Housing the North American City

Housing the North American City

MICHAEL DOUCET AND
JOHN WEAVER

McGill-Queen's University Press
Montreal & Kingston • London • Buffalo

© McGill-Queen's University Press 1991
ISBN 0-7735-0825-2

Legal deposit second quarter 1991
Bibliothèque nationale du Québec

Printed in Canada on acid-free paper

This book has been published with the help of a
grant from the Social Science Federation of Canada,
using funds provided by the Social Sciences and
Humanities Research Council of Canada; and with
the aid of grants from McMaster University and
Ryerson Polytechnic Institute.

Canadian Cataloguing in Publication Data

Doucet, Michael J.
    Housing the North American city
    Includes bibliographical references and index.
    ISBN 0-7735-0825-2
    1. Housing – Canada – History. 2. Housing –
    United States – History. 3. Home ownership –
    Canada – History. 4. Home ownership – United
    States – History. 5. Land use, Urban – Canada –
    History. 6. Land use, Urban – United States –
    History. I. Weaver, John C. II. Title.
    HD7292.A3D69 1991   363.5'0971   C91-090096-5

This book was typeset by Typo Litho composition inc.
in 10/12 Baskerville.

Contents

Tables

Figures

Preface

When embarking on this project in 1981, we had a preconceived set of goals. Drawing upon geography and history, we intended to describe the city-building process through an empirical study of shelter. We thought that we might finish in four or five years. In both cases we misjudged. Theoretical issues and methodological questions proved abundant and tempting, while the source materials were more extensive than imagined.

We had also not considered that contemporary events, other disciplines, and theory would affect our thinking despite our empirical bias. The very decision to write an academic work about housing brought complications that demanded continuous reassessment and rewriting. Many scholars treating the past are insulated against the need for relevance. Early on, we found we could not be so disengaged. At a Mennonite hostel on the edge of the Washington slums, "Irma from Indiana," a missionary on furlough from West Africa, asked how our research might help the poor. We had no answer. As students of shelter and capitalism, we could not ignore the question. However, this book will not respond directly to Irma's concern that scholarship ought to serve human betterment. It is not a prescriptive work. In fact, it is encumbered by the kind of academic paraphernalia that could limit its accessibility to concerned citizens. Nevertheless, we hope that it will have a place in promoting discussions about shelter, as well as about some of the central values and institutional arrangements in North American society.

Writing books, like building houses, brings surprises, for in both exercises problems are encountered from planning to realization. Writing and construction are prone to unexpected upsets: unanticipated shortages of materials, pieces that will not fit, dilatory subcontractors, hidden costs, and funding problems. There is also the satisfaction of completion and the anxiety of having to live with the results. Most important, on the job there is gratification. This was a happy collaboration.

The research took us from city-hall and courthouse vaults to an annex of the New York Public Library and to that model of collecting and hospitality, the Library of Congress. We were assisted by many people. At least a dozen fine undergraduate students have contributed through seminar papers on housing or property development and their work has been cited. David Moore of Hamilton made available an exceptionally rich collection of family papers and D'Arcy Lee of Dundas shared his recollections about mortgage brokerage. George Emory of the University of Western Ontario provided us with a copy of his father's memoirs, which detailed the shifting circumstances of a building contractor. Sociologist Jayne Synge of McMaster shared her research on family life in early twentieth-century Hamilton. The Hamilton-Wentworth regional planning office gave us access to contemporary data and Virginia Mattuzzi of the Hamilton-Wentworth land registry office facilitated our work with subdivision maps and legal documents. David Cook and the staff at the government documents section of the Mills Library were helpful, as were Brian Henley, Margaret Houghton, and many others at the Hamilton Public Library. The abundance of historical resources in Hamilton has been a boon, but the absence of a city or regional archive jeopardizes further collecting and even the maintenance of what has been assembled.

The project that concludes with this book, a project funded by the Social Sciences and Humanities Research Council of Canada and also supported by McMaster University and Ryerson Polytechnical Institute, has seen a number of student researchers come and go. Several have become urbanists. We acknowledge the work of Devi Caussy, Christine Churchill, Peter Clark, Mary Catherine Duvall, Gordon Kerr, Abel Messore, John MacGillis, Janet Nageo, Jordy Robertson, and Stephanie Hill, née Wood. Stephanie provided continuity from summer to summer and was vitally important to our coding of assessment rolls. Carlene Blanche entered the data with great care. The maps were drawn by Reg Woodruff.

Graduate students who have shared information and commented on chapters include John Bacher – a housing scholar in his own

right – Carolyn Gray, W. Thomas Matthews, and Stephen Thorning.
David Burley of the University of Winnipeg contributed criticism,
suggestions, and source materials. Colleagues read portions of our
work or provided data and ideas. Sociologist Peter Pineo did all of
these things. Economist Byron Spencer made several important recommendations. Larry S. Bourne and Judith Kjellberg of the Centre
for Urban and Community Studies at the University of Toronto,
Marc A. Weiss of the Lincoln Institute of Land Policy, and Barbara
Wake Carroll of McMaster University generously provided either
data or access to their own manuscripts in the housing field. Manon
Ames, Jackie Collin, and Pat Goodall from the Humanities Word
Processing Centre at McMaster helped produce the text.

We are grateful to the peer reviewers of our articles and of the
manuscript. We are also thankful to have had shrewd editors who
have allowed us considerable freedom. The copy-editing skill of
Lesley Andrássy smoothed rough edges. At conferences at the University of Winnipeg, the University of Edinburgh, the Australian
Studies Centre in London, the Cité des Sciences et de l'Industrie in
Paris, and the University of Toronto Centre for Urban and Community Studies, as well as at meetings of the Canadian Historical
Association in Guelph and Quebec City, we received advice and
criticism. Many of the concerns of commentators and readers have
not been worked into the book. To write is to choose. We alone are
responsible for our choice.

Many scholars in Canada, the United States, and the United Kingdom have been writing about shelter and society; a number are
acknowledged in the endnotes. The flow of new articles and books
reminds us to end with a caution and an aspiration. There will soon
be works with fresh insights and conclusions contrary to our own.
Nevertheless, we believe that the processes and patterns described
on the following pages will be confirmed for many North American
cities. We hope that our detailed account will assist with the broad
enterprises of understanding how cities have been assembled and
how elements of a North American culture have transcended a border.

For many years, Natalie Doucet and Joan Weaver have understood
and supported our obsession with a study of shelter. Christine and
Brian Doucet and Adam Weaver have been fortunate to live at a
time and place with fine housing. Thinking about their future has
made us alert to potential housing crises and to the nagging fact
that most of the world's children have not been so fortunate. May
Canadians never take their good fortune for granted; may others
profit from our successes and failures.

Housing the North American City

City Building and Shelter History

We are Building for Tomorrow
In a strong, aggressive way,
All the olden beauties borrow
And embody them today.
We are ploughing up and planting
Cutting streets and making plots
We are heaving, pushing, painting,
Putting homes on vacant plots.

We are growing plants and flowers
Velvet lawns and shading trees,
Making most of all the hours,
Building homes for Wealth and Ease.
Homes where people like to tarry,
Homes where workingmen may dwell;
Making nests for those who marry,
Building strong and building well.

We are terracing and parking,
Making all the landscape smile,
Every eminence are marking
With some structure well worth while.
Building Church and school and mansion,
Cottage, factory and store;
Working ever for expansion,
Building better than before.

G. Herbert Palin, "The City Builder"[1]

Some years ago, Roy Lubove reminded students of the city that urbanization was an abstraction. Cities, he wrote in 1967, "are created by concrete decisions over time."[2] He was reacting against the more rarified methods and theories of social scientists. There may be a tinge of irony in introducing an often quantitative study with Lubove's affirmation, but the connections between his brief scheme for studying the city and our data outweigh any apparent conflict. Statistics have been employed here to direct attention to decisions over time; they drive no model and they are not meant to be the objects of manipulative exercises to test theories or the associations of variables. Data are for us intrinsic features of a review of city building. Lubove considered the city as an artifact shaped by peculiarities of site and by probable differences in attitudes among key leaders or city builders. This accent on place and variety in leadership also makes Lubove a seemingly improbable guide for our opening comments, for the following historical account strives to relate the happenings in one North American city to processes throughout the continent. Nevertheless, Lubove's brief essay has influenced our discussion of housing because he proposed that historians should focus on city-building, meaning "the whole range of city-building mechanisms: architecture and landscape architecture, housing and housing finance, the real estate market and realty institutions, transportation, communications, public health and sanitation, industrial technology, and business organization." Not long after these remarks appeared, Sam Bass Warner Jr provided one of the best examples of what Lubove had in mind in *Streetcar Suburbs*, which dissected city building in three Boston suburbs from 1870 to 1900.

The history of any one of a number of human elements to a city can relate much about the city-building process. But the depth and quality of what is related depends on the historian's vision. For streetcar buffs, the topic of traction and power is a matter of precise reporting that stays on track. For historians H.V. Nelles and Christopher Armstrong, the subject has been a means for brilliantly relating an understanding of civic politics and the twists and turns that characterized the building of important sectors of Canadian capitalism. Railways too have had scores, if not hundreds, of historians, but it took John Stilgoe to conceive of a *Metropolitan Corridor* and analyze what this unusual built landscape had meant to Americans.[3] The city has abundant subjects awaiting careful and ingenious treatment. But shelter is special.

Dwellings cover much of the urban landscape. A 1968 report on land use in large American cities indicated that residential uses accounted for an average of 26 per cent of the total land area and 36 per cent of all developed land.[4] Most urban neighbourhoods, at one time, were suburbs. Their history in the United States has been recounted in Kenneth Jackson's highly acclaimed *Crabgrass Frontier*. Jackson set out to explain "the strange, ever-changing physical arrangement of the United States" through a history of suburban development.[5] In some locales, shelter or housing is more than a feature of sprawl, it dominates the skyline. Clearly a necessity, shelter has subtle as well as powerful and obvious social qualities beyond the significance of its dimensions.

Theodore Hershberg and his associates have demonstrated in their Philadelphia Social History Project (PSHP) how, in the late nineteenth century, shelter was connected spatially to employment opportunities. Urban space is not neutral – its social consequences have been profound and, for the unskilled and victims of discrimination, negative.[6] Culturally, shelter forms have crystallized class-based outlooks. Robert Fishman has interpreted the origin of the modern suburb as a bourgeois invention first fully realized in Manchester in the 1830s. His explanation of a class desiring to segregate itself from the working class by abandoning town houses presents social-spatial counterparts to those spelled out by the PSHP. Fishman also attempted to unravel the puzzle of why the English and the North American bourgeoisie abandoned the core while the European and Latin American bourgeoisie retained central-city residences.[7] Class, space, and the consequences of different residential environments are described at many points in the following chapters. Economically, housing has posed interesting theoretical questions about consumer behavior. Although it has use value for the occupant, the owner-occupied dwelling also has exchange value. The owner is a potential shelter producer and therefore mindful of the impact of neighbourhood changes on property values. The residents' and ratepayers' associations in every North-American city represent a barrier to most change. Politically, therefore, shelter has been a challenge in itself and a tangential issue for almost all crucial urban undertakings. Whether one charts change or continuity or both, housing provides serviceable artifacts, and each of the above broad categories of analysis presents an important temporal feature.

The artifacts are not obscure and inaccessible, though if considered alone they relate an incomplete story. It is people, as Lubove suggested, who make cities. Sheer abundance has assured the survival of types of dwellings occupied by many social groups over many generations. In Hamilton, Ontario, our subject city, it is possible to locate examples of housing from the older major eras of city building. An assortment of dwellings – but barely that – has survived the urban renewal projects of the 1960s. Photographs from the 1860s to the 1960s help to show what was lost. A sketchbook of Hamilton streetscapes around 1856, prepared by artist Edwin Whitefield, has helped to confirm assumptions about early housing quality that were based on lifeless census records and mute assessed values. Assessment rolls have been most useful and a note about sampling and coding procedures appears in the appendix. Long used by social historians to explore topics on social stratification, concentration of wealth, and population mobility, assessment rolls consist principally of annual entries on household characteristics, tenancy, and – though in complicated fashion – property values. Other sources confront housing decisions and housing quality more directly: from the 1870s to the 1920s, there was an extraordinary outpouring of carpentry and building journals; from time to time, reformers, social workers, and journalists have exposed the city's worst housing conditions; and beginning in the 1930s, the statistical agencies of the federal and provincial governments have provided a number of useful studies asking ever more sophisticated questions about shelter quality and its costs. Although surprised by the abundance of sources even on unlikely subjects – mortgage brokerage and rental management, for example – the scarcity of testimony by householders has necessarily forced reasoned conjecture on controversial issues. While it is possible to apply internationally accepted standards for housing quality and for debt burden, and to infer broad support for what we call the will to possess by examining several rare private statements and public comments by leaders of worker's movements, there may never be sufficient evidence on these important matters to validate our reasoning conclusively. We think that our data and the conventional sources not only present an account of city building from the top down, but also provide a comprehensive look at what city building meant to a representative variety of households. But, yes, more voices from the past would have been welcome.

THE HOUSE AS TOPIC

Housing or shelter – the terms are used here interchangeably – comes in several architectural forms: the apartment in a high rise, the apartment over a store or in a low rise, the flat, the tenement, the row house, the duplex, and the single-family detached house. Housing also has social forms: the market rental dwelling, the subsidized rental dwelling, the co-operative, the condominium, and the owner-occupied unit. The social and the architectural descriptions can be combined to categorize shelter types. Many of the combinations, especially the rented and owned detached house and the market rental apartment, are discussed on the following pages. The detached house figures as the paramount architectural form, housing the majority of North Americans and projecting an unusual lustre. Indeed, North American culture has maintained as one of its defining characteristics the single-family detached dwelling. Not all of the continent's urban areas have expressed this preference, for geographical restraints, distinct traditions, and investment practices have created cities notable for row housing (Philadelphia, Baltimore, and Washington), the duplex (Montreal), and apartments and tenements (New York). In younger and smaller cities, however, the single-family detached house has been and remains the mainstay of shelter. About three-fifths of all Hamilton dwelling units took this form in 1986, although for nearby Toronto the figure for the same year was just 43 per cent. As a principal emblem of success – whether one likes that attribution or not – the house and its cultural meaning has recently inspired academic antagonists and popular eulogists.

A half-dozen or more revisionist studies have presented the house as an architectural fixation unsuited to modern households, which have become smaller and less family-oriented than those prevailing when the house became refined into its modern form. The house has also been criticized as an unwise investment for working-class families. Revisionist historians, contemporary-housing policy analysts, and feminists have carried present social visions with them into their explorations of the past. They have already given us novel ways of looking at the house, and more results will appear. Generally, they have portrayed the house as a commodity pressed upon a consumer society by a property industry reaping extraordinary profits and by governments hoping to foster a conservative population of proprietors. Developers and governments may well have had these aims and worked for them, but this does not make the case for a manipulated public taste. Kenneth Jackson, appalled though he was by inducements for urban sprawl and the absence of true land-use controls in the United States, gave home buyers credit for intelligence. "Most moved into a single-family house because it maximized their utility from a stable set of preferences. In other words, low density housing was a good deal."[8] Whereas Jackson does not develop historically an explanation for the consumer demand and apparently treats government sponsorship of home buying as autonomous, he at least took the position that the private dwelling is "a more universal aspiration than suburban historians have previously been willing to admit."[9] Recent forecasts of housing taste and demand, which have presented the single-family detached house as either a passing fancy or a luxury unaffordable to the majority, are not alone in having missed the mark. In 1944, just a few years before a housing boom, a panel of Canadian urban experts reported to the Advisory Committee on Reconstruction that "there is little doubt that a preference for one-family houses continues among Canadians, but distinct changes in taste have been registered in recent decades."[10] In fact, the taste for house buying vigorously reasserted itself in better economic times. The following chapters explain this continuity of taste with reference to the radical nineteenth-century roots of a will to possess, and to the specialized industries that help to sustain it by synchronizing with the demand. House buying was and remains a popular ambition that caused land developers, the building industry, and financial service agents to evolve with the demand.

One of the basic appeals of house buying, as it became established in North American culture in the nineteenth century, was independence, but as legions of commentators have affirmed, independence is not equality. Critics of the house, coming to the subject from a concern for those who have been left out, correctly see it as an article of and artifice for inequality. Nothing could be further from their social ideals than Tracy Kidder's *House* (1985), a popular work that painstakingly chronicled an upwardly mobile young couple's quest for self-gratification through the commissioning of a 3,000 square foot residence.[11] Kidder's subjects here and in his Pulitzer prize-winning account of computer designing, *The Soul of a New Machine*, are contemporary independently minded and affluent professionals. They are well-educated, upper-middle class New Englanders similar to the subscribers to *The Atlantic*, where excerpts from both of Kidder's books appeared. Beginning in the mid-1980s that magazine published extensively in the field of housing, presenting superb features on contemporary architecture and informative briefs about housing technology and bourgeois tastes in shelter. Provocative criticisms of the owner-occupied house, owing something to the flourishing state of social criticism twenty years ago, have not dented the predilections of the consumer society. Writing about "The American House" of the 1980s for *The Atlantic*, Philip Langdon concluded that while there no longer was a typical American house, ownership remained firm. "There has been an explosion of variety in housing to correspond with the growing heterogeneity of the population. The housing industry has coped. In an era of spa baths and dual master suites, nearly two thirds of America is connected to a conservative, stabilizing institution: homeownership."[12] Langdon, a design writer, probably exaggerated the variety found in new housing, because he naturally looked for the extraordinary, but he was basically correct about ownership. Since many radical academic assessments about the house have misconstrued the roots of the demand and have manipulated data in order to exaggerate its failure, their lack of influence is understandable. You cannot change what you do not understand. Kidder's couple scrambling toward comfort and status may present an egotistical coda for a conservative phase in recent history. However, they may have acted out of a long-standing cultural urge.

Whereas scholarly monographs are unlikely to dent profound cultural outlooks or self-indulgence, economic circumstances have threatened to do so. Soaring interest rates during the recession of the early 1980s stirred pessimism about the future of the house and prompted discussions about what level of debt load households could bear for shelter. The 1981 Canadian census missed most of the interest-rate escalation, but it established that across the country about 80 per cent of mortgagors should have had no difficulty in meeting their house costs, since housing costs absorbed less than 30 per cent of their disposable income, the accepted guideline in determining housing affordability.[13] It is impossible to say how the brief period of high interest rates (21.5 per cent at the peak) that followed the census would have affected such a calculation, but residential foreclosures rose significantly. Given the great number of house buyers who already occupied their own homes before 1982–1984, the addition of a few new ones paying high rates probably would have had relatively little impact overall. In any case, the rates eventually declined to around 10 per cent. Recent concern about housing affordability in North America has shifted to the prices of houses. The ratio of the average value of owner-occupied dwellings to median household income remained constant at 2.6:1 in Hamilton in 1961 and 1971. It had risen to 2.9:1 by 1981.

Even rising house prices have not convinced the authors of a recent Brookings Institute study that house buying soon will wither. Because most urban Canadians now reside in non-nuclear family arrangements there is some question as to whether the single-family detached dwelling will remain an appropriate housing form in the face of this dramatic change, which American commentators have called the end of "the Norman Rockwell family," but this underestimates the adaptability of the structure and of suburban land design. In fact, in many Ontario cities the legal and illegal conversion of existing single-family dwellings to multi-family structures has been an important response to the twin trends of rising household costs and shrinking household size. Moreover, predicting household traits for even the near future is a risky enterprise. A recent study of what 45,000 Ontario high school students felt about themselves and their future discovered that over three-quarters of each gender expected to marry.[14] We can no more forecast how that generation will decide to mesh household formation with dwelling form and tenure than the experts of the baby-boom years could forecast today's trends. It is not entirely a matter of property-industry obduracy that the single-family house continues to be built in great numbers. Recent American data presented by Robert Fishman led him to the same conclusion: "In 1981 a median American family earned only 70% of what was needed to make the payments on the median priced house; by 1986, the median family could once again afford the median house. Single family houses still constitute 67 per cent of all occupied units, down only 2 per cent since 1970 despite the increase in costs; moreover, a survey of potential home buyers in 1986 showed that 85 per cent intended to purchase a detached, single family suburban house, while only 15 per cent were looking at condominium apartments or townhouses. The 'single', as builders call it, is still alive and well on the urban periphery."[15]

The difficult topics pertaining to the house recommend it for historical analysis. A controversial icon whose fascination for North Americans can be related to the social and economic aspirations of immigrants entering a new society, it is a stable article that, paradoxically, has inspired technological changes all about it. As North America's original large consumer purchase – and a container for most of the others that the continent's industries have produced – the house has driven refinements of consumer credit. Not all of the complex themes and their supporting details were foreseen when we began our research. Our conventional research aspired to recount how land and shelter were assembled, and our data-based inquiries were equally straightforward. They centred on assembling empirical information about the business of city-building, the social attributes of home owners and tenants, the interaction of urban space and society, and aspects of housing quality. What originated as a data-driven investigation – ambitious enough in conception – spread out and drew upon studies in architectural and technological history, the law, and semiotics. The diversification was necessary in order to raise sensible questions about patterns arising from the data. It has also enriched the narrative stream, making it possible to imagine how city-building, in its residential features, has been conducted for approximately 150 years.

Good fortune in locating source materials has enabled us to populate the narrative with land developers, contractors, lumber dealers, mortgage brokers, home buyers, rental estate owners, and disgruntled tenants. A widow threatened with foreclosure makes a brief appearance. Housing is a social matter and common lives belong in the thick of its history, yet it also has enormous economic consequences, so much so that it has been important to the public – and often the private – careers of politicians. Therefore, the narrative introduces John Beverley Robinson as a controversial agent in the making of early Canadian mortgage law. It presents John A. Macdonald as a trust company promoter who collected legal fees from borrowers, and quotes George Brown as a shrewd forecaster of capitalist restructuring in carpentry. Another generation of nation builders also appears, for the civil servants who strengthened the federal government in the 1940s and 1950s concerned themselves with wartime housing, mortgage finance, and social housing. Social, cultural, business, economic, and political history converge upon shelter. A history of shelter also makes local history potentially continental and *vice versa*.

HAMILTON AND THE CONTINENT

American historians have left no Boston cobblestones unturned and, along with social scientists, have reconstructed the social map of nineteenth-century Philadelphia with painstaking research and brilliant interdisciplinary analysis. Hamilton, Ontario, has neither the richness of past nor the concentrations of universities that have inspired and supported the peerless work on the Puritan and Quaker centres. Nevertheless, this industrial community in Toronto's metropolitan shadow is the most-studied urban society in Canada. Michael Katz and his research associates initiated Canada's first quantitative history project with Hamilton data and subsequently published two books. The first, *The People of Hamilton, Canada West*, was a milestone, while the second offered a particular and controversial treatment of the concept of class.[16] Labour historians Bryan Palmer and Craig Heron found the city's iron and steel industries – venues for technological and management innovations, conflicts, and radical declarations and deeds – rich in episodes and social patterns.[17] There are advantages in perpetuating this scholarly effort. It will, for a time, offer students of social, labour, and urban history a familiar focal point, and, "the Hamilton experience," as either model or strawman, will have served a number of disciplines well and will have promoted interdisciplinary research. Within Canada, given the constantly declared concern for regional and local differences, it should provoke more investigations of Atlantic, Quebec, and western cities. It might even stimulate work on Toronto, which is overdue for major social-history research.

Pursuing the universal through the particular is a common problem in the humanities and the social sciences. In his classic article, "If All the World Were Philadelphia: A Scaffolding for Urban History, 1774–1930," Sam Bass Warner Jr made a bold stab at establishing indicators that would fix periods of great urban change.[18] He shrewdly used the subjunctive mood in his title to express an unreal situation. The city of brotherly love is not all the world; neither is Hamilton. In the nineteenth and twentieth centuries, however, their city-builders shared continent-wide practices. But in searching for parallels for Hamilton, a special case like Philadelphia – a great metropolis with a remarkable founder's design – should be set aside. Nor will it do to look at Hamilton as a typical Canadian city – there are none. Examples of diversity, however, abound. Given that historical census data, which are about all we have as a basis for inter-city comparisons, pose considerable problems, it remains absolutely certain that the City of Montreal's housing-tenure history is quite unlike that of Hamilton's.[19] Despite the difficulty of evaluating housing quality, it is probable that most Canadian cities have had generally poorer housing than Hamilton. Conceivably, Hamilton shelter history would fit better into a set of smaller industrial centres settled at about the same time. Middletown would be a convenient city for a comparative study, because the 1929 and 1937 studies by Robert and Helen Lynd and a 1982 follow-up report included observations about housing that also apply to Hamilton.[20] However, the real Middletown – Muncie, Indiana – is too small for comparison. Peoria, Illinois, which was incorporated around the same time as Hamilton and likewise developed from a judicial and farm service town into a city with industry, has potential, although it too is much smaller than Hamilton. Nevertheless, as newer cities in the prosperous heartland of the continent, Hamilton, Muncie, and Peoria have been cities of homes and home owners. Perhaps questions of comparability are beside the point. The universal may be reached through the particular by recounting how bricks and people came together in one city and by noting references to particularities and common practices. In using this strategy, as we have, the goal is to collect examples of how things were done and their social outcomes. The mix of the procedures and of their consequences may vary from place to place, but the procedures and consequences should be highlighted.

There are good reasons for taking the position that there is a North American way of city-building. Land-development strategies, suburban designs, marketing gimmicks, house construction techniques, building materials, and house and apartment plans have passed easily over the border. Only the exceptions make news. Most Canadians learned about cedar shakes and shingles when their export was jeopardized by a 1987 trade dispute. Several of the men that we present as city-builders lived and worked in cities in both countries, while others had international business dealings related to housing. Capital for builders' loans and for home-buyers' mortgages in Canada and the United States was collected by similar practices from the same sources, and distributed by individual agents or institutions on terms that varied regionally while evolving in similar ways. When governments began to intervene in housing, especially through the mortgage market, Canadian authorities watched and frequently adapted American practices, beginning in the 1830s. When Upper-Canadian legislators established a Court of Chancery to deal with mortgage foreclosure, they followed New York legislation. One hundred and twenty-five years later, Central Mortgage and Housing Corporation officials monitored and recommended certain Washington practices. Many contemporary social indicators highlight differences between Canadian and American cities. Michael Goldberg and John Mercer have assembled a splendid collection of these disparities.[21] The very long history of overlaps in fundamental practices of city building in the residential sector suggests, however, that alleged cultural differences between Canadians and Americans encompass only part of the history of the continental counterparts.

TEMPORAL CONSIDERATIONS

Motives and objectives in historical writing should be frankly expressed. We combined because of an interest in residential space and the people who have fashioned it. We believed that intrinsically interesting things could be said about land development, mortgage lending, house construction, and apartment design. Earlier independent work had convinced us that not only were the city-building activities of nineteenth-century Canada poorly understood, but also contemporary studies on land development, social housing, and housing technology were poorer for this lack of a historical record. Frequently, as research advanced, we found claims made or questions raised by housing experts that indicated misconceptions about how the housing practices of their times fitted into a broader scheme. Our interest and our eagerness to set down a record of events led us to begin to write about the housing of a city without a guiding theory. Recapturing past thoughts and conduct is an honourable enough calling, but one crucial ingredient in the research plan made much more than this possible. Believing that topics in a social and economic history of shelter − home ownership for example − required a long-term analysis, we set our sights on 100 to 150 years. Later, the other broad dimension of the study − its fusion of continental and local observations − came into sight when multiple sources laid bare the many strong connections among a set of jurisdictions that grew outward from Hamilton to easily embrace Ontario, to extend to Canada, and to take into account the United States.

Always, however, we worked toward the *longue durée*, but without direct inspiration from Fernand Braudel and the *Annales* historians who popularized the long-term perspective. They have written very little about the connections between housing and culture. As we spelled out in "Town Fathers and Urban Continuity," Hans Blumenfeld directly influenced our searching for the weights of habit that a city-building process carries forward.[22]

Looking for continuity – the weights of habit – makes the identification of change more convincing. All too often housing commentators have asserted change and even revolution without seeing the underlying preservation or without having the historical sense to recognize a cyclical occurrence. In the chapters that follow, temporal concerns direct many passages and order most tables. Many of these concerns are flagged in the text and the book is organized topically, not chronologically. The temporal rubric comprises four temporally conditioned organizing principles: abrupt epochal shifts, evolutionary developments, cyclical patterns, and continuity. The epochal shifts, which bring a long-lasting reorganization of methods, are rare. The spread of the balloon-frame house qualifies, for that mode of construction has dominated house building from virtually its first appearance in Chicago in the 1830s. In Montreal, builders in the nineteenth century used a heavy frame and plank construction, and perhaps other centres also deflected the innovation, but the light balloon frame was overwhelmingly popular. Although its simple components have been refined and mass-produced, its overall ingenious form has remained. The disappearance of rental houses and their replacement by assorted types of apartments, which happened in Hamilton in the 1950s and 1960s, may also qualify, although the disappearance likely occurred at earlier dates in larger older cities. A major and probably more universal post-war housing phenomenon could also have the makings of an epochal shift, but only time will tell since the trend could be reversed. The capacity of fairly young household heads in large numbers to buy houses really only started in the 1950s, but both the young home owners or buyers and the demographic politics behind their purchasing power are still with us.

Minor examples of evolution in housing abound, particularly if we were to narrate changes in building materials in minute detail. A more significant evolutionary path traced in this book follows the transformation of and additions to the organizations responsible for city building (see figure I.1). To provide housing, land must be prepared, structures built, and financing arranged. It is true that in Upper Canada, as well as in many American centres, governments,

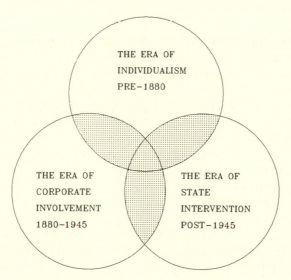

Figure I.1
A three-stage model of city building

idealists, and the ingenious agents of land corporations tried their hand at community building. However, their experiments tended to be just that and their long-term influences, where their plans actually affected surveys, have been confined to peculiarities in the street layouts of a very few centres. The ingenious planners of eighteenth-century and early nineteenth-century town sites represent the path not taken. For the most part, especially by the 1820s in the United States and the 1830s in Canada, the principal builders of urban sites were individuals and families. Initially they performed the basic functions of land development, building, and finance, although governments throughout North America certainly had a hand in city building in this age of individualism. Indeed, local government often seemed to act as an extension of private development interests, and colonial or state law frequently aided the private market. City building proceeded under the direction of individuals, and governments required little in return for their aid. Land development practices, the easing of ancient restrictions on money lending, and the weakening of mortgagors' rights indicate that governments assisted but did not regulate. Truly, then, the first era of city building that has a powerful connection with our own times is an era of individualism. In Hamilton this era may have lasted beyond 1900, but by the late nineteenth century loan and trust companies were well entrenched and by the early twentieth century corporate involvement in land development was significant. The phase of corporate involvement that followed the era of individualism was succeeded in turn by the contemporary one of state intervention (see figure I.1). Governments continue to assist private enterprise; however, they also impose regulations and maintain some objectives that clash with individual or corporate enterprise. It would be tempting to suggest that a contemporary land developer could scarcely imagine the unregulated world of a Victorian counterpart, but we suspect that, restricted as they are by an astonishingly regulated environment, they could rise swiftly to the challenge.

Never neat and tidy at their chronological borders, the phases of evolutionary development nevertheless capture the real sense of innovations in the institutional context of shelter provision. What has occurred in housing merely exemplifies the evolution found in capitalism more broadly. Capitalist change occurred earliest in finance, followed by land development, the rental market, and house building. Incremental developments have refashioned city-building artifacts while their primary kernel remains identifiable over time. This evolutionary quality will be traced in our treatment of the mortgage, the suburban lot, the house, and the apartment building. Industrial manufacturing has encompassed different eras of technology and of organization concurrently, and so it has been with shelter. In the era of state intervention, strains of individualistic organization remain, although the trends of corporate involvement and government action are clear.

The third temporal pattern is cyclical. Over twenty years ago J. Parry Lewis published his richly detailed *Building Cycles and Britain's Growth*, which explained building cycles of roughly twenty years' length from 1700.[23] Approximately the same length characterized Hamilton's cycles of shelter construction, whose growth periods occurred around 1850–1857, 1880–1890, 1905–1913, 1924–1930, 1950–1957, and 1964–1972. Another began in 1985. Mortgage repayment arrangements, which have recently come full circle, also show a cyclical pattern. Three-year and five-year renewals were favoured in the second half of the nineteenth century and into the early twentieth century. Institutional lenders now have returned to short-term agreements – but have retained longer-term amortization – after decades of fixed-term mortgages running twenty years. History repeats itself in other ways. Inflated real-estate appraisals precipitated mortgage-lending crises in North America during the 1850s, the late 1920s, and the late 1980s. A more localized problem arose in Alberta during the early 1980s.

Finally, there stands continuity. Grey matter, alleged Blumenfeld, has been more impervious to change than concrete. In support of this contention, one popularly held idea binds together much of the following chronicle. Home ownership as a goal among North American skilled labourers – not just the bourgeoisie – probably antedated 1850 and cannot be treated as a purely urban idea, for it likely drew strength from agrarian antagonism toward tenancy. In the radical 1880s, home ownership became an obsession for Hamilton's skilled labourers and those in many other cities, if building trade journals are to be believed. It was not an ambition in itself; cultural obsessions never are. Ownership meant many things, including an idea of independence and a potential for satisfying labour, as opposed to the work done for wages in an industrial capitalist economy. Buying or building a house gave working families self-esteem and freedom from the landlord, and signalled their material success to peers. Tied up with these sentiments was the need for household planning. The lot and the house always meant an enormous burden. If the lot were bought alone with the intention of building a house through sweat equity, the working man had to schedule his spare hours. If he proposed to buy a ready-built house or a prefabricated kit, savings had to be set aside. Provident conduct and age, more demonstrably, helped. The will to possess, the persistent urge to own a house, addresses continuity, but so does the relationship of age and ownership. The age or life-cycle feature showed clearly in the Hamilton data from the 1860s to the 1940s. Even among professionals and business proprietors, house buying was most strongly associated with several middle-aged groups. A young businessman usually put capital into his stock or partnership. Profits might later be taken out for the purchase of a family home. Only the widespread introduction of the low down payment or high-ratio mortgage, supported by the interventionist state, altered the pattern. For approximately a century, although most families in Hamilton eventually bought houses, the will to possess existed in tandem with struggle, calculation, and deferral. The high-ratio mortgage may or may not have created more home buyers, but it certainly dropped the age threshold and conceivably helped to erode acceptance of waiting or self-denial in North American culture.

Continuity over time and from place to place can also be seen in the persistence of segregation. Anglo-American cities, on the one hand, and continental and Latin American cities, on the other, evolved contrasting arrangements of residential space to accommodate segregation. Paris exemplified one route. According to

Robert Fishman, the Parisian middle class was on the verge of a
suburban exodus. "Instead, a massive government intervention into
the French housing market totally changed the rules of the game
and made possible the construction of apartment houses near the
centre of Paris on a scale that no one could have foreseen."[24] Louis
Napoleon's and Baron Georges-Eugène Haussmann's reconstruction
of central Paris established an attractive model for bourgeois residential space in many other cities where the state chose to create a
stunning capital. Vienna's Ringstrasse development of the 1860s was
another example of a government-directed scheme to create "a zone
of privilege in the urban core based on monumental apartment
houses whose facades imitated Baroque palaces."[25] From there the
invention spread through central and eastern Europe. In Anglo-
American cities, the middle class and later skilled labourers achieved
segregation by means other than authoritarian planning. Indeed,
suburban home ownership expressed a liberal ideology at odds with
the dynastic pursuits of grandeur. Nevertheless, both nineteenth-
century models of segregation had a common impulse. First the
bourgeoisie and next the skilled labourers wished to place themselves
apart from features of urban life that they had grown to dislike. In
order to escape the poor and the crowded, chaotic, and unsanitary
city cores the bourgeoisie withdrew to their apartments or their
suburbs.

By the 1920s the bourgeoisie in Hamilton had both suburban tracts
and apartment buildings especially designed for their occupation.
Position in the life cycle tended to determine who went where. The
environmental havoc caused by the railway and factory in the North-
American city drove many skilled workers into their own suburban
enclaves. In Hamilton, throughout 150 years, one social group constantly lost out in the locational stakes. Common labourers always
filled the meanest housing stock in the city's least desirable and most
unhealthy residential sectors. Attempts to rectify this situation, by
placing social housing units in market housing areas, met with opposition in Hamilton during the 1970s and 1980s. While this may
be, in part, a consequence of home ownership, which tends to make
households equate changes with a lowering of property values, the
old and powerful instinct of separation has doubtless been at work.
In whatever combination, the will to possess and the desire to keep
apart from poverty and blighted landscapes maintain the continuity
of a segregated city.

Despite the preceding introduction to themes, some readers may
find the transitions in subject matter and method from chapter to
chapter a little jarring. However, we believe that an analysis of city

and society requires a variety of approaches. Geographers and historians may fail to illuminate the processes of city development if the former adhere to exercises restrained by theory or if the latter insist on a consistent narrative voice. Some readers may find the conclusions tentative. After a decade of research, we have found that questions worth asking remain difficult to answer, particularly when attempting to treat a great many social categories and processes, rather than concentrating on a class or a period.

This book is organized topically. The first six chapters take readers through decisions and actions that conclude with people moving into shelter. Our initial three chapters consider land development as design and business exercises. Chapter four speculates on the sources of the widely based urge for home ownership. The next chapter establishes in detail the clustering in time of innovations in house construction in relation to economic cycles, and indicates where and why habits and tastes persist. Home financing, the topic in chapter six, is treated in much the same analytical way as shelter construction. The mortgage is scrutinized, and change and continuity highlighted. Beginning with chapter seven, a study of home-ownership patterns, material and social history are united as we consider people in their accommodations and neighbourhoods. Chapters eight and nine deal with rental houses and apartments, landlords and tenants. The concluding chapter explores a number of ways to evaluate shelter quality and environments historically.

1 Residential Land Development before 1880: The Era of Individualism

Fundamentally, cities are composed of four basic elements – the natural environment, people and their culture, buildings, and infrastructure. These four are melded together in many different ways during the evolution of individual urban places. One such nexus of interaction is the urban land-development process. Most simply, urban land development can be defined as a series of stages whereby agricultural, wild, or otherwise vacant land is converted to some sort of urban land use (figure 1.1). There are many possible paths through this seemingly simple process. If dwellings are the end product, for example, then residential development has occurred, with neighbourhoods as the final outcome. Since residential areas account for the largest single block of urban space, the home-building process cumulatively determines both the form and character of urban communities. Three major stages – subdivision, construction, and marketing – characterize urban residential land development. These stages may be followed in an orderly and rapid sequence, as is usually the case today, or they may be pursued in the disjointed, haphazard fashion typical of earlier times.

The urban land-development cycle begins whenever someone recognizes urban potential, no matter how prematurely, in a parcel of non-urban land. Subdivision, the first stage in the process, entails the partitioning, by means of a survey, of some large block of land into smaller parcels. This immediately intensifies the use of the land. A one-hundred-acre farm lot that was home to a single rural household has the potential to be converted into building lots that, when

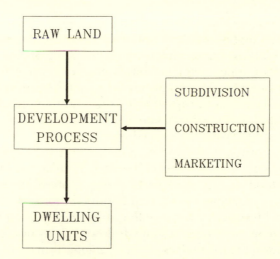

Figure 1.1
The residential development process

developed, can house more than three hundred urban families. Very early in the history of Ontario, politicians and landowners decided that professionals, the provincial land surveyors, were needed to lay out such surveys. By 1847 the registration in local registry offices of their work, the so-called plans of subdivision, had been made mandatory. This brought a greater degree of certainty to land transactions by providing a means for identifying the precise location of any given land parcel, though it by no means guaranteed that the title to the parcel was clear.[1] For decades, these would be the only controls over the land-development process.

The construction or physical development stage of residential development involves the design and erection of some type of shelter on a surveyed building lot. The variables associated with this phase of development include the identity of the builders, the locations chosen for their activities, and the materials and techniques that they have employed. At various times, construction has been either the second or the third stage in residential land development. Housing forms and tenures, as this volume documents, especially in chapter 5, have displayed both continuity and change over the course of North American urban history.

In the marketing phase of the process a surveyed land parcel is sold, either alone or with an already completed dwelling on it, to a consumer who may or may not be a land speculator or investor.[2] During most of the nineteenth century, the marketing of lots was split into two parts that enveloped the construction stage. More recently, marketing has occurred only once, at the end of the process. Over time, both the people involved in the selling of real estate and the tools that they have employed have become ever more specialized and sophisticated. Today computerized data bases are used both to record the inventory of properties for sale and to monitor trends in the real-estate market. Auction sales similarly have given way to more formalized property-buying procedures.

Residential property development in North American cities in the early 1990s is a complex process, but it is familiar and reasonably well-understood. It is dominated by large, sophisticated, vertically-integrated corporations whose final product, more often than not, is as much a lifestyle as it is a dwelling unit.[3] Subdivision remains as the first stage in the process, usually followed today by construction. The dwelling units themselves are technological marvels and come in an ever-increasing range of shapes, sizes, and densities – single-family houses, duplexes and the like, granny flats, rowhouses, stacked townhouses, and low-rise and high-rise apartments. Tenure forms have also proliferated to encompass freehold, leasehold, condominium, and co-operative arrangements. Contemporary dwelling construction can best be described as the on-site *assembly* of components (windows, stairs, roof trusses, and others) that have been manufactured elsewhere. Ideally, once the necessary approvals have been secured it takes 16 to 20 weeks to construct a single-family residential unit in Ontario today. Most such dwellings are built for owner occupancy. A myriad of rules and regulations are in place in most jurisdictions to control the development process. These run the gamut from zoning by-laws to building codes to requirements for the registration of both property boundaries and building plans with the appropriate authorities (see chapters 2 and 3).[4]

In marked contrast, nineteenth-century residential development was dominated by very small, craft-oriented firms that typically produced low-density forms of housing – single-family dwellings and townhouses (see figure 1.2) – most of which would be rented by their occupants (see chapter 8). Apartments, even in low-rise form, were virtually unknown outside of New York City (see chapter 9).[5] Dwelling construction at that time was characterized by the on-site *fabrication* of the components needed for the specific housing unit. Few rules and regulations governed the development process, and entry into the business required relatively little capital. As a result, there were many players and the development industry was unregulated, fragmented, and technologically and organizationally unsophisticated.[6]

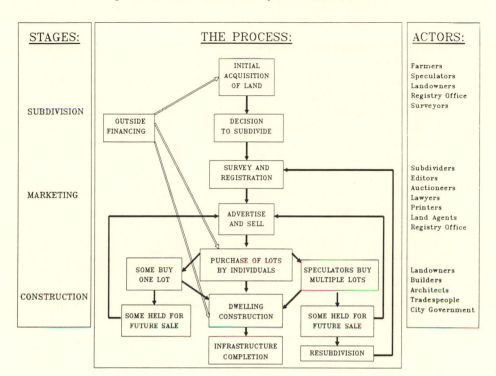

STAGES:	THE PROCESS:	ACTORS:

STAGES:

SUBDIVISION

MARKETING

CONSTRUCTION

THE PROCESS:

INITIAL
ACQUISITION
OF LAND

OUTSIDE
FINANCING

DECISION
TO SUBDIVIDE

SURVEY AND
REGISTRATION

ADVERTISE
AND SELL

SOME BUY
ONE LOT

PURCHASE OF LOTS
BY INDIVIDUALS

SPECULATORS BUY
MULTIPLE LOTS

SOME HELD FOR
FUTURE SALE

DWELLING
CONSTRUCTION

SOME HELD FOR
FUTURE SALE

INFRASTRUCTURE
COMPLETION

RESUBDIVISION

ACTORS:

Farmers
Speculators
Landowners
Registry Office
Surveyors

Subdividers
Editors
Auctioneers
Lawyers
Printers
Land Agents
Registry Office

Landowners
Builders
Architects
Tradespeople
City Government

Figure 1.2
Residential development during the era of individualism

   While the residential-development industry has always been involved in the production of dwelling units, both the nature of these
units and their construction have changed over the course of North
American urban history. It is useful to think of the history of North
American housing construction in terms of the three periods or
paradigms that we developed in the introduction: 1) the era of individualism (pre-1880); 2) the era of corporate evolution (1880–
1945); and 3) the era of state intervention (post-1945). In residential
development, these paradigms evolved because of the changing
inter-relationships among a number of key factors that bear upon
the decision to build housing – legislative and regulatory influences,
construction technology, corporate organization, home-financing
methods, and consumer demand. Each paradigm produced different housing and neighbourhood forms and each is characterized by
a specific and dominant variant of the residential-development process, though the degree of dominance has never been absolute. The
shift from one paradigm to another was neither abrupt nor complete.

They present one of several useful ways of temporally ordering the features of city building, and we employ them here and in the next two chapters to structure our analysis of the residential-development process, as well as in later chapters dealing with home construction and with the mortgage. Many researchers have examined one or two of the three major components of the residential-development process. Most have studied residential development under a single paradigm. We believe that the strength of our examination lies in its breadth. Over the remainder of this and the subsequent two chapters, we portray subdivision, home construction, and marketing under three development paradigms, beginning with the least well-researched and temporally most remote period – the era of individualism.

SIGNIFYING URBAN INTENTIONS: THE SUBDIVISION PROCESS

It was largely through the creation of residential subdivisions that the spatial form of North American towns and cities was established. By the mid-1780s this process could no longer occur in a vacuum. The alignment of new urban subdivisions was usually related to an already existing survey system such as the rectangular concession-lot and concession-line arrangement that had been superimposed on much of Southern Ontario by the end of the eighteenth century, in preparation for the speedy and efficient agricultural settlement of the area. This had produced a basic grid of one-mile squares in some areas and one-and-one-quarter mile squares in others. These were further subdivided into rectangular farm parcels, which were granted to both settlers and speculators by the Crown. In the Hamilton area the farm parcels were one-quarter mile wide by five-eighths of a mile deep and contained 100 acres; in Toronto they were one-quarter mile by one and one-quarter miles and contained 200 acres.[7]

Unless influenced by severe topographical variations or by an already established system of Indian trails, the urban surveys in this period were invariably drawn up on the basis of the grid system that had become popular in Renaissance Europe. This idea was introduced to North America by William Penn in the late seventeenth century, and rapidly became the most common urban form.[8] Penn's dominance of the design of early Philadelphia was not unusual. The early years of subdivision activity in many widely scattered cities were usually controlled by a few individuals. Central Hamilton, for example, had only about ten Crown grantees, while that part of central Toronto lying between Queen and Bloor Streets and known as the

Park Lots (an area of more than 12 square kilometres) had fewer than three dozen original owners.[9]

The case of Hamilton's Mills family illustrates the role that prominent early settlers often played in North American cities during the era of individualism. James Mills became one of the handful of early property owners who controlled the land that would form the nucleus for modern-day Hamilton. Over about four decades he rose from tailor and merchant to wealthy landlord. His influence on the local real estate market began in earnest in 1816 when he and his brother-in-law Peter Hess purchased roughly 500 acres in Barton Township from the widow Margaret Rousseau for £750. At first the two partners grew crops and raised livestock, but agriculture was not their primary objective. Mills and Hess hoped that land values, especially for building lots, would rise quickly because of the designation of George Hamilton's town site as the future courthouse town for the Gore District in March 1816. In plain terms, Mills and Hess, like most of their wealthy contemporaries, were land speculators. They signed their transaction with widow Rousseau on 24 June 1816 and registered the legal instruments recording the sale on 13 July 1816.[10] Such a registration was purely voluntary. Many colonial land sales, all being private business transactions, were never registered officially. That created confusion in the land market, for there was a risk associated with the purchase of land whose title was unclear. Mills and Hess avoided carelessness, knowing that a registered transaction would increase a potential buyer's trust. Shortly after their land was registered, they divided it in half, leaving the two men free to pursue separate courses of action.

Low prices for Upper-Canadian agricultural products and low rates of immigration meant that Hamilton grew very little during the 1820s. In contrast, the 1830s brought a seven-year boom and Hamilton's locus of growth favoured the Hess and Mills lands. The former's tract was closer to the original town site and, like most large landowners of the day, Hess sold blocks of property to other land speculators. Investors like Hess generated their profits through a relatively quick turnover of their holdings. Mills, on the other hand, pursued development tactics that promised higher long-term profits but at a much slower pace. Mills certainly made sales to individuals and, in the case of a 10-acre transfer to Abraham Smith in 1824, to speculators.[11] Rather than selling all of his land in this fashion, however, Mills partitioned much of his property into small lots. He had cheap houses as well as some fine commercial buildings built on some, while others were sold directly to small-scale builders and investors. A shrewd marketer and property evaluator, Mills had advertised his building lots as early as 1834, and took care to stress their favourable topography. Elevated parcels provided more than a scenic view in the decades before the advent of an elaborate infrastructure system of water mains and sewer lines. Another early Hamilton subdivider, Hugh B. Willson, proudly proclaimed of his newly platted east-end Hamilton property: "these Lots [are] unsurpassed by the most favored neighbourhood ... The soil is dry and warm, and of the finest description for Gardens, and requires little or no drainage – while the most abundant supply of pure water is obtained in every part of the property, by sinking wells to a depth of only twelve feet."[12] A meandering ridge cut diagonally across the middle of the Mills estate and it was here that he had built his own brick residence, "The Homestead," in 1835. The fifty lots situated along this ridge were thought to have the best potential for prestigious development, and by mid-century they housed a higher social class than those in the lower-lying sections of Mills's survey, while more than 200 lots had been sold across the entire Mills tract (see chapter 8).[13]

How much profit did Mills make? Prices for lots were seldom advertised during his lifetime, and only a few transactions at the registry office recorded a sale price. From 1840 to 1850, eight land transfers involving Mills's lots cited fairly uniform prices, averaging £55 per lot. It is conceivable, therefore, that by mid-century if Mills had sold about half of the land, for which he originally had paid £375, he had netted £6,000 to £12,000, or better than twenty times his initial investment. Granted, it had taken twenty-five years, but property taxes were very low and promotional and developmental outlays had been minimal.[14]

The influence of the Mills family on Hamilton lasted well beyond the subdivision stage of the residential development process. Family members were memorialized in the city's street names, and unlike most of their peers, the Mills became property developers and landlords. Their real-estate holdings were passed down through the generations. At the time of his death in 1852 James Mills probably owned about fifty residential and commercial units that together would have provided him with an annual rental income of some $2,500 to $3,000. By 1861 the family's yearly rental income from its ninety-five dwellings and commercial properties would have been between $7,000 and $10,000. A typical unskilled urban labourer would have earned $250-$350 per year at this time.[15] For a land-rich family like the Mills, urban real estate could prove to be very profitable. Our portrait of the Mills family suggests both continuity and concerted effort with respect to Hamilton's land market. Such families helped to determine the course of urban development. Others who were involved in land subdivision during the era of individualism took a path similar to that followed by Hess.

Building lots were seldom in short supply during the era of individualism. Between 1850 and 1880 more than one hundred surveys, containing nearly five thousand lots, were registered for Hamilton in the Wentworth County registry office. These plans meant that enough new building lots were created during those three decades to accommodate almost thirty thousand new Hamiltonians, although the city's population increased by about only twenty-two thousand people over the same period. The new lots were added to those remaining undeveloped and even unsold from earlier subdividing booms in the 1830s and 1840s. Hamilton's assessment rolls for 1852 identified 2,525 vacant lots within the city. By 1881 there still were 2,166 undeveloped building lots in Hamilton, enough to house more than thirteen thousand additional inhabitants.[16] Even this understates the excess of subdivided land within easy access of the centre of the city, for platting had gone on beyond Hamilton's boundaries, especially to the east of the city. In the 1850s alone, surveys containing at least 676 lots had been registered for those sections of Barton Township adjacent to the city. The lots in these subdivisions were often touted for their modest prices and low taxes. Settlement proceeded very slowly in these peripheral areas, for they lacked many of the hallmarks of civilized urban existence, such as access to piped water.[17] While inexpensive to buy, ex-urban property did not escalate in value very quickly. In 1868 Hamilton land agents Moore and Davis told John Billings of Galt, the owner of some lots in Gilkison's Survey, Township of Barton, that "single lots are of little value, being so far from the city." Five years later they urged Robert Walker of Toronto to accept an offer of $100 for lot 291 in the same survey and suggested that other lots in the area had fetched only $50 to $60. At an 1874 auction sale of villa lots in suburban Burlington Beach, seventy properties sold for an average of $46 each.[18]

Subdivision in and around Hamilton during the era of individualism underscores the cyclical nature of the urban land market (see table 1.1).[19] As in other centres, Hamilton subdividers did not always respond with equal enthusiasm to the siren song of property investment. Plan registration dates alone suggest a fluctuating interest in subdividing, producing cycles that fell into line with broader North American economic trends, a point underscored in Isobel Ganton's study of the 794 subdivisions registered in Toronto between 1851 and 1883.[20] Platting was vigorous in Hamilton through the middle

Table 1.1
Economic Cycles in Nineteenth-Century North America

Boom Years	Bust Years	Recovery Years
1816–1819	1819–1823	1823–1834
1834–1837	1837–1842	1842–1854
1854–1857	1857–1862	1862–1865
1865–1873	1873–1879	1879–1886
1886–1891	1891–1896	

Sources: Murray G. Rothbard, *The Panic of 1819: Reactions and Policies* (New York: Columbia University Press, 1962); Reginald C. McGrane, *The Panic of 1837: Some Financial Problems of the Jacksonian Era* (Chicago: University of Chicago Press, 1964); George W. Van Vleck, *The Panic of 1857: An Analytical Study* (New York: AMS Press, 1967); and Samuel Rezneck, *Business Depressions and Financial Panics: Essays in American Business and Economic History* (New York: Greenwood Publishing, 1968)

of the 1850s (see table 1.2). Between 1850 and 1857 an average of 3.8 surveys and 212 lots were placed on the market annually. Comparable figures for 1858–67 were one survey and 51 lots per year; for 1868–74, 5.3 surveys and 200 lots; and for 1875–80, 4.5 surveys and 197 lots. After 1857, the same year as the devastating international Panic, subdividing activity was much reduced, especially during the early 1860s, with no registered surveys at all in 1862 or 1863. Platting rose from this nadir to a new peak between 1868 and 1873 when 34 surveys were registered. The latter year marked the onset of another international financial crisis and subdividing did not pick up again in Hamilton until 1877 or 1878. Only a few Hamilton subdividers acted counter to these trends, probably because of either imperfect knowledge of market conditions or over-commitment to a particular project.

The term accretion best describes Hamilton's growth during the era of individualism (see figure 1.3). Most parts of central Hamilton were platted prior to 1847, when the mandatory registration of subdivisions came into effect. These old, unregistered surveys often produced messy land titles.[21] The addition of new subdivisions to the core area after 1847 was not always a neat and orderly process. As early as the 1850s several plans had been registered for subdivisions in the area beyond the city's eastern limits. In contrast, some areas within the city limits were not subdivided until well after 1880. The role of individual decisions in the timing of platting was important – someone had to see urban potential in a given parcel of land. The property, in turn, had to be obtainable for this purpose. In other words, the owner had to be willing either to subdivide the

Table 1.2
Characteristics of Hamilton Registered Surveys, 1850–1880

	No.	%
DATE OF REGISTRATION		
1850–54	16	15.4
1855–59	16	15.4
1860–64	4	3.8
1865–69	12	11.5
1870–74	29	27.9
1875–80	27	26.0
NUMBER OF LOTS		
fewer than 6	7	6.7
6–10	10	9.6
11–25	34	32.7
26–50	27	26.0
51–100	13	12.5
101–200	8	7.6
more than 200	5	4.8
MODAL FRONTAGE (IN FEET)		
fewer than 30	10	9.6
31–40	9	8.7
41–50	49	47.1
51–60	18	17.3
more than 60	6	5.8
not listed/illegible	12	11.5

Source: Calculated from all plans of subdivision in the Wentworth County Registry Office for the years 1850–1880

land or to sell it to someone who would. A reluctant owner, or a battle over the ownership of the property if the original owner died, could delay the process interminably, often causing development to leapfrog the dormant parcels. Only later was it understood that subdivision in the era of individualism had been premature, haphazard, and costly. Writing in the 1930s, the noted planner Thomas Adams observed that

speculative methods ... result in subdivision excessive in relation to existing and even to prospective requirements. Much of the subdivided land would have retained its value for agriculture had it remained in acreage, but this value is destroyed by the process of subdivision. As prospective building land, it may have no value at all ... On the other hand, if the land is sold at a price profitable to the speculator, then all the expenses due to wasteful speculation have to be charged against the prospective user. Enormous financial losses to communities and owners are final consequences of these methods.[22]

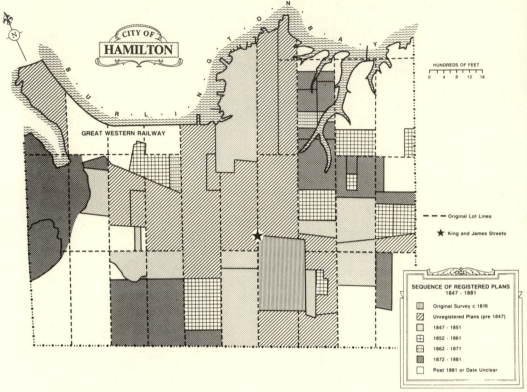

Figure 1.3
Sequence of registered plans in Hamilton, 1847–1881

Accessibility was important to astute land subdividers. Property near the main thoroughfares of the city (King, James, John, Main, and York Streets) was usually subdivided before land some distance from a major street (see figure 1.3). This was certainly true in Hamilton's east end where, by 1861, subdivisions fronted the entire length of King Street, and extended well beyond the city limits into Barton Township, while property only half a mile to the north or south remained unsurveyed for considerably longer. Topography could also influence the timing of residential surveying, as illustrated by the Mills family's holdings. Low-lying, marshy areas, such as those in south-western and north-eastern Hamilton, usually took longer to become subdivided, especially if they also suffered from poor access to the centre of the city.

In at least three respects patterns established in Hamilton's early unregistered surveys persisted throughout the era of individualism.

As in the pre-registration years, few plans laid out after 1847 crossed over the lot lines of the original one-hundred-acre parcels of the 1790s. Land assemblers usually seek to minimize the number of property owners with whom they must deal. The generous size and the grid plan of the original Crown grants made such negotiations largely unnecessary. Those registered plans that did cover portions of more than one of the original Crown lots were found in areas where the property had already been in the hands of one owner. A second persistent feature in the post-1847 plans was a grid alignment regardless of topography. All of the plans examined adhered to this format, and Hamilton did not get alternative configurations until after the turn of the century (see chapter 2).[23] The grid was adopted without question in most North American cities at this time. As Isobel Ganton has noted, the grid "was easy to lay out and extend, easy to describe in deeds and portray on plans, and readily comprehensible in terms of the size, location, and utility of lots by buyers."[24] The final feature of the Hamilton registered plans that seems to have been inherited from the earlier surveys was a relatively generous attitude toward the size of building lots. From the first, lots with frontages of fifty or sixty feet were common (see table 1.2). The typical lot in the typical Hamilton survey (50 by 125 feet) contained one-seventh of an acre of property. Such lots were quite large when compared to those being platted in most American cities at this time, though much narrower lots began to appear in Hamilton after 1880 (see chapter 2).[25]

The earlier, pre-1847 surveys differed from the later ones in one important respect. Almost without exception the former plans had encompassed relatively large numbers of building lots. George Hamilton's survey contained 229 lots and most of the other early plans were composed of at least half this number. James Mills's largest survey covered 19 city blocks and contained 240 building lots, another defined 12 blocks and encompassed about 200 lots. One 1834 plan, drawn up for Sir Allan Napier MacNab, covered 40 blocks in Hamilton's north end and contained 429 lots, the largest survey laid out before 1881. Many of the earliest subdivisions had been platted by Hamilton's pioneers who, like James Mills, held at least one original one-hundred-acre Crown grant. After 1847 there was more variation in the size of surveys, with greater emphasis on smaller plans. Some of the smaller surveys in Hamilton and elsewhere represented resubdivisions of earlier schemes, a process that sometimes produced rather narrow lots.[26] The size of the surveys was a function of the land that was available. When large plots remained undeveloped, large surveys could be laid out; when they did not, smaller plans were registered. Large-scale plans did not disappear as the era of individualism progressed, they merely became less common, except at the edge of the city. Transportation improvements and changes in the characteristics of the subdividers themselves would combine in the years after 1880 to bring larger subdivisions back into prominence (see chapter 2).

Who were Hamilton's subdividers during the era of individualism? The 104 plans registered between 1850 and 1880 were laid out by 94 different people and groups. Seventeen of them involved partnerships, usually consisting of two or three people. Nine other surveys were registered by the trustees for the estates of deceased persons, and one survey was registered by a bank. The rest (74 per cent) were registered in the names of individuals. Not only was the registration of plans dominated by individuals, but land subdivision must have been a mere sideline for most of those who were involved in the process. Seventy-six of them were mentioned in only one plan, twelve in two plans, and five in three plans. Archibald Ferguson, who inherited the better part of two east-end Crown grants from his father, Peter Ferguson, was the only one who had made anything approaching a career in land development during this period – he was involved in the registration of ten plans.

Hamilton's mid-nineteenth-century subdividers looked after their land interests through participation on bodies that dealt with property matters, especially if the decisions made by these organizations could influence land prices. For example, twenty-five, or 30 per cent of those eligible (eliminating estates, companies, and women) served on the Hamilton City Council at some time between 1847 and 1881. Put another way, one of every eight who sat on the council during the era of individualism was directly involved in land subdivision.[27]

Manuscript census records for the period reveal something about the age, occupation, religion, and birthplace of Hamilton's mid-Victorian subdividers. Most lived in and around the city, and fairly complete census information exists for seventy-two people.[28] These data show that at the time when their first plans were registered their average age was 42.8 years, so they were probably well-established members of society. Unfortunately, their length of residence in Hamilton is not known. Hamilton's subdividers held a fairly restricted range of occupations. Merchants and lawyers dominated, accounting for 43 per cent of the group, while a further 17 per cent were gentlemen and physicians. Only 8 per cent were listed as builders or contractors. These figures are similar to those calculated by Ganton for Toronto's subdividers during the same period.[29] The religious affiliation of those involved in the registration of plans in Hamilton was overwhelmingly Protestant. Only 4 per cent were Roman Catholics. Finally, those born in either Canada or England predominated, accounting for slightly more than three-fifths of the names on the plans. The occupational, religious, and ethnic characteristics of the subdividers mirror the contours of the city's mid-century elite delineated by Michael Katz.[30] Given the unequal distribution of property ownership during the era of individualism, this was hardly unexpected. Subdividers were potentially powerful people, but the process that they initiated was a remarkably decentralized one.

The records available for the study of residential land development in Ontario are circuitous and voluminous. For this reason a sample, composed of all plans of subdivision registered for the City of Hamilton between 1847 and 1861 that contained at least fifty building lots, was drawn from the Wentworth County registry office records. Complete records could be located for 14 surveys, covering 115 city blocks and containing 1,668 building lots (see table 1.3). This sample contained two-thirds of all lots registered between 1847 and 1861, and one-third of all lots recorded between 1847 and 1881. Collectively, the 14 subdivisions encompassed 23 per cent of the blocks within Hamilton's city limits and they provided good areal representation of those portions of Hamilton that were being subdivided during the era of individualism (see figure 1.4).[31]

Their developers, too, seem representative. All eighteen were men of high occupational status, and the vast majority resided in Hamilton at the time when their survey had been registered.[32] In terms of organization, two men each were associated with two different surveys and four of the plans involved partnerships. The individuals responsible for these surveys must have known each other well. Five sat as members of Hamilton's City Council, and two others, MacNab and Cameron, were members of the Provincial Legislature, the former serving as Prime Minister from 1854 to 1856. We know for certain that ten of them served as directors of business concerns such as railways and other transport interests, utilities, and insurance companies, or incorporated organizations such as libraries and the Mechanics' Institute.[33] In the 1850s Hamilton's large subdividers were strangers neither to each other nor to the levers of power in their mid-nineteenth-century city. They were the cement that held the place together in the face of a tremendous rate of turnover in the population as a whole. And one of the most important factors in their persistence must have been their involvement with land.

Table 1.3
Characteristics of the Sample Surveys and Their Subdividers

Survey Number[1]	Date of Plan	City Blocks	No. of Lots	Modal Frontage (Feet)	Sales before Registration
1	03/58	8	98	47	No
2	10/48	20	220	63	No
3	01/54	12	106	56	Yes
4	10/57	10	129	66	Yes
5	08/48	9	109	50	Yes
6	?/54	6	100	40	No
7	10/47	12	206	52	No
8	10/47	8	164	52	No
9	03/53	2	47	66	Yes
10[2]	01/55	4	55	30	Yes
11	06/55	2	49	48	No
12	?/57	3	68	50	Yes
13	?/47	4	83	57	Yes
14	11/54	15	234	50	Yes
Totals		115	1668		

Survey Number[1]	Subdivider(s)	Occupation	Elected Office	Directorships	
				Rail	Others
1	Stinson, T.	Banker	No	–	–
2	Hamilton, P.H.	Gentleman	No	1	1
3	Ferguson, P.	Gentleman	No	–	–
4	MacNab, A.N.	Gentleman	MPP	3	2
5	Beasley, R.G.	Gentleman	No	–	–
6	Tisdale, V.H.	Merchant	No	–	–
7	Willson, H.B.	Gentleman	City	2	–
8	Willson, H.B.	Gentleman	City		
9	Cameron, J.H.	Gentleman	MPP	–	–
10[2]	Pringle, J.D.	Lawyer	City	2	1
	Logie, A.	Lawyer	No	–	–
	Griffin, W.	Crown Clerk	No	–	–
11	Billings, W.L.	Physician	City	1	1
	Lister, J.	Merchant	City	1	–
12	MacNab, A.N.	Gentleman	MPP		
13	Moore, E.	Lumber Merchant	No	–	–
	Moore, J.F.	Lumber Merchant	No	–	2
14	Kerr, A.	Merchant	No	1	4
	McLaren, W.P.	Merchant	No	3	4
	Street, R.P.	Actuary	No	2	1

Source: Wentworth County Registry Office records, Hamilton City Council Minutes, and the *Canada Gazette*

[1] see figure 1.4 for the location of each survey

[2] A resubdivision of lots 77–88

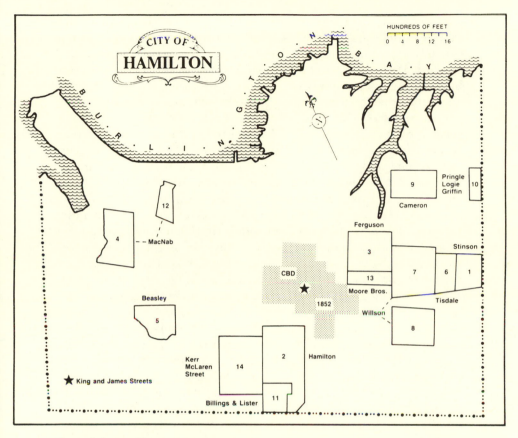

Figure 1.4
Location of the fourteen sample surveys

MARKETING THE SUBDIVISIONS

Popular sentiment seems to hold that those who buy and sell urban property make a profit. Few detailed studies exist for the era of individualism, but several careful analyses of the purchase and sale of urban land for later periods have pointed to a surprising number of individuals who actually lost money.[34] As one recent newspaper article proclaimed, "Making a Profit on Real Estate Means Hard Work."[35] It also took some luck. Profit on land cannot be regarded simply as the selling price minus the initial purchase price, for a variety of different costs are associated with the purchase, retention, preparation, and sale of urban property. Prudent analysis, therefore, is required to accurately assess the profitability of land subdivision for the period prior to 1880.

It would be instructive if the charges incurred for each of the sample Hamilton subdivisions could be precisely indicated. Unfortunately this cannot be done for any of the fourteen surveys under scrutiny. Detailed account books simply do not exist. Those involved in the development industry have often displayed a secretive bent, an attitude that led John Herzog to comment that, even today, "builders are notorious for their failure to keep more than very crude records of their operations."[36] Residential building, according to a recent study by James McKellar, "is not an industry that has been well studied and much of what is documented is anecdotal; information resides primarily in the memories of industry participants."[37] During the era of individualism, to judge from surviving business and civic records, accounting practices were often idiosyncratic. Losses due to periodic moves or even to routine housecleaning have eroded the already notoriously slim base of original material recording the balance sheets of mid-nineteenth-century members of the development industry. Nevertheless, a few traces have survived.

Some significant expenses could mount up in the purchase and sale of urban land, as the surviving account books of a Toronto lawyer, George Taylor Denison, reveal. In December of 1867, Denison's brother, Fred, purchased twelve Toronto lots from the family estate for a total of $495. The brother kept these lots for two years and three months, at which time he sold eight of the lots, gave two to George to pay off a debt, and kept the remaining two for himself. The eight lots sold for $710, a gross profit of $215. This seemingly handsome return was reduced by almost two-thirds due to a variety of expenses such as mortgage interest, property taxes, and miscellaneous items that were meticulously recorded by George Denison in his ledger.[38] Thus, even when the initial purchase price was low, carrying charges on vacant land could be relatively high, especially if the would-be investor did not have sufficient capital to finance the purchase without a mortgage. The Denison example and most published studies concerning the profitability of urban land relate to instances in which already platted lots were bought and sold. What was the situation when the property being offered for sale was newly created? Unfortunately, complete and reliable data were available for only ten subdivisions. However, by estimating their profitability we hope to show that even though the subdividers could face substantial carrying charges and other expenses, most of them did very well indeed (see table 1.4).

The initial price that a subdivider paid for land had a considerable impact on the ultimate profitability of any platting venture. A low purchase price, in conjunction with the rapid resale of the property,

Table 1.4
The Success of the Subdividers in Selling Their Lots: The First Decade of Sales

Survey Number[1]	First Sale	No. of Lots	Lots Unsold after:				% Not Sold	Taxes[2]	Best Year
			1 Yr	3 Yrs	5 Yrs	10 Yrs			
1[3]	02/52	98	98	97	91	40	41	$1,086	55–6
2[4]	11/48	220	183	172	96	0	0	1,116	51–2
3	09/53	106	67	42	30	20	19	522	53–4
4	05/45	129	49	34	26	12	9	392	45
5	10/48	109	108	106	12	1	1	573	51–2
6[5]	10/55	100	100	77	77	0	0	908	55–6
7	07/48	206	203	155	146	9	4	1,656	55–6
8	08/48	164	0	0	0	0	0	246	48
9	02/53	47	47	2	2	0	0	83	53
10	10/53	55	26	12	6	5	9	137	53–4
11	07/57	49	49	46	33	4	8	435	62–3
12	10/48	68	17	12	4	4	4	113	48–9
13	05/47	83	68	62	31	1	1	500	50–1
14	09/53	234	144	19	16	16	7	441	54–5

[1] For locations of each survey see figure 1.4.
[2] Estimated on the basis of $1.50 per unsold lot per year over the course of ten years.
[3] Thomas Stinson built 41 cheap houses on his lots and rented them to members of Hamilton's working class. The dwellings were still owned by family members in 1881.
[4] Peter Hamilton died in 1854 and his unsold lots were divided among his heirs.
[5] Valentine Tisdale lost 60 of his lots through foreclosure in 1862.

almost always resulted in a substantial profit for the subdivider. Of course, a lucky few paid nothing for their land, receiving it either through an inheritance or as a grant from the Crown. Most of the subdividers in our sample, however, paid something for their property.

A vital factor that determined the magnitude of a subdivider's profits was the length of time it took to dispose of all of the building lots that had been carved out of the original raw block of land. Taxes had to be paid each year on the unsold lots, and while the rates (about $1.50 per lot per year) were quite low, this cost diminished earnings as the years passed. Success in the marketing of building lots was related to the demand for land, which closely mirrored the economic or business cycle. Those who released land after the cycle had begun to turn down often incurred high total tax payments on their properties. In some cases the city even seized land because subdividers would not, or perhaps could not, pay their taxes.[39] Furthermore, while building lots remained unsold, many subdividers had to make mortgage payments, thus forfeiting any interest that

their equity could have made elsewhere. Robert Swierenga has suggested that, at this time, US investors could expect rates of return
of between 4 and 7 per cent from such relatively safe sources as
government, commercial, and railway bonds, and savings banks.[40]
In 1878 Hamilton property agents Moore and Davis informed local
MPP James M. Williams that in certain situations it would be much
better for people "to invest their means in Bank Stock, Bonds, and
Mortgages than [to] build houses and pay taxes upon them, even if
they were compelled to loan money at 7% interest."[41] Even though
there apparently were safer avenues of investment during the era
of individualism, these were not without their own pitfalls, and land
remained a very attractive arena for capitalists throughout the period.[42]

The primary goal of subdividers during the era of individualism
was the marketing of vacant building lots. First, their raw property
had to be "prepared" for sale. Few regulations governed this procedure during these years, but several types of land or site preparation costs were incurred by all urban subdividers. A surveyor had
to be hired to divide the land into building lots. Records indicate
that in the late 1850s and early 1860s surveyors were paid about
$4.50 per day. Given that at least two men were required and that
a proper survey for a project with more than fifty lots probably took
two to five days to complete, a figure of between twenty and forty
dollars would appear a reasonable estimate for this component of
the land preparation costs.[43] Once the survey crew had completed
its task, the resultant plan for the subdivision had to be both approved by the local provincial land surveyor and registered in the
local registry office. The fixed charges for each hardly burdened the
subdivider. In 1865 an Ontario registrar could charge $1.25 for the
registration of a plan of subdivision.[44] If the subdivider employed
the services of a lawyer to oversee the registration of the plan, then
legal fees, likely under five dollars, had to be paid. In 1849, for
example, Toronto lawyer George Taylor Denison charged Septimus
Tyrwhitt 2s 6p (about 50 cents) to register an instrument in the York
County Registry Office. Miscellaneous expenses, such as the printing
of the plan and related forms, probably brought the total cost of
survey approval and registration to between ten and twelve dollars,
a modest sum.[45]

To make the property more attractive to prospective purchasers,
subdividers could improve their holdings. In the 1850s the only such
betterments to be considered were the installation of streets and
sidewalks and such aesthetic embellishments as the planting of fruit
and ornamental trees on the vacant lots. According to Ann Durkin

Keating, few Chicago subdividers sold improved lots prior to 1880.[46] Judging by the long list of memorials to the Hamilton Board of Works in which the residents of new areas of the city petitioned their municipal government for roads and sidewalks, few early Hamilton subdividers invested very much in the physical improvement of their holdings either. The representatives of one group of east-end residents in an 1865 memorial to the Chairman of the Board of Works noted that "we are agents for a block of buildings on the corner of Victoria Avenue and Robert Street, called Piper's Row, and we have great complaints from the tenants on account of not having sidewalks to them. The sidewalk is about half way from Wellington Street and we only ask a one plank walk to the houses so that tenants may be able to get to them in a muddy season of the year."[47] At this time, most Hamilton subdividers probably did no more than hire a ragged crew of day labourers (at one dollar each per day) to clear a rough path in order to demarcate the road allowances in the subdivision. Some did not even do this much, for tree stumps remained in the middle of some Hamilton streets well into the 1850s.[48]

Most of Hamilton's subdividers ignored aesthetic features during this period. Only one of the group under study here, Hugh B. Willson, ever advertised that he had planted trees on his lots. Yet such adornments could be important in a tight real-estate market for as Moore and Davis, the Hamilton land agents, reported to Hamilton-area MP James Rymal: "there is no handsomer ornament can be put about dwellings in cities and towns than shade trees on the streets, and to encourage the planting of them there should be a stringent clause inserted [in the Vagrant Bill] for their protection."[49] On average, it seems unlikely that many subdividers spent more than one or two dollars per lot on improvements. For developed properties, expectations and expenses probably were higher. George Taylor Denison, for example, recorded expenses of six dollars for the planting of spruce trees, six dollars for advertising, and six dollars for the preparation of deeds in conjunction with the sale of a brick house in Toronto in the 1860s. These are minuscule figures compared to the large sums that developers now must spend to provide the serviced lots, requisite park land, and amenities required to satisfy both government regulations and market demands (see chapters 2 and 3).[50]

In the mid-Victorian city, services such as gas, water, and sewer lines were left to be negotiated between residents and the bodies providing the infrastructure. Parks would appear as afterthoughts, if at all. Electricity, telephone lines, and public transportation did not arrive in Hamilton until the 1870s, well after the subdividers in question had disposed of most of their holdings. In the case of water supply, Moore and Davis spelled out the alternatives open to Hamilton property owners in an 1861 letter to Joseph Sudborough of Toronto: "As to making the well deeper, it would probably cost from ten to fifteen dollars. As to putting the water in your houses, the company put them in as far as the line of your lot and you would require to pay the balance. It would probably cost some five or six dollars each and it would require to be put in each house most likely."[51] While hardly onerous in terms of charges, the achievement of a fully-serviced building lot in a mature neighbourhood during the era of individualism required much perseverance and the advice and services of professionals.[52]

The final category of costs associated with the subdivision process related to the actual disposition of the building lots. Initially, at least, virtually all of the subdividers attempted to sell their building lots at public auction, for as long as they had at least ten lots remaining in a survey. During the period in question, the auctioneer's fee was between 1 and 3 per cent of the selling price of the property.[53] Advertising was an important part of the residential land-development process, and newspaper editors played a helpful role in the sale of urban land. All of the subdividers examined advertised auction sales of their properties at one time or another in the pages of the Hamilton *Spectator*, other local journals, and even some out-of-town publications. Advertising expenditures frequently bought more space than might have been expected, for the editors of the local papers also "hyped" the upcoming auctions in their local news sections and the auctions themselves were nearly always treated as newsworthy events. Some subdividers supplemented their newspaper advertisements with the distribution of inexpensive handbills and broadsides that were usually printed in a newspaper office. Advertisements ran for long periods and probably cost the average subdivider at least eighty dollars.[54] While most contained straightforward descriptions of the property and the details of its disposition, some were accompanied by elaborate and costly maps and a number of them intimated other costs that had to be borne by the entrepreneur. A few subdividers offered prospective buyers inducements such as free lunches and a limited number of free or cut-rate lots (see figure 1.5). At a conservative estimate, it would appear that most subdividers could count on spending at least one hundred dollars on various forms of advertising in the marketing of their holdings.

Finally, a lawyer usually prepared the deeds that the subdivider would use to transfer the lots to their new owners. Very often a printed form was used for this purpose (see figure 1.6), thereby

SCHEME

FOR THE

DISPOSAL OF ONE HUNDRED LOTS!!

KING, STEVEN, TISDALE, AND EMERALD STREETS,

ON

IN THE CITY OF HAMILTON.

Five Lots at £15 each, Eighty-three Lots at £120, & Twelve Lots at £200 each.

Payable, £15 on day of purchase; balance, at the expiration of Five Years. The interest only being required half-yearly, say, on the 1st days of July and January in each year, with the option to purchaser of paying the whole principal sum at any time, for which a Deed in fee simple will be granted.

FIVE LOTS.—Nos. 1, 8, 31, 56, 71, ... **£15 each.**

The Drawers of which will receive a Deed immediately.

TWELVE LOTS.—Nos. 47, 48, 49, 50, 51, 52, }
95, 96, 97, 98, 99, 100. } .... **at £200 each.**

The balance, say EIGHTY-THREE LOTS, ... **at £120 each.**

As per the accompanying Plan. The Choice of Lots to be decided by Ballot.

The drawing to take place on the .... *day of* .... *next.*

V. H. TISDALE.

WE, the Undersigned, hereby agree and bind ourselves to become purchasers of Lots, in the City of Hamilton, upon the above-stated conditions.

HAMILTON, 21st May, 1855.

Figure 1.5
Valentine Tisdale's scheme for the disposition of lots, 1855

CONDITIONS OF SALE

OF PART OF LOT 18, IN THE 3RD. CONCESSION, BARTON, THE PROPERTY OF

JAMES C. MACKLIN, Esq.

1. No person shall advance less than $5 at any bidding, and no person shall retract his bidding.

2. The highest bidder shall be the purchaser; and if any dispute arise as to the last or highest bidder, the property shall be put up at the last undisputed bid.

3. The vendor reserves the right of making one bid for each lot.

4. The purchasers shall at the time of sale pay down a deposit in the proportion of $10 for every $100 of their purchase money to the vendor or his solicitor, and the balance of one-third of the purchase money in one month from the day of sale, the remaining two-thirds to be secured by mortgage, payable in two annual instalments without interest. Ten per cent. off the whole purchase if paid down.

5. The vendor will prepare and execute deeds, at his own expense, to the purchasers. The purchasers to pay the expense of preparing and executing the mortgages and of registering both deeds and mortgages.

6. The title of the property has been thoroughly investigated several times during the last few years, and the Bank of Montreal, who sold to the vendor, foreclosed a mortgage upon the property in the Court of Chancery, and had a direction inserted in the decree to ascertain who were the owners of, and encumbrances upon, the land: the inquiry was answered by the officer of the court after investigation, and all the persons whom he found to be interested in the property were foreclosed. The title shall therefore commence with the conveyance from the Bank of Montreal to the vendor, and no purchaser shall call for any evidence of the prior title, but may at his own expense have copies of the abstract from the Registry Office and copies of the deeds in possession of the vendor.

7. The purchaser of the largest part in value of the property shall be entitled after the completion of the sale of all the property to the custody of the muniments of title in the vendor's possession, and shall at the expense of the purchasers of the other portions of the property requiring the same enter into usual covenants with them respectively for the production and furnishing copies of such muniments. Until all the property shall be sold, the vendor shall retain such muniments and the purchasers of the portions sold shall be entitled in the meantime at their own expense to the production of such muniments and to copies of them but not to a covenant to that purpose.

8. If the purchaser shall fail to comply with the conditions aforesaid or any of them, the deposit and all other payments made thereon shall be forfeited and the premises may be re-sold: and the deficiency, if any, by such re-sale together with all charges attending the same or occasioned by the defaulter shall be made good by him.

I agree to purchase from James C. Macklin, Lot No._____ in Block_____ on_____

_____ Street, in the survey of part of lot eighteen in the third concession of Barton as laid out

into town lots by David C. O'Keefe, P. L. S., for him, for the sum of $_____

_____ and upon the terms mentioned in the foregoing conditions

of sale.

Dated this _____ day of _____ A. D. 1873.

Witness

}

Figure 1.6
Conditions of sale in Macklin's survey, 1873

reducing the overall legal fee. These charges varied from solicitor to solicitor, but a figure of between ten and twenty dollars would be a reasonable estimate for this expense, especially if standardized contracts were prepared. In 1849, for example, George Taylor Denison charged Septimus Tyrwhitt £1 5s (about $5.00) to draw up a memorial of a deed and £1 10s (about $6.00) to fashion a mortgage.[55]

Most of the expenses outlined above were negligible in themselves and there was surely considerable variation in the costs paid by individual subdividers. It seems unlikely, however, that many spent more than five hundred dollars on the preparation of their land, advertising, and legal fees. The major expenses, aside from the purchase of the raw land, were the fees paid to the auctioneer who sold their building lots and the cost of holding the property until it was sold. The former was a direct function of the size of the survey, while the latter was inversely related to the subdivider's success in selling land.

Few subdividers were able to dispose of their building lots overnight during the era of individualism. Development lagged even more. Town founder George Hamilton had been able to sell only about a fifth of the 229 lots in his 1813 survey during the subsequent decade, and about 10 per cent of the parcels remained vacant in 1852. Ebenezer Stinson, a brother of Thomas, had sponsored a thirty-one-block survey containing more than 360 lots in Hamilton's north end in 1839. About fifty of the lots remained in the family's hands in 1852 and nearly two hundred of the properties still were vacant at that time. Of the subdividers examined in detail (see table 1.4) only one, Hugh B. Willson, was able to sell all of the lots in a survey within one year, and Willson probably sold out for reasons of financial distress. Most had land left to sell five years after entering the market and ten of the fourteen were still selling their building lots after a decade. In only half of the surveys examined had the subdividers been able to dispose of at least half of their platted lots within three years. To a large extent the pattern of sales reflected trends in the real-estate cycle. Activity in the mid-Victorian Hamilton real-estate market peaked around the time of the Panic of 1857. In seven of the surveys the best year for sales fell between 1853 and 1857. Only one subdivider experienced his greatest marketing success after 1857. Those who placed their lots on the market either early, before 1850, or late, after 1855, often had to wait until the glut of surplus lots had been cleared from the market and the boom had begun to pick up again before they were able to record large numbers of sales. But there were some exceptions to this pattern, a reflection no doubt of the varying ability of individual subdividers either to take advantage of market opportunities or to create their own sales niches. For example, Sir Allan Napier MacNab quickly disposed of the lots in his two surveys by selling them to his colleagues in the Provincial Legislature, a market much less accessible to most other subdividers. Those who put their lots up for sale in 1854 or 1855 had to have some luck on their side. By that time the number of vacant lots on the market had reached almost four thousand. The Panic of 1857 caught a number of the latecomers with a large proportion of their holdings still unsold. Not everyone could ride out the lean years ahead.

Most subdividers, however, reaped quite handsome profits from their enterprises. This was especially true for those who had purchased their raw land at an early date. Even though they had been forced to sit on their holdings for relatively long periods, the initial prices that they had paid for their properties had been so modest that impressive profits could still be realized. And although they had done very little except to divide their land into building lots, they were selling property that now had a potential for much more intensive use, so large profits were to be expected. The intensification of use gave the subdividers a considerable advantage over those who speculated in already platted building lots. The latter were more susceptible to the vagaries of the land market. As we saw in the case of Fred Denison, holding costs were much more significant for these investors, since wildly inflated selling prices could rarely be justified by any further potential intensification of use. Nor did they share the subdividers' ready access to mortgage funds (see chapter 6). The building-lot speculator could garner profits comparable to those enjoyed by subdividers only under ideal market conditions. A speculative boom could provide the appropriate investment atmosphere, but the profits of the lot speculator could also be boosted through a combination of low purchase prices and rapid turnover. As we noted earlier, current research suggests that a significant number of those who speculated in building lots during the era of individualism lost money. Our evidence indicates that the same fate rarely befell the property subdivider (see table 1.5).

Financial statistics were complete enough for ten of the surveys to allow us to construct crude balance sheets for these projects. Only Peter Ferguson did not have to pay for the block of land from which his subdivision was carved. Eight of the remaining nine subdividers paid amounts ranging from $18 to $40 per lot created (Stinson, Hamilton, and MacNab) to $73 to $117 per lot created (Tisdale, Willson, and Pringle, Logie, and Griffin). In 1853 Kerr, McLaren,

Table 1.5

Estimates of Profits Realized in Ten Sample Surveys Based on Sales during Their First Decade[1]

Survey[2] Number	Purchase Price	Value of Mortgage	Holding[3] Costs	Prep.[4] Costs	Sale[5] Costs	Sales[6]	Profit	Average Annual Return
1	$2,684	$4,000	$1,686	$147	$314	$19,395	$14,564	54.3%
2	4,000	1,440	6,212	330	576	45,595	34,477	24.6
3	nil	nil	1,379	159	540	41,968	39,890	–
4	5,040	6,000	4,184	194	412	29,184	19,354	16.7
6	20,000	16,000	3,308	150	288	16,787	−6,959	−4.4
7	24,000	20,000	4,656	409[7]	813	69,344	39,366	16.4
8	16,500	nil	nil	246	274	15,350	−1,670	−10.1
10	4,000	nil	137	83	313	19,260	14,727	36.8
12	1,680	2,000	1,646	102	328	20,800	17,044	39.0
14	54,200	54,200	6,537	351	1,128	100,803	38,587	7.1
Totals:	132,104	103,640	29,745	2,171	4,986	378,486	209,380	15.8

[1] All values listed in the Registry Office records in pounds sterling have been converted to dollars at the rate of £1 = $4.
[2] See figure 1.4 for the locations of each survey.
[3] Based on the following assumptions: unsubdivided land was taxed at the rate of 20 cents per acre per year, unsold lots were taxed at the rate of $1.50 per year, mortgage interest was 6 per cent per year (with the principal retired in four equal, annual payments). Costs also include interest, at 4 per cent per year, lost while the land remained unplatted.
[4] Based on the assumption that $1.50 per lot was expended.
[5] Assuming that auctioneers received 1 per cent of the value of all sales, that expenses for advertising amounted to $100, and that legal fees associated with the sale amounted to $20.
[6] Based on the actual listed selling price for each parcel.

and Street paid $232 per lot for their large and well-located property at the base of the Mountain in Hamilton's south end. Land for seven of the ten subdivisions was purchased with the aid of a mortgage. At least three of these were held by banks, a source of capital that was not open to ordinary land buyers at this time. All of the mortgages were for at least 36 per cent of the purchase price of the raw land; three of them were for amounts that actually exceeded the purchase price. Such high-ratio mortgages were unusual within the real-estate market at this time (see chapter 6). They bear witness to the power and influence of the subdividers. In spite of their financial obligations, collectively the subdividers were able to parlay initial investments totalling $132,104 into sales of $378,486 and estimated profits of $209,380. Over the first ten years of sales, the average annual rate of return on investment was 15.8 per cent.

Losses were incurred on two of the surveys, at least on paper. In fact, probably only one of the subdividers failed to make a profit.

Hugh Willson was responsible for two east-end surveys containing a total of 370 building lots, both of which were placed on the market late in 1848 (see figure 1.4). The land for these subdivisions had cost Willson the substantial sum of $40,500. Despite extensive newspaper advertising and an unusual concern for the aesthetic appearance of his properties, public response to the lots in his northernmost subdivision was slow, with only three lots selling in the first year (see table 1.4). Apparently there were no nibbles for the lots in the southern-most subdivision. Probably in need of cash, Willson accepted an offer of $15,350 from Russell Prentiss of St Louis, formerly of Hamilton, for all 164 of the lots in the smaller, southern survey. Given the apparent sluggishness of the market, it seems that Willson preferred to take a small loss to rid himself of the financial burdens of one of his subdivisions. In the end this may have been unwise, since Willson was able to realize a handsome profit from his other survey and since Prentiss turned a profit when he sold out for $22,800 two years later to Ebenezer Stinson. On balance, however, Willson came out ahead on his two platting schemes.

The second tale of a subdivision that failed to produce a profit had a much sadder ending. In March 1854 Valentine H. Tisdale purchased fifteen acres of land in Hamilton's east end and had it surveyed by the dean of Hamilton's Victorian surveyors, Thomas Allen Blyth, PLS, into one hundred building lots.[56] This property had cost $20,000 and Tisdale had to take out a mortgage from the previous owner of the land, local banker Andrew Steven, for 80 per cent of its value (see table 1.5). In spite of frequent newspaper advertisements, and a proposed lottery scheme to dispose of the lots, from which Tisdale expected to realize almost fifty thousand dollars (see figure 1.5), he had little luck in selling the properties. Only 23 of the lots had sold in the first two years, and none at all during the next five years. The timing of his survey proved to be very poor. There already were too many vacant lots on the market in 1854, especially in the east end, where Tisdale had to compete not only with Hugh Willson and Peter Ferguson (see figure 1.4), both of whom had been on the scene earlier and still had property to sell in 1854, but also with speculators who had laid out subdivisions beyond the city limits and who offered their lots at lower prices than did the city subdividers. Moreover, the topography in Tisdale's survey may have made his lots unsuitable for immediate residential development. As late as 1870 the land agents Moore and Davis reported to the then-owner, J.R. Steven of New York, that sixty-four lots remained unsold and "as there are so many low lots it is not easy to get them sold."[57] The depression following the Panic of 1857 caught up with Valentine Tisdale and the mortgagee took back sixty of the lots through a final order of foreclosure in 1862. Bad luck, poor timing, and international financial crises could combine to crush even an innovative entrepreneur such as Tisdale. Subdividers burdened with large mortgages often were in jeopardy during the era of individualism. Yet, given even reasonable luck and sufficient funds to invest, the sale of previously non-urban land for future urban use was a rewarding proposition.

The mid-Victorian subdivider played a lesser role in shaping urban development than does the present-day land-development conglomerate. After analysing more than fifteen hundred plans registered for the San Francisco area between 1860 and 1970, Elizabeth Kates Burns concluded that

the completeness of a developer's control over development has varied. The subdivider of the nineteenth century often only laid out street and lot lines, with a minimum of improvements for streets and public utilities. The consumer was left to contract separately for the building of a house on his lot. Later subdividers prepared both land and house for the consumer, setting the initial character of the tract with more certainty. More recently "community builders" provide residential areas for a range of income levels and include commercial and public facilities in a single large development.[58]

Nevertheless, the actions of early subdividers had several implications for both the physical development of the city and the formation of what might be termed an "urban-development mentality." On one level, real-estate platters provided the physical context for urban growth. By converting their visions about the future of their city into reality, they determined such lasting features as the width and spacing of streets, the presence and location of amenities, and the dimensions and orientation of building lots. In Jerome Fellmann's words, they provided the "framework within which the expansion of the built-up area of a city occurred."[59] In many instances, subdividers provided the names for the streets in the newly platted areas, frequently using their own surnames – as was the case with MacNab, Stinson, Ferguson, and Tisdale – and the names of family members.[60]

During the era of individualism, subdividers went about their business in various ways, much like the period's artisan manufacturers (see chapter 4). Even though all of the Hamilton subdividers who were studied in detail employed the grid system, with its rigid, rectangular arrangement of streets and building lots, the plans they sponsored were by no means identical. Lots varied in both their size and their orientation around the blocks in the surveys – the modal frontage, fifty feet, was used in only 32.8 per cent of all of the lots examined. The next most common frontage, sixty-six feet (the length of a surveyor's chain and a common width for road allowances), accounted for another 8 per cent. Frontages, then, varied a great deal. The average width of the building lots in the sample surveys was 54.7 feet. Blocks, too, differed in size from subdivision to subdivision, often later resulting in a rather discontinuous street pattern if the visions of adjoining subdividers did not coincide.[61] Over time, however, subdividers slowly developed crude rules of thumb for subdivision design. Hugh B. Willson advertised in 1848: "the survey has been made upon the most liberal basis. The streets are wide, and every lot has an alley passing in the rear. The principal streets run North and South, thereby giving the lots Easterly and Westerly aspects, which in City residences are always considered preferable to any other."[62] In the 1930s Homer Hoyt suggested that at least five different ways had been used to subdivide forty-acre tracts in Chicago. For San Francisco, Elizabeth Kates Burns noted that "the diversity of [the] subdivisions is far more obvious than [their] similarity. Subdivision size, the intended social status of future residents, lot and street design, the character of the tract's initiator, date, and location are only some of the differentiating characteristics."[63] All these features would help to shape the character of the neighbourhoods that developed with the aging of the subdivisions.

On a more subtle level, the subdividing practices of the era of individualism helped to foster the twin and persistent notions that continued urban expansion was a good thing and that there was money to be made in urban land. Although many of their plans were premature, most of Hamilton's early subdividers profited from their experiences. Unfortunately, only a limited number of people could be involved in the lucrative land-subdivision process, so most individuals seeking real-estate investments put their money into a much riskier kind of property – the building lot. As long as subdividers continued to show a profit and as long as newspaper editors fuelled the "will to possess," people must have been tempted to invest their money in land. After all, subdividers and editors were visible community leaders who doubtless knew what was good for themselves and their city, so others followed their lead. Unfortunately, they seldom had the luck, the power, the access to capital, or the long local experience of the original land platters.

With very few exceptions, the subdividers of this period were not property developers. They were content to purchase large blocks of land, have plans of subdivision prepared for their holdings, and sell that land as building lots. Not all of these land parcels would be developed quickly, so it is instructive now to focus on what occurred after the subdividers withdrew. The picture is one of a piecemeal process of residential development. Many people were involved in the various stages of dwelling construction and it often took many years and repeated changes of ownership before completed neigh-bourhoods emerged. Development depended on whether the prop-erty owner wanted either to build or to have built a dwelling unit or some other permanent urban structure on a vacant piece of land. Speculative interests first had to work their way through the mar-keting phase of the land-development process.

Between 1847 and 1881 the 1,668 lots selected for intensive study changed hands a total of 7,154 times, an average of 4.3 transactions per lot. This average, by its very nature, masks the full details of the spectrum of activity in the Hamilton land market at that time. About 9 per cent of the lots changed hands at least eight times over the years examined. At the other extreme, about 6 per cent of the lots remained in the hands of their original purchasers for the entire length of the trace. Most of the properties in the sample (85.3 per cent) changed hands between two and seven times in the first thirty-five years after their creation.[64] The frequent turnover of property and the rapid escalation in land value that, according to Homer Hoyt, characterized Chicago's mid-Victorian real-estate market was much less evident in Hamilton, partly because it was both smaller and growing more slowly than Chicago during this period. Then, too, Hoyt's study included a large number of lots in Chicago's Central Business District (CBD), while such lots were excluded from this study. In fact, many of the Chicago lots that sold very frequently were in the CBD, where the speculative mania reached dizzying heights. Land within a one-mile radius of that city's main intersec-tion, State and Madison, increased in value from $810,000 in 1842 to $50,750,000 in 1856. The Hamilton land market was not in the same league.[65] Certainly, an examination of the transaction histories of Hamilton's building lots shows that as time passed the lots turned over less frequently. The next sale after 1881 was identified for all lots in five of the sample surveys.[66] In each case, the most common date for the next sale was sometime after 1900. Stability of ownership came with the passage of time.

The characteristics of those owning property in the sample surveys changed during the era of individualism (see table 1.6). Initially, members of the development industry, the merchant/industrial bourgeoisie, professionals, or the landed gentry had owned 81.3 per cent of the building lots.[67] By 1881 the male members of these

Table 1.6
A Comparison of the Initial and Final Lot Owners in the Selected Surveys,
1850–1880

Variables	Initial Owners*	Final Owners
OCCUPATIONAL GROUP		
Development Industry	25.6%	19.1%
Builders	12.3	12.0
Servicers	13.3	7.1
Government Officers	3.9	1.2
Merchant/Industrial Bourgeoisie	27.7	19.4
White-Collar Workers	1.4	7.9
Skilled Workers	4.8	15.8
Unskilled and Semi-skilled	1.5	6.7
Professionals	1.4	2.0
Landed Gentry	26.6	9.6
Women	3.0	13.0
Unclassifiable	4.1	5.4
PLACE OF RESIDENCE		
Hamilton	75.5	91.6
Elsewhere	24.5	8.4
NUMBER OF LOTS	1668	1811
NUMBER OF OWNERS	456	937
AVERAGE LOTS PER OWNER	3.7	1.9

Source: Wentworth County Registry Office records
* defined as those who purchased land from an original subdivider

groups controlled just 50.1 per cent of the properties, not an in-
consequential amount but a substantial drop over the period in question. Women, who initially had owned a mere 3 per cent of the lots,
nominally controlled 13 per cent by 1881. When a married woman
acquired land, often so that her husband would not lose it to creditors, the legal instruments usually recorded the name and occupation of her spouse. The percentage of the lots in the hands of
developers and the elite in 1881 must be raised by at least 6 per
cent, for about three-quarters of the women who were listed as the
final owners of the lots could be identified as the wives of members
of the development industry (20 per cent), the merchant/industrial
bourgeoisie (33 per cent), professionals (3 per cent), or the landed
gentry (19 per cent).[68] Nevertheless, land ownership was becoming
more dispersed and there is evidence that Canada's elite were losing
some of their investment interest in land, a point that is supported
by the research of both Gustavus Myers and Tom Naylor, and Jock
Bryce, who examined the Hamilton records for the credit-reporting
firm of R.G. Dun.[69]

Some real property-ownership gains had been made by Hamilton's less affluent citizens by 1880. Skilled, unskilled, and semi-skilled workers held more than four times as many lots as they had initially. Those engaged in the building arm of the development industry, primarily small-scale, skilled craftsmen, had held their own in terms of properties owned. The robustness of the property-ownership levels for the working-class groups stands in marked contrast to the relative declines that they had experienced within the overall occupational structure of Hamilton – 67.8 per cent in 1852 and 61 per cent in 1881.[70] Further evidence of the trend toward a more equitable control over the properties under examination can be seen in the ownership figures for white-collar workers, primarily clerks. Members of this group owned almost six times more building lots in the early 1880s than they had in the early 1850s. The elite still controlled most of the sampled lots in 1881, but property had begun to filter down to lower levels in the socio-economic hierarchy. Perhaps the land market was more efficient than many anecdotal studies of speculation have suggested.

Other changes also were evident in the characteristics of the property owners over time. After the first round of sales, almost one-quarter of the lots had been in the hands of nonresidents. Thirty-five years later, they held only about one-twelfth of the properties. This provides further evidence that speculative fever cooled considerably as time passed. Land purchases became more calculated as more of those involved became aware of the realities of urban land markets.[71]

Land ownership became less concentrated in the selected subdivisions as time elapsed. During the first round of sales the properties had passed from the 14 subdividers to some 456 new owners (see table 1.7). While most of these (57 per cent) acquired only one lot, they actually controlled only 15 per cent of the properties. The building lots were still under relatively concentrated control after the first round of sales, for 53 per cent were in the hands of just thirty-three people (7.2 per cent of all purchasers). This pattern of control altered over the ensuing three-and-a-half decades. While the number of building lots in the surveys increased through resubdivisions by about 7 per cent, the number of individual landowners more than doubled. By 1881 those who owned only one or two properties controlled 54 per cent of the lots, more than twice as many as at the outset of the marketing stage of the land-development process. Furthermore, both the number of large lot owners and the number of lots that they owned decreased dramatically. By the end of the era of individualism there were only two owners who held

Table 1.7
Changes in the Distribution of Lot Ownership in the Sample Surveys:
A Comparison of the Initial and Final Owners, 1850–1880

| | Initial Owners | | | | | |
| | Individuals Involved | | | Lots Controlled | | |
No. of Lots Owned	No.	%	Cum. %	No.	%	Cum. %
1	259	56.8	56.8	259	15.5	15.5
2	83	18.2	75.0	166	10.0	25.5
3–5	61	13.4	88.4	224	13.4	38.4
6–9	20	4.4	92.8	137	8.2	47.1
10–19	22	4.8	97.6	295	17.7	64.8
20–49	7	1.5	99.1	168	10.1	74.9
50+	4	0.9	100.0	419	25.1	100.0
Totals	456			1,668		

| | Final Owners | | | | | |
| | Individuals Involved | | | Lots Controlled | | |
No. of Lots Owned	No.	%	Cum. %	No.	%	Cum. %
1	585	62.4	62.4	585	32.3	32.3
2	198	21.1	83.5	396	21.9	54.2
3–5	110	11.7	95.2	385	21.3	75.5
6–9	29	3.1	98.3	204	11.3	86.8
10–19	13	1.4	99.7	192	10.6	97.4
20–49	2	0.3	100.0	49	2.6	100.0
50+	0	0.0	100.0	0	0.0	100.0
Totals	937			1,811		

Source: Wentworth County Registry Office records

more than twenty properties. One was lawyer Thomas H. Stinson, a son of Thomas Stinson (who had died in 1864), the only subdivider in the sample group to build rental dwellings on his holdings, and the other was the City of Hamilton itself. Over time, more and more of the lots had passed into the hands of people who intended to develop them. Land speculators were still operating in Hamilton at the beginning of the 1880s (and are to this day), but most had turned their eyes to greener subdivisions.[72] As industrialization and re-source exploration activities progressed in Canada, the importance of property as an avenue of investment for the elite diminished. Thus a reduction in the concentration of property ownership over the course of the era of individualism is hardly unexpected.[73]

During this period the fluctuations in the prices paid for vacant land in Hamilton resembled those described by Hoyt for Chicago.[74]

In both cities land prices followed recognized North American economic cycles (see table 1.1). Average lot prices in Hamilton increased continually throughout the speculative boom of the early 1850s, rising from $163 in 1852 to a peak of $1,024 in 1857. Financial panic struck North America in that year, and some of the consequences can be seen in the precipitous decline in average lot prices to $69 in 1867, when considerable numbers of vacant lots were sold for back taxes.[75] In 1859 Moore and Davis, the Hamilton land agents, responded to a land valuation request from Benjaman Van Norman of Tillsonburg, Ontario, by noting, "as to the value we could not tell what it is worth, there not being any property selling of any consequence."[76] Lot sales also fluctuated, falling from 698 in 1854 to just 30 in 1864. Sales picked up in 1865–66, first as a result of the flood of cheap, tax-sale lots and then in response to improving economic prospects and conditions. Prices, too, rose from their nadir, but even in the best years (1873 and 1878) they barely approached half the value of the 1857 peak. Slowly, a wariness about the investment potential of urban building lots worked its way into the popular psyche. Even the editor of the boosterish Hamilton *Spectator* had been chastened by the impact of the depressions that always seemed to follow periods of intense speculative activity. In 1872 he cautioned readers that "we are not by any means anxious to see real estate go up to fancy prices. The natural increase in size and prosperity of our city will send up property as fast as it ought to go. If prices are forced up by *unnatural speculation*, or by combinations among property holders, the result will be disastrous: rents will increase and wages must follow."[77] Vacant lot prices were more stable in the 1870s, averaging between $200 and $500 for most of the decade. The memory of hundreds of vacant lots being seized for back taxes was fresh in the minds of most Hamiltonians. Speculators were forced to seriously re-examine the profitability of investing in urban building lots. In 1869 Moore and Davis warned their client Arthur W. Smith of London, Ontario, that while "we may say that property is rather looking up in the City ... there is not a great deal changing hands, particularly vacant and unproductive property."[78] In the same year they advised Toronto barrister James Maclennan: "we have an offer of $2,000 for lots, 1, 2, 3, 7, 8, 11 on King, George and Ray Streets from Rev. Geddes, payable $500 down and balance in five yearly payments @7% yearly ... Please let us know if you will accept it. We think probably the interest will be more than the increase in value."[79] Geddes's offer amounted to about $333 per lot, which was very close to the average price of $293 recorded for the 1,339 vacant-lot transactions registered in the sample subdivisions between 1870 and 1879. Written, considered offers became more common as the era of individualism waned, but auction sales remained an important means of selling several properties simultaneously. Here the prices could vary considerably, even for similar properties in the same subdivision. In 1873, for example, eleven properties on Stinson Street in Hamilton's east end sold for prices ranging from $160 to $490. Later in the same year the press reported that lots on Markland and Concession Streets in the city's south end had been sold for from "$250 to $500 per lot according to position and size."[80] Many examples from the period emphasized that location within the city strongly influenced the sale price of Hamilton properties. Moore and Davis reported to Captain Edward Harrison of Belleville, Ontario, that his "house is vacant and it appears there is very little chance of renting it for one tenement [it] being in a poor locality and so far off."[81] Later, they warned J.G. Harper of Toronto that "we are sorry there is very little chance of selling your lots at present. They are in the poorest place in the city for selling, being near the cemetery."[82]

CONSTRUCTION DURING THE ERA OF INDIVIDUALISM

The final stage in the residential development process during the era of individualism usually was the erection of housing units (see figure 1.2). Between 1851 and 1881 the number of occupied dwellings in Hamilton increased by about 255 per cent, more than keeping pace with the 155 per cent increase in the population and the 184 per cent increase in the number of families (see table 1.8). By modern standards the pace of housing construction was quite slow, but the mean number of houses built per year did show an increase from decade to decade. On average, 132 houses were added to the building stock annually during the 1850s, 156 in the 1860s, and 197 in the 1870s.[83]

The composition of the construction industry in any nineteenth-century city largely mirrored the skills needed to build a dwelling. Donald Adams has suggested that the major labour costs in the construction of a house in Philadelphia between 1785 and 1830 were carpentry (43 per cent), brick and stone work (20 per cent), plastering (11 per cent), and painting and glazing (9 per cent). Adams found that in 1859 labour accounted for 56 per cent and materials 44 per cent of the total cost of a dwelling, a ratio that he argued changed only marginally over the course of the next century.[84] Figures for the composition of Hamilton's mid-Victorian construction industry

Table 1.8
Demographic and Housing Statistics: Hamilton, 1851–1891

Characteristic	1851	1861	1871	1881	1891
Population	14,112	19,096	26,716	35,961	47,245
Families	2,471	3,332	5,084	7,016	9,282
Dwelling Types					
Stone	150	322	n.a.	n.a.	324
Brick	414	789	n.a.	n.a.	4,541
Frame	1,349	2,160	n.a.	n.a.	4,356
Log	10	0	n.a.	n.a.	n.a.
Shanty	27	n.a.	n.a.	n.a.	1
Occupied Dwellings	1,950	3,271	4,830	6,802	9,222
Vacant Dwellings	50	91	44	351	552
Under Construction	32	11	62	69	75
Total Dwellings	2,032	3,351	4,936	7,222	9,849
Occupied Dwellings/ Family	0.79	0.98	0.95	0.97	0.99
Average No. of New Dwellings per Year by Decade		132	156	197	242

Source: Census of Canada, various years and volumes

faithfully mirror Adam's estimates (see table 1.9). With the notable exceptions of the introduction of both the balloon-frame method of construction and slit nails in the 1830s, technological advances in dwelling construction were modest until the end of the era of individualism.[85]

Hamilton's construction industry did not remain static during the mid-nineteenth century. Employment expanded by 93 per cent. While this was less than two-thirds of the 155 per cent increase in the population in general between 1852 and 1881, it must be remembered that in the former year the construction industry was still trying to cope with a severe housing shortage. Over time it succeeded, for the average number of persons per dwelling dropped from 7.2 to 5.3 between 1852 and 1881.[86] Changing economic conditions and an increasing desire to own better-finished houses acted to modify the composition of the building industry as the years passed. The Panic of 1857 had a devastating effect on the economic health of many of Hamilton's skilled trades, including those related to building, and in 1861, employment in construction was marginally lower than it had been in 1852. While brick workers, carpenters, gasfitters, painters and glaziers, and plasterers increased in number during this decade, all of the other construction occupations suffered ab-

Table 1.9
The Composition of Hamilton's Construction Industry, 1852–1881

Construction Occupations	1852		1861		1872		1881	
	No.	%	No.	%	No.	%	No.	%
Brick Layer	2	0.4	5	1.1	28	3.3	51	5.9
Brick Maker	4	0.9	5	1.1	15	1.8	22	2.6
Builder	13	2.9	11	2.5	12	1.4	28	3.3
Cabinet Maker	73	16.4	36	8.2	61	7.2	64	7.5
Carpenter	197	44.3	233	52.8	482	57.0	389	45.3
Contractor	8	1.7	4	0.9	6	0.7	11	1.3
Gasfitter	1	0.2	3	0.7	2	0.2	3	0.3
Joiner	17	3.8	4	0.9	–	–	2	0.2
Mason	65	14.6	55	12.5	64	7.6	52	6.1
Millwright	–	–	1	0.2	–	–	2	0.2
Painter and Glazier	31	7.0	43	9.8	101	12.0	104	12.1
Plasterer	16	3.6	30	6.8	37	4.4	53	6.2
Plumber	–	–	2	0.5	16	1.9	28	3.3
Renovator	–	–	–	–	1	0.1	1	0.1
Stone Cutter	13	2.9	5	1.1	10	1.2	32	3.7
Turner	5	1.1	4	0.9	10	1.2	19	2.2
Total Workers	445		441		845		858	

Source: Hamilton Assessment Rolls

solute declines as a result of the downturn in the economy. And well they might have, for in 1861 the assessment rolls revealed that fully 21 per cent of Hamilton's houses stood vacant. During the 1860s construction employment almost doubled and most occupations grew, though the share of total construction employment accounted for by individual designations continued to change – a partial reflection, no doubt, of the gradually increasing sophistication of both the construction process and building technology. The position of ordinary carpenters, numerically the largest group in the hierarchy of the building trades, eroded in the 1870s in the face of a slow but steady wave of technological developments and shifting market demands. Most notable among these were the incipient prefabrication of dwelling components, the standardization of lumber sizes, the advent of housing amenities such as electricity and central heating, and the increased use of brick for dwelling exteriors (see chapter 5).

If the pace of development during the years in question was relatively slow, the scale of organization in the building industry was even more modest. Our Hamilton data complement the earlier findings of H.J. Dyos for the London suburb of Camberwell, R.M. Pritchard for Leicester, Peter Aspinall for Sheffield, and Sam Warner Jr for suburban Boston.[87] Vestiges of the Georgian building practices, described so admirably for British cities by C.W. Chalkin, persisted well into the nineteenth century.[88] The houses erected in Hamilton during the era of individualism were the products not of a highly integrated construction industry, but rather of the skills and efforts of literally hundreds of craftsmen and ordinary labourers, many of whom worked independently, though they often came together momentarily in any one of a seemingly endless series of loose, job-based associations with fellow tradespeople (see chapter 5).

In Great Britain, according to E.W. Cooney, four types of building firms existed during the era of individualism. The simplest, most traditional, and least sophisticated were headed by master craftsmen (such as carpenters, masons, or bricklayers) who worked only in their own trade and employed only a few journeymen and apprentices. Three other types had evolved by the early nineteenth century: master craftsmen who undertook responsibility for the construction of all parts of a building, but who employed directly only those in their own trade and contracted with other master craftsmen for the remainder of the work; builders (often architects or merchants) who erected entire buildings on the basis of contracts with master craftsmen in the various trades; and, finally, master builders who built entire buildings and employed more or less permanently a relatively large body of labourers and workmen in all of the principal building crafts. London builder Thomas Cubitt is widely acknowledged to have pioneered this fourth method for organizing the building business sometime between 1815 and 1820.[89] In Hamilton, the first three of Cooney's building-firm types co-existed throughout the era of individualism, and the final category emerged in the late 1880s (see chapter 2). The balance, however, was already shifting away from the simpler forms of organization and toward the builder category by the 1870s. As brick houses became the norm, brick workers had to be integrated into the construction process, and there is some evidence that home buyers had begun to demand that more attention be paid to the aesthetic and sanitary details of house design. Thus electricians, painters, pipe fitters, plasterers, plumbers, stone cutters, and wood turners increasingly had to be blended into the ever-expanding construction team.[90] Gwendolyn Wright has demonstrated that this trend continued to grow during the era of corporate evolution because of the increasing industrialization of building. The need for a builder/contractor to coordinate the men and materials required to construct a house, especially one of any size, increased as time passed.[91]

Despite the more prominent role of the builder-contractor and a moderate increase in technology, the scale of most house-building firms remained quite modest both in numbers employed and in annual output, even in the 1870s. Some insight into this issue is provided by the industrial schedule of the 1871 *Census of Canada* (see table 1.10). Of the nineteen Hamilton building firms enumerated, thirteen employed fewer than a dozen men, and in fifteen of them, total capitalization was under $4,000. The largest firm, that of C.W. and T.L. Kempster, appears to have been very much out of step with the rest of the industry. It had 281 employees at its separate building and brick-making operations, total capitalization of $64,000, annual wages of $83,800, and annual production valued at $116,500. According to some sources, its chief business was in the manufacture of sashes, doors, and window blinds, so the figures for this company are quite difficult to assess.[92] Even if the employment figure for the Kempster firm is accepted at its face value, the leading building, painting, and plumbing firms in Hamilton in 1871 employed only 414 people. This represents just 49 per cent of those identified in the assessment rolls as being employed in the construction industry at that date (see table 1.9). There still were large numbers of independent craftsmen in Hamilton's building industry toward the end of the era of individualism.

Most builders told the census enumerator that they worked the whole year round. While this may have been true for the employers, few of their workers normally enjoyed such full employment. There was a very definite construction season in Canada and much of the northern United States during the mid-nineteenth century. It began with the onset of spring and usually was over by the time of the first snowfall. The editors of Hamilton's newspapers knew this pattern well, and each March they would anxiously anticipate the prospects of the coming season. In 1872 the *Spectator* reported that "a walk through the streets of the city shows that the building prospects of the coming season are excellent. If the so far surly March would begin to assume the lamblike character which traditionally belongs to his later days, the work would be actively commenced."[93] For the better part of four or five months each year little or no construction took place in Hamilton. Basements could not be excavated, mortar would not set properly, nor would paint and plaster dry when temperatures remained below the freezing point. It is no wonder that employment in this industry failed to grow as rapidly as the population as a whole.

The annual seasonal lull in building was mirrored in the cyclical nature of the land market itself. In February 1868 Moore and Davis replied to J.G. Harper of London, Ontario, that "on the opening of

Table 1.10
The Hamilton Building Industry, 1871

	Capital					
	Fixed	Float	Months Worked	Number Employed	Annual Wages	Value of Product
BUILDERS						
McGill & Van Allen	$2,000	$2,500	12	15	$6,000	$8,000
W.W. Summers	300	500	12	3	1,400	3,400
Peter Brass	1,000	3,000	12	15	6,000	8,500
Yates & Garson	1,000	3,000	12	16	8,000	25,000
Alex. Milne	600	400	12	3	1,500	2,450
Geo. Murison	400	2,500	12	10	5,000	7,500
Robt. Chisholm	800	1,000	12	20	8,500	18,700
Wm. Richardson	–	500	12	4[1]	1,400	2,000
Robt. Waugh	–	2,600	12	5	1,560	3,750
Chas. Britten	250	100	4	3	450	1,500
Allen & Co.	1,000	1,000	12	10	4,000	10,000
Geo. Hellig	–	400	8	6[1]	2,050	7,000
Jeff. Houlden	–	–	2	–	1,500	2,400[2]
Geo. Sharpe	200	100	12	2	900	2,000
C.W. & T.L. Kempster	25,000	20,000	12	250	75,000	103,500
David Edgar	3,000	4,000	12	10	4,500	12,000
Wm. Clucas	600	5,000	12	19[1]	5,000	13,000
Stephan Searle	–	1,000	–	1	–	1,500[3]
Geo. White	1,000	500	12	10	4,000	9,000
Sub-Totals	37,150	48,100		402	136,760	240,400
BRICK YARDS						
Henry New	5,000	6,000	7	15[1]	1,000	6,000
Alfred Little	6,000	8,000	6	36[1]	6,000	12,000
Kempster	7,000	12,000	6	31[1]	5,800	13,000
Aaron Bawden	2,000	2,000	6	10[1]	1,200	4,800
Joseph Faulkner	500	500	4	8[1]	200	800
Sub-Totals	20,500	28,500		100	14,200	36,600
PLANING MILLS						
Geo. Sharpe	2,000	2,000	12	10	5,000	15,000
Travaskis Bros	200	100	6	1	–	625
John Semmons	4,790	2,000	12	29[1]	5,000	8,000
Geo. Bacheldon	–	–	12	25[1]	7,800	13,950
Michael Brennan	3,000	2,000	12	22[1]	7,800	28,000
Sub-Totals	9,990	6,100		87	25,600	65,575
PLUMBERS AND PAINTERS						
Donald McPhie	160	900	12	4[1]	1,000	4,500
Henry Harding	3,000	1,200	12	4[1]	1,300	5,000
V.W. Edgecomb	300	500	12	4	1,500	4,000
Sub-Totals	3,460	2,620		12	3,800	13,500
TOTALS	71,100	85,300		601	180,360	356,075

Source: Industrial Schedule, *Census of Canada*, 1871, MSS

[1] Employed boys under the age of 16

[2] Marginal notes indicated that Holden built 2 houses ($300) and 5 cottages ($1,800) and did general jobbing ($300).

[3] Marginal notes indicated an output of 2 frame cottages.

Spring we intend putting up posts and notices on a large number of properties."[94] Later in the same year they informed Thomas Walker of New York City that "there will be little chance of selling [your property on Victoria Avenue] this winter. We will, however, attend to it and put up notices in the Spring. If put up now they will likely be torn down."[95] Land transactions, too, had periods of high and low intensity each year, with 13.9 per cent of the sales registered for the 14 sample subdivisions dated between January and March, 40.5 per cent between April and June, 23.9 per cent between July and September, and 21.7 per cent between October and December. Almost 69 per cent of all the transactions examined for the era of individualism were signed in the months of May through October.[96] When people spoke of the real-estate season, it was with good reason.

Data relating to building operations in Hamilton in any given year during its early history are very sporadic. Neither building permits nor City Council reports on construction activity were issued in Hamilton at this time. On occasion, however, the local press made some attempt to spell out the details of the past season's construction activities. Unfortunately these reports were published irregularly and they probably never were complete. Nonetheless, the information contained in the reports for 1873 and 1881 has been summarized to gain more insight into activity in the construction industry toward the end of the era of individualism (see table 1.11). These surveys reveal that the city's builders and contractors were engaged in a variety of construction activities, including a type of minor repair work known as "jobbing." They put up new dwellings and they added to and modified the existing stock. In addition, in both years they built, enlarged, and renovated factories, commercial structures, and institutional buildings. Many of the institutional buildings were architecturally designed and were so complex that they could be tackled only by the city's largest firms. In some cases the architect himself appears to have functioned as the overseer of the project. William Leith, for example, played this role in the erection of the $11,000 St Joseph's Convent at the corner of Park and Colborne Streets in 1873. Leith employed local tradespeople on this project: H.E. Bush (masonry); Alderman Sharp (carpentry); Alderman Fitzpatrick (painting); W. Farmer (plumbing); and Mackay and Boyd (plastering).[97] Some structures, however, were beyond the capabilities of the local members of the building industry. In 1881 an office building valued at $150,000 was put up for the Canada Life Insurance Company. The architect was a Mr Wells of Buffalo, while the masons were Brown and Love of Toronto.[98]

Table 1.11
Building Activity in Hamilton, 1873 and 1881

Type of Construction Activity	1873	1881
RESIDENTIAL CONSTRUCTION		
Houses Built	75	78
Additions	3	10
Repairs/Renovations	7	–
Medium of Construction		
Frame	29	2
Brick	43	47
Stone	2	–
Roughcast	1	–
Other/Not Listed	–	29
Listed Values		
less than $500	8	–
$500–$999	20	8
$1,000–$2,499	23	31
more than $2,500	24	16
not listed	–	23
Most Houses by One Builder	7	12
No. of House Builders Listed	14	13
Avg. No. of Houses/Builder	2.5	4.5
NONRESIDENTIAL CONSTRUCTION		
Factories	7	12
Commercial Structures	9	17
Institutions	3	5
Other Additions	6	7
Other Repairs/Renovations	5	3
TOTAL PROJECTS	115	132
TOTAL VALUE OF PROJECTS	$835,000	$599,000

Source: Hamilton Spectator, Friday, 21 November 1873; Monday, 24 November 1873; Thursday, 27 November 1873; Saturday, 13 December 1873; and Saturday 29 October 1881

Few Hamilton builders put up more than a dozen houses in any given year during the era of individualism. In 1873 the listed builders averaged 2.5 dwellings, and 57.1 per cent of them worked on no more than two houses. For 1881 the comparable figures were an average of 4.5 dwellings per builder, with 38.5 per cent of those listed working on no more than two houses. Only one firm, that of Robert Cruickshank, built more than six houses in 1873, and only two builders, Thomas and Jocelyn and Michael Pigott, accomplished this scale of activity in 1881.[99] At the other extreme, there is evidence that some of the houses mentioned in the year-end press reports were built, or at least contracted, by the families who came to occupy them. Editors took care to name the builders and architects involved in the houses detailed in their reports. Indicative of the enduring heritage of self-building in Hamilton, no builder or architect was listed for 45 houses (56.3 per cent) in 1873 and for 30 houses (38.5 per cent) in 1881.

Associations amongst the various building trades were extremely fluid at this time. The building projects of Hamilton's 1881 season provide ample evidence of the shifting nature of these relationships, even over the course of a single year. For example, A. Peene, a mason who often served as a contractor, worked with Burns (carpenter), Matthews (painter), Kennedy (plasterer), and Clark and Squibb (plumbers) on a house built on Concession Street for W.C. Waltham. For another house for John Alexander on the same street he employed the same plasterer and plumbers, but used Burrows for carpentry and Barker for painting. Peene worked with yet another carpenter, T. Smith, on a pair of two-storey brick residences on Catherine Street for a Mr Brown. Smith, in turn, also worked for Thomas and Jocelyn on a pair of two-and-a-half storey, ten-room dwellings on Herkimer Street that were built for a Mr Balfour. Matthews and Kennedy worked together on the $34,000 City Hospital on Barton Street, which was designed and coordinated by architect L. Hills. They were joined on the project by J. Beer (masonry), R. Cruickshank (carpentry), Young and Brothers (plumbing and steam fitting), and J.B. Bishop (slating). The $75,000 Ontario Cotton Mill at the corner of Simcoe and MacNab Streets brought together yet another band of tradespeople: Thomas and Jocelyn (masonry); R. Chisholm (carpentry); Young and Brothers (plumbing); Clark and Squibb (steam fitting); Matthews (painting); and Wallace and Son (roofers).[100] Clearly, there were many permutations and combinations to the building teams that worked in Hamilton during the era of individualism. As Cooney suggested for London, associations of master craftsmen and firms led by builders were the most important forms of corporate organization at this time.[101]

The construction activity that we have described for Hamilton was typical of that found throughout North America and Britain during the nineteenth century. In Baltimore forty builders each erected at least one hundred dwellings between 1869 and 1896. This accounted for no more than one-ninth of the new housing built during this period, and probably averaged no more than seven units per year.[102] Research by David Hanna has shown that nearly six thousand housing units were erected by about two thousand builders in Montreal between 1868 and 1877. Only one-quarter of the new units were put up by the 3 per cent of the builders who constructed at least ten units. Half of Montreal's builders were involved in no more than two houses over the decade.[103] Between 1865 and 1900 three-quarters of the builders active in Sheffield were involved in the construction of no more than eight houses, while only 2 per cent were able to put up at least one hundred dwellings.[104] What of house building in larger cities? C.G. Powell found that in 1845 it took 31 firms to build 137 houses in North Kensington and four-fifths of all London builders active in that year put up six or fewer houses.[105] The scale of house building did increase slowly in London during the last half of the nineteenth century, though, as Aspinall argues, probably not until later in other British cities.[106] In 1872, 5,911 houses were built in London. According to H.J. Dyos, 85.4 per cent of these dwellings were erected by builders involved in no more than two dozen houses each, and 37.2 per cent were built by firms involved in six or fewer units each. Only two builders put up more than sixty London houses in 1872, averaging about seventy units each. In 1899, 7,135 dwellings were put up in London by 614 builders – 48.8 per cent by builders of two dozen or fewer units, and 14.3 per cent by builders of fewer than seven houses. The number of firms involved in the construction of more than sixty houses grew to seventeen, accounting for 29.9 per cent of the total production, and averaging about 126 units each.[107] In Sam Bass Warner Jr's Boston suburbs, 74 per cent of the 9,030 builders active between 1872 and 1901 erected no more than one house per year, collectively accounting for one-third of the 23,073 dwellings put up during this period. At the other extreme, 122 firms erected more than 20 houses over the period, but they accounted for just 23 per cent of the new stock. The largest of these firms was involved in 328 houses, an average of 11 units per year. While it was easy to get enough capital to put up one or two houses, problems in obtaining sufficient and steady financing for larger schemes unquestionably dictated the slow pace at which most builders put houses on the market during the era of individualism.[108] Limited technological and organizational sophistication further restricted the size of building firms at this time (see chapter 2).

The building reports for 1873 and 1881 shed some light on the type of housing that was being built in Hamilton at the end of the era of individualism. For example, there is clear evidence that brick dwellings, which accounted for about 60 per cent of the homes built in each year, were becoming the norm, especially in the better class of houses. In 1852 the *Census of Canada* reported that 71.1 per cent of Hamilton's houses were of frame construction. This figure dropped to 66.0 per cent by 1861, and to 47.2 per cent by 1891 when building material again was reported in the *Census* (see

Table 1.12
Timing of Lot Development in the Sample Surveys*

Survey Name	Percentage of Lots Developed by Time Period					Total Lots
	Pre–1852	1852–61	1861–72	1872–81	Post–1881	
1 Stinson	n.a.	49%	15%	12%	24%	98
2 Hamilton	12	24	15	32	17	220
3 Ferguson	n.a.	72	18	0	0	106
4 Macnab	2	16	11	20	51	129
5 Beasley	1	19	20	19	41	109
6 Tisdale	n.a.	0	25	57	18	100
7 Willson	17	2	20	43	18	206
8 Willson	1	15	34	41	9	164
9 Cameron	n.a.	40	9	32	19	47
10 Pringle	n.a.	24	18	34	24	55
11 Billings	n.a.	41	20	16	23	49
12 MacNab	0	54	21	6	19	68
13 Moore Bros	58	10	17	11	4	83
14 Kerr et al.	n.a.	18	20	29	33	234
All Surveys	7	23	20	28	22	1668

Source: Wentworth County Registry Office records and Hamilton Assessment Rolls
*Refer to figure 1.4 for the locations of each survey
n.a. = surveys platted after 1852

table 1.8).[109] Houses in a variety of price ranges were being erected during the 1870s and early 1880s. Several dwellings valued at less than $1,000 were put up in both 1873 and 1881, as well as roughly the same number of large homes valued at more than $2,400, some of which were mansions worth more than $8,000. The typical new house in this period, however, probably sold for somewhere between $1,000 and $1,500, a range that corresponds with Dorothy Brady's estimates for United States urban house prices at this time. This spectrum of prices was also substantiated in both real-estate advertisements of the period and the Moore and Davis letterbooks. The average price paid for the 1,373 houses that changed hands in the selected surveys prior to 1882 was $1,099.[110]

Building on most of the lots in the 14 sample subdivisions was slow. By 1861 about three-tenths of the 1,668 lots had been developed, and only one-half had been built on by 1872. More than one-fifth of the lots were not developed until sometime after 1881 (see table 1.12).[111] In eight of the subdivisions more lots were developed either between 1872 and 1881 or after 1881 than during any other period. None of the surveys experienced its most active development period during the economically depressed 1860s.

Development was most rapid in the Ferguson and Moore Brothers surveys (see figure 1.4), which were just to the east of Hamilton's CBD and immediately adjacent to other portions of the city that already had been developed by 1852. The pace of development was also relatively rapid in the more remote Stinson survey, where it had received a boost from the active participation of the proprietor, banker Thomas Stinson, in the building process. He had erected some 41 rental dwellings on his survey, being the only subdivider in the group studied to become involved in construction. (The role of the subdivider-landlord is explored in chapter 8.) The northernmost MacNab survey also developed rapidly, because its proximity to York Street, the main road to Toronto, and the yards and shops of the Great Western Railway attracted working-class buyers. At the other extreme, development was very slow in two west-end surveys owned by MacNab and Beasley. These subdivisions not only were relatively remote, but also were on rough and poorly drained land. Dyos, too, has noted that poorly drained subdivisions in the London suburb of Camberwell developed more slowly than those on higher and dryer ground.[112] Of course such patterns underscore the premature nature of many of the surveys of this era.

Who was responsible for the development that did take place in the selected subdivisions during the era of individualism? Most of the construction must have been carried out by the members of Hamilton's building trades, but since they each owned relatively few of the city's vacant lots, usually they were not the ones who made the actual decision to develop any particular piece of land. It is useful to approach the question of who decided to develop land from two perspectives – which occupational groups owned lots at the time of their development and how involved each group was in property development. C.G. Powell, writing about the British building industry, neatly referred to control over the decision to develop as "building sponsorship."[113] In Hamilton the behavioural patterns of building sponsors varied by occupation and almost seven hundred individuals had decided to develop property in the sampled subdivisions by 1881 (see table 1.13).

Some 2,183 different people had owned the 1,668 building lots in the selected subdivisions by 1881. Fewer than one-third of these, however, were involved in a decision to develop any of the properties prior to that date. Many of the people who purchased land during the era of individualism did not intend to develop their holdings, a fact that can be supported by the figures for the proportion of lot developers among the members of different occupational groups.[114] Lot owners associated with the building arm of the property-devel-

Table 1.13
Lot Developers by Occupational Category

Occupational Category	No. of Buyers	% Developers in Group	No. of Lots Developed					Total Lots Developed	Average Lots Developed
			1	2	3–5	6–9	10+		
Development Industry	418	43.8%	110	29	29	10	5	422	2.3
Builders	313	46.0	94	22	23	5	0	248	1.7
Servicers	105	37.0	16	7	6	5	5	174	4.5
Government Officers	38	21.0	5	2	1	0	0	12	1.5
Merchant/Industrial									
Bourgeoisie	386	29.5	57	20	25	6	6	313	2.8
White-Collar	84	39.3	23	6	3	1	0	53	1.6
Skilled Workers	303	35.3	83	18	6	0	0	138	1.3
Semiskilled and Unskilled	154	33.8	42	6	4	0	0	68	1.3
Professionals	56	23.2	8	4	1	0	0	19	1.5
Landed Gentry	326	23.3	43	14	12	6	1	171	2.3
Women	297	24.2	53	8	10	1	0	107	1.5
Unclassifiable	121	17.4	21	2	3	1	1	61	2.2
Totals	2183	31.4	445	109	94	25	13	1364	2.0

Source: Wentworth County Registry Office records and Hamilton Assessment Rolls

opment industry had by far the best performance record in this regard. Fully 46 per cent of landowners from this category were involved in a development decision prior to 1881, a figure that was more than 50 per cent higher than that for the members of the merchant/industrial bourgeoisie and roughly double the figures for government officers, professionals, the landed gentry, and women. Of the elite groups, only those who serviced the development industry (agents, financiers, lawyers, and the like) came within 10 per cent of the figure for builders. White-collar workers and members of the working class, whether skilled or not, displayed a much greater predilection for building than any of the elite groups that were not associated with development. The white-collar workers even surpassed those who serviced the development industry. At least one-third of the white-collar and working-class owners of land in the sample subdivisions had decided to develop their property by 1881. The conflicting goals of extracting use-value or exchange-value from property are evident in these figures. In the former case land is purchased for its consumption or potential occupancy purposes, while in the latter case it is obtained for its potential value as an investment. The pursuit of exchange value is not always compatible with housing production.[115]

Nonresidents were far less likely to be lot developers than were Hamiltonians. Only 20.8 per cent of nonresident purchasers developed lots in the sample surveys, compared to 33.1 per cent of purchasers who lived in Hamilton. It is tempting to conclude that since so much land had initially been controlled by both the local and nonresident elite, this must have slowed the process of development. But many of the subdivisions that were laid out in Hamilton and elsewhere during the era of individualism were simply much too premature. There was no real demand to develop all of the lots under examination, even by 1881. The early trading of land among the elite really did not hamper the overall pace of residential construction, though it undoubtedly fostered an unnecessarily patchwork pattern of development. As the need to build on the lots increased, more and more of them filtered down into the hands of people whose only concern was development and the use-value of property. Because of the vagaries of human behaviour, this took longer in some properties than it did in others.[116]

Not unexpectedly, those associated with the development industry owned more lots at the time of dwelling construction (31 per cent) than did the members of any other occupational group. Next in importance came the members of the merchant/industrial bourgeoisie, at 23 per cent, followed by the landed gentry, 12 per cent, and skilled workers who were not in the construction industry, 10 per cent. Again, the small scale of property development during the era of individualism stands out. Only the members of the development industry, the merchant/industrial bourgeoisie, the landed gentry, and the group labelled others averaged more than two development decisions within the sample subdivisions. Most striking, fully 65 per cent of those who owned lots at the time of their development were responsible for the decision to develop only one property in the selected surveys. Among skilled workers and those in the unskilled and semi-skilled category, 78 and 81 per cent, respectively, of the developers made but a single development decision. Such individuals were tapping the use-value of their property. For others, land was coveted for its exchange value. Among those who serviced the development industry, multiple property developers were in the majority (59 per cent). They accounted for exactly half of the members of the merchant/industrial bourgeoisie and 43 per cent of the members of the landed gentry. In all, however, only thirteen individuals, firms, or institutions were responsible for the development of at least ten properties. None of these belonged to the building arm of the development industry, which is further evidence of the small scale of most such enterprises. Moreover, not all of the larger developers were involved in the erection of dwellings, for this list included the City itself (twenty-five lots, mostly for parks and schools) and industrialist Richard M. Wanzer (eighteen lots for a sewing-machine factory). Yet a number of landowners did decide to build several dwellings in the surveys under study, often in at least two different subdivisions.[117]

The leading owner-developers from the building arm of the development industry were uniformly small-scale operators and included brick maker Joseph Faulkner (8 lots in 1 survey), painters Nehemiah Ford (5 lots in 1 survey) and Thomas Freeborn (6 lots in 2 surveys), carpenters John Little (6 lots in 1 survey) and James Milne (6 lots in 2 surveys), and builder George Murison, who served on Hamilton City Council for more than fifteen years and was Mayor of the City in 1870, (9 lots in 3 surveys). As the press reports on building activity discussed earlier show, builders were more active than this in the construction of the dwellings in our surveys. By and large, however, they did not instigate a great number of the decisions to begin construction. In the sample surveys such decisions were made by a wide variety of people. Furthermore, in the vast majority of cases (81 per cent), those involved in such determinations influenced the development process for one or at the most two (frequently

adjoining) parcels. The development of adjacent properties, perhaps
even through the erection of semi-detached houses, allowed a small-
scale landlord to keep a close check on his or her investment (see
chapter 8). House building, whether for rental or owner occupancy,
remained a remarkably decentralized undertaking in Hamilton during the era of individualism.

The fragmented nature of the decision to build houses usually
resulted in variegated visual patterns along the streets of Hamilton.
In his widely acclaimed book, *Streetcar Suburbs*, Sam Bass Warner Jr
placed considerable emphasis on the uniform visual result of the
residential development process in the late nineteenth-century Boston suburbs that were the focus of his study. He even went so far
as to state that while 9,030 people built the 23,070 homes in his three
suburban towns, "somewhat paradoxically, from the extreme indi-
vidualization of agency in the building process came great uniformity
of behaviour, a kind of regulation without laws."[118] Such uniformity
was not a characteristic of the development process in the Hamilton
surveys under examination for the era of individualism, but Warner's time period largely fell within the period that we have defined
as the era of corporate involvement, so direct comparisons may be
unfair. Nonetheless, both the scale and the pace of development
combined in Hamilton to produce very ragged streetscapes, more
akin to those in H.J. Dyos's Camberwell than to those in Warner's
Boston.[119] Few of the Hamilton blocks in the subdivisions examined
were fully developed during the same decade, let alone the same
year. For the 234 lots in the Kerr, McLaren, and Street survey in
Hamilton's south end the following pattern of development occurred: 1850s – 15.8 per cent, 1860s – 16.7 per cent, 1870s –
34.2 per cent, and 1880s and beyond – 33.3 per cent. In the 370
lots contained in Hugh B. Willson's north and south surveys, respectively, the timing of development was: pre-1852 – 14.6 and 1.2
per cent, 1852–1861 – 20.4 and 14 per cent, 1862–1872 – 17 and
33.5 per cent, 1873–1881 – 34 and 41.6 per cent, and post-1881 –
14.1 and 9.8 per cent (see figure 1.7). Houses that had been built
two or even three decades apart often sat side by side on the streets
of these surveys. In many of these tracts a significant number of
vacant lots remained even thirty years after the development process
had commenced. While there were some rows of virtually identical
houses in Hamilton, the degree of uniformity was by no means as
striking as Warner's Boston work led us to expect. The era of individualism was a time of significant change in the architectural tastes
of urban North Americans, and Hamiltonians provided no exception

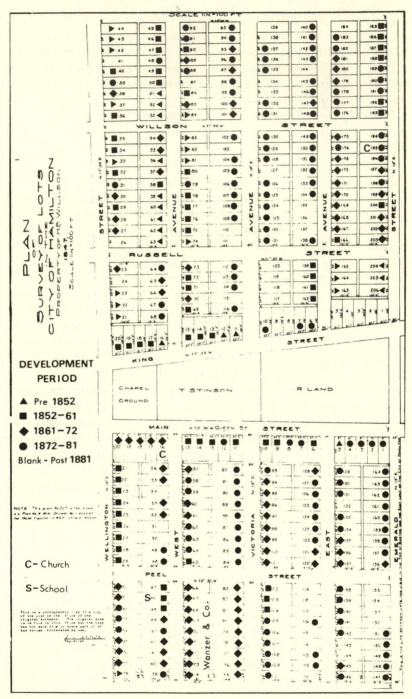

Figure 1.7
Development patterns in Hugh B. Willson's surveys

to this trend. In the years leading up to the 1880s, the neo-classical style of the late 1840s and 1850s gave way to the Gothic treatment of the 1860s which, in turn, was superseded by the Italianate styles of the 1870s and 1880s. This interval witnessed a shift from the historically pure and symmetrical designs of the Early Victorian period to the picturesque eclecticism of the later stages of High Victorian architecture.[120] The differences among these approaches to dwelling design were by no means small, and the juxtaposition of houses from the different periods produced anything but uniform streetscapes.

Nearly 1,500 buildings could be found scattered throughout the 14 sample subdivisions by 1881. The overwhelming majority (92 per cent) were low-density residential structures. Perhaps this outcome was the Hamilton equivalent of Warner's "uniformity without laws," because the Kerr, McLaren, and Street survey was the only one in which restrictive covenants had been used to force developers into residential land use (see chapter 2).[121] But there were some notable intrusions into the emerging neighbourhoods, not all of which were compatible with housing. On the complementary side, a few of the lots that fronted along King Street in the Stinson, Tisdale, and Willson surveys, and even a few of the interior lots, had been developed for grocery stores, inns, and taverns. Furthermore, by 1881 churches had appeared in five of the surveys, and a boys' home had been established in MacNab's northern survey. The Hamilton, Ferguson, and southern Willson surveys held primary schools and the central school had been placed on two acres of land in Hamilton's subdivision. A curling club had also been built.

Less fortunately, industrial and transportation facilities had been developed in six of the subdivisions. Ferguson's survey, near the eastern edge of the CBD, was the most popular target for these enterprises. The Shedden Company, a cartage firm, had its facilities here, along with the freight office of the Hamilton and Northwestern Railway. To the horror of many local residents, Samuel Nash had erected a pork factory in the same subdivision in 1872.[122] Other firms that had been constructed in the selected surveys by 1881 included the Copp Brothers foundry in the Moore Brothers survey, some yards and buildings of the Great Western Railway in MacNab's northern survey, a right-of-way for the Hamilton and Port Dover Railway in Cameron's survey, and the Sewer Pipe Company in Beasley's survey. By far the largest industrial enterprise in any of the subdivisions was R.M. Wanzer and Company's huge sewing-machine factory, which occupied all eighteen of the lots in the block bounded by Peel Street, Victoria Avenue, Stinson Street, and West Avenue in Willson's southern survey (see figure 1.7). By 1878 this firm em-

Table 1.14
Occupation and Tenure by Subdivision, 1881: Percentages

Survey Number[1]	Prof./ Prop.	White-Collar	Skilled	Semi-Skilled	Unskilled	Unclass.	Owned	Total Dwellings
1	5.5	12.3	51.4	4.1	14.4	12.3	39.9	138
2	32.0	40.1	12.2	3.4	1.4	10.9	57.5	127
3	4.0	18.5	50.8	6.5	8.3	12.0	29.0	214
4	–	–	38.5	7.7	38.5	15.4	76.9	13
5	–	11.1	40.7	5.6	27.8	14.9	39.6	53
6	5.4	17.9	44.6	14.3	12.5	5.4	28.3	53
7	7.5	17.3	53.7	5.6	7.5	8.4	32.5	203
8	20.6	53.4	16.0	1.5	–	8.4	48.4	122
9	5.5	7.3	50.9	5.5	21.8	9.1	29.6	54
10	7.1	–	42.9	7.1	28.6	14.2	42.9	14
11	36.4	18.2	27.3	–	4.5	13.6	78.6	14
12	–	2.1	72.3	12.8	8.5	4.3	42.6	47
13	4.6	16.1	41.4	13.8	12.6	11.5	25.6	78
14	8.8	26.4	39.6	6.9	11.3	7.0	53.8	145
All Surveys	9.7	22.1	42.1	6.2	9.8	10.1	40.2	1275
Entire City	6.9	18.5	40.4	7.0	13.4	13.8	34.4	6646

Source: Hamilton Assessment Roll, 1881
[1]See figure 1.4 for the locations of the surveys

ployed five hundred people, had up to fifteen thousand machines in various stages of production at any one time, and shipped its products all over the world.[123] Most of these industries blended poorly with the residential neighbourhoods in which they were located, but in a period of largely pedestrian movement between home and workplace, the proximity might have resulted in a countervailing benefit for some residents. The worst intrusion of all – one without any advantages to the residents – was created by the Hamilton and Lake Erie, later Hamilton and Northwestern, Railway. In the early 1870s this company was allowed to lay its tracks along Ferguson Avenue in the city's east end, thus bisecting both the Ferguson and the Moore Brothers surveys. In spite of rhetoric to the contrary, proximity to a railway certainly depressed the value of urban property. Moore and Davis bluntly told Alfred Hooker of Prescott, Ontario, that "the reason of [his] property being depreciated is that as soon as the Railway was built all property below the Railway became very much reduced in value."[124] While the sight and sound of a belching, thundering steam locomotive might have symbolized progress for some of Hamilton's most ardent boosters, it must have been a daily annoyance to those who lived near the tracks. A 1917 study into Hamilton's railway facilities termed the Ferguson Avenue line the "worst offender of all" and declared that it "destroys in a great measure, the value of the surrounding property; it is a serious and increasing obstruction to the free movement of street traffic, and should be eliminated or depressed."[125] More than seventy years later this line is still in use in the same location, and it continues to depress land values in the immediate area.

Spatially, patterns in the selected subdivisions underscored the trends in neighbourhood development that were emerging in the city as a whole during the era of individualism. Areas to the north of King Street tended to be working class, and those south of King Street generally were upper class (see table 1.14).[126]. Later, these distinctions would become more pronounced, and an east-west split would also develop. In 1881, however, there was still some inter-mixture among the members of the different occupational categories in most of the sample surveys. As we demonstrate through a case study of the Mills estate in chapter 8, and again in our discussion of the quality of shelter in chapter 10, residential patterns were more homogeneous at finer levels of spatial analysis such as the block or the block-face. The rapid industrialization fostered by the harnessing of electric power in the 1880s caused job opportunities to shift from the CBD toward the north-end waterfront, thus further crystallizing residential patterns.[127]

The emerging social-structural distinctions among Hamilton's neighbourhoods were mirrored in the property values in different areas of the city. These provide a crude, but useful, surrogate for housing quality (see chapter 10). Properties in the northern group of surveys had an average assessed value of $773 in 1881; those in the southern group had an average assessed value about 3.3 times higher – $2,526. These figures can be compared to the 1881 city-wide average of $1,115, and provide some additional evidence of the emerging residential bifurcation in Hamilton. Since assessed values usually reflected a myriad of property-related factors – general location or situation, favoured positions, size of lot, site quality, land use, and building material – the differences in assessed value above and below King Street suggest that by 1881 the quality of neigh-bourhoods in the city varied widely.[128]

Levels of home ownership generally reflected the patterns found for both social structure and property values. Overall, 40.2 per cent of the houses in the sample surveys were owner-occupied by 1881. This exceeded the city-wide figure of 34.4 per cent by a good margin, but the owners matched or outnumbered the renters in only five surveys, all but one of which again lay to the south of King Street. In contrast, in the subdivisions north of King Street, levels of home ownership were below the average for Hamilton at that time by as much as 9 per cent.

CONCLUSIONS

During the era of individualism, the residential development process in urban North America was largely unregulated, highly decentralized, and not very well coordinated. Literally hundreds of individual development decisions culminated in the patchwork of neighbourhoods that made up most cities and towns in Canada and the United States c. 1880.[129] At the same time, the period produced an ironic urban morphology characterized by the monotonous use of the grid for street allowances and building lots. *Laissez-faire* urban development practices had created mental and physical tensions in many cities by the later 1870s. Individualism would have to be curtailed. It had provided no mechanism to prevent the building of railway tracks along residential streets or a pork factory or a shanty on the lot next door. Soon, full-blown attempts at urban planning would emerge and be applauded. Citizens, in turn, would come to expect better-serviced neighbourhoods and better urban design.

The dwelling units within the neighbourhoods of the era of individualism, while almost always low density in form, varied vastly in size, orientation, and style, often within the same city block. They were built by countless individuals and small-scale firms that came together in an ever-changing array of business associations. Architectural tastes shifted dramatically over the period, and these changes were visible in the form of the built environment that evolved along most residential streets. While looked upon with favour today, the raggedness of mid-Victorian urban streetscapes might not have been so highly regarded by the residents of the period. Historian Walter Van Nus has argued that by the late 1890s, in the minds of many Canadian architects, "haphazard mixture along a street of architectural style and of building size was held to create a disturbing effect on the minds of passers-by, while coherence of line among buildings along a street fostered serenity."[130] Most of the housing that was built in Hamilton during the period was constructed for known purchasers. To use Powell's term, it was largely "custom built."[131] The economics of home construction mitigated against speculative building in Hamilton, though certainly not everywhere.[132] Families often built their own homes. Those without the time or necessary skills turned to the small-scale enterprises that dominated the building industry in every North American community (see chapter 5). In times of rapid growth, the industry had trouble coping with the demand for new housing units. Then, as had been the case in the early 1850s, Hamilton families were forced either to double up or to live in substandard accommodations (see chapter 10).

Urban services were provided after the fact. Residents cajoled and even paid private companies to extend gas, electricity, and street railway services to or past their homes. They petitioned the civic government to provide water and sewer facilities, sidewalks, and paved streets.[133] Extra assessments were levied by the City to defray the costs involved, a practice that would continue for many decades after 1880 (see chapter 2). Parks and other forms of public open space usually came as an afterthought, if at all. Major urban parks, such as High Park and Allen Gardens in Toronto, often originated as gifts from munificent land owners, rather than as the results of civic foresight.[134] All of these factors – timing, technological barriers, individualism, the level of urban services and amenities, and market demands – combined to produce urban neighbourhoods that were varied in their completeness, physical appearance, and social structure.

During the era of individualism most residents lived in dwelling units that they rented from a landlord (see chapter 8). The building industry itself was oriented toward the production of rental dwellings. David Ward has suggested that home ownership levels typically hovered around 25 per cent in large US cities in 1900.[135] The situation in Hamilton was just slightly better. Home ownership in this period was a dream that remained beyond the financial reach of most families, though the prospects for ownership did improve considerably with age. Even so, until 1911 most household heads under the age of 55 were tenants (see chapter 7). Generations of experience with tenancy in both Europe and North America had conditioned most household heads to expect no more than good rental accommodations for much of their lifetimes. Although North America had beckoned to many immigrants as an open continent in which they might escape tenancy, it was never an easy goal to achieve, as Elizabeth Blackmar's recent study of property ownership in Manhattan during the early nineteenth century has demonstrated.[136] Even after political struggles against landed estates had been won in many North American jurisdictions, the whole system for financing and constructing dwellings during the era of individualism worked against widespread home ownership. Costs, not laws or a powerful landholding aristocracy, were the culprits. It took years of effort to accumulate the sum needed for a down payment, and many of those who could get past that hurdle did not have the employment stability to ensure that they could pay off a mortgage. The growth of trade unions and the expansion of white-collar occupations in the years after 1880 would serve to rectify this problem for some fortunate households. From the perspective of the building industry, however, housing affordability could be addressed in two ways. First, the real costs of home construction could be assaulted. Here the form of the building industry could be altered in order to promote the greater use of labour-saving technology in home construction and to encourage greater efficiency. Master builders like London's Thomas Cubitt provided the model for this corporate reorganization.[137] Smaller homes and new and better designs could also be used to control price rises (see chapter 5). Second, the perceived costs of home ownership could be manipulated. New ways of paying for homes could be developed so that they would appear to be more affordable, thus broadening the potential market of home buyers (see chapter 6). These developments, along with the emergence of mechanisms for ensuring more homogeneous neighbourhoods, were the most important hallmarks of the era of corporate involvement in North American urban residential development. It is to that period and beyond that we now direct our attention.

2 The Transition to Modern Development: The Era of Corporate Involvement, 1880–1945

Eras seldom begin or end abruptly, rather the characteristics of adjacent epochs overlap. Such is the case with the property-development periods defined in this book. In the previous chapter we examined, in great detail, residential development during the era of individualism, not only because of the paucity of scholarly reporting on the way in which housing was built in urban North America prior to 1880, but also to establish a basis from which to monitor changes in traditional development practices. Resource limitations and a much larger urban area will not permit the same level of detail for the post-1880 eras. However, through the use of published sources and case studies of selected developments we can point out both the constants and the changes in residential development over time.[1] Secondary sources are much more abundant for the years following 1880, and government reports and serial publications on housing and the residential-construction industry also appear for the later time periods.

Our eras describe major trends in the residential-development process. They should not be viewed too rigidly. For example, in our discussion of the era of individualism we did not highlight the important changes in property development that occurred during the latter years of the period. Both enterprises that serviced the property development industry and house-building firms themselves were feeling the impact of new trends by 1880. Of course, not every firm quickly recognized that a fundamental restructuring of the residential-development process was taking place. It would take time for the revolution to crystallize, and some firms never would feel the need to adopt the new methods. Several practices and philosophies of residential development would persist well after 1880. Small house-building firms, for example, continue to thrive in urban North America in the early 1990s.[2] Nevertheless, when viewed collectively, the numerous and varied changes that began during the era of individualism provided the seeds for a distinctly different paradigm for dwelling construction – the era of corporate involvement.

CHANGES IN THE ORGANIZATION OF THE PROPERTY-DEVELOPMENT BUSINESS

The housing-development paradigm changed slowly at first, but the shifts began well before 1880. For one thing, the era of individualism had witnessed the emergence and proliferation of specialized property-handling agencies, the forerunners of modern real-estate firms. In Hamilton one of the first of these was the Moore and Davis Company, founded in 1858 and operating today, primarily in the insurance business (see chapter 8). We make extensive use of the records of this company throughout this volume. By 1864 Hamilton supported six such house, land, and estate agents, and by 1881 there were twelve land and estate agents and dealers listed in the city directory.[3] Over time, as we document for Hamilton (see chapter 8) and as Marc Weiss has recently illustrated for the United States, real-estate firms became more numerous, took on increasingly sophisticated roles, and became active participants in the residential-development process. Pearl Davies, an authority on real-estate history in the United States, has argued that during the 1880s and 1890s "[the] standing of the occupation went up notably because *real estate business became identified with community growth and community prestige.*"[4] By 1913 forty-two real-estate firms were reported for Hamilton. While most conducted what was termed a "general real estate business," a number claimed to specialize either in the type of property that they handled (industrial, subdivisions, commercial) or in the area that they covered (city, suburbs, east end, mountain).[5] One such agent was Alex S. Dickson, whose motto, "I have faith in Hamilton," underscored the growing symbiosis between realtors and their locales. Dickson specialized in inside residential, business, and manufacturing sites. In a gesture symbolic of real estate's increasing professionalism, Dickson spent over two thousand dollars on advertising in 1912–13 alone, including a pamphlet that extolled real-estate opportunities in Hamilton to English investors.[6]

According to Weiss, the move toward specialization and the changes in methods of operation in the real-estate business were accompanied by the emergence of both larger firms and industry-wide organizations that came to have a great influence on the course of residential development. Indeed, realtors were the driving force behind much of the early planning legislation in the United States.[7] These trends were continent-wide and they spread quickly. The Hamilton Real Estate Board was formed in 1922 and local realtors Norman Ellis and Stuart Chambers helped to found the Ontario Association of Real Estate Boards in December of 1922.[8] As real-estate firms became more numerous, auction sales to dispose of urban property declined. Real-estate agents began to sell most city property at an advertised price, for which they received a fixed rate of commission. For many years there was true competition over the percentage of the sale price charged.[9] By 1900 Hamilton newspapers typically contained a full page of such advertisements. At the same time, and often with the impetus provided by members of the real-estate industry, a number of new financial institutions – most notably the building societies and trust companies that we discuss at length in chapter 6 – emerged to cater, at least in part, to activities in the real-estate market.

Many changes had begun to affect the home-construction process by 1880. Technological breakthroughs in building materials revolutionized construction through the production of house components in standard sizes. Industrialization in the planing mills resulted in more elaborate, yet cheaper, products and ultimately led to the manufacture of sophisticated prefabricated house components, and even complete dwellings, that could be shipped to the building site and installed or erected with a minimum of labour and skill (see chapter 5). All of these changes were visible, though by no means dominant, by 1880. Improvements in plumbing and heating systems, the introduction of telephones and electricity, and the adoption of public transportation systems also had occurred by the end of the era of individualism, and some scholars have pointed to an incipient "feminization" of housing during the closing decades of the nineteenth century. Each of these had an impact on either dwelling design or location.[10]

The organization of the residential-development industry also began to change. On 27 November 1874 a provincial charter was granted to six Hamilton men to incorporate the Hamilton Real Estate Association, with an authorized capital stock listed as the substantial sum of $100,000 (4,000 shares at $25 each).[11] This association was established to enable its members to explore several different facets of property development. Their vision of the city-building process was broader than was typical of the era of individualism. In the statement of purpose filed with the association's papers of incorporation, the founders declared an interest in "buying and selling, building upon, and leasing Real Estate in the City of Hamilton and County of Wentworth ... using portions of said Real Estate as a Public Park and Gardens and as a Hotel and carrying out all business incidental thereto." In the 1875 *Prospectus* issued by the company prospective investors were enticed with three possible sources of profits: "the difference between the cost of houses built in numbers and the price of the same sold singly, which could reasonably be estimated at twenty-five per cent; the interest on mortgages held by the Company for purchase money of houses sold; [and] the advance in price which could be obtained for land bought by the [C]ompany in large blocks, when sold *by the foot*, as well as the natural increase in the value of all land held for building purposes."[12] The Hamilton Real Estate Association was not unusual by the mid-1870s – for example, Ann Durkin Keating notes that similar organizations were at work in the suburbs of Chicago.[13]

The embryonic form of the modern property-development corporation can be seen in such associations. For instance, the company took a vertically-integrated approach to the process of land development. Its directors had strong ties to organizations such as banks, insurance companies, and utilities that could help the corporation to succeed. Moreover, its founders had a direct influence on municipal decision-making – two of the original six backers served on the Hamilton City Council during the 1870s and early 1880s. While this was hardly novel for people with an interest in property development, the thrust and organization of the Hamilton Real Estate Association were more sophisticated than was typical of the time. And the enterprise lived up to at least some of its founders' expectations, for as we note in chapter 1 it was one of the most active residential-development enterprises in the city in the 1870s, building houses on twenty-two lots in three of the fourteen closely studied subdivisions alone.

Later others would come to share the vision and the methods of this organization, but there was no enthusiastic dash to restructure Hamilton's residential-development industry. Few builders and contractors, the backbone of the industry during the era of individualism, appear to have moved in this direction. As Marc Weiss has reported for the United States, even in the late 1930s

builders ... were very numerous, and most of them operated on a very small scale with few if any regular employees. The majority of general contractors and special trade subcontractors were actually building trades journeymen who moved back and forth between hiring construction crews and working on them. The bulk of building construction was performed under contract ... The US Bureau of Labor Statistics calculated in 1938 that even in the largest cities, the typical housebuilder constructed less than four houses a year, with most building only one or two per year. Very few firms built more than 10 houses annually.[14]

The Bureau of Labor Statistics had surveyed nearly fourteen thousand builders in seventy-two cities. Fewer than 2 per cent of them put up more than twenty-five houses in a single city in 1938, collectively accounting for 30 per cent of the houses built by the entire group. Only thirty-three built more than one hundred houses a year in a single city, but these large builders accounted for 11 per cent of the houses constructed by the builders included in the Bureau's survey. All of them operated in cities with populations of more than half-a-million people.[15] If the scale of building firms was related to city size, then Canada presented entrepreneurs with few opportunities for growth. Respected housing analyst Humphrey Carver, bemoaned the lack of scale in the Canadian house-building industry in the late 1940s, and even suggested that large-scale firms were necessary if Canadian housing costs were to be lowered.[16] But economies of scale are not easily achieved in home construction. Building enterprises, even today, do not have to be capital intensive. This ease of entry into the building business encouraged the formation of small firms and meant that the number of participants remained high throughout the era of corporate involvement.

The organizational structure of most of the elements of Hamilton's home-construction industry remained fragmented for many years. Building tradespeople in Hamilton first incorporated in 1886, when ten local carpenters banded together to form the Hamilton Building Contract Company, capitalized at just $3,000. Members of the company had subscribed for stock worth only $155, but their ambitions were clearly formulated: "carrying on the trade or business of Builders, Carpenters, and General Contractors for building houses and other structures and for the purposes above mentioned to acquire lands by purchase or otherwise with power to sell or otherwise deal with same as may be advisable according to the exigencies of said business."[17] In November of 1898 about forty builders and contractors met to organize Hamilton's first builder's exchange.[18] For the most part, however, the construction industry in Hamilton, as in other North American cities, remained small scale and quite weakly organized until well after 1900. The traditional contracting system, with its loose and ever-shifting sets of associations, persisted in home construction. Typically, when Hamilton real-estate agent

William Moore decided to become a developer, he contracted with a builder who, in turn, subcontracted some of the tasks.[19] Terry Naylor uncovered a similarly diffuse pattern of building along Ravenscliffe Avenue, near the base of the Mountain, during the first two decades of the twentieth century.[20] But a few city builders had begun to understand that corporations like the Hamilton Real Estate Association would have an increasingly important role to play in residential development.[21]

One of the most prescient city-building organizations was Patterson Brothers. The brothers came to Hamilton from County Tyrone in 1867. John Patterson worked for three local firms, the Wanzer Sewing Machine Company, E. & C. Gurney and Company, and the Spring Brewery, before travelling to Ohio and Illinois where he managed lumber concerns. He returned to Hamilton in 1878 and with his architect brother, Thomas, established a lumber business that was dissolved in the early 1900s, by which time other enterprises had diverted his attention from real estate. In 1896 Patterson and four others (known collectively as the five Johns) helped to form the Cataract Power, Light, and Traction Company, which brought cheap hydro-electric power to Hamilton by 1898 and served to stimulate industrial development in the early twentieth century. The company acquired control of the Hamilton Street Railway and several radial lines in the late 1890s and John Patterson promoted two electric radial lines – the Hamilton Radial Electric Railway and the Hamilton, Waterloo, and Guelph Electric Railway. He was instrumental in the establishment of the first blast furnace in Hamilton in 1895 and by 1900 he was investing in Pennsylvania coal fields.[22]

The brothers' lumber firm soon spread into speculative construction, erecting an "estimated 150 houses annually" in the mid-1880s and registering a dozen small residential surveys.[23] They were major shareholders in some Hamilton radial lines and assembled tracts of land close to the extensions of the streetcar system – a business integration of sorts.[24] Local real-estate agents Moore and Davis described the brothers as "pushing men [who] do an immense amount of building every year."[25] In 1887 the farsighted John Patterson proposed the development of a manufacturing suburb to the northeast of the city limits on some fifty acres of his land at Huckleberry Point.[26] By the time of his death, in late January of 1913, his vision had become a reality. His enterprises had been rewarding, for he left an estate valued at almost $440,000.[27]

While Hamilton had had a few subdivider/builders during the era of individualism, firms like the Patterson Brothers were different. Not only did they participate fully in all three stages of the residential-development process – subdivision, marketing, and housing construction – but also they expanded the responsibilities associated with each of these previously separate roles. Patterson Brothers supplied the lumber and other materials for the dwellings that they built from their 75,000 square foot yard at the corner of Robert and Barton Streets in Hamilton's north end. And they operated on a scale that was unheard of prior to the 1880s, boasting in one slick, eight-page brochure that "we never have less than twenty to thirty buildings in course of construction at once, and we can get work done for far less than would be possible were it let in single houses."[28] They were, in E.W. Cooney's terminology, "master builders" who took every advantage of the limited opportunities for economies of scale presented by the late-nineteenth-century home-building industry (see chapter 5).[29] The brothers also apparently took pains to internalize their business wherever feasible and, at the same time, to keep their operation as economically lean as possible: "We make our own plans and have no architects' fees. We buy our lots for cash and only when they are bargains. We pay cash for all our building material, and only buy when it is cheap. Our expenses are light. We have no president, secretary, manager or board of directors to pay, and we do not require enough profit to keep a long list of shareholders."[30]

The Pattersons and counterparts throughout North America displayed an uncanny grasp of the opportunities in the residential real-estate market that were opening up because of changes in late-nineteenth-century North American society, namely, urbanization, the growth of middle-class occupations, concern about urban design and planning, and the desire for exclusive residential areas.[31] They sensed, too, that the market for owner-occupied dwellings could expand. Their business plan was simple and straightforward and emphasized quality: "We are doing and intend to do the largest building business ever done in Hamilton, and to do it we must build at a price that others cannot touch, and we think we can do so. We wont [sic] build what are called thrown-together-houses. There is no money saved by it, and a house badly built will do us more harm and leave an impression that fifty well-built ones will not efface. Our success in selling we attribute to the fact that every customer finds his house exactly what we represent, and we can refer them to any from first to last."[32]

By the late 1880s it was no great feat to build twenty, thirty, or even fifty houses a year. The difficulty was to sell a high volume. This marketing problem persisted in Canada until some time after the end of World War II, for as Humphrey Carver observed in 1948: "house-building has not attracted large-scale construction firms and the field has been left almost entirely to the small home-builder whose organization can quickly adapt itself to the fluctuating demand for housing. The small speculative builder undertakes the erection of only a few houses at a time, saving the cost of managerial overhead and office space and protecting himself from market uncertainties by keeping his output in line with the immediate demand of prospective purchasers."[33] Like many entrepreneurs of the period, beginning with William E. Harmon in Cincinnati, the Patterson brothers realized that the marketing of homes had to be revolutionized. Harmon is acknowledged as the first person in North America to apply the instalment selling idea to real estate. The genius here was to focus the buyer's attention on small, manageable, regular payments and away from the entire cost of the purchase. Under Harmon's purchase contract system, ownership of the property was transferred to the purchaser only after a certain portion of the sale price had been paid. This reduced the financial risk for the subdivider and the need for the large down payments formerly required to offset the costs of possible foreclosures. It was the latter point that made Harmon a true pioneer. In 1873 Hamilton real-estate firm Alanson and Hilton advertised the sale of so-called "inferior" lots at the old race course property in the city's west end for as little as five cents a day. Under the terms of sale, purchasers were required to put one-third of the $50 price down and then paid $1 per month for two years. In 1875 the Hamilton Real Estate Association proposed to sell the houses that it built "at a reasonable profit, *on the instalment plan*, the purchase money being paid in equal monthly or quarterly sums for a term of years." The instalment plan was described as "the only system on which a person of limited income and without capital available for the purpose can ever *become his own Landlord*." Purchasers, however, were still expected to be able to manage down payments equal to at least 25 per cent of the house price. Such marketing practices were becoming common, for the Hamilton Real Estate Association claimed that "companies working on this plan are active in nearly every city in the United States; also in Montreal and Toronto, and, in each instance where their reports have been obtained they are found to be making very satisfactory profits."[34] Fine tuning to the instalment idea – even lower down payments and tiny weekly repayment schemes – was needed to complete this real-estate marketing revolution. It occurred south of the border and took about a decade. Beginning in December 1887, Harmon and associates began selling building lots in the Cincinnati area for as little as ten cents down and ten cents a week. Their first subdivision, Branch Hill, contained 303 (25 foot by 125 to 162.5 foot) building lots priced at $25 each. Purchasers needed to put just two dollars down and make weekly payments of twenty-five cents. Located in an area that John Stilgoe has evocatively labelled as the "Borderland,"[35] Harmon's survey lay some seven-and-a-half miles from the Cincinnati city limits and required a monthly commuting expenditure of $7.00 to reach the central city. Branch Hill nonetheless sold out in four days. Hazelwood, a second subdivision, larger and more distant from Cincinnati than Branch Hill, was equally successful. Not surprisingly, the combination of a low down payment with the instalment idea spread quickly during the 1890s; the Harmon firm alone marketed subdivisions in more than twenty cities.[36] In 1896 John Patterson advertised the sale of lots near Hamilton's new iron smelting works in the north-east part of the city as "The Chance of a Lifetime." These properties were priced from $125 to $160. Purchasers were required to pay two dollars down and fifty cents weekly, with no interest and no taxes to be paid for five years. At these terms, the cheapest lots would be paid for and less than thirty dollars would be owing on the most expensive properties after just half a decade.[37] By the early years of the twentieth century, easy-payment plans for building lots were the norm in Hamilton.

For large-scale builders to succeed, the market for new homes had to be expanded. One way to accomplish this was to assault the question of affordability directly. So some house-building firms such as the Patterson Brothers adopted easy-payment plans based on low, monthly, blended instalments that were amortized over long periods (for that time) of from fifteen to twenty years (see chapter 6). In the words of Gwendolyn Wright, the easy-payment plan "seemed to offer an opportunity to expand the ranks of the [North] American middle class by making inexpensive suburban homes more widely available. A family with a yearly income of only $1,000 could consider becoming suburban homeowners and taking on middle-class status."[38] The greatest documented user of this marketing scheme prior to 1900 was Samuel Eberly Gross, Chicago's foremost late-Victorian builder. By 1892 he had sold over 40,000 lots, built and sold some 7,000 houses, and developed more than 150 subdivisions and 16 towns in the Chicago area.[39] The Pattersons were not in the same league as Gross, but like the Chicagoan they offered open mortgages and a crude form of mortgage insurance that would "allow you four months in which to pay up should you be unable through sickness or any other cause to make your payments on time, and if you have been paying for any length of time we will arrange your mortgage afresh should you still be unable to make up what had fallen behind."[40]

The thrust of the Pattersons' marketing effort compared the economics of renting and owning a home. People had to be convinced that it was cheaper to buy than it was to rent. The brothers' arguments support our contention that the easy-payment scheme and the rhetoric associated with it provided one of the greatest stimuli to home ownership in the history of urban North America. Their promotional literature enthused:

We sell our houses for prices that are very little more than the rent they would bring, and we want only a small payment down; but if this payment equals one quarter of the price to be paid we will make your payments LESS than the rental, and you always have the privilege of increasing them or paying the whole mortgage at any time, without having to pay interest for any more than the money has been owing.

... Say that you have been renting a house for the last twelve years at $10.00 a month. That money is completely gone; you have nothing for it. But in addition to this, say you have at the same time been able to save $2.00 per month, which you have regularly put in a savings bank, and which has been drawing four per cent interest (the highest they will pay) and it has grown and grown and compounded till you have now nearly $380.00 of principal and interest to your credit. This is very good, but suppose instead of that you had bought a house from us equally as good, if not better than the one you live in, and your payments, taxes and all are $12.00 per month, how would you stand. You would simply have a home completely paid for, that would be worth not less than $1200.00, and possibly much more, showing a clear gain of more than $800.00 over and above what you could possibly get from a savings bank.

We have sometimes heard the statement that it would be better to save money till you could buy a house for cash, as by our plan you have to pay a good deal of interest. We will look at this simply from a business point of view. It cannot be expected that a house can be bought on time without having to pay interest, but which way will benefit you most. We think no sane man will say that three or four hundred dollars in a savings bank equals a house that will sell for twelve hundred dollars.[41]

Not wishing to leave anything to chance, the Pattersons adorned the final page of their brochure with two tables – one compared the instalments needed to repay $1,000 yearly, half-yearly, and quarterly over periods ranging from 2 to 20 years, while the other illustrated the monthly instalments required to repay loans varying from $50 to $1,500 when amortized over 2 to 15 years. To underscore the eminent economic sense of home ownership, they entitled one section of their brochure "The Best Savings Bank in the World is one of Patterson's Houses bought on the Instalment Plan."[42]

The market eventually responded in a substantial and positive way to the alteration in the perception of housing costs created by builders like the Pattersons, but the depressed economy of the 1890s kept owner-occupancy in Hamilton at the traditional one-third level in 1901. By 1911, however, it had risen to 51.4 per cent. Richard Harris found an ownership boom of similar magnitude in Toronto between 1899 and 1913.[43] The correspondence between increased owner occupancy and greater mortgage debt in this era is quite easy to demonstrate. Weiss notes that in the US the number of residential mortgages grew by 350 per cent between 1890 and 1920.[44] Unfortunately, we have no comparable figures for Hamilton, but in Canada as a whole the value of mortgage loans on real estate held by life insurance companies increased from $13.2 million to $125.7 million over these same three decades. This represented an increase of more than 850 per cent.[45] In the entire history of Hamilton, only the 1941–1951 period, when owner-occupancy rose from 48.7 to 65.0 per cent, would surpass the decadal gain in ownership recorded between 1901 and 1911. Between 1940 and 1950 the value of mortgages held by Canadian life insurance companies grew by just 164 per cent.[46] It was during the era of corporate involvement that the dual concepts of widespread home ownership and routine mortgage debt became part of the Canadian urban experience. Before homes could be owned, however, they had to be built; and before dwellings could be put up, subdivisions had to be laid out. In spite of the radical changes just described in the organization of the residential-development industry, the actual process associated with housing development did not break entirely from past practices during the era of corporate involvement.

SUBDIVIDING ACTIVITY

The initiation of the residential-development process retained several characteristics from the era of individualism. Hamilton subdividers continued to focus on the periphery of the city throughout the era of corporate involvement. After 1890 this often meant that the new subdivisions were being platted in recently annexed areas of Hamilton (see figure 2.1). In 1890 Hamilton's city limits encompassed some 3,050 acres. Between 1891 and 1914 the area of the city increased by 134 per cent, and by a further 45 per cent between

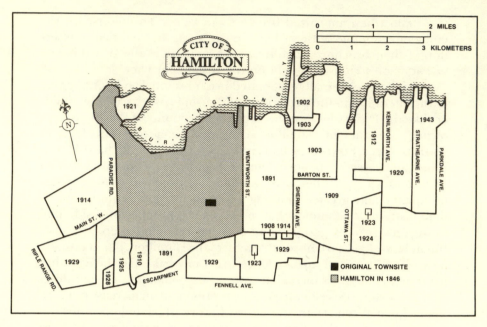

Figure 2.1
Annexations by the City of Hamilton to 1945

1915 and 1945, by which time it stood at 10,332.7 acres, more than three times its original extent.[47] The subdivision of land continued to occur largely without government regulation, though Ontario's City and Suburbs Planning Act [1912] provided some regulatory control over the number and width of streets, their direction and location within a subdivision, and the size and form of the lots. The law must have had an impact on Hamilton-area platting because it applied to all subdivisions within a five-mile radius of any Ontario city with a population of at least 50,000 people.[48] Now it required at least four signatures for suburban subdivisions to be registered; namely, those of the head of the local government (usually the Reeve of the Township), the Mayor of Hamilton, an Ontario land surveyor, and a representative of the Ontario Railway and Municipal Board, the body charged with administering the 1912 Act.

Subdivision registration continued to reflect broader economic cycles during the era of corporate involvement (see table 2.1).[49] Almost 45 per cent of the 281 surveys drawn up for the Hamilton area between 1881 and 1945 were registered during one spectacular burst of activity that lasted from roughly 1906 to 1913. This was the greatest period of platting mania in Hamilton's history. Thirty-two

Table 2.1
Subdividing Activity: Hamilton Area, 1881–1945

Period	Surveys		Lots		Average Lots Per Survey
	No.	%	No.	%	
1881–85	13	4.6	535	1.5	41
1886–90	16	5.7	604	1.7	38
1891–95	6	2.1	605	1.7	101
1896–00	5	1.8	451	1.3	90
1900–05	16	5.7	1,416	3.9	86
1906–10	45	16.0	5,759	16.0	128
1911–15	96	34.2	18,950	52.7	197
1916–20	13	4.6	2,412	6.7	186
1921–25	20	7.1	2,097	5.8	105
1926–30	6	2.1	519	1.4	87
1931–35	5	1.8	102	0.2	20
1936–40	7	2.5	490	1.4	70
1941–45	33	11.7	2,027	5.6	61
Totals	281		35,967		128

Source: Registered Plans, Wentworth County Registry Office

surveys, averaging 175 lots, were registered in 1911 alone, twenty-five subdivisions averaging 178 lots in 1912, and a year later, the twenty-four registered surveys averaged 220 lots. One 1913 subdivision, Kenilworth, laid out by J. Walter Gage, the "dean of Hamilton real estate promoters," contained eleven hundred properties.[50] So great was the subdividing activity that the local surveying firm of Mackay, Mackay, and Webster could estimate, probably without exaggeration, that in the spring of 1913 alone they had "laid out between six and seven thousand town lots in the neighbourhood of Hamilton, employing a staff of about forty men."[51] In the three decades preceding 1880, fewer than six thousand lots had been surveyed and then registered for the Hamilton area (see chapter 1). Between 1910 and 1913, however, more than 16,500 lots were subdivided and registered. Sixty-four per cent of these building lots lay outside of the city limits, primarily in Barton Township immediately to the east and south of the city and in Ancaster Township immediately to the south-west, "districts which but lately were regarded as being in the country."[52] The outward march of subdividing activity was truly impressive. As real estate broker and subdivider Thomas Crompton noted: "in 1906 there was nothing east of Wentworth Street but open fields. When F.B. Robins and I started the Crown Point subdivision, centered on today's Ottawa Street, people thought we were crazy. Crown Point lay a mile and a half east of the city, there were no conveniences, no roads worthy of the name, for sidewalks only a few hundred feet of planks that washed away in the rainy season, and the land was clay, greasy and sticky as paint. We always kept rubber wading boots for prospects to wear in spring and rainy weather."[53] Subdividers began to turn their attention to the Mountain area south of the city in 1909 when J. Walter Gage and G.A. Turner registered plans for the East Mount and Wycliffe subdivisions respectively.[54] By 1913 platting was occurring in western Saltfleet Township, which, though contiguous with Barton Township, began some four miles to the east of the centre of Hamilton. As P. Obermeyer, the Hamilton correspondent, reported in June of 1910 to the Department of Labour's monthly publication, the *Labour Gazette*, Hamilton was "experiencing the nearest approach to a real estate boom in its history, and prices are advancing rapidly."[55] Using Helen Monchow's rule of thumb of twenty-two lots for every one hundred of population, Hamilton's subdivision boom between 1910 and 1913 provided enough land for 75,000 additional people.[56] Impressive as this appears, the efforts of Hamilton's subdividers before World War I paled in comparison to the exertions of platters in other Canadian cities, especially on the Prairies.[57]

For every boom, there is a bust. By early 1915 there was open talk of the "depressed condition of business" and an "industrial depression."[58] Reports suggested that building tradespeople departed from Hamilton to search for employment elsewhere. In October 1914 the Hamilton United Relief Association opened a labour bureau, and within two or three months about four thousand people had registered for work.[59] Up to 75 per cent of the members of some branches of the building trades were unemployed in December 1914. Given the downturn in the economy, the glut of building lots on the market, and the outbreak of war, further subdivision development would have been risky, even foolish. Only twenty-four surveys were registered between 1916 and 1930 and merely ten, containing 592 lots, between 1931 and 1940.[60]

The early-twentieth-century platting boom was possibly more intense in many growing Canadian cities than in larger American centres such as Chicago and Los Angeles. Aggressive boosterism, frequently associated with railway development and extraordinarily high levels of immigration, played a definite role in the stimulation of subdividing activity in the Canadian West.[61] In Hamilton it was fueled by rapid and massive industrialization, which in turn attracted thousands of people. By the centennial of the city in August 1913, Hamilton's population was reported to have topped 100,000, almost double the total enumerated by the city's assessors in 1906. This represented an average annual addition of some 6,400 people – 11,890 in 1912–13 alone – who needed to be housed, fed, clothed, and employed. Some predicted that Hamilton would be home to 150,000 citizens by 1924.[62] People flocked to Hamilton because of the city's astounding industrial growth, which was stimulated by Hamilton's good location, incentives, and cheap electric power, and largely financed through American investment.[63] A brochure issued around 1913 by Hamilton's Commissioner of Industries, H.M. Marsh, proudly listed thirty-seven US-based firms that had located in the city. It was claimed in a 1930 publication from the same office that "American Industry Favors Hamilton ... And They Still Come." By then the number of American-based plants had reached one hundred.[64] From a base of about two hundred industrial plants in 1903, Hamilton's industrial sector expanded to about four hundred factories by 1913, twenty-eight of which employed more than one hundred people.[65] Thirty-eight factories, valued at $1,080,650, were erected in 1911 alone and during 1912 twelve new industries, valued at more than $2 million and expected to employ more than 26,000 people, were in various stages of construction.[66]

Entrepreneurs rushed to accommodate the new Hamiltonians through the creation of massive numbers of pristine building lots. Most of the subdividers of the period continued to employ the grid-plan and to serve simply as land preparers, a process that often led to inefficient development patterns and other residual problems. In the words of historian Walter Van Nus: "a developer's desire to extract the maximum number of lots from his [holdings] often led him to ignore the location and/or width of projected or existing streets nearby, if by doing so he could squeeze lots out of the property. This lack of coordination forced expensive road relocation and therefore higher local taxes. The desire for quick and maximum profit also encouraged speculators to instruct surveyors merrily to lay out the [property] on the familiar grid pattern, regardless of topography."[67] Even ardent Hamilton boosters cast a critical eye on the lack of imagination employed by local subdividers. The editor of the *Herald*, writing near the height of the 1912–13 platting boom, suggested that "what we want ... is a little more variety in laying out subdivisions to get away from the stereotyped plan of laying out all surveys in squares. How much better it would look to have them laid out in boulevards and crescents."[68] Nonlinear models for subdivision design were readily available both on paper and on the ground well before the outbreak of World War I. After all, architect Alexander Jackson Davis had experimented with nonlinear streets at Llewellyn

Park, New Jersey, as early as 1853, and in 1869 the landscape architects Frederick Law Olmsted and Calvert Vaux had laid out their
pioneering curvilinear subdivision, Riverside, on 1,600 acres of land
along the Des Plaines River near Chicago.[69] Streets in Toronto's
Rosedale area began to take on their meandering form in the
1880s.[70] In 1912 and 1913 the Canadian Northern Railway had had
plans drawn up and registered for very large, curvilinear suburbs
at Mount Royal (near Montreal), Port Mann (east of Vancouver),
and Leaside (near Toronto). At roughly the same time, the Canadian
Pacific Railway was developing the Shaughnessy Heights area of
Vancouver and the Dovercourt Land, Building, and Savings Company was marketing several nonlinear subdivisions in North Toronto.[71]

John Reps argues that by 1900 most large American cities had at
least one fashionable, nonlinear suburb. C.N. Forward and John
Weaver make the same case for Canadian cities, though in Canada
the process lagged by about a decade.[72] These areas came to be
known as "romantic or picturesque suburbs" because of their winding
streets and their conscientious adaptation to the existing topography.[73] Ravine areas and shorelines were favourite sites for such
enterprises because they presented a variety of elevations. Romantic
suburbs were carefully designed to be in harmony with the landscape
rather than superimposed on it, and their creators drew on the
growing corpus of literature about city planning for their inspiration.
The history of the North-American urban-planning movement has
already been well documented. There is no need to recount it here
except to note that early planning thought was influenced by the
work of British, German, and American landscape architects and
planners; the Parks and Boulevard movement; Ebenezer Howard's
Garden City movement; and the City Beautiful movement long associated with the World's Columbian Exposition that was held in
Chicago in 1893.[74] Apparently these accomplishments had very little
influence on Hamilton's subdividers, for romantic suburbs were slow
to appear in the Steel City.

While a number of ravines cut through Hamilton, the area's first
nonlinear subdivision was platted on the Mountain (Niagara Escarpment) in then still remote Ancaster Township. There the Home
and Investment Realty Company, a corporation whose directorate,
according to its promotional literature, was "made up of some of the
most successful and well known Real Estate men in Canada," subdivided Mountainview, a 132-acre parcel, into 430 building lots in
1913. Mountainview was intended to be "one of the finest suburbs

to be found in any city in Canada." A park was designed and laid out along the edge of the Mountain and, like many romantic suburbs of the period, Mountainview was only a short distance from the new "Country Club."[75] Mountainview, however, was a hybrid of the old and new subdivision forms. Those of its streets near the brow of the Mountain took on the graceful crescent form, but those to the south end of the subdivision were laid out in grid fashion and lined with 50 by 125 foot lots. Just over one-quarter of the building lots in Mountainview deviated from the conventional rectangular shape. Mountainview was no Riverside, but it did provide the requisite pleasant view to some, and parts of it were carefully landscaped.[76]

Hamilton's subdividers probably stuck to the grid plan in the early years of the twentieth century because so much of what they were doing was intended to meet the demands of the city's burgeoning industrial labour force. As the editor of the *Herald* observed of the city's new subdivisions in 1913: "they have been developed in the best possible way – homes have been erected upon them: homes, in many instances, owned by the men who fill the factories."[77] Then, too, there were expenses to consider. During the boom, real estate in Hamilton was a valuable commodity, but as Herbert Lister, author of Hamilton's official centennial volume argued: "where, however, the largest and quickest returns are made is in the buying of acreage in the suburbs, subdividing into building lots, developing the estate by grading and levelling the streets, putting in sewers and sidewalks, and selling the improved lots with building clauses suitable to the locality. It is not straining the truth to state that there is no form of investment in existence that shows such a sure and handsome return without any of the risks that usually attend a high rate of profit."[78] For those motivated by profit, the grid plan remained the speediest method for placing building lots on the market.

In Hamilton as elsewhere, slowly but surely new wrinkles were added to the process of creating residential building lots. By 1914 boulevards had been incorporated into four surveys, the earliest example being the Union Park subdivision, which had been laid out in the city's east end in 1900. Mountainview and four other subdivisions contained crescents.[79] Taken together, these nine surveys embraced 1,324 building lots, but far less than half of them were influenced by nonlinear street designs. Beginning with the Wentworth Court survey in 1925, cul-de-sacs began to be introduced into some of the smaller subdivisions, but again the impact was slight.[80] The nonlinear movement did gain some momentum after the end of World War I. Of the 38 surveys registered between 1920

Table 2.2
The Hamilton Platting Boom, 1906–1913

Subdivider Type	Surveys	Lots	Average No. of Lots
Individual – one survey	35	3,473	99.2
Individual – multiple	4	420	105.0
Partnerships	19	2,271	119.7
Corporations	59	11,920	202.0
Illegible/Unknown	11	2,877	261.5
Totals	129	21,238	164.6

Source: Registered Plans, Wentworth County Registry Office and Vernon's *City of Hamilton Directory*, various year

and 1940, 14 contained at least some non-grid features. Yet Hamilton remained an overwhelmingly gridded city throughout the era of corporate involvement.[81]

Other important changes occurred in the subdivision process between 1881 and 1945. Beginning with Thomas Bush's "Beulah," which was registered in 1881, surveys were increasingly given names other than the subdivider's surname. By 1900 the practice was more or less universal. And what a collection of appellations spread across the city's landscape – "Fairleigh Park" (1890), "Crown Point" (1904), "Ravenscliffe" (1909), "Bloomsdale" (1909), "Orchard Hill" (1910), "Normanhurst" (1912), "Rosemount Gardens" (1912), to list but a few examples. In spite of these grand names, the building lots in the subdivisions of the era of corporate involvement tended to be narrower than they had been in the era of individualism. By 1909 most Hamilton subdivisions contained lots that had frontages of between 25 and 30 feet, which, according to planning texts of that period, was typical of North American cities at that time.[82] Sixty-four per cent of the Hamilton subdivisions that were registered between 1909 and 1913 contained lots with frontages that were thirty feet or less.

Two other changes in subdividing practice were significant. First, as in suburban Chicago, there was a growing trend toward the involvement of corporations, or at least full-time real-estate professionals, in the subdivision process (see table 2.2).[83] At least 56 per cent of the lots created in Hamilton between 1906 and 1913 had corporate origins. Those who dominated Hamilton's market for building lots during this period had often entered the business through a kindred enterprise. J. Walter Gage came from a background in fruit growing and rural real estate, Alex Metherell from stock brokerage in the United Kingdom, George Armstrong and

Alex Dickson from insurance agencies, and W.D. Flatt, like John Patterson, from lumber and building supplies.[84] The rise to prominence of the land-corporation/real-estate professional stemmed from both escalating land costs and growing consumer expectations. In 1912 the *Herald* reported that virtually all the vacant land along the city's streetcar routes was in the hands of speculators.[85] One four-acre parcel in the east end was purchased for $50,000 in 1913. The buyer intended to erect forty houses, valued at between $5,000 and $7,500 each, on the land.[86] Few small entrepreneurs could compete in this economic climate.

After 1900, as Keating and others have shown, consumers came to expect more from the land market than just a raw building lot.[87] Discerning purchasers wanted improvements to the property prior to the completion of the transaction. These included water and sewer facilities and graded streets as the bare minimum, but ideally encompassed paved streets, cement sidewalks, hydro, gas, and telephone lines, landscaping, open space, and public transportation. Only well-funded corporations could afford to market improved lots, as the costs involved were substantial. Adrian Theobald found that in the Chicago area around 1930 it cost between $10 and $20 per front foot to install sewers, sidewalks, and paved streets. By the mid-1930s planner Thomas Adams was using a figure of $16 per front foot as an estimate of the costs of local improvements.[88] While precise figures are hard to discover for Canadian cities during the era of corporate involvement, it is known that Toronto subdivider Robert Home Smith incurred a cost of $170,000 just to clear his 3,000 acre tract in Etobicoke in 1910.[89] The Dovercourt Land, Building, and Savings Company paid $14,591.50 to install sidewalks, sewers, water mains, gas mains, and concrete roadways in its fourteen-acre Lawrence Park West subdivision in North Toronto around 1917. Another $822.25 was spent on landscaping. Assuming a density of six building lots per acre, these expenditures translated into preparation costs of about $183.50 per lot.[90] Between 1908 and 1911 it cost the Canadian Pacific Railway's Land Department an average of $3,943 per net residential acre to clear, grade, and service 268.8 net acres of land in the Vancouver area known as Shaughnessy Heights. Around 1925 the British Columbia government paid $6,210 per net residential acre, or about $2,019 per lot, to clear, grade, and service a parcel of 79 net acres on Vancouver's University Endowment Lands. The raw land had cost $4,000 per net residential acre, so servicing was becoming an important factor in the cost of housing, and surely dissuaded many from becoming subdividers.[91] Costs per unit of measurement were modest, but spread over an entire subdivision they quickly mounted. In Hamilton in 1910 macadam roadway cost the City 25 cents a square yard and cement sidewalks averaged 10.7 cents per square foot.[92] Yet the pace of servicing expansion was impressive. By 1911 Hamilton had 170 miles of cement sidewalks, 146 miles of water mains, and 107 miles of sewer lines. These figures represented increases of 21.4 per cent, 21.7 per cent, and 50.7 per cent, respectively, in Hamilton's infrastructure network since 1905.[93]

If people did buy lots in unimproved subdivisions, a majority of the new owners then had to band together and petition the City to have the improvements installed. This was time-consuming, especially during the boom before World War I, when the demand for services was great. It also meant that a local improvement levy would be applied to the property taxes of those owners who benefitted from the new services. The two Local Improvement Books extant for Hamilton for the period between 1902 and 1943 contain summaries of hundreds of petitions mostly for, but sometimes against, the installation of more than a dozen different types of improvement including sewers, cement walks and curbs, street grading, road widening, and several types of pavement.[94] Charging citizens for improvements was raised to the level of a science in most municipalities. When Hamilton decided to add roadways to its list of local improvements sometime around 1905, City Engineer Ernest G. Barrow was commissioned to ascertain the costs for various types of pavement and to determine how other municipalities in the Great Lakes' region paid for such work. His analysis revealed that in most of the fourteen cities studied, roadways were paid for by a local improvement levy, with the property owners covering at least two-thirds of the costs, but with the municipality paying for the cost of intersections. Barrow recommended that citizens be allowed to pay either all at once or over the expected life of the pavement, which varied from three years for gravel to fifteen years for brick.[95] Costs, then, quickly became well-known and residents were charged a fixed rate per foot of frontage that they owned. Rates charged per front foot in 1908 included: cement walk and curb – $0.101, cement curb – $0.073, brick pavement – $0.429, street grading – $0.065, and macadamizing – $0.151.[96] While these rates were quite modest because they were amortized over fairly long periods, the costs of providing the services tended to translate into substantial sums when they were accumulated for the entire city. In 1914 the City of Hamilton spent $430,312.16 on roadways, sewers, sidewalks, and curbs, all but $175,000 of which had to be paid for directly by the affected property owners.[97]

This outdoor work was a great, if unpredictable, source of employment to many of Hamilton's working-class citizens, but there was no guarantee that property would be improved in a reasonable amount of time under the local improvement system. Sometimes funding for the improvements could be obtained only if the voters approved special by-laws that authorized the sale of municipal bonds. It was not always possible to secure the signatures of the majority of residents for a petition to the city. We estimate that about one-quarter of the petitions for local improvements did not result in the passage of a local improvement by-law.[98] Labour supply, too, could be uncertain. In the summer of 1912 the city employed 1,200 on outside construction, but needed at least 200 more to meet the demand for services. As the economy weakened in 1914 and 1915, the city could not afford to carry out much improvement work and laid off workers because "property owners were not petitioning for local improvement work to the same extent as in previous years."[99] So the purchase of an already serviced lot must have seemed like a wise idea for would-be land buyers. Increasingly, this is what the corporate subdividers provided.

MARKETING

As the nature of the property being placed on the urban land market changed, so did the marketing of that property. Building lots began to be sold at a fixed price per foot of frontage. This rate varied according to such things as the size of the lot, its drainage characteristics, and its location. Around 1913 the developers of the affluent Lawrence Park Estate in Toronto divided their 190 lots into seven different quality types, with prices varying from $25 to $85 per front foot.[100] In Hamilton in 1910 J. Walter Gage was selling lots in the east-end Crown Point survey for $4 to $8 per front foot and in East Mount, on the less accessible Mountain, for $3 to $6 per front foot, with corner lots priced at $8 per front foot.[101] Prices rose dramatically during the boom. In 1913 lots in the Orlando Heights survey on the Mountain were being sold for $6 to $12 per front foot, while those in the Hamilton Realty Company's Kingsvale survey in southeast Hamilton were priced from $15 per front foot.[102] Moore and Davis estimated that between 1903 and 1913 the price of building lots had at least doubled, and in some places quadrupled. Herbert Lister suggested that the price of lots in the Maple Leaf Park and Westmount surveys more than doubled between 1910 and 1913.[103] Unsubdivided parcels increased in value even more rapidly. Land in the north-east end, near the corner of Ottawa Avenue and Barton

Street, that had sold for $250 per acre in 1903 was selling for $18 to $40 per front foot in 1913. If in the shape of a square, an acre contains just under 209 feet per side, but the need for road allowances would reduce the marketable frontage to perhaps 169 feet. Since two frontages would be saleable in such a block of land, property sold at from $15 to $40 per front foot would fetch between $5,070 and $13,520.[104] As Edel, Sclar, and Luria observed for Brookline, Massachusetts, at roughly the same time, "the real profits in real estate are to be made by buying land at the fringe of the urbanized area, providing transport to that land (or buying just before someone else provides the transport), and then selling out relatively quickly and capturing the initial steep rise in land values."[105]

Subdividers had two alternatives for disposing of their building lots – they could handle the sales themselves or they could allow real-estate agents, often through exclusive contracts, to do it for them. Alexander Metherell was the exclusive agent for at least a half dozen surveys in 1913, while H.G. Ogg & Company handled the Roxborough Gardens subdivision in the east end.[106] Whether the lots were sold through an internal sales department or contracted out, temporary site offices, which were open evenings and on Saturdays, were frequently set up at the subdivisions themselves to facilitate sales. Real-estate agent Robert E. Jarvis even opened an east-end office that catered "exclusively to the Industrial classes."[107]

Printed advertisements, employing both old themes and new techniques, continued to be important elements in the marketing of new building lots. Virtually all the advertisements during the 1906–13 platting boom mentioned "easy terms" for payment, with down payments of only five dollars requested by major subdividers such as J. Walter Gage and W.D. Flatt. Inducements, a time-honoured real-estate strategy, continued to be highlighted in many advertisements during the era of corporate involvement. Some subdividers refunded taxes and interest payments for a certain period of time. Enticements to get people to see the lots included band concerts, free transportation, and contests for free lots and even free houses.[108] The improved condition of the building lots was prominently featured in the era's promotional literature. It is not known precisely when the first improved subdivision came on the market in Hamilton. What is clear, however, is the importance of services in the annexation movement that swept suburban Hamilton around 1909. Outlying areas throughout North America were willing and eager to join established cities as long as they could expect speedy connections to the city's service infrastructure.[109]

By 1910 most new Hamilton subdivisions contained some hard services. The range of services offered, as Ann Durkin Keating found in suburban Chicago, was quite varied.[110] Joseph Pim's Queen's Park survey on Hamilton's Mountain boasted graded streets, shade trees, electricity, telephone and natural gas lines, and cement sidewalks.[111] J. Walter Gage advertised graded streets, sewers, and cement sidewalks in his Barnesdale, Crown Point, and East Mount surveys.[112] Shade trees on every lot, graded streets, and cement sidewalks were features of W.D. Flatt's Brightside survey.[113] In 1913 the developers of the St James Park survey, on the Mountain, advertised sidewalks and natural gas lines.[114] The promoters of the up-scale Mountainview subdivision offered graded streets, sidewalks, sewers, waterworks, telephones, and electric lighting.[115] Some subdividers also began to include so-called softer services in their projects. By the early years of the twentieth century aggressive land developers began either to set land aside to sell to the city for parkland or to include parks and boulevards within their subdivisions, often sharing the costs with Hamilton's Board of Park Management, which had been created in 1899.[116] To advertise their servicing efforts, the corporate subdividers used slick brochures for high-class subdivisions like Mountainview, and elaborate newspaper advertisements for more modestly priced properties. These advertisements often covered more than half a page and nearly always featured fancy artwork, sometimes in the form of photographs or drawings of houses. Around 1911 Alex Metherell of the Hamilton Realty Company even took to publishing a monthly bulletin on the local land market entitled *Opportunities*.[117] All such efforts cost money. Toronto's Dovercourt Land, Building, and Savings Company spent $4,362.33 in 1914 alone to promote its Lawrence Park subdivision.[118] Entry into the subdivision world was no longer as easy as it had been during the era of individualism.

To protect and enhance property values in their subdivisions, the corporate developers made extensive use of restrictive covenants in their deeds of sale. The origin of these property-control devices is not clear, but Charles Ascher suggests that while they were probably first inserted into deeds in medieval England, their use as urban planning tools is less than two centuries old.[119] By the early years of the twentieth century many developers had found an enticing rationale for their use. In the words of Patricia Burgess Stach: "realizing that the activities occurring on one parcel could affect the value and condition of those nearby, developers imposed restrictions to stabilize and enhance land values, thus protecting their investments. Profit, derived from marketability, was the realtors' primary concern ... [Developers] could make more money and better protect [their] property investment with planned and restricted subdivisions."[120] Restrictions had existed in Hamilton since at least the mid-1850s, and the earliest known examples were found in the deeds associated with the Kerr, McLaren, and Street survey in Hamilton's south end (see chapter 1). Purchasers of property in this subdivision were "covenanted not to build any dwelling or other houses (except out houses) unless the same were built of stone or brick, or to carry on any noxious or offensive trades on said premises."[121] The widespread use of restrictions in new subdivisions began in the 1890s when, as Ascher argues, "far-seeing real estate subdividers sensed a market in our rapidly growing cities for communities of homes whose extra value would be partly in amenity of planning and partly in assurance that the graciousness of environment would be protected. Improved transportation made readily accessible tracts on the fringe of the city, large enough to create their own quality and atmosphere of neighborhood and community."[122] One of the earliest examples of a North-American subdivision with restrictive covenants in its deeds was Roland Park, an exclusive suburb of Baltimore designed by Frederick Law Olmsted in 1891. The idea of the restricted subdivision spread quickly. By 1928 Helen Monchow was able to identify eighty-four major examples, including twenty-nine surveys laid out by the Olmsted Brothers, J.C. Nichol's 5,000-acre Country Club District in Kansas City, Frank Vanderlip's 3,200-acre development at Palos Verdes, California, the Van Swerigen brothers' massive Shaker Heights project near Cleveland, and two Canadian examples – the Uplands in Victoria and Sunalta in Calgary. She included numerous examples of actual clauses in her monograph.[123]

The restrictive covenants found in the deeds of the era of corporate involvement fell into three categories. Each limited the owner's free use of the property. Some restrictions, such as those found in Hamilton's Kerr, McLaren, and Street survey, dealt primarily with the issue of *land-use control*. This usually meant the exclusion of non-residential uses, but later came to be used to limit the density of residential development as well. Once the residential tone of the area had been established through the use of this first device, a second type of deed restriction could be imposed to establish *physical control* over the appearance of the subdivision. These restrictions governed a variety of housing parameters – a time frame within which construction should begin, the location of the house on its lot, its size and value, its architectural style, the nature and even colour of the materials used on the facade of the dwelling, and whether or not an architect had to design the house. Robert Home Smith used a very comprehensive set of deed restrictions in the 1920s and 1930s in an attempt to create "A bit of England far from England" in what is now the City of Etobicoke (near Toronto). In the Kingsway Park area these included a minimum lot frontage of forty-three feet, the right of the company to approve the plans for the house, a minimum allowance of not less than three feet on each side of the house, the bulk size of the house to be no less than 16,000 cubic feet, the house to be set back from the street line by twenty feet, garages to be attached, and the type of architecture to be "white painted brick, stone, and clapboard, stucco, or other forms of white architecture, conforming to more or less the same general design."[124] Deeds received from the Home Smith Company stipulated that the restrictions were to be in force for a period of thirty years.[125] The end result in Kingsway Park was an affluent neighbourhood of spacious, Tudor-style houses. Carefully crafted, combined, and enforced, the physical and land use covenants should have been sufficient to ensure the emergence of exclusive, single-family residential areas, at least initially.

Some subdividers included a third, insidious type of restriction in their deeds that was designed to bring about *social control* over future residents within their subdivisions, by specifying groups to whom the properties could not be sold. In American suburbs, this usually meant non-Caucasians. But, as we will illustrate shortly through our discussion of Hamilton's Westdale area, some rather lengthy lists of undesirable purchasers could be concocted. Happily, social-control covenants were ruled unconstitutional by Supreme Court decisions in both the United States and Canada, but not until after the end of World War II. The importance of the widespread adoption of the use of deed restrictions during the 1881 to 1945 period cannot be overstated, and their corporate origins are unmistakable. As Marc Weiss has correctly argued, "that they were willingly and in many cases eagerly accepted by purchasers opened the wedge for the introduction and extension of public land-use controls. Deed restrictions, an innovation of community builders and their attorneys, served as both the physical and political model for zoning laws and subdivision regulations."[126] Often the restrictions found in deeds imposed far more limitations on the use of property than zoning laws ever could. As John Delafons has suggested, "if the controls exercised by public authorities over land-use in America seem excessively detailed and capricious, the controls happily adopted by private citizens are positively sadistic."[127]

Hamilton's corporate subdividers began to make extensive use of deed restrictions during the 1906–13 platting boom. Few advertisements specified the precise nature of the restrictions, but most cor-porate subdividers mentioned that they were present. As early as 1907 homes worth at least $1,500 had to be erected on the lots in W.D. Flatt's Beulah survey.[128] J. Walter Gage boasted of "building restrictions for moderate-priced homes" and "building restrictions to ensure a good class residential property" in his 1910 advertise-ments for the East Mount and Barnesdale surveys.[129] The promoters of the St James Park subdivision promised that their lots were "all 50 feet or over, which insures the purchaser against any crowding of houses, and the restrictions are such that the purchaser is so protected that no building erected will cheapen his property."[130] In 1912 the Ambitious City Realty Company promised purchasers in its Forest Lawn survey that "with the building restrictions placed on each lot you can rest assured that [the homes] will be of the better class."[131] Restrictions on the exclusive Mountainview survey were more detailed. It was claimed that Mountainview was "the only Prop-erly Restricted District in Hamilton, and ... the future home of our millionaires."[132] No apartments, tenement houses, stores, or mer-cantile businesses were to be allowed on any part of the lots. Dwell-ings had to have minimum values of between $2,500 and $5,000 depending on their location. Fencing and building lines varied from 20 to 30 feet from the property line according to the location and width of the street along which the dwelling fronted.[133] Deed re-strictions had a positive impact on the quality of housing being built in Hamilton. In a 1922 paean on housing developments in the city, the editor of the *Herald* suggested that "one reason for such fine houses is the building restrictions which were not in effect when the older portions of the city were built up."[134] Over time, as our analysis of the Westdale area reveals, deed restrictions in Hamilton became both more widespread and more complicated.

HOUSING DEVELOPMENT

During the era of corporate involvement, the number of dwelling units in Hamilton increased almost sixfold. Most of this growth was related to new family formation trends, since population increased only four-and-a-half times compared to a sixfold increase in the number of families (see table 2.3). On average, about 550 units were added to the city's housing stock annually between 1881 and 1941. By decade, this varied from a low of 158 units per year during the 1890s to a high of 1,383 units per year during the 1910s. Two-thirds of the new dwellings erected during the entire period had been completed by 1921. So the counterpart to Hamilton's platting boom

Table 2.3
Population and Dwelling Characteristics: Hamilton, 1881–1941

Year	Population	Dwellings	Families	Average No. of Persons per Dwellings	Average No. of New Dwellings Per Year, by Decade
1881	36,661	6,800	7,016	5.4	–
1891	48,959	9,222	9,282	5.3	242
1901	52,634	10,802	10,909	4.9	158
1911	81,969	15,157	16,812	5.4	436
1921	114,151	28,984	28,984	3.9	1,383
1931	155,547	35,117	40,363	4.4	613
1941	166,337	39,889	42,267	4.2	477

Source: Census of Canada, various years and volumes

was an equally impressive rise in construction. Neither would be matched again until the years after the end of World War II. How was this notable accomplishment in residential construction achieved? We contend that changes in scale and efficiency were not the most significant determinants of the increased housing production levels during this period – rather, the size of the construction industry itself fluctuated with market demands. Put crudely, when demand was great more people entered the house-building business, a point further developed in chapter 5. Fluctuations in the numbers of real-estate agents and builders between 1881 and 1945 illustrate this point adequately for the moment (see table 2.4).[135] Neither moved directly with population growth. Rather, the loosely organized land-development business expanded and contracted in step with market conditions. Almost 40 per cent of the forty-six real-estate firms listed in a 1913 review had been in business for fewer than four years.[136]

Operational scale for builders changed little over the period. An admittedly incomplete 1913 analysis of building activity in Hamilton identified only two major building firms – John Poag & Company, Limited, a lumber and building firm with seventy-five employees that specialized in homes with an average value of $4,800, and Richard Press & Son, which had been in business for over forty years and had put up entire blocks of houses in the factory district.[137] If we calculate the ratio between the number of residential building permits issued and the number of builders listed in the City Directory we find an average number of dwellings per builder of 5.1. This figure varied from 1.3 in 1897 to 10.2 in 1913, so modest changes in scale were occurring. But some houses unquestionably were built

Table 2.4
Building Contractors and Real-Estate Agents: Hamilton, 1881–1945

Year	Building Contractors	Real-Estate Agents	HREB[1] Members
1881	19	11	–
1885	14	4	–
1889	35	15	–
1893	82	24	–
1897	88	21	–
1901	57	14	–
1905	63	30	–
1909	79	41	–
1913	138	113	–
1917	132	75	–
1921	100	89	–
1925	260	110	38
1929	146	90	24
1933	130	83	16
1937	78	73	19
1941	94	77	26
1945	90	90	47

Source: Vernon's City of Hamilton Directory, various years
[1]Hamilton Real Estate Board

by the owners, and it was still considered newsworthy when a builder put up more than thirty houses in a year. In September of 1906 Samuel Landers reported to the Labour Gazette that one unnamed Hamilton builder had taken out building permits, valued at $76,000, for the erection of 32 dwellings.[138] Six years later it was observed that contractors Bennet and Thwaites intended to erect some 75 houses on land that they had recently purchased in Hamilton, along with 25 summer houses at Burlington Beach.[139] In March of 1913 there was a report of the intention of one builder to put up forty houses in the east end.[140] While building permits were required after November of 1891 whenever the cost of new construction or repair work exceeded one hundred dollars, these documents have not survived.[141] We know very little about the identity of the builders of specific Hamilton dwellings and their scale of activities during the era of corporate involvement. Fortunately, some summaries of the building permits that were issued exist in the form of the City Building Inspector's annual reports. These help to shed some light on the outcome, if not the origin, of building activity for the years between 1897 and 1925.

Permits were issued for the erection of almost seventeen thousand dwellings in Hamilton between 1897 and 1924 (see figure 2.2). This

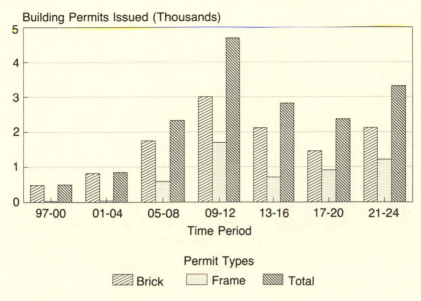

Figure 2.2
Residential construction in Hamilton, 1897–1924
Source: Building Inspector's Reports

represents an underestimation of building activity in the Hamilton area. Not every house builder would apply for a permit, and the documents were not required in those areas that lay outside the city limits. Before 1905 the houses were almost exclusively (95.9 per cent) brick structures. Late in 1903 the Master Carpenters' Association successfully lobbied to reduce the extent of the city's so-called fire limits.[142] Frame structures accounted for about one-third of the residential permits issued between 1905 and 1925. Residential building activity peaked in 1912 when permits were issued for 1,128 brick and 631 frame dwellings. Generally, the permit values for brick houses were double those for frame dwellings, with the average figures for 1912 being $2,000 for brick and $1,021 for frame houses. As with platting activity, residential construction trailed off after 1913, first in response to the depression that began late in that year and then as a result of the outbreak of World War I. More than 2.6 times as many residential permits were issued between 1909 and 1913 as between 1914 and 1918. Construction activity did not cease, even during the worst years, but it was cut back to what might be termed necessary building. In June of 1915, for example, it was reported that in Hamilton "the building trades were not nearly as active as is customary at this season, there being practically no speculative building going on this year."[143]

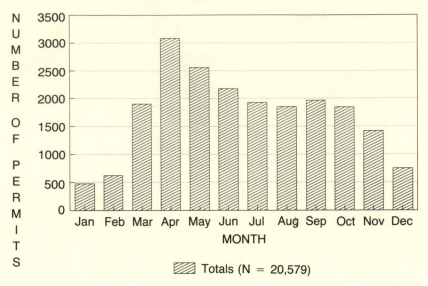

Figure 2.3
Building permits by month, Hamilton, 1897–1924
Source: Building Inspectors' Reports

As in the years prior to 1881, Hamilton's construction industry continued to be plagued by the vagaries of the Canadian climate. The eight-month or nine-month building season remained a reality during the era of corporate involvement, especially for outside work.[144] In February 1907 Samuel Landers reported that the building trades were "rather slack owing to the extremely cold weather. Painters and plumbers employed on interior work were fairly busy, but bricklayers, masons, carpenters, etc. were dull."[145] In January 1909, however, P. Obermeyer observed that "the building trades were fairly well employed, favorable weather enabling operations to be carried on continuously."[146] Building-permit summaries, however, suggest that over the long haul there was a marked seasonality to building activity (see figure 2.3). These monthly reports include all types of permits, not just those for new construction. It seems certain that many of those issued during the colder months must have been for inside repair work to homes, shops, and factories. Nevertheless, the figures reveal that, by quarter, there was a definite cycle to the issuance of building permits. Only 14.5 per cent of the permits were dated during the first three months of the year, compared to 38 per cent for the second quarter, 27.9 per cent for the third quarter, and 19.6 per cent for the final quarter. While the

Table 2.5
The Progress of Utility Construction: Hamilton, 1886–1921

| | Per Cent of Building Lots Passed by: | | | |
| | Water | | Sewers | |
Year	Original City	Annexed Areas	Original City	Annexed Areas
1886	65.0	–	41.3	–
1891	67.7	–	49.2	–
1896	72.9	28.6	61.2	28.6
1899	82.1	4.5	68.8	4.5
1906	86.2	30.9	79.8	49.0
1911	94.0	59.6	89.3	54.4
1916	98.4	86.2	91.1	89.3
1921	97.9	72.0	97.5	76.6

Source: Hamilton Assessment Rolls. For a complete discussion of this source see Appendix A.

Labour Gazette could report that a 1920 study by a large Chicago contracting firm found that there was little if any difference in the amount of time lost between winter-season and milder-season work, winter construction remained atypical in Hamilton during the era of corporate involvement.[147]

If weather conditions influenced building, so too did the completeness of the infrastructure network that fronted a given building lot. Remarkable progress was made during the early years of the twentieth century toward completion of Hamilton's utility systems.[148] By 1921 virtually every lot within the original city limits was passed by water and sewer lines, and coverage extended to more than 70 per cent of the properties in the annexed areas of the city (see table 2.5).[149] Development was less likely on lots that were not passed by water and sewer lines, especially in the suburbs (see table 2.6). In 1921, for example, only 19.3 per cent of the suburban lots without water and 16 per cent of the lots without sewer lines had been developed, compared to 77.6 per cent of the suburban lots with water and 75.2 per cent of the properties with sewers. Peter Moore found a similar lag between infrastructure development and home construction in Toronto's Annex area in the early twentieth century, as did Roger Simon for Milwaukee.[150] This underscores the growing value of serviced building lots as the era of corporate involvement progressed.

Cities such as Hamilton did more than merely provide infrastructure. Concern over civic design and housing conditions had

Table 2.6
Local Improvements and the Pace of Land Development

	Per Cent of Properties that Were Developed:									
	Original City					*Annexed Areas*				
	Water		*Sewer*			*Water*		*Sewer*		
Year	*Yes*	*No*	*Yes*	*No*	*N*	*Yes*	*No*	*Yes*	*No*	*N*
1886	91.5	87.4	90.9	89.5	1,298	–	–	–	–	–
1891	91.8	87.4	93.0	89.5	1,402	–	–	–	–	–
1896	96.9	96.3	96.2	97.6	1,496	100.0	28.0	100.0	28.0	35
1899	96.2	86.8	96.1	91.0	1,353	87.5	20.7	87.5	20.7	177
1906	96.1	82.9	97.0	83.7	1,398	60.0	18.0	55.8	6.9	314
1911	97.6	69.4	97.4	83.7	1,428	57.3	9.2	55.6	16.6	1,239
1916	98.3	32.4	98.4	85.2	1,748	65.2	14.3	63.1	17.2	1,358
1921	98.1	51.4	98.1	59.1	1,728	77.6	19.3	75.2	16.0	1,598

Source: Hamilton Assessment Rolls

emerged in Hamilton by the middle of the era of corporate involvement. Both the Medical Officer of Health and Fire Department officials began to display an interest in housing conditions.[151] In 1912 a meeting was held at the Board of Trade offices to debate the merits of a limited dividend housing company like the one that was then being formed in Toronto.[152] In July 1913 City Council debated a new building by-law that was intended to guard against slum districts.[153] The City Council passed Hamilton's first attempt at enacting a comprehensive set of building code regulations in May 1914. By-Law 1630 took up 110 single-spaced pages in the *City Council Minutes* of that year and contained 142 sections. Section 9 required that all dwellings have a yard with an area of not less than 400 square feet and that each room of the dwelling should have a minimum height of eight feet and at least one window, the size of which should be equivalent to at least 10 per cent of the floor area of the room. Other sections of the by-law contained detailed specifications for foundations, chimneys, wooden columns, stairs, mansard roofs, apartment houses, reinforced concrete construction, cement and concrete, and eavestroughs. The by-law that it replaced had had a mere 42 sections and consisted of only 14 pages.[154] By autumn 1915 references were being made to a Hamilton City Planning Commission, a by-product, no doubt, of the passage by the Province of Ontario of the 1912 City and Suburbs Plans Act.[155] The noted planner Thomas Adams spoke to the members of this Commission in November 1915.[156] All of

Table 2.7
Home Ownership by Area: Hamilton, 1886–1921

Area	1886	1891	1896	1899	1906	1911	1916	1921
ORIGINAL CITY								
North East	38.6	34.5	36.3	35.2	37.7	41.2	36.8	40.2
North West	30.6	27.7	26.6	·28.5	31.7	32.2	32.0	35.7
South East	34.0	37.4	37.0	39.1	38.3	44.5	50.0	51.3
South West	48.1	48.0	36.0	49.4	46.3	52.0	49.6	49.8
Port	37.6	35.9	32.7	28.6	34.7	31.6	27.7	38.1
Total	37.7	34.4	43.4	34.8	37.1	39.0	37.7	41.6
ANNEXED AREAS								
1891 East	–	–	28.6	39.4	48.6	52.1	50.2	65.2
1891 S. West	–	–	–	100.0	69.2	84.0	78.7	69.2
1902–12 East	–	–	–	–	–	50.6	49.7	72.4
1916 West	–	–	–	–	–	–	–	–
Total	–	–	28.6	47.4	51.8	53.2	52.2	70.3
ENTIRE CITY	37.7	34.4	33.4	35.2	38.1	42.5	42.4	52.4

Source: Hamilton Assessment Rolls

this represented a move away from the *laissez-faire* philosophy that had prevailed during the era of individualism.

Earlier in this chapter we suggested that a significant increase in home ownership took place in Hamilton during the era of corporate involvement. If we look closely at an areal breakdown of the home-ownership figures, it is easy to see that the growth was brought about largely through the construction of dwellings for owner-occupancy in the suburbs. As Edel, Sclar, and Luria recently argued in their book, *Shaky Palaces*, "the development that made homeownership available to a large proportion of the American working class without upsetting basic capitalist institutions was suburbanization. The linking of large hinterland areas to cities by cheap commuter transportation allowed individual homes to be built more cheaply."[157] For the purposes of this part of the discussion, Hamilton has been divided into two broad areas – the original city and its later annexations – with the former region further broken down into five smaller areas about Hamilton's main intersection of King and James Streets – four quadrants plus the port area. Within the limits of the original city, levels of home ownership remained quite stable between 1886 and 1921 in all areas except for the south-east quadrant (see table 2.7). In this section, which encompassed several prestigious, elevated mid-nineteenth-century subdivisions – by James Mills, Peter Hunter Hamilton, Kerr, McLaren, and Street, and Billings and Lister – owner-occupancy levels rose to 51.3 per cent, an increase of almost

Table 2.8
Occupational Structure by Area: Hamilton, 1921

Area	Prof./ Prop.	White- Collar	Skilled and Semi-Skilled	Labour	Unclassified
ORIGINAL CITY (N = 1,552)					
North East	13.1	6.0	54.2	9.3	12.1
North West	9.1	11.7	51.9	14.1	13.0
South East	20.8	14.7	37.5	12.7	14.2
South West	15.1	8.8	48.6	10.4	17.1
Port	6.3	6.0	50.7	24.6	12.3
Total	12.1	10.6	50.7	13.8	13.5
ANNEXED AREAS (N = 933)					
1891 East	13.6	20.4	47.2	8.4	10.6
1891 S. West	23.2	30.4	36.2	2.9	7.2
1902–12 East	10.0	13.5	50.6	19.4	6.4
Total	12.0	16.6	48.5	15.2	7.5
ENTIRE CITY	12.1	12.8	49.4	14.3	11.3

Source: Hamilton Assessment Rolls

51 per cent over the period. No other original-city area increased its rate of home ownership by more than 17 per cent, and in three of them owner-occupancy levels remained at or below the 40 per cent level even in 1921. Most of the increase in home ownership had occurred in the burgeoning suburbs, where almost from the beginning of their settlement, and certainly after 1906, the majority of the dwellings were built for owner-occupancy. Neither the depression of late 1913 nor the outbreak of World War I were sufficient to halt this trend. By 1921 home ownership levels in the occupied suburban areas averaged just over 70 per cent.

Hamilton's socio-economic structure continued to crystallize as the era of corporate involvement progressed. The growing area under the control of restrictive covenants placed limits on the quantity of land available for inexpensive housing. For 1921 statistically significant differences in occupational structure were observed for the same small areas that we employed in the foregoing analysis of home ownership changes (see table 2.8).[158] Insofar as occupation can be used as an indicator of wealth and class, this translated into a city with an increasingly sharply defined socio-economic structure. Two definite working-class areas, with concentrations of skilled, semi-skilled, and unskilled workers well in excess of the city-wide figure of 63.7 per cent, had emerged by 1921 – the port area in the north end of the original city (75.3 per cent) and the 1902–12 east-end annexations (70 per cent). These were more or less the same areas

Table 2.9
Building Permit Values: Large Ontario Cities, 1905–1925
(in Thousands of Dollars)

Year	Ontario	Hamilton	(%)	London	Ottawa	Toronto
1905	$17,599	$1,511	8.6	$539	$1,534	$10,348
1906	23,043	2,125	9.2	1,200	1,729	13,160
1907	26,175	3,030	11.6	875	2,365	14,326
1908	22,012	1,331	6.0	866	1,794	11,795
1909	32,488	1,547	4.8	850	4,528	18,139
1910	33,603	2,605	7.8	805	3,023	21,127
1911	39,669	4,256	10.7	1,037	2,998	24,375
1912	50,022	5,492	11.0	1,136	3,622	27,402
1913	49,475	5,110	10.3	1,790	3,991	27,039
1914	38,558	3,704	9.6	1,838	4,398	20,694
1915	14,353	1,522	10.6	1,207	1,605	6,652
1916	20,230	2,410	11.9	926	1,530	9,882
1917	17,408	2,747	15.8	838	1,041	7,164
1918	18,477	2,472	13.4	878	2,636	8,535
1919	40,585	5,087	12.5	2,455	3,252	19,618
1920	47,175	4,340	9.2	2,146	3,305	25,737
1921	49,277	4,639	9.4	2,527	2,716	23,878
1922	67,247	4,928	7.3	2,606	5,022	35,238
1923	59,889	5,453	9.1	3,261	3,521	30,609
1924	57,330	3,310	5.8	2,114	2,541	23,926
1925	74,673	2,676	3.9	2,390	4,942	25,797
Totals:	781,689	70,295	9.0	32,284	62,093	405,444

Source: Labour Gazette, various years and volumes

as those identified by Rosemary Gagan as having the highest mortality rates in Hamilton in 1910.[159] In contrast, professionals and proprietors and white-collar workers were highly over-represented, compared to the city-wide figure of 24.9 per cent, in three more salubrious areas – the south-east quadrant of the original city (35.5 per cent), the area in the south west that had been annexed to Hamilton in 1891 (53.6 per cent), and the area to the east of the original city limits that also had been annexed in 1891 (34 per cent). A fuller account of the relationship between housing quality and social structure is presented in chapter 10.

While Hamilton's construction efforts compared favourably with those of other large Ontario cities, typically accounting for about 9 per cent of the building permits issued annually in the province (see table 2.9), there is some evidence to suggest that by the early 1920s Hamilton's residential construction industry could no longer supply the demand from two perspectives – quantity and affordability.[160]

As early as 1903 the editor of the *Hamilton Times* suggested that as "A Solution to the Problem of Houses – Large Number will be Built by Individual Owners."[161] Olivier Zunz uncovered a similar dual home-construction market in Detroit for this period, as did both Deryck Holdsworth for Vancouver and Roger Simon for Milwaukee. According to a recent study by James McKellar, owner-built or owner-contracted housing remains important in rural areas of the Maritime Provinces and even in some small cities such as Sault Ste Marie and Prince George.[162] Periodic references in the *Labour Gazette* pointed to a scarcity of moderately priced rental houses in Hamilton.[163] By 1909 reports of overcrowding and other housing-quality problems were becoming regular entries in the *Annual Reports* of the City's Board of Health.[164] As the era of corporate involvement wore on, the housing industry turned its attention to the demands of the middle and upper classes. Modestly priced houses were not very profitable. Less affluent urban residents were left to fend for themselves. John Saywell suggests that their strategies included doubling up, worker co-operatives, the acceptance of company-built dwellings, rental of poor quality housing, and the erection of shack towns in the industrial north-east of the city.[165] No philanthropic housing, common in some American and British cities, appears to have been built in the Ambitious City.[166] One Hamilton housing achievement in the pre-World War I period was notable, however. As part of the city's centennial celebrations in August 1913, Alderman James Bryers orchestrated four hundred of the city's construction trades-people in the task of erecting a twelve-room brick house in just twenty-four hours, a feat that was later reported in Ripley's "Believe It Or Not." Ironically, as hundreds watched this event in east-end Britannia Park (each paid twenty-five cents for the privilege), many Hamiltonians had never been so poorly housed.[167] However, many others had never lived so well. Overall, as we show later in this chapter and in chapter 10, Hamiltonians were probably better housed than most urban Ontarians by the end of the era of corporate involvement. Unquestionably, they enjoyed better residential conditions than their contemporaries in Glasgow and New York.[168]

THE ERA OF CORPORATE INVOLVEMENT, 1921–1945: THE CASE OF SUBURBAN WESTDALE

In the year following the end of World War I Canada came as close to experiencing massive civil unrest as ever in her history. More than 600,000 people were demobilized after the war, and returning soldiers and others faced rapidly rising prices and expensive and insufficient housing stock in most Canadian cities. Events culminated in the bitter Winnipeg General Strike during May and June 1919. We are not suggesting that the unrest that swept Canada in 1919 was rooted solely in housing issues – clearly this was not the case, but housing costs and conditions were an often-cited part of the problem.[169] In response, in 1919 the federal government made a direct loan to the provinces of $25 million for home construction, and 6,244 dwellings were completed under the program by 1923.[170] This small effort proved to be a sufficient involvement for the federal government because private-sector builders quickly came to its rescue. The building industry responded in two ways to rising labour, material, and land costs during the early 1920s. Builders put up houses on cheap suburban land for those who could afford to become owners and they erected apartments for those who could not afford or were not interested in home ownership (see chapter 9). From virtually a zero share at the end of World War I, apartments increased to about 15 per cent of urban housing stock in Canada by 1931.[171] Hamilton did not escape these trends. Between 1921 and 1931, the percentage of households in our assessment roll samples that resided in apartments increased from 1.6 to 8.4 per cent (see chapter 9). About midway between these years, a guide book identified 253 apartment buildings in the city.[172]

The inter-war years in Hamilton and elsewhere were characterized by the growing importance of urban planning and corporate developers. Both would leave their mark on the structure and form of new residential neighbourhoods. In order to explore the nature of residential development in Hamilton during this period, we have chosen to focus on the Westdale area, for the saga of Westdale is very much the story of residential development in Hamilton and, we suspect, much of urban North America between 1921 and 1945. When completed in the early 1950s, Westdale would be home to more than 1,700 households.[173] Its 16 subdivisions and 2,112 building lots represented 36.4 per cent of the surveys and 32.4 per cent of the lots placed on the Hamilton real-estate market between 1920 and 1945 (see table 2.10).[174] As a private enterprise, Westdale's development was governed by commercialism, but like the best planned North American suburbs of the era of corporate involvement – such as the exclusive Country Club District of Kansas City or Vancouver's Shaughnessy Heights – it balanced aesthetic and environmental concerns with financial ones. This made Westdale different from many of the commuter suburbs of the period, but its history still progressed in step with national and even international

Table 2.10
Westdale's Component Surveys, 1920–1944

Plan Name	Year	No. of Lots	Typical Lot Area (Sq. Ft.)	Use of Grid Plan	Minimum Building Values
Woodlawn	1920	229	3,000	All	$2,500
Clinelands	1921	181	3,750	Most	$3,000
Paisley Gardens	1921	206	4,500	No	$3,000
Westmoreland	1921	135	4,500	All	n.a.
Crescentwood	1921	325	4,750	No	$3,000
Elmhurst	1923	309	4,500	Most	$3,000
Crestwood Addition	1924	12	4,750	No	$3,000
Princess Heights	1925	137	3,250	All	none
Oak Knoll	1925	27	8,000	No	$7,000
Oak Wood	1927	21	10,500	No	$8,000
Crescentwood Exten.	1929	19	4,750	No	$3,000
Forest Hills	1929	65	8,500	No	$8,000
Collegiate Park	1930	132	3,150	All	$3,000
Parkside Vista	1939	56	3,100	All	$2,500
Parkside Gardens	1941	187	3,450	Half	$2,500
Parkside Gdns Add.	1944	71	3,450	No	$2,500

Sources: Registered Plans, Wentworth County Registry Office and Weaver, "The Suburban Life of Westdale," 418

urban trends. Westdale's developers, builders, and residents shared experiences with their counterparts across Canada and the United States. Their philosophy of residential development was similar to that espoused by a leading contemporary American developer, Jesse Clyde Nichols:

The best manner of subdividing land should not necessarily mean the quickest sale. The destiny and growth of your town is largely affected by the foresight of the man who subdivides the land upon which you live. The most efficient manner of platting land should be the plan which gives the greatest value and security to every purchaser, adds the greatest amount of value and beauty to the city as a whole, yet produces a big profit to the man who plats the land. To follow this method, one must have supreme imaginative confidence in his city and its future.[175]

The story of the development of Westdale neatly illustrates the way in which large North American subdivisions of the 1910s, 1920s, and 1930s represented a bridge between residential-development processes during the era of individualism and during the post-1945 period. Westdale's development differed from both the "free-ride and barbecue days" of the early twentieth century and the packaged suburbs of the present time.[176] The area's developers are best understood as part of a group that Marc Weiss has labelled "community builders":

subdividers who changed the nature of [North] American land development during the early decades of the twentieth century. They did this initially by taking very large tracts of land and slowly improving them, section by section, for lot sales and home construction. Strict long-term deed restrictions were imposed on all lot and home purchases, establishing uniform building lines, front and side yards, standards for lot coverage and building size, minimum housing standards and construction costs, non-Caucasian racial exclusion, and other features. Extensive landscaping and tree planting were emphasized to accentuate the natural topography and beauty. Public thoroughfares included curved streets, cul-de-sacs, and wide boulevards and parkways. Often special areas were set aside for retail and office buildings, apartments, parks and recreational facilities, churches, and schools. Private utilities and public improvements were coordinated as much as possible with present and future plans for subdivision development and expansion.[177]

"Someone has said that there is only one crop of land, but there is an endless crop of babies and every baby on the face of the earth makes every foot of land more valuable."[178] This article of faith, which was employed in an artistically-striking brochure by Westdale's developer, has been promoted by the land-development industry past and present. Discerning patterns of urban growth, both spatial and temporal, developers such as those involved in Westdale have attempted to interpret and influence public taste. Since the early years of this century, they have acted as instruments for drawing together political and legal acumen, capital-raising facilities, planning talents, and the building trades. Value judgments about land developers, therefore, must be constructed carefully so as to permit areas of ambiguity. Westdale, for example, was a well-conceived community with a fine array of amenities. It also encompassed racial discrimination and "clever" tactics on the part of its developers.

At the turn of the century, the thrust of Hamilton's residential growth was being channelled by topography. By 1910 the escarpment to the south, Burlington Bay on the north, and a wide ravine in the west had turned most platting and building activity eastward. As we saw earlier, east-end surveys stood at a considerable distance from downtown Hamilton by the onset of World War I. Moreover, Hamilton's factories, with their smoke and noise, were concentrated in the east end. For several reasons, then, the level plateau that stood at the far side of the ravine forming part of Hamilton's western border caught the eye of Toronto contractor J.J. McKittrick, who in 1910 began to promote a 100-acre plot known as "Hamilton Gardens."[179] His venture lacked urban services and McKittrick did not have the necessary resources to secure them. As a consequence he became associated with local partners whose careers had been meshed with the development of Hamilton. Legal talent, for example, came from Sir John Gibson, a former Lieutenant-Governor of Ontario, and one of Hamilton's famed Five Johns. Gibson and other members of what came to be known as the McKittrick Syndicate were connected with both Hamilton's pioneer electric utility firm, Cataract Power and Light, and the Hamilton Street Railway. (Had John Patterson, lived long enough, he too would have been joined the other Johns in Westdale.) The Southam family, publishers of the influential *Spectator*, acquired a major interest in the land syndicate.

Soon the new group had expanded McKittrick's original 100-acre parcel to 800 acres and had successfully negotiated an agreement whereby in January 1914 Hamilton annexed its holdings.[180] The agreement set down conditions that forced the syndicate to construct and maintain a bridge over the ravine separating Westdale from the rest of Hamilton. Early in 1914 a contract for the steel superstructure valued at $76,625 was awarded to the Hamilton Bridge Works, while another for grading and excavating valued at $33,794.90 was granted to the Mackay Paulin Construction Company. The work was to be completed by December.[181] In return for building the bridge, the low, pre-annexation rural assessments on the entire property were to be frozen until 1919. Yet even with the syndicate's entrepreneurial talent and the obvious benefits associated with annexation, the endeavour proved unpromising. The 1913–15 recession followed by the strictures of the war economy retarded property sales until the early 1920s.

The cash flow anticipated by McKittrick Properties dried up for about eight years, during which time its tax-concession cushion expired. It was subsequently renegotiated for a period extending to 1926. This provided some relief, though a further commitment entered into during the palmy days before the war returned to plague the syndicate. To secure a key parcel of 100 acres belonging to the Hamilton Cemetery Board, McKittrick Properties had purchased another site and exchanged it, along with a promise of $40,000 in compensation, for the coveted parcel. Financial stringency forced the syndicate to default on one of the compensation instalments. Eventually, the Cemetery Board agreed to a settlement, but not before the influence of the syndicate had been rebuked by Hamilton's electorate. A "McKittrick man," contractor and Board of Control member William Henry Cooper, ran for Mayor in 1916. Despite the avid backing of the *Spectator*, which, in a steady stream of front-page pieces, lauded him as a prudent administrator and the friend of both unions and small property owners, Cooper was defeated in the election that was held on New Year's Day 1917. He came third behind both a bitter rival, fellow Controller Thomas Skinner Morris, and the dark-horse winner, merchant Charles Goodenough Booker.[182] The early experiences of the McKittrick syndicate suggest that even a powerful alliance among the civic elite could sometimes be subject to fiscal and political embarrassments.

Facing difficulties, the developers of Westdale sought fresh management. One of the investors, John Moodie, President of Eagle Knitwear and another of the famous Five Johns, invited his son-in-law, F. Kent Hamilton, to guide the fortunes of McKittrick Properties. A Winnipeg lawyer, Hamilton had learned a great deal about the planning and promotion of suburbs from his western experiences. Arriving in 1918, Hamilton quickly established both downtown and Westdale offices and a sales staff that climbed to eight members in good times. During the next seven years he also took a direct hand in the design of the publicity campaigns used to market the Westdale surveys. Kent Hamilton rapidly gained stature in both the local and provincial real-estate communities. He was instrumental in the formation of the Hamilton Real Estate Board in the early 1920s and in 1925 he became the first Hamiltonian to serve as president of the Ontario Association of Real Estate Boards.[183] He was alert to the nuances of development south of the border. It was Hamilton who commissioned New York landscape architect Robert Anderson Pope to prepare a street plan for the syndicate's holdings.[184] Pope was one of many urban planners of the period to have been influenced by German and British planning concepts. As early as 1910 he had recommended German-style urban deconcentration, urging a shift away from urban designs that concentrated lines of transportation on the core of the city. Pope argued that this led to overcrowding and high residential land costs. New suburbs, more or less self-contained, promised a remedy.[185] When Pope designed Westdale, his plan included an area that was set aside as a central shopping district and contained a few sites for apartment buildings (see figure 2.4). In 1921 Westdale's promoters could truthfully boast that "we have sites for homes, stores, and apartments, but we don't mix them."[186]

Kent Hamilton, like his Toronto-developer contemporaries Robert Home Smith and Wilfred Servington Dinnick, had a full understanding of the housing issues that had concerned North American

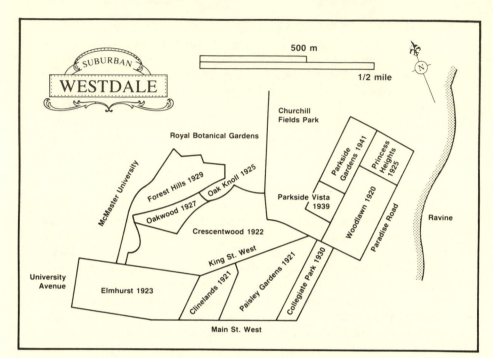

Figure 2.4
Component surveys in the planned suburban community of Westdale

reformers between 1900 and 1920, having read articles and attended lectures on most of the era's remedies for the housing crisis. His expertise would later bring him before the Lieutenant-Governor's Committee on Housing Conditions in Toronto.[187] By 1919 Hamilton was aware of the English garden city movement, limited-dividend housing companies, tax incentives for builders and home buyers, company housing, and even the wartime housing constructed by the United States government. He supported the creation of the Hamilton Town Planning Commission and for a while considered supporting the still novel idea of public housing. Kent Hamilton recognized that studies by the Town Planning Commission would necessarily benefit McKittrick Properties because their land was the last major undeveloped parcel near the central city and, therefore, would be a prime candidate for any experiments in public housing. In the economic uncertainty of 1919, a sale, even to a public housing corporation, was to be welcomed. Kent Hamilton had one further reason for supporting action on the housing problem. He had been advised that "failure to take care of the returned soldiers, not only from the stand point of housing, but also from the stand point of opportunity for earning a decent livelihood, may result in a social uprising that in the end would be far more expensive to the city of Hamilton than a theoretical excess cost of providing adequate housing."[188]

The turbulence of 1919 frightened some civic leaders into repression, but it moved others towards an expedient consideration of reform. With the return of social and economic stability, public housing proved a will-o'-the-wisp. Indicative of a more conservative approach was a week-long Better Homes Exhibit in the Hamilton Armories that was sponsored by Kent Hamilton and the newly created Hamilton Real Estate Board. Their aim was "to educate the average renter into the method and means of ownership."[189] As a tangible move in that direction, the first and easternmost Westdale survey – Woodlawn – had come onto the market as worker's parcels with thirty-foot frontages arranged on a grid layout. Woodlawn was touted for its proximity to central Hamilton compared with contemporary east-end surveys, its lack of smoke, its favourable elevation, and its building restrictions.[190] This subdivision, however, had a lower potential value than other holdings in Westdale, for along the opposite slope of the ravine the City maintained a garbage dump. It also was close to the City Isolation Hospital and a brickyard that Kent Hamilton described as "a boil on the thumb so far as our property was concerned."[191] As late as 1928, a syndicate official could write that these lands remained unattractive: "I do not expect that they ever will be very desirable."[192] One direct incentive had hastened the surveying of worker's lots. The Ontario Housing Act of 1919, applying federal funds, assisted the raising of mortgages for modest six-room homes costing under $4,000. The "Hamilton A-1 Plan" that provided a home for $3,850 was a government-approved brick-dwelling design that was used extensively in Westdale.[193] The eastern portion of the syndicate's holdings would continue to be developed on the grid plan into the 1940s (see table 2.10).

From 1920 until 1926 the syndicate faced lean times. Local conditions formed an element in a national interlude of slow residential construction after the modest boost provided by the 1919 Housing Act. Moderate improvement occurred in 1922 by which time streetcar, water, and sewer lines had reached Westdale, but construction suffered a setback in the next two years. After 1924 the McKittrick group joined other development ventures in a subsequent steady improvement in business circumstances that continued unbroken until 1928, a peak year for residential construction in Canada.[194] As conditions improved, the syndicate responded by altering its tactics. Several small, curvilinear subdivisions containing lots that were double and even triple the size of those in its first surveys were placed on the market to appeal to more affluent home buyers (see table 2.10). These elite surveys – Oak Knoll, Oakwood, and Forest Hills – clung to a narrow, ravine-indented fringe at the western extremity of the syndicate's holdings. Here high minimum building values were written into the deeds of sale, many of which included specifications for building materials and established broad architectural guidelines. Still, the attempt to cater to a prosperous clientele brought no sudden windfall. Twenty years after the initial land assembly, ten years after the registration of the first survey, only half of the residential lots had been built on. Nonetheless, this represented a much more rapid pace of development than that in another contemporary, though smaller, development – Victoria's Uplands area, where it took almost forty years to accomplish a similar rate of home construction.[195] While some of the 830 vacant Westdale properties appearing on the 1931 assessment rolls were held by contractors, the property developers still held 570 lots, or almost 70 per cent. Free bus trips for inspection tours, gold certificates buried on a few lots, the innovative opening of a model home (probably Hamilton's first), a contest to name Westdale, and the barnstorming of Jack V. Elliot in his *Canuck* aeroplane could not stir sufficient interest. McKittrick Properties went into bankruptcy.

With the prospect of the tax assessment freeze being lifted in 1926 and the suburban bridge, which the syndicate had financed, now serving a major highway, the developers requested an amended agreement with the city. In fact, the McKittrick group faced serious cash flow problems because of their fixed annual payments. To meet expenses, the syndicate turned to an expedient. Since land had been mortgaged to finance the initial assembly of their holdings, some additional collateral had to be discovered. The company borrowed against contractors' agreements-to-purchase, with loans coming from the major shareholders. As contractors repaid the company, the loans were retired.[196] The sticky cash flow problem became a critical worry, for with time running out on the assessment freeze, the City of Hamilton and the Ontario Railway and Municipal Board refused to amend the 1913 agreement. Caught in a pinch that required dramatic proof to the city that relief was required, the Board of Directors met in June 1926, refused to pay a minor bill, and precipitated bankruptcy proceedings.[197]

Two major unsecured creditors were affected by the bankruptcy: those shareholders who had loaned funds against the contractors' agreements-to-purchase, and the city, which was owed $200,000 in tax arrears and service charges.[198] As real as the financial crisis was, certain features of it suggest that this bankruptcy had tactical dimensions. Shareholders paid to the full all creditors except the city.[199] It soon dawned on the Mayor and the members of the Board of Control that their obduracy had drastic implications. Future tax revenue, the city's credit rating, and Hamilton's reputation as the "Ambitious City" were jeopardized. The episode had a familiar ring. A major private endeavour had become so closely identified with commitments to expansion that a government hard line was precluded despite some hard-nosed posturing to convince voters that City Hall was not "soft" on developers. The Public Trustee appointed by the syndicate's shareholders expressed the situation well enough when he noted: "The city asked us to come, as a matter of fact, because they were worried about the situation ... I told them that there was now no money to pay taxes ... and out of that arose the suggestion that they might take some of our land for parks, to clear up those arrears of taxes."[200] The city subsequently took 377 acres of ravine land for parks and erased the syndicate's tax debt. Since this rugged land was unsuited for development and the creation of a park would bolster land values, the agreement was a rewarding one for the developer.[201] With the debt lifted and the Ontario Railway and Municipal Board approving a relaxed agreement that released the developers from maintenance of the bridge, a newly chartered company, Westdale Properties, composed of shareholders in the former syndicate, sprang to life and bought out McKittrick Properties. Except for the sacrificed park land and a legal bill of $7,573, the manoeuvre had cost nothing, but it had succeeded in forcing the better terms that the syndicate had been seeking for five years.[202]

Just as the new company took shape, it fell heir to a boon secured by the old syndicate. Late in 1921 Kent Hamilton had begun courting McMaster University, which had been considering a move from its Bloor Street location in Toronto.[203] After lengthy negotiations and generous contributory pledges to McMaster from John R. Moodie, William Southam, G.H. Levy, and Sir John Gibson – all members of the original and new land companies – McMaster located in Westdale.[204] Immediately, property in two Westdale subdivisions was "expected to be more attractive than anything at present on the market as a result of the McMaster University location."[205] The developers also calculated that the university would carry the burden of expense for a major water main, purchase electricity from their power company, and lure a "colony of Professors" to their property. For all of its apparent ingenuity and sinister cunning, the syndicate was neither cohesive nor an instant success. Kent Hamilton sued for payment of commissions owing during the bankruptcy; the Southams found certain legal charges of their syndicate partners, Gibson and Levy, far too high; and as for professors, most were too impecunious to buy Westdale property. The Depression dashed whatever expectations had been raised by the new agreement and the enticing of McMaster. Only in the building boom after World War II would vacant fields disappear.

The creation of Westdale preceded the era of totally integrated property development. In a sense, Westdale emerged at a mid-point in the evolution of the residential land-development industry. It had the benefit of sophisticated planning ideas but, unlike recent packaged suburbs, construction was not a branch of the developers' activities. Given this absence of corporate integration, Westdale's completion drew upon an array of traditional crafts and specialists whose relationships tended, for all their diversity, to be intimate and coordinated. Mortgage brokers, contractors, and sub-contractors worked independently, but their ability to deviate from the developers' vision of the finished community was constrained.

By the early twentieth century it was firmly held in real-estate circles that unregulated growth, non-conforming land uses, and certain ethnic groups offended middle-class and upper-class home buyers. No longer were building lots sold by auction, a system that minimized the developers' control over the tone of the subdivision. Conformity was now deemed to be important. As we have already seen, its implementation in the face of a multitude of individual actors led to the continued evolution of the real-estate profession and to the appearance of both restrictive covenants and urban planning.[206] Deed restrictions gained wide application in Hamilton during the early years of the twentieth century. As noted, those used in the industrial east end specified minimum building values and required brick construction. Westdale's deeds were written to a higher standard. Newspaper advertisements designed by Kent Hamilton in the early 1920s reminded prospective homebuyers to consider "whether the location is restricted so that it will increase rather than decrease in value." Westdale was described as being "well restricted."[207] It was. Westdale's covenants contained controls not only over land use and the physical character of its homes, but also over the social characteristics of its residents. Minimum house values were established for most of the component surveys (see table 2.10). In the more exclusive subdivisions, like Oak Knoll, Westdale's developers, like their Toronto counterparts in Lawrence Park and Kingsway Park, retained the right to approve the "location, plan and specifications, exterior elevation, and type of construction." Social control was exercised through a single, blunt, all-encompassing clause: "None of the lands described ... shall be used, occupied by or let or sold to Negroes, Asiatics, Bulgarians, Austrians, Russians, Serbs, Rumanians, Turks, Armenians, whether British subjects or not, or foreign-born Italians, Greeks, or Jews."[208] In describing the advantages of Westdale in a brochure, Kent Hamilton felt it sufficient to employ the phrase "restricted."[209] Regulation in the early years was enforced. A real-estate agent for the syndicate warned a contractor not to sell to an interested Italian-born greengrocer: "Tom, we don't want people like that in here."[210]

Builders, dependent on credit and a sound reputation with subdivision developers, lacked the security to risk breaking the covenants. Relying on quick sales, they dared not risk architectural innovations either. Most of Westdale's builders operated as family combinations without their own design facilities. They used standard blueprints with a demonstrated popularity. This meant that despite Westdale's distinctive physical planning, the bulk of its housing stock resembled much of what was being built in the rest of contemporary Hamilton. The appearance of large, custom-built homes in the 1940s altered the picture somewhat, for even smaller houses constructed in these years attempted to incorporate imitations of expensive flourishes: leaded windows, decorative stonework, bay windows, and Tudor-style wood and stucco treatments on the second storey.

Westdale's builders were a mixed group. For many men anxious to improve themselves, an opportune route out of the labour pool was to become a small-scale builder. Like street peddling or the operation of a corner store, contracting provided a few urban labourers with access to an independent lifestyle. A few, such as Hamilton-based builder Michael Pigott, even carried success beyond their original locality, but for most it was a struggle. As J.M. Mattila observed, "since many contractors have moved up from trades or carpenters, painters or similar artisans, they often lack the necessary business and engineering training for making accurate estimates of cost."[211] Some insight into the hustle and flexibility of the builders is provided by merely considering the fragments of information conveyed in city directories, advertisements, and assessment rolls. One family, the Theakers, demonstrated a few fairly common characteristics of the approximately thirty builders involved in Westdale. The men worked on homes in both the east end and Westdale, using the same plans. Gladys Theaker kept the books and worked for a mortgage broker. Family members lived within walking distance of Westdale. Occasionally, such family builders formed transitory partnerships, for the building trades remained almost as fluid in their associations during the 1920s and 1930s as they had been during the era of individualism. Joseph Vickers and his son held twenty Westdale lots in 1931, but the father also retained several in the east end with plasterer Fred Beldham. These were father and son operations, but certain family-builders established modest construction dynasties. The Mills family had participated in home construction and property development since the mid-nineteenth century (see chapter 8). The Armstrongs had both depth and experience, having constructed some six hundred dwellings in Hamilton by 1930. With William C. Armstrong as realty and financial agent and William D. Armstrong heading the architectural department, they were the most ambitious domestic contractors in the city during this period.[212] Thomas "Carpenter" Jutten and his son Charles were builders, but Jutten senior, Mayor of Hamilton from 1923 until 1925, and later a Member of the Ontario Legislature, was also a significant politician. The father, in his role as Mayor, cut the opening ribbon on a Westdale economy home in 1924, and the son was erecting houses in Westdale as late as 1948.[213] Whatever the scale of their operations, the builders had a parochial orientation. They depended on the establishment of a local reputation for credit and customers. Unlike many of today's builders, they were not working for large suburban builders with an impersonal corporate name. Building technologies have advanced since the 1920s, but there was something of value in having contractors mindful of their community's esteem. One builder, Thomas Casey, could easily identify homes he had constructed fifty years earlier, and, as he did so, rhapsodize about the quality and distinctive character of his brickwork.[214] The careers of Casey and several other builders are detailed in chapter 5.

Several major Westdale property owners had no interest in dwelling construction, but instead purchased lots and held them for future capital gain. In 1931, outside of those involved in the syndicate, the largest such investor, with twenty-five lots, was a doctor whose brother managed the local-real estate branch for National Trust and was president of the Hamilton Real Estate Board in 1930. The degree of speculation in Westdale lots and the background of the participants are difficult to establish because of the existence of a complex set of "straw companies." Kent Hamilton is a case in point. Besides his connection with McKittrick Properties, he managed at least two other companies that dealt in real estate: Blackstone Realty Securities Limited and Gorban Land Company Limited. Hamilton and similar property entrepreneurs placed some of their activities at a remove in order to limit personal liability, but the numerous corporate labels imply additional motives. At the least they suggest the involvement of different combinations of shadow investors and the desire for anonymity. This would prove to be an enduring and widespread feature of the residential property-development industry.[215]

In many instances those who dealt in Westdale real estate had assorted business interests across the city and often engaged in several facets of the housing industry, providing a foretaste of the elaborate corporate integration that later would come to characterize large segments of the residential-development business. Kent Hamilton, for example, also functioned as a mortgage broker. Forty Investors, the largest owner of commercial property in Westdale, was managed by W.C. Thompson, a real-estate agent and mortgage broker. His other operations included Forty Associates, Hamilton Home Builders Limited, Hamilton Improvement Company, Traders Realty, and Thompson and Thompson Realty. Less diverse agents understandably located their offices close to such complementary services. Long-time realtor Norman Ellis owned three Westdale lots in 1931, and to arrange financing for builders or prospective purchasers he only had to go next door to the Hamilton Finance Corporation.[216] Realtor J.W. Hamilton, who also held three properties in 1931, sold building lots to contractors. At the same time he served as secretary of a major wholesale lumber company with offices on the same office-tower floor as his real-estate agency. These and a host of comparable connections helped to guarantee Westdale's disciplined development. The realty agents and investors who owned lots were not likely to sell to contractors who might introduce structural or ethnic non-conformity for fear of depressing the value of their remaining properties. Moreover, speculation in Westdale lots had few amateurs. Almost three-fifths of those who owned two or more vacant lots in Westdale in 1931 were involved in real estate or building.[217]

From the outset, Westdale developed a distinctive residential character in spite of the variety of its surveys and their building lots. Like other Hamilton suburbs of the era of corporate involvement, it was developed primarily for owner-occupancy. In 1931, 72 per cent of its dwellings were owner-occupied, compared to just 48 per cent in Hamilton as a whole. The Westdale figure would rise to 82.5 per cent two decades later, in comparison to a 68.5 per cent rate for all of Hamilton. It was an overwhelmingly Protestant area – 88 per cent in 1931 and 77.7 per cent in 1951, figures that were more than 11 per cent higher than those for the entire city at both times.

Generally speaking, status increased from east to west in Westdale. This was reflected in assessment values, occupational categories, and dwelling sizes and styles.[218] The central blocks of Westdale were made up of surveys with moderate deed restrictions on building values and unpretentious survey names – Crescentwood, Clinelands, and Paisley Gardens. The streets here did not follow the simple grid pattern found in Westdale's easterly surveys, but neither were they protected from the noise and commotion of through traffic like the subdivisions along the elite western fringe. Geographically and figuratively, Westdale's centre was truly mid-way. Contractors responded most frequently to the developer's characterization of these central surveys by constructing the two-and-a-half storey "square plan" homes that had proven popular all across Hamilton.[219] Slight variations in the style of brick, porch, and window details did little to break the monotony or, as some preferred to think, the continuity. The centre emerged as something of a compromise between the simplicity apparent in the worker's surveys and the elegance arrayed along the fringe. Since Westdale's builders responded to the prosperity of 1925–29 with a frantic construction of middle-class dwellings, this central portion had a number of completely developed blocks by 1931. After the 1930s it continued to have more architectural cohesion than any other part of Westdale. For example, while Thomas Casey built the whole side of one block face, one of his rivals, Joseph Vickers, built a row of houses behind him using similar materials, blueprints, and embellishments.

Westdale's completion as a prime and distinctive development was fully realized by 1951. It had taken about three decades. In comparison to development in the era of individualism, this was remarkably swift, especially considering the prolonged negative effect on residential construction of the Depression and World War II.[220] Most Westdale homes had been erected in two short bursts – 1925–29 and 1945–51. The subdivision development cycle, then, was significantly shortened during the era of corporate involvement. Over time, Westdale gained an image of convenience and prestige that dispelled the reservations of the 1920s about raw and unattractive lots in surveys such as Woodlawn and Princess Heights. Measured against the new suburbs of the 1950s being laid out in Stoney Creek, on the Mountain, and on land to the west of Westdale, the district had considerable appeal. Westdale represented the best efforts of private enterprise in the housing field, but it had been a slow and fragmented process. Impressive as they were, especially when compared to those during the era of individualism, the forces marshalled in the building of Westdale would not be sufficient to meet demands for new housing in post-war urban Canada. A rapid rate of new family formation, coupled with massive foreign migration into many Canadian cities in the years after 1945, created an unprecedented demand for new housing. To meet this demand, a new development paradigm would be needed.

3 Modern Residential Development Practices: The Era of State Intervention

An acute housing problem troubled wartime and post-war Canada. It may be defined as an immense unsatisfied need for accommodation that derived from housing supply shortages, replacement requirements, and overcrowding associated with the depression and the war. Individuals in all income groups wanted housing, but low and medium income tenants felt the need most keenly.

Jill Wade, *Wartime Housing Limited*[1]

THE CONTEXT FOR POST-WAR RESIDENTIAL DEVELOPMENT

By 1945 Canadian housing production had become more regulated, more coordinated, and more integrated. Increasingly, either corporations or real-estate professionals brought building lots onto the market. These lots were sold for a stated amount per foot of frontage rather than by auction, and they were nearly always marketed on easy-payment terms. No longer raw land parcels, newly platted lots came with an impressive array of services already in place. By 1941 the infrastructure system within the City of Hamilton was virtually complete – 99.7 per cent of all dwellings had electric lighting and the same percentage had running water – though figures were usually lower than this in exurban areas.[2] The increasing sophistication of the residential development process in urban Canada in the years prior to 1945 was signalled by the birth of several national housing-related organizations – the Town Planning Institute of Canada (1918), the Canadian Association of Real Estate Boards (1943), and the National Home Builders Association of Canada (1943) – paralleling similar trends in the United States.[3] Residential construction, however, continued to be dominated by small building firms even as World War II ended, but larger enterprises were becoming less unusual. As the era of corporate involvement progressed, governments came to exercise greater control over most aspects of residential development. In 1945, however, two important elements in the present-day arsenal of development practices had yet to emerge fully: the rise to prominence of large, vertically integrated development companies and the active involvement of government in dwelling construction.

As had been true in the early 1880s, the transition between residential development paradigms around 1945 was anything but abrupt. Many old philosophies and methods would persist from earlier times, yet some important seeds of change had been sown prior to 1945. Among the early harbingers of change, the Patterson Brothers and a few other "master builders" had taken a stab at corporate integration well before 1945, but they never achieved anything approaching market dominance. Builders of all sizes began to face growing regulatory control much earlier. By 1914 Canadian municipalities were passing building codes and zoning by-laws with great zest.[4] Control over the subdivision of land had begun in 1912 with Ontario's passage of the City and Suburbs Planning Act (see chapter 2). By the mid-1940s subdivision approval in communities like Hamilton required the signatures of the Mayor, the City Clerk, the City Engineer, a Provincial Land Surveyor, the Registrar, and the Minister of Planning and Development.[5] Housing conditions and costs in British, Canadian, and US cities began to come under the scrutiny of the press and early social scientists before 1900.[6] Concern about housing matters had been expressed by all three levels of government at various times during the era of corporate involvement, especially around 1919. Even before that, affordability had emerged as an issue that needed to be addressed. A limited-dividend housing company had built the Spruce Court project in Toronto in the early 1910s and the idea had been discussed in Hamilton. Loans for home building had been guaranteed for a brief period through temporally limited federal and provincial legislation enacted in 1918.

Mirroring trends south of the border, steps had also been taken in Canada to begin to flesh out housing policies, goals, and standards. The country's first National Housing Act was passed in 1935, again providing loan guarantees for residential construction. According to David Hulchanski, the Dominion Housing Act of 1935 was significant, not for its effect on housing, but because it set a precedent by "defining an 'appropriate' role for the federal government in Canada's housing sector."[7] Unlike the experience in the 1920s, this Act was renewed and the government's role was expanded in both 1938 and 1944. By 1945, 21,708 mortgage loans with a value in excess of $87 million had been approved.[8] In a separate and equally interventionist initiative, the federal government guaranteed 125,652 loans, worth almost $50 million, between 1936 and 1940, under its Home Improvement Program, which was viewed as being the "saviour of the building trades."[9] This would not be the last time that the national government would use housing as a mechanism to stimulate a moribund economy. A national building code was developed by the National Research Council in the early 1940s. In 1945 the Central (now, Canada) Mortgage and Housing Corporation (CMHC) was created by the federal government to administer the National Housing Act.[10] So by 1945 "the principal elements of federal postwar housing policy had already been tried out in some sense."[11] Not to be outdone, provincial governments, which were responsible for municipal matters by virtue of the British North America Act of 1867, also began to exercise their regulatory might in areas that had direct consequences for the residential development process. Ontario, for example, passed the Department of Planning and Development Act in 1944 and the Planning Act in 1946.[12]

Little in the way of dwelling units resulted from any of these early government interventions until 1941 when the federal government created a Crown Corporation, Wartime Housing Limited (WHL), to provide inexpensive rental accommodations for war workers and veterans. By 1949 almost 46,000 modest dwelling units had been produced by WHL in 73 different Canadian centres, with about 1,700 of the units built in Hamilton, largely in the east end and on the Mountain.[13] In the words of Jill Wade,

WHL houses provided unpretentious but suitable living accommodation. Yet, their construction, design, and site planning may be characterized as progressive, experimental, and distinctive ... Because it was expected to remove its housing at the war's end, it had to build temporary, not permanent units. The houses were to rest upon posts of blocks, rather than basements, and they demanded a construction method that would facilitate their eventual dismantling and possible re-assemblage elsewhere. WHL employed an innovative semi-prefabricated [building] technique ... WHL house plans were plain and practical, yet curiously distinctive. Across Canada, the company used the same standard house types for both its two-bedroom and four-bedroom bungalows; later it added a third two-bedroom type.[14]

While Wartime Housing had been created under the most trying of conditions, the federal government would not be able to escape as easily from its new-found residential construction obligations as it had from its tentative initial foray into the housing field, namely, the loan guarantees of the early 1920s. This time, small-scale, private-sector building enterprises would not be able to remove the necessity for state involvement, even initiative, in housing.

In 1945, then, the stage was set for the emergence of a new residential development paradigm. Both property development corporations and government agencies would expand their roles in the housing field in the post-war era. Government policies and programs related to housing matters would proliferate and would be focused on two broad goals – the improvement of market efficiency in residential construction and the promotion of social justice and equity.[15] In the process, the importance of small builders would first rise and then diminish, but it would certainly not be eliminated. Finally, the sequence of the stages within the residential development process would change, especially as companies became increasingly vertically integrated. We have termed the post-war development paradigm the era of state intervention to signify the area of greatest change in the field of residential development.

The increasing influence of the state on North American housing development in the years after 1945 has been well documented and need not be repeated here in any great detail.[16] In Canada the activities of CMHC are noteworthy, however, for its role has grown well beyond its original mandate to merely administer federal housing legislation. CMHC does not just guarantee mortgages. Over the years it has developed sites and built housing, established housebuilding standards, provided hundreds of designs for modest homes, set site planning criteria, and published manuals for would-be home builders.[17] One CMHC design, a one-and-a-half-storey dwelling known as the "Type C" unit, has even been described by James McKellar as "the quintessential Canadian house."[18] Through its Research Division, CMHC has monitored housing trends regularly, issued both studies and statements on Canadian housing matters, and funded graduate-student and other scholarly research in the housing field. There is little doubt that it has had a positive influence on the housing circumstances of the Canadian population.[19] By now it is difficult to imagine a Canada without the CMHC.

Our understanding of the current workings of the residential development process in urban North America is remarkably complete.[20] Housing data are routinely gathered and reported, making analyses of residential development less difficult for the present than for earlier periods, though we have to agree with Melvin Charney that the information, for all of its apparent richness, remains "partial and in no way reflects the [Canadian house-building] industry as a whole."[21] Most researchers continue to identify subdivision, marketing, and construction as the major stages along the path from raw land to the production of housing units. Today, however, the second and third stages are usually combined. Housing consumers now purchase dwelling units, not building lots. Government and scholarly interest in housing issues in post-war Canada has been keen, resulting in numerous task force reports, articles, and monographs that have dealt with a multitude of topics.[22] In part, no doubt, this attention has been fuelled by both the importance of residential construction to the national economy and the remarkable accomplishments of Canada's house-building industry since the end of World War II.[23] Residential development during this period has occurred within a dynamic demographic setting. Between 1941 and 1981 Canada's population doubled and the proportion living within urban areas increased from 55 to 76 per cent.[24] Over the same period Canada's housing stock grew from 2.6 to 8.3 million dwelling units, a more than threefold increase. Housing production outpaced population growth in post-war Canada for two main reasons – to provide for the 382,000 families who, by 1945, had been forced to double up with other families because of a shortage of housing units and to meet the unprecedented rates of new household formation brought about because of high volumes of immigration, increased longevity, and growth in the number of families. In spite of the so-called "Baby Boom" of the 1950s and 1960s, all of these trends combined to reduce the average number of persons per Canadian dwelling from 4.5 in 1941 to 2.9 four decades later.[25]

Canada's housing stock not only increased in quantity, but also improved markedly in quality during this period. The Canadian census reported that 27 per cent of all dwellings were in need of major repair in 1941, a figure that dropped to just 7 per cent by 1981.[26] Housing units also became larger and better equipped during this interval. The average number of rooms per dwelling rose from 5.3 to 5.7, while the proportion of homes heated by a stove or space heater fell from 61 to 7 per cent. By 1971 more than 90 per cent of all Canadian homes had a refrigerator, piped running water, an inside flush toilet, and an installed bath or shower, compared with just 21, 61, 56, and 45 per cent, respectively, in 1941. Perhaps the most striking change within Canadian homes related to the type of fuel used for heating. In 1941 coal, coke, and wood accounted for 93 per cent of the total, a figure that had dropped to just 4 per cent by 1981. Coal chutes and wood piles, features of Canadian homes for generations, had been replaced by oil tanks and natural gas meters and lines.[27]

Neither of the earlier residential development paradigms could have produced these results in such a short time span. Both increased scale and an economic climate conducive to heightened dwelling-construction activity were necessary for Canada's post-war housing achievements. It has been estimated that Canada's housing industry doubled its capacity between 1946 and 1955. Cheap suburban land and especially the availability of high-ratio, long-term, government-backed mortgage funds were the keys (see chapter 6). Between 1945 and 1955, 249,231 mortgage loans, worth about $2.4 billion, were approved under the provisions of the National Housing Act (NHA). Compared to the 1935 to 1945 period, these figures represented more than an elevenfold growth in the number of loans approved and an almost twenty-sevenfold increase in their value.[28] In spite of the vast sums being expended on house building, most building firms remained small during the immediate post-war period. CMHC, the agency that served as the federal government's window on housing, argued in the mid-1950s that, "the residential building industry is neither cohesive nor homogeneous. It comprises some large general contractors neither continuously nor wholly engaged in residential construction; a much larger group of merchant builders operating chiefly, but not wholly, on a speculative basis; a group of builders who limit their operations to work on a contract basis; and a very large group of others who fall into none of even these ill-defined categories."[29] In 1955 some 1,700 builders operated under the provisions of the NHA, averaging almost 22 housing completions each in that year. Just over 78 per cent of them built no more than 24 houses in 1955. Only 5 per cent of the NHA builders put up at least one hundred houses in 1955, while 42 per cent were involved in no more than six dwellings. Indicative of the momentum developing within the house-building industry, 37.4 per cent of the firms had entered the business after 1950, with almost one-fifth starting up in 1954 or 1955.[30]

James McKellar argues that CMHC's Integrated Housing Plan (IHP) actually fostered this pattern of small firms because, under the IHP, "speculative builders undertook to sell houses at a price previously agreed upon, and in turn CMHC undertook to buy back [any] unsold houses [at a guaranteed price] ... The integrated plan gave encouragement to small builders who were entering the house-building industry for the first time. Enthusiasm, and a few hand tools, were convenient substitutes for skills and experience.[31] Some builders, such as the Campeau Corporation, Quality Construction, and Consolidated Building Corporation, thrived on this state-supported environment, using a standardized product, the ready availability of serviced lots, and a huge demand for new housing to achieve impressive scales of operation.[32]

With scale increases came an increased propensity toward vertical integration. Melvin Charney surveyed Canadian home builders of all sizes in the early 1970s and found that "there is an important threshold of output around 100 units per year. Firms of this size tend to have a more "bureaucratic" form of organization, 80 per cent of them are incorporated [and] 39 per cent of them consider [land development] a most important aspect of their operations ... Large firms undertake practically no rehabilitation or improvement [work] ... The majority of construction work is sub-contracted by builder/developers. The larger the firm, the more work is sub-contracted." Almost 90 per cent of large firms, compared to only about 40 per cent of small (less than 25 units per annum), were builder-developers.[33] By 1971 some 80 per cent of Canadian housing starts were being carried out by just twenty builder-developers.[34] After 1975 the supply of new housing units began to outpace demand. This combined with the recession of the early 1980s to push many large building companies out of the residential-construction field, thus, according to McKellar, "leaving the field to myriad small home-builders who had survived [the earlier push towards bigness by concentrating] on larger custom homes and the home renovation market."[35] In 1985 only 34.2 per cent of Ontario's building firms were incorporated and 76 per cent of the residential contractors had total output of less than $250,000.[36] At roughly the same time, a CMHC report concluded that "the typical homebuilding firm in Canada is small, building fewer than ten houses per year. Even the largest firms, which may build up to 2,000 housing units per year, are small in scale compared to the average firms in such other goods-producing industries as automobile or consumer-appliance manufacturing. Few homebuilders operate in more than one market area and those that do so operate only in selected markets. There are no national homebuilding firms active in all major markets from coast to coast."[37] Yet while small firms continued to predominate, even a recession could not wipe out all of the gains made by larger enterprises. The 44 largest builders in 1983 (just 0.5 per cent of all builders), collectively accounted for 29 per cent of the revenue generated by Canada's new single-family home builders. In 1985 Canadian builders averaged just 9.9 houses each, but half of all the dwellings completed in that year were attributable to only 241 firms.[38] For the most part, we agree with James McKellar when he argues that, except for the very largest firms, "the residential construction industry in Canada has evolved since 1945 with a minimum of capital investment, little standardization, varying skill levels in the labour force, an aversion to technological innovation, and a reliance on a myriad of subcontractors, suppliers and material producers. It is an industry that has a complex organizational structure, is fragmented, is subject

Table 3.1
Subdividing Activity: Hamilton Area, 1946–1987[1]

	Surveys		Building Lots		
Period	No.	%	No.	%	Average Lots/Survey
1946–50	60	6.1	4,206	7.8	70.1
1951–55	220	22.3	17,629	32.7	80.1
1956–60	48	4.9	3,595	6.7	74.9
1961–65	136	13.8	4,138	7.7	30.4
1966–70	81	8.2	3,429	6.4	42.3
1971–75	112	11.3	6.067	11.2	54.2
1976–80	88	8.9	5,689	10.5	64.6
1981–85	138	14.0	4,924	9.1	35.7
1986–87	104	10.5	4,265	7.9	41.0
Totals	987	100.0	53,942	100.0	54.7

[1]Figures for 1946–50 have been established from an analysis of all registered plans. Figures for 1950–75 have been estimated from a 10 per cent random sample of registered plans. Figures from 1976 onwards apply to the entire Regional Municipality of Hamilton-Wentworth.

to major cycles in the economy, and is regional in character."[39] As elsewhere, however, private firms are not the only producers of housing in Canada today. There are, as Cardoso and Short have recently suggested for Great Britain, four distinct modes of housing production – self-produced, individual contract, institutional contract, and speculative production.[40] The same set of housing producers exists in North America. It is, therefore, against a backdrop of dramatic demographic, bureaucratic, and technological change that we must now evaluate the residential development process in Hamilton during the era of state intervention.

LAND SUBDIVISION

While new rules and participants characterized residential development in the years after 1945, the subdivision of land remained the initial stage of the process. For major urban centres like Hamilton the post-war demand for new building lots drove subdividing activity to heights that were only rarely achieved earlier. Between 1850 and 1945 some 400 subdivisions containing about 41,000 building lots had been registered for the Hamilton area. These figures were dwarfed by the totals of nearly 1,000 surveys and 54,000 building lots registered for Wentworth County between 1946 and 1987 (see table 3.1). In fact, the subdividing activity for the entire period between 1850 and 1880 was virtually duplicated in the combined efforts of 1986 and 1987 alone. Platting patterns for this most recent era have been quite consistent – 3,400 to 6,000 new lots placed on the market every five years – with the major exception of the 1951–1955 interval when an average of more than 3,500 lots were registered annually, a subdividing boom that nearly rivalled the mania of 1911–1915 (see chapter 2). During the post-1945 period, subdivision approval became both a formal and a formidable process (see figure 3.1). Thirty-six copies of the draft plan were required by the 1980s, and the approval process often took much longer than the suggested minimum of 102 days, a far cry from the simple registration procedure of a century earlier. In 1976 Hamilton builder Heinz Seebeck, then chair of the Ontario council of a builders' lobby group known as the Housing and Urban Development Association of Canada (HUDAC), claimed that it took "up to 50 hearings, appeals, presentations, and submissions to get a housing subdivision approved, [all of which] adds thousands of dollars to the cost of a lot."[41]

As always, the new surveys were being registered for areas on the periphery of Hamilton, which for the post–World War II period meant that the subdividers generally moved even farther afield – east to Stoney Creek, west to Ancaster, and especially south to property on the Mountain. Growth on the Mountain was spectacular. The population there increased from an estimated 10–12,000 in 1945 to 25,000 by 1952, 50,000 by 1960, and 137,275 by 1986, an eleven-and-a-half fold increase in just four decades.[42] In the 1910s the edge of subdividing activity had been about eight kilometres from the core of Hamilton. By the mid-1970s the locus of subdivision development, under the influence of widespread automobile ownership, had moved out to a radius of thirty or forty kilometres and more.[43] It then encompassed such long-established municipalities as Grimsby and even St Catharines to the east, Burlington and Waterdown to the north-east, and Galt and Brantford to the south-west.[44]

Two currents of change that began during the era of corporate involvement picked up speed during the years after 1945: the use of non-grid plans and the participation of corporations in survey registration. Between 1946 and 1975, 53.5 per cent of the plans examined contained at least some non-grid features, though the grid continued to dominate many of the surveys registered for working-class areas in north-east Hamilton and on the Mountain.[45] Changes to the norms in subdivision design for middle- and upper-class areas came quickly and almost completely. Of the seventeen plans analysed for the years between 1968 and 1975, only one small survey for just seven lots was linear. Three-quarters of the plans registered between

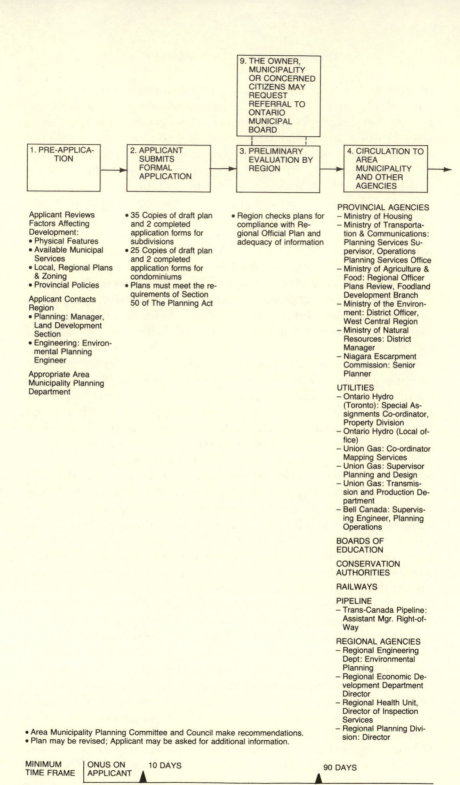

9. THE OWNER, MUNICIPALITY OR CONCERNED CITIZENS MAY REQUEST REFERRAL TO ONTARIO MUNICIPAL BOARD

1. PRE-APPLICATION

2. APPLICANT SUBMITS FORMAL APPLICATION

3. PRELIMINARY EVALUATION BY REGION

4. CIRCULATION TO AREA MUNICIPALITY AND OTHER AGENCIES

Applicant Reviews Factors Affecting Development:
• Physical Features
• Available Municipal Services
• Local, Regional Plans & Zoning
• Provincial Policies

Applicant Contacts Region
• Planning: Manager, Land Development Section
• Engineering: Environmental Planning Engineer

Appropriate Area Municipality Planning Department

• 35 Copies of draft plan and 2 completed application forms for subdivisions
• 25 Copies of draft plan and 2 completed application forms for condominiums
• Plans must meet the requirements of Section 50 of The Planning Act

• Region checks plans for compliance with Regional Official Plan and adequacy of information

PROVINCIAL AGENCIES
– Ministry of Housing
– Ministry of Transportation & Communications: Planning Services Supervisor, Operations Planning Services Office
– Ministry of Agriculture & Food: Regional Officer Plans Review, Foodland Development Branch
– Ministry of the Environment: District Officer, West Central Region
– Ministry of Natural Resources: District Manager
– Niagara Escarpment Commission: Senior Planner

UTILITIES
– Ontario Hydro (Toronto): Special Assignments Co-ordinator, Property Division
– Ontario Hydro (Local office)
– Union Gas: Co-ordinator Mapping Services
– Union Gas: Supervisor Planning and Design
– Union Gas: Transmission and Production Department
– Bell Canada: Supervising Engineer, Planning Operations

BOARDS OF EDUCATION

CONSERVATION AUTHORITIES

RAILWAYS

PIPELINE
– Trans-Canada Pipeline: Assistant Mgr. Right-of-Way

REGIONAL AGENCIES
– Regional Engineering Dept: Environmental Planning
– Regional Economic Development Department Director
– Regional Health Unit, Director of Inspection Services
– Regional Planning Division: Director

• Area Municipality Planning Committee and Council make recommendations.
• Plan may be revised; Applicant may be asked for additional information.

MINIMUM TIME FRAME	ONUS ON APPLICANT	10 DAYS		90 DAYS

Figure 3.1
The land subdivision process c. 1988

10. THE OWNER OR MUNICIPALITY MAY REQUEST REFERRAL OF THE CONDITIONS OF APPROVAL TO ONTARIO MUNICIPAL BOARD			ONTARIO MUNICIPAL BOARD DECIDES

5. DRAFT APPROVAL OR REFUSAL	6. PREPARATION OF SUBDIVISION AGREEMENTS	7. CLEARANCE OF CONDITIONS	8. FINAL APPROVAL
• Region evaluates comments and suggested conditions from Area Municipality and Agencies • Regional staff prepare report (with conditions) for a decision by Regional Planning and Development Committee and Council • Subject to conditions and Regional Council's approval, the plan is endorsed "Draft Approved" by Regional Chairman after 21-day appeal period, if there are no appeals • Application may be refused	APPLICANT (OR AGENT) • Contacts Regional Engineering and Area Municipality for details on requirements for respective Subdivision Agreements REGIONAL AND AREA MUNICIPALITY ENGINEERING DEPARTMENTS • Issue approvals for engineering drawings, specifications, etc. • Prepare reports for approval of cost-sharing and requesting authority to enter into agreements • Prepare agreements and forward copies to applicant (or agent) • Execute Subdivision Agreements, deposits cash payments and securities along with tax certificates, executed Certificates of Title, executed Notice of Application to Register Agreement, and executed Easements or Deeds (if required)	• Applicant satisfies conditions imposed by contacting each Agency • When conditions have been satisfied, Agency sends a clearance letter to Regional Planning describing how condition was satisfied • Applicant submits final plans with form "J" from Land Titles to Regional Planning staff for approval • Region reviews final plans checking for compliance with approved draft plan	• Region forwards final plans endorsed by Regional Chairman for registration to Examiner of Surveys for final check • Examiner submits final plans for registration to Land Titles Office

11. LAND TITLES MINISTRY OF CONSUMER AND COMMERCIAL RELATIONS EXAMINER OF SURVEYS

ONUS ON APPLICANT   2 DAYS

1976 and 1987 had been drawn up for corporations or building firms, which fell into three categories: holding companies; building, development, and construction companies; and numbered companies. The role of individuals in the residential land-subdivision process, then, continued to diminish. Significantly, one-fifth of the non-corporate surveys registered between 1976 and 1987 had been prepared for public bodies such as the Ontario Housing Corporation and the Ontario Land Corporation.

A few large surveys were laid out in the east end and on the Mountain in post-war Hamilton, but most subdivisions platted since 1945 have contained no more than one hundred lots.[46] Given the increased participation of corporations, the rather small average size of the subdivisions registered after 1945, about 55 lots per survey, might seem paradoxical. This peculiarity was linked to the growing trend for subdividers to become developers as well. Large parcels were still purchased by such builder-developers, but they were placed on the market in stages that reflected the ability of the company to erect houses within the building season.[47] This also meant that surplus land could be retained in a non-urban form so that it would be assessed at a lower rate than that applied to urban building lots. Using strategies typical of this staggered approach, in 1963 R.A. Garside Construction Limited registered a six-lot subdivision on the Mountain called Lawfield Manor No. 4, and the Queenston Development Company registered Glendale Estates No. 2 (Phase 4), a 115-lot east-end project, in 1972.[48] Between 1981 and 1987 Toronto-based development giant Bramalea Limited registered three phases and a total of 152 lots in Stonegate Estates in Ancaster.[49] Cardinal Heights, a Mountain-area project of Abbotsford Homes, was the focus of nine separate additions or phases, totalling 248 lots, between November 1980 and July 1985.[50] While survey size declined in the post-war period, the dimensions of the typical building lot increased. Forty feet became the minimum frontage required by CMHC for NHA-financed homes. In Hamilton subdivisions, the fifty-foot lot quickly became the norm. Two explanations can be provided for this trend. Larger lots were needed to accommodate both septic tanks (in areas not yet linked to municipal sewer systems) and the sprawling, ranch-style bungalows that grew in popularity during the 1950s and 1960s.[51] In 1958 CMHC estimated that eight out of ten new single-family houses in Canada were bungalows.[52]

The increases in area, population, families, and dwelling units that occurred in Hamilton in the years after the end of World War II dwarfed earlier growth (see table 3.2). Annexations between 1948 and 1966 expanded the area within the city limits by more than a

Table 3.2
Population and Dwelling Characteristics
Hamilton Census Metropolitan Area, 1951–1986

Year	Population	Dwellings	Families	% Owner-Occupied Dwellings	% Single-Detached Houses	Average No. of Persons per Dwelling	Average No. of New Dwellings per Year
1951	281,908	68,640	68,820	68.0	68.9	4.1	–
1956	338,294	86,990	84,941	n.a.	n.a.	3.9	3,670
1961	395,189	105,240	98,837	73.5	73.0	3.8	3,650
1966	449,116	123,352	110,005	68.2	67.8	3.6	3,622
1971	498,510	146,285	125,030	63.7	63.5	3.4	4,586
1976	529,371	172,510	138,730	63.7	58.8	3.1	5,246
1981	542,095	190,245	146,885	63.4	58.5	2.9	3,547
1986	557,029	201,330	153,165	64.6	59.8	2.8	2,200

Source: Census of Canada, appropriate years and volumes

factor of three, from 10,332.7 to 34,846.2 acres.[53] In January 1974 Hamilton became part of the provincially created Regional Municipality of Hamilton-Wentworth, which established a two-tiered system of local government for the area and effectively eliminated both the opportunity and the need for additional annexations. Between 1951 and 1986 population in the Hamilton area nearly doubled, while the number of dwelling units almost tripled, closely matching national trends.[54]

THE UNIFICATION OF MARKETING AND CONSTRUCTION

Beginning in the early 1950s more and more subdividers became housing developers. Homes and even lifestyles were being marketed now rather than mere building lots. This trend, like many others during the era of state intervention, added to the final cost of housing. According to James Lorimer, by the mid-1970s most developers built a cushion of between $1,000 and $1,500 into the price of their new homes to cover their marketing expenses.[55] Subdivision developers still did not actually build the dwelling units in every case.[56] Ever cautious and conservative, some continued to follow the model used earlier in areas like Westdale and made arrangements with established building firms to erect dwellings on their subdivisions. In several instances, especially in small projects, a definite symbiosis between the subdivider, builders, and a real-estate firm emerged. The most careful subdividers usually called on several builders to work on their projects. For example, in 1952 thirteen building firms were involved in the west-end University Gardens Survey.[57] According to John Sewell, a similar strategy was used between 1953 and 1962 in the development of Don Mills, a 2,062 acre area of North York, near Toronto.[58] Such an approach to subdivision development ensured a variety of housing styles and also gave subdividers a great deal of control over their builders. If they did not do a good job, they would not get any more building lots.[59] Five principles underlay the plan for Don Mills: the development of neighbourhoods individually focused on an elementary school and collectively focused on a shopping centre, the separation of vehicular traffic and pedestrians, the provision of green space, the provision of employment opportunities in the shopping centre and a separate industrial park, and the exercise of control over architectural and design elements. With its controlled construction, elaborate plan of development, and financial success Don Mills quickly became established as the model for community building in Canada and "an

inspiration for large-scale developers in the US and the world."[60] In
the words of John Sewell, "It is difficult to overestimate the influence
of Don Mills on urban development in Canada. Don Mills defined
the basic design elements and the business practices now used in
contemporary suburban developments."[61] Its descendants include
Erin Mills, Meadowvale, and Bramalea in the Toronto area, Bow
River in Calgary, Kanata in Ottawa, Highridge Estates and Heritage
Green on Hamilton's Mountain, and countless other Canadian suburbs.[62]

Nothing to match the scale of Don Mills or its most famous offspring appeared in post-war Hamilton until at least the mid-1980s.[63]
In the late 1960s the Ontario Housing Corporation assembled almost
1,700 acres of land on the Mountain in what is now the City of
Stoney Creek. This property, known then as Satellite City and now
as Heritage Green, was to be developed by the Ontario Housing
Corporation (OHC) under Ontario's Home Ownership Made Easy
(HOME) program.[64] Five separate medium-density neighbourhoods
and a population of 43,000 were envisioned. Ten per cent of the
lots were to be reserved for rent-geared-to-income housing. Ownership of the project was transferred to the Ontario Land Corporation (OLC) by the early 1970s, and the holdings were expanded to
include some 2,060 acres. The OLC, which became by far the largest
developer in the Hamilton-Wentworth region, renamed the project
Heritage Green in the late 1970s to remove any associations with
what many had incorrectly perceived as a subsidized housing project.
New plans for the area were drawn up in the mid-1980s featuring
a ring road, curved streets, and lower densities. These plans recast
the focus of the development toward a definite lifestyle – the accommodation of young families – primarily because both schools
and an elaborate park system already were in place. A total of 11,590
dwelling units and a population of about 32,500 were projected for
Heritage Green in the new plans.

Dwelling construction, however, has proceeded quite slowly. By
1985 the OLC had serviced and sold to builders some 1,355 lots,
while a private developer, Albion Estates, had developed another
750 lots on property that it owned in the north end of the project.[65]
Between 1985 and 1988 only two new subdivisions, containing 258
lots, were approved and registered. This latest delay was caused both
by market conditions and by the desire of the OLC to see the Red
Hill Expressway approved before continuing with the project. In
1989–90 three hundred multi-family units were constructed in the
project, largely to fulfill the provincial government's directive to
include more affordable housing units in all new urban residential

developments. For the immediate future draft approval was obtained for the construction of 456 more housing units on 58 acres of land. These lots were to be sold to builders in the Fall of 1990. So, while Heritage Green matches Don Mills, Bramalea, and other Canadian corporate suburbs in size, its genesis in the public sector and its slow pace of development are quite different. After more than a decade of OLC stewardship, only about one-quarter of the lots in Heritage Green had been developed and the area was home to just 9,000 people in mid-1990.[66]

In the Steel City, most residential growth has occurred through the steady, often piecemeal, addition of rather modest subdivisions. Most of the building and development that took place after 1945, as before, was accomplished by local firms and entrepreneurs. Bramalea Limited notwithstanding, few external private-sector players were attracted to Hamilton, and they probably accounted for no more than ten or fifteen per cent of new dwelling construction.[67] Between 1951 and 1986 annual population growth in the Hamilton area averaged about 7,861 people. This was simply not sufficient to draw the largest Canadian builders to the Hamilton new-home market, except for rather small and, as yet, tentative projects.[68] All of this could change if house prices continue to climb in the Toronto area, forcing first-time home buyers especially to look farther afield to places like Barrie, Peterborough, Guelph, and Hamilton, where land prices remain much lower.[69]

Not only were people buying homes rather than building lots during the era of state intervention, the new houses generally came fully serviced.[70] Stimulated by the powers given to municipalities by the Ontario Planning Acts of 1946, 1950, and 1955, in the mid-1950s Hamilton began to demand that subdivision developers sign agreements with the City regarding services as a requirement for its approval of the plan of subdivision.[71] Mitchell Contractors, for example, was charged $58,059, or almost $1,117 per lot, for the servicing of their 52-lot Tower Heights survey in 1956.[72] The "developer's agreements" were usually quite specific about not only the costs but also the nature and timing of the services to be installed. The 1956 accord between the City and contractor William Grisenthwaite regarding the 55-lot Fennell Gardens Addition was typical. By virtue of this agreement, Grisenthwaite was charged a total of $82,327.84, or almost $1,500 per lot – $34,060 for street services and street improvements, $1,703 as a fee for the use of the services of the City Engineer, $1,090 for the City's inspection charge, $21,850 for the owner's portion of roadways, $22,024.84 for local improvements, and $1,600 for maintenance costs for one year. The agreement also spelled out requirements for landscaping. Trees had to be "of species and varieties hardy in the Hamilton area," at least 6 to 8 feet tall, and "of number 1 quality in accordance with the standards of the Canadian Nurseryman's Association." Any trees that died within the first year had to be replaced. Lawns were to be seeded on a base of six inches of topsoil, and the turf was to be watered. All phases of the landscaping work were subject to the approval of the Commissioner of Streets and Sanitation.[73]

Government regulations regarding servicing had the same effect as had the mandatory registration of subdivision plans a century earlier – both led to the creation and expansion of a service industry. In 1966 Sunshine Homes used a private engineer for some of the preliminary work on its 45-lot Berrisfield Gardens Addition, but still needed to agree to pay the City $67,425.25, or just over $1,498 per lot, for services.[74] Edgemount Developments paid the City $33,865 in 1966, or almost $3,079 per lot, to service its 11-lot survey on Atwater Crescent.[75] Within the city limits, then, servicing became much more predictable. The "front-loaded" costs were passed on directly to home buyers and probably added between $2,000 and $4,000 to the price of a new Hamilton single-family home in the mid-1960s. James Lorimer estimated that by the mid-1970s in major Canadian cities it cost between $4,000 and $13,000 to fully service a building lot.[76] Recently John Miron has suggested that the introduction of municipal control over servicing provision and standards created a "forward shifting" of servicing costs that may well have reduced the affordability of suburban housing. Such controls forced buyers to purchase a given package of servicing standards at the outset, rather than staggering the costs as the services were installed.[77]

In the Hamilton area, builders were complaining by the mid-1970s about the "gold-plated standard and Cadillac approach" taken by local municipalities with regard to services, and estimated that servicing costs per lot ranged from $6,000 in Hamilton to $8,400 in Ancaster.[78] Some builders began to call for smaller lots, the elimination of storm sewers, curbs, and sidewalks, and a lowering of other servicing standards in order to reduce costs.[79] Lot levies for services were not the only charges inflicted on residential developers by local and regional municipalities. By the 1970s subdividers throughout Ontario were forced by the Planning Act to dedicate land for parks and playgrounds or pay 5 per cent of the value of their land in lieu of such a land transfer. For one 29-lot Hamilton survey registered in 1975, this amounted to $50,180, or $1,730 per lot. Subdivision charges and servicing costs brought the total pre-building expenditures for this project to $148,771, or $5,130 per lot.[80] In 1983 the new Ontario Planning Act changed the base figures used to calculate the parkland charges owed by subdividers. Under the old system, the municipality collected 5 per cent of the value of the undeveloped and unserviced lot. Now the cash-in-lieu payment is based on 5 per cent of the market value of the land on the day before the building permit is issued. According to George Barclay of the Hamilton and District Home Builders' Association, this adds about $2,000 to the cost of a serviced building lot with a market value of $45,000, and roughly $175,000 to the cost of developing an 8-hectare (20-acre) subdivision containing 100 houses.[81] All these charges, have been passed on directly to home buyers, thus increasing the price of new homes. Special levies to pay for new schools are anticipated in the 1990s. This, plus the federal government's new Goods and Services Tax (GST), which is expected to come into effect in January 1991, will add several thousands of additional dollars to the cost of new homes.[82]

While builders and developers have blamed escalating costs on servicing standards and vandalism, analysts such as James Lorimer and Peter Spurr have pointed their fingers at land speculation and excessive profit-taking. For the mid-1970s Lorimer estimated that the profit per lot realized by developers in seven major Canadian cities ranged from a low of $5,000 in the Forest Hills area of Halifax to a high of $24,600 in the Marlborough Park area of Calgary. Only in Thunder Bay did servicing costs exceed the developer's profit totals on a lot. In most of the cities analysed, Lorimer found that profits exceeded service charges by a ratio of more than two to one.[83] While little was done to curb or even effectively tax land speculation in Ontario, the Province did respond favourably to the builders' requests for new standards for servicing.[84] At the same time, it began to encourage the design of more energy-efficient subdivisions.[85]

Slowly but surely a movement towards larger scale building firms occurred in the Hamilton area, as it did throughout urban North America in the years after 1945. In the United States, Sherman Maisel could write about a transition in the production capacity of firms in the house-building industry by the early 1950s.[86] Models of development organization and practice began to emerge. The Washington-based Urban Land Institute (ULI) issued the first edition of its comprehensive *Community Builders Handbook* in 1947. In 1950 it prepared a *Home Builders Manual for Land Development* for the National Association of Home Builders. About a decade later they jointly issued a study of concepts and innovations in residential land development. A separate manual on land development was published in 1969. In the late 1970s the ULI published a treatise on the pros and cons of large-scale development.[87] All of these publications would have been available to Canadian builders. By the mid-1950s some US builders were producing homes at unheard of rates – such as Eichler Homes with an annual production of 300 to 400 units and Levitt and Sons with 3,000–5,000 homes per year.[88] Even in the United States, this was considered exceptional. In Canada only a few Toronto-based firms would ever come close to this scale of house building.[89]

Most Hamilton builders remained small, even after 1945, but the Steel City's new housing market eventually came to be dominated by a handful of larger firms. In 1949 the Elliot Construction Company established a new record for Hamilton when, on a single day, it secured building permits for 61 houses in its 163-lot Brandon Hill Survey on the Mountain. The four-room and six-room frame dwellings were priced to sell for between $5,900 and $7,800, with down payments ranging from $600 to $1,000.[90] George Sinclair Construction, one of Hamilton's oldest builders, was founded in 1949 and over the course of the next four decades built about 2,000 dwelling units of all types.[91] Typical of the Hamilton builders of the early post-war period, though probably larger than most, was the Grisenthwaite Construction Company Limited. Started by William H. Grisenthwaite in 1936, it remained in business for more than three decades, until he retired in 1962. Like many builders, he started in a small way. Grisenthwaite "sold his first house at a profit and built two others with the return. The business grew from there."[92] By the late 1940s the company was involved in land assembly and development and between the time of its first survey, Gary Park, in 1947 and the demise of the firm in 1962, Grisenthwaite Construction likely built an average of more than 100 homes per year.[93] Over its more than three decades of existence, the firm built almost 4,000 dwelling units in Hamilton. At the time of William Grisenthwaite's death in 1980, Ron Fraser, who joined the firm in 1950, simply noted, in what amounted to a testament to both the prowess of the company and the notoriously cavalier attitude of builders to detailed record keeping, that "we kind of gave up keeping track of the number of houses we built."[94] By 1952, when Grisenthwaite was elected president of the National Home Builders Association, his company was said to have been associated with ten other firms that were engaged in land development, rental properties, building materials, lumber, and the manufacture of hot air heating equipment, thus foreshadowing the vertical integration that soon would characterize large Canadian building firms.

The Grisenthwaite Construction Company had built houses, apartments, factories, schools, stores, and warehouses by the early 1950s and it was active, at various times, not only in Hamilton but also in Oakville, Acton, Ottawa, and Edmonton. It took full advantage of the strategies and opportunities that the era of state intervention presented. In the urban Canada of the post-war period, this initially meant building modest homes that purchasers would be able to finance through NHA guaranteed mortgages. Grisenthwaite was the first Hamilton builder to put up a house under the old Dominion Housing Act and by the mid-1950s he billed himself as "Hamilton's leading NHA builder." The firm specialized in laying out and developing surveys on the Mountain, at the rate of about one per year, where it built inexpensive bungalows and one-and-a-half-storey houses. His brother Jim claimed in 1980 that William Grisenthwaite had "built half the houses on the Mountain."[95] In 1954 a Grisenthwaite home required a down payment of $990 to $1,580 and monthly payments of $47.63 to $56.16.[96] The architect-designed "Brentcraft," a 1,295 square-foot bungalow, sold for $16,850 in 1961, with $1,186 down.[97] Low prices and easy terms apparently were no longer sufficient to sell real estate. Buyers needed new enticements. To market their new homes, Grisenthwaite and other Hamilton builders began to make extensive use of model homes, a strategy that Ned Eichler claims US builders also started to employ in the early 1950s.[98] In 1956 Grisenthwaite invited Hamiltonians to visit his collection of eight furnished and decorated model homes on the east Mountain, open seven days a week from 1 p.m. until 9 p.m.[99] Other large Hamilton builders such as Abbotsford Homes and Sunshine Homes employed the same strategy, and by the early 1960s it probably was universal.[100]

The tradition of long-established, locally based building firms, like those of George Sinclair and William Grisenthwaite, continues in Hamilton into the early 1990s.[101] Two examples illustrate the roles of both apprenticeship and innovation in the establishment of successful building firms. Starward Homes Limited, a company now capable of building 90 to 100 houses a year, was founded by Ward Campbell and his brother-in-law Jan Szostak in 1979. Campbell, like many contractors, came from a building background.[102] A grandfather had put up his first Hamilton home around 1905, and his father had founded a successful building company, Abbotsford Homes.[103] Both Campbell and Szostak had learned the home-construction business by working at Abbotsford, but their own firm embarked on some new initiatives. Starward Homes built Hamilton's first super-energy-efficient R-2000 home in 1983, and by 1988 had erected about 30 more. Ward Campbell was elected president of the Hamilton and District Home Builders Association in 1987. In 1988 his company was involved in two residential projects – Rexford Gardens in the east Mountain area and Scenic Woods in Ancaster.[104] Typical of the tangled web of enterprises that has characterized Hamilton's building industry since at least the time of Westdale, the plan for Rexford Gardens had been registered by Abbotsford Homes Limited, while two plans, totalling 240 lots, had been registered for Scenic Woods by 551908 Ontario Incorporated. To further confuse matters, a plan for the 14-lot Brigadoon Village Extension had been registered by Ward Campbell.[105] The use of multiple companies remains very common in the development industry. Sometimes a different corporate name is used for every project. According to one Hamilton development expert, there are two reasons for this practice – to reduce taxes and to maximize publicity.[106]

Another prominent contemporary Hamilton builder, Peter DeSantis, has a history reminiscent of many immigrant success stories in urban Canada. DeSantis came to Hamilton from Italy in 1958 and, like many immigrants, he started slowly, first establishing a small carpentry business and later expanding into house building.[107] His company, Homes by DeSantis, was incorporated in 1980, and by 1988 had built more than 2,000 dwelling units, including more than 500 single-family homes, in the Hamilton area. Peter's brother Aldo migrated to Hamilton in 1960 and now acts as the exclusive realtor for the firm. In 1986 Homes by DeSantis was awarded, through the tendering process, the right to build on 152 of the 194 lots released by the OLC in Highridge Estates, a 540-lot project on the Mountain. By mid-1988 the firm had built and sold more than 475 houses, often using the name Empire Developments (Hamilton) Limited, in Highridge Estates and a second OLC project, Heritage Green.[108] Something of an innovator, Peter DeSantis was one of the first Canadian builders to incorporate a sophisticated security system into his homes. His firm also pioneered in the effort to down-size houses in response to changing market conditions. In 1987 he developed a small townhouse project containing 41 units on the Mountain. Priced between $97,900 and $115,900, all the dwellings were sold within 26 hours. He was named builder of the year by the Hamilton and District Home Builders Association in 1980, and in 1984 he captured the Beaver Award presented to Canada's builder of the year by the Canadian Home Builders Association.[109]

Grisenthwaite, Campbell, and DeSantis managed to establish businesses that were both innovative and successful. Their firms typified those that came to dominate the house-building business in the Steel

Table 3.3
Size of NHA Operations of Builders Obtaining NHA Loans for New Housing
Hamilton, 1960–1973

| | Dwelling Unit Ranges | | | | | | | | | |
| | 1–25 | | 26–50 | | 51–100 | | 101+ | | Totals | |
Year	Bldrs	Units	Bldrs	Units	Bldrs	Units	Bldrs	Units	Bldrs	Units
1960	91	351	3	81	–	–	–	–	94	432
1961	118	572	4	161	2	134	–	–	124	867
1962	109	568	5	167	3	176	–	–	117	911
1963	127	713	13	431	3	214	1	124	144	1,482
1964	98	699	7	280	3	190	1	101	109	1,270
1965	83	527	8	257	4	340	3	329	98	1,453
1966	74	549	6	257	1	59	2	318	83	1,183
1967	72	701	11	401	3	188	1	147	87	1,437
1968	33	252	3	104	3	208	1	103	40	667
1969	13	95	3	109	3	244	1	115	20	563
1970	48	250	8	309	4	257	4	988	64	1,804
1971	68	484	15	499	6	507	4	669	93	2,159
1972	53	432	17	555	5	321	8	1,237	83	2,545
1973	25	164	6	203	1	60	3	390	35	817

Source: CMHC, Canadian Housing Statistics, appropriate years

City, and we suspect elsewhere. Figures for the construction of NHA-financed homes in Hamilton between 1960 and 1973 neatly illustrate the scale shift that took place in the city's residential construction industry during the era of state intervention (see table 3.3).[110] Only rarely did small builders (1–25 houses), who collectively never averaged more than 10 dwellings per year, fall below three-quarters of those involved in the construction of NHA-financed houses. Yet their share of total production declined from 81.3 per cent in 1960 to 20.1 per cent in 1973. In 1960 none of the Hamilton NHA builders had put up more than fifty houses. By 1973 the four largest builders each averaged about 112 homes and accounted for more than 55 per cent of the total NHA-backed production. Over the entire period 17,590 NHA homes were constructed, with an average production of 14.8 dwellings per builder per year. Builders of more than fifty houses per year accounted for 42.2 per cent of this total. By the early 1970s there were more than a dozen firms operating in Hamilton that were capable of building on that annual scale. In 1972 the thirteen largest NHA builders each averaged almost 120 houses. While larger firms did come to dominate Hamilton's NHA-financed new-home market, their supremacy was not as complete as it was in some other large Canadian cities. Hamilton, in the words of Joachim Schwarz, the present Manager of Subdivision and Condominium Administration for the Regional Municipality of Hamilton-Wentworth, has maintained "a tradition of small, local building and development firms."[111] For Canada as a whole, the proportion of NHA-financed dwellings put up by builders who erected more than fifty houses per year increased from 26.2 per cent in 1963 to 64.0 per cent in 1973.[112] In Toronto, by contrast, between 80.5 and 94.1 per cent of the NHA-financed houses built from 1968 through 1973 were erected by firms that built more than fifty dwellings a year.[113] By 1973 builders of more than fifty NHA-financed units a year accounted for 54.8 per cent of production in Montreal, 90.8 per cent in Ottawa, 77.8 per cent in Winnipeg, 71.1 per cent in Calgary, 50.6 per cent in Edmonton, and 79.9 per cent in Vancouver.[114] By the mid-1970s these levels of concentration became a major concern for housing analysts such as James Lorimer and Peter Spurr, with Lorimer arguing that government intervention had fostered private-sector consolidation and large-scale undertakings.[115]

Data collected by Ontario's New Home Warranty Program for the period between 1978 and 1984 indicate that 37 per cent of the 19,136 new housing units constructed in Hamilton between those years were erected by firms that built more than fifty dwelling units in a year. While this might appear to be a decrease in concentration from the 1960–1973 interval, figures for the two periods are not directly comparable. Those for the earlier years included only NHA-financed homes, while those for 1978–1984 incorporated all dwelling units built for sale. It seems likely that the modestly-priced, NHA-financed homes were the most amenable to mass-building efforts. By 1955 CMHC had issued a 129-page booklet outlining the minimum standards that had to be met by such homes. John Miron argues that these regulations contributed to "a standardization of housing constructed in the postwar period. In choosing a new dwelling, households were faced with either an NHA-financed unit built according to these standards or a conventionally financed dwelling with perhaps different characteristics but at less favourable financial terms. In effect, the subsidies implicit in NHA programs helped to tip the consumer demand toward housing based on those standards."[116] The effects of housing standardization on Hamilton did not go unnoticed. By early 1972 the *Spectator* warned that Hamilton's Mountain area, the focus for much of the city's NHA-inspired construction, "could easily evolve into one of the dullest places to live in the Toronto-centred region. Here we are faced with the prospect of acres and acres of boring identical subdivisions, dotted with flat uninspiring ball parks, and racked with walls of apartment blocks."[117] A dozen years later, Tim McKay, president of both Lounsbury Real Estate and the Metropolitan Hamilton Real Estate Board, complained that "some mass-produced housing in Hamilton is so repetitious that [I have] actually become lost in a subdivision because everything looked the same."[118] Not unexpectedly, this similarity in style had counterparts in other housing variables. Over time the assessed value of Hamilton's houses became more closely grouped around the mean assessed value (see chapter 10).

Scale increases, then, did not always translate into societal gains, so perhaps it was fortunate that not every Hamilton builder opted for a major change in scale. A recent analysis of the Ontario New Home Warranty Program data by Barbara Wake Carroll confirms Hamilton's continuing tradition as a haven for smaller builders, but with a solid core of larger firms as well.[119] For each of the seven years and for each of the sixteen Ontario cities included in her data base, Carroll computed the proportion of new residential construction accounted for by the leading four building firms. The figures for Hamilton ranged from a low of 17.6 per cent in 1978 to a high of 28.1 per cent in 1981. In each year Hamilton was well below the average for all sixteen cities, which ranged between a low of 35.1 per cent in 1978 and a high of 56.2 per cent in 1982, and displayed the second to the fourth lowest percentages on this measure. At the same time, however, the output of Hamilton's four largest builders was 40.1 to 69.6 per cent higher than the average for all Hamilton builders. On this latter measure, the leading firm index, Hamilton was consistently above the average for all sixteen cities and had the third to the sixth highest figure in each year. Forty-eight different firms built more than fifty dwelling units in a single year in Hamilton between 1978 and 1984 (see appendix B). While two-thirds of the firms only accomplished this feat once, those who built more than fifty homes at least twice accounted for 57.2 per cent of the output from the large builders. So in the 1980s Hamilton's residential construction industry was a mix of the old and the new. Most firms remained small in their scale of operation, but a solid cadre of larger enterprises had gained a sizeable share of the Steel City's residential construction.[120]

What led to the increasing market segment enjoyed by large-scale building enterprises in urban Canada? We share the contention of many others that the greatest stimulus to concentration of housing production in the post-1945 period came from the housing and planning policies developed by the various levels of government.[121] These have already been summarized and analysed by Albert Rose, James Lorimer, Larry Bourne, John Bacher, John Miron, and others.[122] In 1948 Humphrey Carver had argued that large-scale building firms would not evolve in Canadian cities unless there were both: "opportunities to develop projects large enough in scale to justify new forms of operation and offset the overhead cost of administration [and] an assured continuity of demand which would make it possible to attract capital."[123] Both conditions had been met by the early 1950s. Beginning with the National Housing Act of 1944, Canada's federal government has enacted a series of measures that have ultimately encouraged the emergence of large-scale building and development companies. Not all of these have been housing-policy acts *per se*. As James Lorimer forcefully argued in 1976, federal tax-deferral policies served to stimulate the expansion of several Canadian development corporations into some of the largest such enterprises in North America.[124] Legislation enacted by provincial and local governments has also encouraged the participation of large firms in the residential land-development process. After 1945 zoning by-laws and Official Plans needed to be digested and, perhaps, contested before the Ontario Municipal Board. Planning bodies now must be consulted as a matter of course.[125] The increasingly careful scrutiny of subdivision plans, coupled with the need for public hearings and elaborate pre-building servicing, have both lengthened and complicated the approval process for new residential developments. This has resulted in better-planned communities in most cases, but only well-financed and expertly staffed organizations can participate today in what a century ago was a very open game. Technological change, too, has encouraged the development of larger firms. Four decades ago Humphrey Carver observed that

any serious consideration of the costs of building houses seems to lead to the conclusion that no really significant changes in the levels of costs are likely to be achieved without some revolutionary changes in the building process itself. In the simplest terms, that process consists of the movement of materials from their primary source into the finished building with labour being applied to the fabrication of these materials at various stages along the line. The ultimate cost of a building is, in fact, made up of the accumulated costs of moving the materials, *plus* the accumulated costs of the labour that has been applied to those materials, *plus* the accumulated overhead and profit of all those who have handled the materials *en route*. It is clear therefore that the efficiency and economy of the building process is most likely to be improved by efforts to smooth the flow of materials and to concentrate the work of fabrication at the fewest possible points along the line.[126]

By the mid-1950s so-called platform wood-frame construction methods were being used by some builders to streamline the sequencing of the building tradespeople needed to complete the typical single-family detached house.[127] It would seem that these efficiencies and economies have been realized most readily by larger firms.[128] Large companies may also have gained some advantage through their bulk purchases of land. A recent study of four Hamilton builders found a general consensus that "larger contracting firms can almost always sell their houses cheaper than small builders who produce identical homes [because] the larger firms buy land on a grand scale, hence making each lot cheaper, [while small builders] must buy serviced lots at high prices from developers."[129] In the mid-1970s Peter Spurr found a direct correlation between the size of a building firm and its tendency to develop land – 87 per cent of large builders, compared to 42 per cent of small builders, were involved in all stages of residential property development.[130]

In 1981 the Hamilton area's 190,245 dwelling units could be categorized into four almost equal age groups – 24.6 per cent had been built prior to 1946, 25.4 per cent between 1946–60, 23.6 per cent between 1961–70, and 26.4 per cent between 1971 and 1981.[131] The annual and total rates of dwelling construction after 1945 pointed to a greatly matured housing industry. It was, as we have seen, an industry that had developed both greater scale and the ability to respond quickly to changes in government policy. While modern residential developers frequently complain about the red tape associated with their house-building endeavours, this has not stopped them from erecting large communities within rather short time spans.[132] Our assessment roll data point to an ever-contracting time frame for the substantial development of Hamilton's city blocks. Blocks annexed by Hamilton in 1891 took, on average, twenty-five years to become 80 per cent developed. Those annexed in 1956 took about two years to reach the same level of development (see table 3.4). Not only have modern developers built up new subdivisions quickly, they have also displayed a remarkable facility for altering their mix of products in response to shifts in both government policy and market conditions (see figure 3.2). As one housing analyst recently noted, "builders are an adaptable lot. If people want luxury homes with 3,000 or 4,000 square feet, they will build them. If they want no-frills townhouses, they will build them, too!"[133]

Table 3.4
Changes to the Residential-Development Process c. 1850 to c. 1990

Stage or Phase	Individualism	Corporate Involvement	State Intervention
Subdivision	– Uncoordinated grids – registration of plans – individual entrepreneurs	– some crescents and cul-de-sacs – coordination of plans – approval of plans – specialists involved – restrictive covenants and some zoning	– mostly non-grid – well coordinated – elaborate approval process – large firms and state involved – zoning controls
Marketing	– auction sale of lots – few or no services – inducements – high down payments – short-term mortgages	– lots priced per front foot – serviced lots – use of covenants – site offices – easy-payment plans – longer-term mortgages	– sale of house and/or lifestyle – use of model homes – use of electronic media – new tenure forms – low down payments – long-term mortgages – fully serviced – state-initiated inducements
Development or Construction	– low density – small, fluid firms – self-building – limited building season – housing fabrication	– low density with some small apartments – a few large, integrated firms – less limited building season – housing assembly – first building codes	– low, medium, and high density units – production dominated by large firms – unlimited building season – housing assembly – state-built housing – rigid building codes

As had been true during the previous era of corporate involvement, dwellings erected in areas annexed by Hamilton during the post-war period were built for owner-occupancy. In 1956 the level of owner-occupancy in the sample blocks in the post-1945 annexations stood at 84.8 per cent, compared to a figure of 57.8 per cent for the entire set of blocks in our data base. A decade later, the respective figures were 85.9 per cent and 66.9 per cent. Starts of

Dwelling Starts by Type
Hamilton, 1958–67

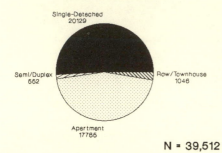

Single-Detached
20129

Semi/Duplex
552

Row/Townhouse
1046

Apartment
17765

N = 39,512

Dwelling Starts by Type
Hamilton, 1968–77

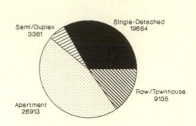

Semi/Duplex
3361

Single-Detached
19684

Row/Townhouse
9135

Apartment
26913

N = 59,113

Dwelling Starts by Type
Hamilton, 1978–87

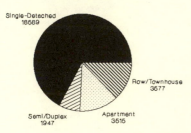

Single-Detached
18589

Row/Townhouse
3577

Semi/Duplex
1947

Apartment
3515

N = 27,628

Dwelling Starts by Type
Hamilton, 1958–87

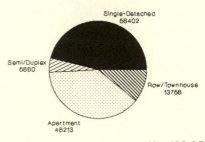

Single-Detached
58402

Semi/Duplex
5860

Row/Townhouse
13758

Apartment
48213

N = 126,253

Figure 3.2
Hamilton dwelling construction by type, 1958–1987

single-family detached dwellings, the widely accepted ideal form for owner-occupied housing (see chapter 4), have remained steady in Hamilton over the course of the three-decade interval ending in 1987, averaging about 1,950 per year. Yet by decade the share of all starts captured by this form has varied from a low of 33.2 per cent between 1968 and 1977 to a high of 67.3 per cent between 1978 and 1987. Between 1958 and 1967 government policies and market conditions encouraged Hamilton's builders to construct an almost equal mix of single-family detached dwellings and apartments. During the next decade, however, a federal program, the Assisted Home Ownership Program (AHOP), and Ontario's HOME program, both designed to expand owner-occupancy to include those with modest incomes, combined to bring townhouse construction into prominence.[134] About two-thirds of the Hamilton townhouses built during the three decades in question were started between 1968 and 1977. Both townhouse and apartment construction received a stimulus in 1967 with the passage of Ontario's first Condominium Act. This brought home-ownership possibilities to new forms of dwellings. In 1975 the Ontario government introduced rent controls that clearly affected the mix of dwelling units started in Hamilton between 1978 and 1987. Only about 7.3 per cent of the apartment units started between 1958 and 1987 were commenced during the last decade of the period. While rent controls undoubtedly slowed apartment construction, the rising cost of money was probably equally important. Interest rates, which had been 11 per cent for first mortgages in early 1976, rose to over 20 per cent in mid-1981.[135]

Government policy during the post-war years more than merely shaped the size of building firms and the nature of the housing stock built by the private sector. A place quickly emerged for public-sector housing as well. The National Housing Act of 1944 contained provisions for the funding of slum clearance and urban renewal schemes. Toronto's Regent Park North, which was started in 1947, was the first of a series of renewal projects that spread across urban Canada.[136] Hamilton's initial public-housing venture, the 496-unit east-end Roxborough Park project, opened in 1952.[137] By 1988 a total of 11,368 social-housing units, spawned by several different programs and policies, and provided to three broad groups – families (60.7 per cent), seniors (36 per cent), and singles (3.1 per cent) – were located in the Hamilton area, with another 442 under development. This represented about 6 per cent of the total housing stock in the area.[138] Only 51.3 per cent of Hamilton's social-housing units, however, were provided under rent-geared-to-income programs at that date. The remainder, especially the units in housing co-operatives, were often occupied by members of the middle class under programs designed to encourage a broad mixture of tenants. This was intended to remove the stigma that had so often been attached to the residents of low-income-only public-housing projects. Nevertheless, the new approach, which provides partial subsidies to people who do not need them, has been criticized by some as a misallocation of scarce housing funds.[139]

WHITHER RESIDENTIAL DEVELOPMENT?

A careful examination of the nature of residential development in Hamilton, and much of urban North America, in the early 1990s reveals what can be viewed as a trifurcation of the housing market. Decades ago, private-sector builders abandoned the poorest housing consumers. Traditionally, such individuals have simply had to make do by building their own accommodations (something that has been made increasingly difficult because of ever more stringent building codes and servicing standards), by doubling up with someone else (another strategy that has been made difficult in some Ontario municipalities), or by paying a high proportion of their limited incomes for dwelling space.[140] Since 1945 more and more of the least affluent members of Canadian urban society have been sheltered in housing built, and usually managed, by state agencies. Co-operatives, a relatively recent but underfunded housing alternative, offer tenants an opportunity to manage their own buildings, thus giving them control over, and a stake in, their housing environment, a sort of bridge between the worlds of pure tenancy and outright ownership. Not enough has yet been accomplished, for throughout the United Nations-sponsored International Year of Shelter for the Homeless in 1987 news poured in from all parts of Canada that the demand for social-housing units substantially outstripped the supply.[141] Outright resistance to social housing has been characteristic of many of Ontario's growing suburban communities, so much of the burden falls on the older central cities.[142]

Poorer people still have to make do in their search for decent housing in urban Canada, and Hamiltonians are no exception. In 1981, 29.1 per cent of Hamilton's tenant households spent at least 30 per cent of their income on rent.[143] A 1986 survey of 5,147 general welfare recipients in the Hamilton-Wentworth Region found that about half of the unmarried recipients spent more than 55 per cent of their income on housing and 43.8 per cent of recipient families spent between 45 and 64 per cent of their income on shelter. "The generally accepted rule-of-thumb is [that no more than] 25 per cent of a person's income should be spent on housing so enough money is left for food, clothes, and other necessities."[144] In 1988 the Hamilton area received joint federal/provincial funding approval for the construction of just 345 new non-profit residential units out of a total of 6,990 such units allocated throughout the entire province.[145] Supply will continue to lag behind demand into the foreseeable future.

Since the early years of this century, the middle class has represented the largest segment of the housing market, and the prime target for the builders of dwellings for owner-occupancy. Today their housing demands are satisfied almost entirely by the efforts of the large-scale builders found operating in the suburbs of all Canadian cities. Their housing products of the early post-war years, so-called suburban, owner-occupied, single-family-detached "tract" developments, now compete with a variety of shelter lifestyles including condominiums, rejuvenated inner-city neighbourhoods, and country and small-town life. For some, the bloom is definitely off the suburban rose. The harshest critics look upon these areas, no matter how expensive and well planned, as sterile and monotonous "Cul-de-Sicks."[146] Early post-war suburban developments often followed a grid plan and were mocked because their houses were "all made out of ticky-tacky and they all look just the same."[147] Today's subdivisions are more cleverly draughted. They certainly look more elegant, but as Toronto-area planning consultant Joe Berridge has noted, "all [their] curves are computer-generated to maximize frontage – you sell by the front foot – and minimize the run [the length of one straight row of houses], so that people won't realize every house looks the same."[148] Beyond such concerns about urban design, there is growing distress about the ability of builders to market homes that can be priced to fall within the budgets of many middle-class Canadians, especially young first-time buyers.[149] In the last century young men were encouraged to "Go West." Today, young couples are being advised to commute.[150] As they did around the turn of the century with the poor, private builders apparently are now abandoning yet another segment of the Canadian housing market. In the words of one housing analyst, writing in 1988, "most [Canadian] builders are continuing to focus on the mid-range of the market."[151] How the housing needs of the less affluent members of the middle class are to be met in the future remains unclear. The final segment of the market, the well-to-do, requires little sympathy. It demands features and a quality of finish that generally can only be satisfied by small-scale firms that still place great emphasis on craftsmanship.[152] There seems to be no shortage of expensive homes in most Canadian markets. In reflecting on the current trifurcation of the housing market, it is apparent that many of the old ways in residential development persist to this day, but now they co-exist with other and quite different forms of housing and housing production.

The modern residential-development process is well understood. Its continuation in this present form, however, is jeopardized by a number of serious problems that were being debated in the late 1980s and early 1990s – most importantly, the recurring and now quite persistent question of affordability.[153] The problems of housing supply and the volatility of land and housing prices are complex. Blame is not always easy to assign, but several avenues of attack have been followed. Larry Bourne, for example, has identified five theories that have been used in this regard – conspiracy, demand-pull, multiple bottleneck, cost-push, and institutional or neo-Marxian. He concluded that the problems were so complex that all five theories were correct in part. High housing prices are the product of some combination of inflation, cost increases, demand pressures, tax policies, and bureaucratic bottlenecks.[154] Remedies will not emerge very readily, but changes in both home financing and housing design again seem possible. Some housing experts, for example, are predicting that Canada may have to follow a recent Japanese strategy – inter-generational mortgages amortized over 100 years – if housing payments are ever to be manageable for average Canadians again.[155] A recent study by the Toronto Home Builders' Association indicated that it would be possible to build a new home in the Toronto area for under $100,000, or roughly half the median house price of $203,044 recorded for Toronto in May of 1988. Such a home – a 670 square-foot bungalow, on a small lot, located about an hour from downtown Toronto, and with minimal services – would fall, however, well below the expectations of most middle-class Canadians, expectations that have been fuelled by developers, the media, and four decades of government pronouncements, programs, and policies.[156] And to date private builders have shown little interest in the erection of this type of dwelling. Nor have municipalities rushed to re-zone land to accommodate such population densities. So while such a house is possible, it is not probable.

Most of the attention in the affordability debate has been focused on the Toronto area, where mayhem has sometimes reigned at subdivision sales openings. In 1986 frantic buyers bought, sight unseen, 259 Markham-area houses, priced at between $122,900 and $156,900, in just one hour.[157] Saner purchasing behaviour has prevailed in Hamilton – it took the DeSantis brothers 26 hours to sell 41 townhouses – but recent studies have shown that the Steel City has not been immune from the affordability problem. For the twelve months ending in May 1988, the Hamilton-Burlington area recorded the greatest percentage increase in the price of new homes (14.1 per cent) in all of metropolitan Canada. Between May 1986 and May 1988, prices for new house in the Hamilton area rose by 31.7 per cent. If history has taught us anything about housing markets it is that what goes up eventually must come down. By the autumn of 1989 the housing-price spiral halted its seven-year ascent as the housing industry in Hamilton and all of urban Ontario became caught in the vice grip of adverse monetary policy (rising mortgage rates) and fear of an impending recession. Both housing starts and prices fell in response to this newest downturn in the building cycle. Once again fears were being expressed about the ability of building firms to ride out a depressed market. Unfortunately, this latest market correction was expected to have little impact on housing affordability in the face of growing uncertainty about the overall state of the economy.[158]

A second problem plaguing the residential-construction process in urban North America is a looming shortage of skilled workers. For example, the average age of Ontario brick layers in 1988 was fifty-five.[159] Already the lack of skilled workers can be seen in the growing number of complaints concerning the poor quality of new houses, especially those put up by large firms.[160] Overall, however, Hamilton builders continue to erect homes of above-average quality. For 1989 the Ontario New Home Warranty Program reported that 14.6 per cent of Hamilton builders, compared to 11.6 per cent in the province and 8.8 per cent in Toronto, received excellent or above-average ratings based on complaints registered against them.[161] The industry's ability to respond to future sharp increases in demand, however, must be questioned. High housing-start totals in both Toronto and Hamilton in 1987 and 1988 precipitated labour and material shortages and long delays in the closing of sales.[162] The aging of the workforce is not the only labour-related problem facing the construction industry. Strikes can also play havoc with building schedules and dwelling production. Each of the many distinct building trades is represented by a separate union, so it is possible to have waves of strikes that can bring building activity to a standstill for long periods.

It is clear that the residential-development industry will have to continue to adapt as the era of state intervention progresses, just as it did throughout the eras of individualism and corporate involvement, in order to meet the housing needs of Canadians in the final years of the twentieth century and beyond (see table 3.4). The challenges to be met include more than the basic and obvious questions of the production of an affordable product and the ability to meet market demands, though the identification of market niches will remain important for builders of all scales. One expert has predicted that this will mean a growing focus on "lifestyle housing, catering to the needs and desires of various non-traditional buyer groups including singles, empty nesters, and seniors."[163] Building-industry members cite fragmentation, declining efficiency, lack of investment in research and development, loss of share in domestic and export markets, and government over-regulation as the major problems that they face for the future.[164] The ability to deal with these issues will determine the nature, size, and organization of Canada's building industry in the years to come. In a study for CMHC, Hussein Rostum recently identified five issues that must soon be faced by the

Canadian residential-construction industry – the influence of new technology, the transfer of technology to the industry, regulatory reforms and standards, industry adjustment to the country's changing demography (smaller families and an older population), and the need for greater productivity and competitiveness.[165] Already some building firms have diverted their attention from new construction repairs and renovations. The rather piecemeal development patterns of the eras of individualism and corporate involvement have left current builders with a great deal of potential for a variety of "renewal" projects. The inner city has become a favoured target for many small entrepreneurs. Most major Canadian cities now contain older housing districts, once run-down and dominated by rooming houses, that have become "gentrified" or "whitepainted," a process that often reduces a city's stock of affordable housing. In Hamilton several old mansions near the base of the Mountain have been converted into luxury condominiums.[166] Sometimes residential renewal has taken builders well beyond repairs and renovations. In some cities small, old homes on large, well-located lots have begun to be bulldozed one by one to make room for what many have called "monster homes."[167] Rostrum found that by 1986, 43 per cent of all residential construction expenditures in Canada were accounted for by renovation activity.[168] Recently Clayton and Hobart have suggested that the supply of new housing in Canada will be directly related to financing conditions, energy price and availability, land price and availability, the supply of labour, and government policies and regulations, factors that are subject to a high degree of both volatility and unpredictability.[169] Canadians will continue to need people who can subdivide and service vacant land and build dwelling units, but perhaps at a diminished level of enterprise as the population ages, family formation rates decline, and the twenty-first century approaches.[170] Who will initiate the process as we move beyond the mixed private and public responsibilities of the era of state intervention remains an unanswered question (see figure 3.3).

At present, most Canadian developers operate in the private sector, but the public sector presence is now a permanent and significant fixture in the process. The state not only regulates the residential-development industry, but also participates in most facets of housing production. As in many other aspects of Canadian life, ranging from gasoline retailing to broadcasting, a balance has been struck in the housing field, with a role for both entrepreneurs and government agencies. It is hard to imagine housing in Canada without the involvement of both economic sectors. As one CMHC publication recently noted,

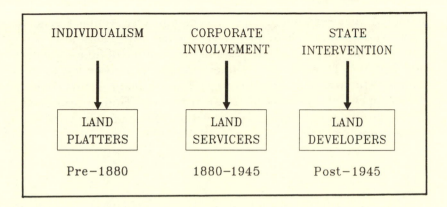

a reliance on the private sector and the powerful incentive of the profit motive, guided by government through well-developed and administered building standards and codes, has provided Canada with one of the highest standards of housing in the world. The public sector has played an important role in the success story. In Canadian experience, governments have recognized and utilized the private-market mechanism to produce housing in the numbers and the types that are affordable to most Canadians. They have also exercised a stabilizing role during times of economic uncertainty or recession. Government itself has established the framework for the provision of housing and to ensure its quality and standards, but has not built the housing. In addition to creating the framework for the effective and efficient functioning of the private market, governments in Canada have played a significant role in housing low-income people whose housing needs are not usually responded to effectively by private-market forces.[171]

Even Canada's Conservative government of the mid-1980s acknowledged a role for the public sector in housing, albeit merely one of "encouraging a climate of stability in which the private sector can function most effectively," which, if nothing else, is consistent with the Tory push towards "privatization."[172] If, however, private developers fail to deal with the needs of the lower-middle class, a large and articulate group, the balance between the two building spheres described so well by CMHC may well shift. Ironically, some observers contend that this transformation would take place with, perhaps, little evident change in either the nature of residential development or the price of housing.[173] So the future of the residential-development process remains murky. Housing choices, for those with money, have never been greater – ownership versus rental, freehold versus condominium, leasehold versus co-operative, central city versus suburban versus exurban locations, a multitude of housing styles and features. Those without money will continue to be left to the whims of the public and, increasingly, the voluntary sectors. Whatever happens, the process of transforming raw land into finished dwelling units will remain an essential and fascinating part of the Canadian way of life.

4 Launching the Will to Possess: North American Property Ownership as a Cultural Phenomenon, 1820s–1980s

In the previous three chapters, land development had been described and analyzed in relation to a group of city builders, the individuals who guided the land-development processes. The creation of a medium of exchange for a lot with a dwelling for use requires another set of players, for there must be households who wish to occupy the dwellings. To complete the picture of urban residential development, a discussion of consumer demand is mandatory. It is an intractable topic, prone to ideologically inspired insights and controversies, and unamenable to the empiricism of the research on land development. The latter, an aspect of business history and historical geography, holds research problems but no final enigmas. Motives are patent and actions tangible. In contrast, consumer behavior is puzzling. The controversies over consumer demand hinge on arguments over the nature of humanity – selfish or social, free or manipulated. These polarities oversimplify, but they indicate the parameters of discussions about consumer behavior. Consider, by way of introduction, the issue of manipulation. In some sense, particularly in the twentieth century, the property industry's advertising may become a part of the influences affecting demand itself. Even the design of certain suburbs can demonstrate a sophisticated integration of law, landscaping, and promotional gimmicks to fashion an appealing image. However, the demand for ownership of land and shelter is not to be comprehended, either in the past or now, as a clear-cut urge aroused by a cunning property industry hiring admen to run specific campaigns for individual surveys or more universally promoting slogans about the wisdom of owning real estate: "there is a finite supply of land and an infinite supply of babies"; "your home is worth more today than it was yesterday."

In this chapter, we summarize various theories about demand, criticize several of them, and posit that the rise of home ownership emanated from protests against the concentration of land ownership in Europe, and the distinctive possibilities in North America. Demand for owner-occupied shelter was connected to concepts of freedom that flourished in the transatlantic world from the 1820s to the 1890s. The consumption of goods as an expression of self-concepts has been recognized for many years as a crucial variable in consumer behaviour. Such self-concepts, as they have related to home ownership, preceded the onslaught of advertising and arose initially from socio-political upheavals affecting relationships between tenants and landlords in Europe and North America. It is conceivable that individual home ownership in the twentieth century has provided capitalists with a convenient and diffuse ideological lever to use against public ownership and extended government services. However, the actual foundations of a will to possess were built from the materials of radical assaults upon tenancy – a legacy of feudalism – and not on a modern ideology propounded for and by capitalists simply to help sustain a socio-economic system.

An increase in home ownership or in home-buying households across all socio-economic groups is detailed for Hamilton in chapter 7. Understanding why and how it occurred presents opportunities to consider the attitudes, technologies, and capital-mobilizing institutions that facilitated and have sustained the trend. On the basis of scattered evidence, it seems probable that this phenomenon of rising home ownership occurred across most of urban North America within roughly the same decades. However, the absence of home-ownership data in the published censuses for Canada and the United States in the nineteenth century and the inherent time-frame bias of the social sciences have truncated analytical accounts of home ownership. Moreover, while an exploration of the past introduces an essential dimension to the study of home ownership, historical inquiry cannot conjure up abundant direct statements on attitudes. Contemporary research unquestionably benefits from methodologies using questionnaires and interviews, which allow greater sophistication and assurance about conclusions. For the nineteenth century we do have the letters and autobiographies of some common men, which are plentiful enough to hint at a many-faceted impulse to own property and at the ease of acquiring it – and losing it – in North America.

Why does one find a large percentage of North American households residing in detached houses that they own or, through mortgage finance, are buying? Why does the percentage of home-owning and home-buying households exceed that in affluent European nations? As Nancy Duncan has demonstrated in a fine review of the literature that sought to answer these very questions, there is no simple way to manage the discussion.[1] The questions engage theoretical and empirical work by urban economists, Marxists, urban geographers, sociologists, anthropologists, and historians. The scope and implications of the studies, as well as their sheer volume, intimidate. Often home ownership is a small point embedded in more comprehensive theoretical constructs. The principal subject has sometimes been residential location or home design, with ownership an appendage. Furthermore, whatever else shelter has become, it is a necessity that implies an allocation of resources, and constitutes a considerable portion of landscape. Therefore ideological and prescriptive outlooks spice most writing, and few discussions are neutral. In addition to divining which visions and insights a particular scholar brings to his or her map of this epistemological swamp, there are further challenges. The links between cause and result are ever complex, sometimes unfathomable. Cause and result relationships can spin into circularity. For example, households seeking shelter are not and have never been purely autonomous decision-making units, but how far have they been hoodwinked? Shelter decisions are made in an environment that has been built, literally, by a complex industry composed of many interests. In the prior chapters, we have outlined the major historical features of land developers. Do their agents and those of building contractors manipulate taste to create a demand or do they react to antecedent goals and desires of shelter seekers? One of the most lucid and thoughtful writers on home ownership, Constance Perin, put the problem succinctly. "The confusion stems from our not really knowing whether producers lead or follow the demand."[2] This chicken or egg puzzle will recur in our presentation of representative literature. We ourselves have introduced one final complication – for reasons that should be clear at the end of the chapter, we believe the result to be explained is something more than the rise of home ownership; it is also the decline of tenancy. In terms of the socio-economic forces at play in the transatlantic world in the nineteenth century, the decline of tenancy dovetails more exactly with major issues and developments than ascending home ownership *per se*.

THE PRINCIPAL THEORIES OF DEMAND

Neoclassical economists have usually considered home ownership in association with locational decisions. The models of urban land-use articulated by William Alonso and those who have elaborated on his work in urban economics assume that people work in an urban core and reside at a distance from the centre. Their models, often based on indirect or surrogate data that rely on available statistics to measure phenomena for which no direct measures exist, are employed to theorize how households would apportion income to commuting costs, land costs, and other goods and services. The models test hypotheses about the location of different income groups, family life-cycle features and shelter characteristics, the influence of age of dwellings on housing expenditures and on the location of households, and a host of other questions. Richard Muth, one of the foremost practitioners of the art, directly addressed home ownership in a classic 1960 article.[3]

Muth set out to help clarify and possibly resolve a long-standing difference in opinion about the price and income elasticities of demand for housing. The problem he set for himself bears a resemblance to our own, although his model-building exercises in 1960 and later were meant to predict how households reacted to marketplace conditions or to rises in income. He did not attempt to dissect the motivations behind general behaviour in the marketplace. Nevertheless, Muth recognized that motivation was closely related to perceptions about what a house meant. On the subject of income elasticity of demand for housing, he cited Walter Morton, who equated housing with a necessity like food. Thus Morton proposed that housing should reveal a low elasticity. "Housing expenditures," he wrote, "do not bear a constant but a decreasing ratio to income." For contrast, Muth quoted the influential *Principles of Economics* by Alfred Marshall. "Where the condition of society is healthy ... there seems always to be an elastic demand for house room, on account of the real conveniences and the social distinctions which it affords."[4] Muth's strawmen had adduced evidence to support Morton. Taking into account that a period much longer than one year is needed for ultimate responses of housing stock to changed demand conditions, Muth found price and income elasticities of unity or higher.[5] In plain terms, housing demand was highly responsive to changes in income and price. Muth did not return to Marshall's proposed tandem explanation about convenience and social distinction. In later work, he has mentioned that variations in taste can generate differences in bidding for shelter, and he quickly linked taste to higher income.[6] For predictive models, factors need to be switched away from debates about how the real world operates and into ingenious searches for data supposedly measuring the essential factors in a rational process. Muth's neoclassical methodology led him only tangentially to the question of why people might want to own a house and to passing comments about distinction and taste. In common with the research thrust of his branch of economics, he primarily pursued deductive model-building with the objective of enhancing the model's ability to predict.

In neoclassical studies, the individual operates freely and rationally, making adjustments in association with fluctuations in income or price. Marxist structuralists operate from a contrasting set of notions about perception and conduct. If Muth and Marshall alluded to buying "more" shelter because of some assumed drive based on the calculations of households, Marxists like David Harvey and Manuel Castells have placed the egg ahead of the chicken, that is, they found the city to be a manipulated environment. Harvey's article, "Use Value, Exchange Value and Urban Land-use Theory" presented a brilliant theoretical account of shelter as a manipulated commodity, managed by vested interests that exploit its exchange value.[7] Michael Doucet's study of the boosting of real estate by the nineteenth-century press in Hamilton, although not couched in Marxist terms, offers an empirical exploration of inducing demand.[8] However, even a researched inquiry into promotional hype and the real property market begs our central question about which came first: the advertising or the public's taste. Advertisements reflect existing attitudes or desires and their information achieves tactical usefulness only to the extent that consumers are responsive to it. The manipulated-city hypothesis presents a top-down portrayal of social dynamics that does not extend enough credit to the will of people. Manuel Castells in *Grassroots in the City* addresses this weakness and deals with the public in circumstances of protest. Castells, though making a claim to unite urban systems and personal experiences, limits the North American personal experiences to studies of community mobilization. Ideological blinkers prevent him from treating the proposition that the public may be part of the very problems that he set out to expose.[9] By the selection of case studies, he gives the impression that grassroots flourish in the counterculture of major cities. It should be recognized that they also flourish in the suburbs.

Harvey's work has seemingly tackled home ownership directly. He and other structuralist Marxists relate the fostering of demand for a particular type of housing to the requirements of capitalism. In a 1978 article Harvey developed this line of analysis. Competition and pursuit of accumulation by individual capitalists, Harvey alleged, generates too much capital in the system relative to the opportunities for employing it profitably. In industrial capitalism, the consequence has been gluts on the market, falling prices, excess productive capacity, and unemployment. One way to resolve such over-accumulation problems for a while, explained Harvey, is to switch investment into a circuit of the economy that will aid consumption. Somehow the capital market mediates between a crisis for capitalism and housing. To avert the threat to capitalism, the mechanisms of capitalism gear up to promote and channel spending.[10] The surburban house is an article involving just such a massive mobilization of capital, and, according to Castells, it is also an engine to drive further consumption. It individualizes food preparation and entertainment and, in turn, it stimulates the consumption of appliances. Owner-occupation thus had desirable multiplier effects for industries producing mass-consumption goods. The promotion of the ideal of home ownership, therefore, surfaces indirectly in consumer-product advertising. This critique of home ownership moves beyond the suggestion that the shelter sector of the economy might supply opportunities that buy time for capitalism at the price of waste and manipulation. Castells went further than Harvey in his account of capitalism and home ownership.[11]

For Castells, the single-family detached dwelling performs even an ideological function. It detracts from collective responses to social issues, including housing, by accenting individualism and privacy. The lawn and yard compensate for the degenerated relationship between people and nature in the industrial workplace. Other Marxist writers have added to the list of advantages to the capitalist system stemming from owner-occupation. The goal of paying off the mortgage, it is alleged, reinforces work discipline and job stability. Whether home ownership induces job stability or *vice versa* is, once again, the type of fundamental question about the sequence of influences that bedevils the home-ownership question at many turns. Finally, a few commentators have proposed that home ownership splits the working class. Although owner occupation appears among all occupational groups, it is clearly less prevalent among less-skilled workers in unstable employment sectors.[12] The validity of this assertion is demonstrated in chapter 7. Both the sweeping general model of housing and capital mobilization and the specific plausible social criticisms present problems. Harvey's model of capitalism reifies structures and leaves out the individual. It finds an escape from the circularity of influences by proposing that individuals have no influence whatsoever on taste. Capitalism confronts a crisis and promptly evolves a remedy by producing a new array of consumer demands. Suburbanization, for Harvey, is the most potent recent tactic for stimulating capitalism. Historical research might illuminate or humanize the discussion of an investment shift, taking it out of a vaguely elaborated model of under-consumption and explaining it in terms of chronicled events that flesh out the reasons for North America's unique path within capitalism.

The introduction of historical perspectives raises two further concerns that challenge the easy claim that home ownership must be regarded as a conservative force. There are no shortages of quotations to the effect that politicians and business interest groups in the twentieth century encouraged home ownership to still radicalism. Frequently, conservative declarations that home ownership will combat Bolshevism are cited to argue that ownership has become the dominant form of housing tenure in many advanced capitalist countries through the ideological and financial backing of the state. That may be happening now in Europe.[13] State action in the United States and Canada only started to promote home ownership during the 1930s.[14] The sequence of events in North America undercuts the emphasis on state conduct as a universal maxim. Home ownership was substantial before state sponsorship programs appeared. It was strong at a time when it was less a handmaiden of conservatism than seems likely today. In the setting of nineteenth-century social relations, home owning and the attack on tenancy it represented figured as radical and progressive. One of the rare testimonials to property ownership written by a workingman, in this case a carpenter in Hamilton in 1837, boasts of obtaining property and achieving the franchise. We will soon consider this document and other historical reconstructions of workers' attitudes. Not only twentieth-century conservatives have praised home ownership.

Harvey's conception of a switch of capital from a primary circuit of industrial production to a secondary one of a consumption fund, with housing as the prime example, fails to explain differences between North-American and European shelter forms. Harvey does not address the principal question of this chapter: why has home ownership been more prevalent in North America than in Europe? Possibly significant cultural and historical differences are central to the issue. Castells has said that the rise of homeownership can not be taken as an illustration of the primacy of culturally distinctive societies.[15] To insinuate the notion of a specificity of societies due to distinct cultures is to undermine the primary dynamic of capitalism that drives the structuralist model. Despite the historically arid approach of Harvey and Castells, the structuralists have made a significant contribution to seeing housing questions in relation to the entire capitalist social order. In denying that North American shelter forms – ownership and suburbanization – were determined by a distinctly new-world culture Castells has suggested as a strikingly similar alternative that they occurred because the continent's cities were shaped by a particular pattern of capitalist development at a particularly critical stage. Doubtless the North American urban economy underwent a critical adjustment in the era of growing home ownership. The list of financial, technological, and organizational innovations that exemplified the adjustment in North America are the substance of much of this book.

Our chore, in accepting leads from the many schools of theoretical analysis, is to select the more convincing arguments and to flesh out their disembodied and timeless rendering of social developments with documented instances of changing behaviour set in historical events. The conduct of investors, thanks to the provocative works by Harvey and Castells, will require thoughtful review. The rise of ownership and its reciprocal, the fall of rental tenure, involved demand and supply considerations. Something occurred to reduce the reputation or desirability of rental-income investment and to strengthen the appeal of mortgage-income investment. In turn, the home-buying and shelter-rental ratios were affected. The idea that home-buying could represent an investment for households does not emerge from the structuralists' presentations about grand capitalist strategy nor does it arise in Muth's effort to determine the elasticity of demand. Global-scale or broadly conceived schemes in urban economics, whether formulated under neo-classical or Marxist auspices, have failed to perceive home buyers as investors. In contrast to the avoidance of this topic by academic theorists, the property industry itself has predictably played on a belief that an owner-occupied home works as an investment for the occupant. Why pay rent to a landlord and have nothing to show? Protect your savings against inflation and buy real estate! Recently, filling a gap in the economic and historical literature, work by Matthew Edel, Elliott Sklar, and Daniel Luria has challenged the commonly pronounced wisdom of real estate as an investment for the majority of households, while acknowledging that the public has accepted the notion of real property as a sound investment. Their book, *Shaky Palaces: Home-ownership and Social Mobility in Boston's Suburbanization*, advances four main arguments concerning home ownership. All are expressive of the new left historiography of the 1960s and early 1970s in that they upset notions about the distribution of economic opportunity and political power in America, seek to enlist history in current policy debates, and find social conflict rather than consensus in the events leading to suburbanization.[16] The first important argument in *Shaky*

Palaces bears directly on home ownership as an investment. The authors, after an analysis of real-estate values in sample communities in and around Boston from 1870 to 1970, conclude that "real estate investments by homeowners were not as remunerative as those investments open to larger-scale investors."[17] By considering investment purely in terms of the capital gains of house sales versus the potential capital gains from Dow Jones stocks, *Shaky Palaces* pursues a single idea too doggedly. Home ownership may have offered other economic benefits such as gains arising from the freedom to improve a property, the freedom to take in boarders, and – in contrast to tenement rental – the commonplace production of food on suburban lots. Also, as a point of historical record though not one that damages their counterfactual argumentation, it is quite unlikely that lending institutions would have permitted common men to borrow for stock investments to the levels permitted for home mortgages. If real estate did not perform as well as the Dow Jones index, it may have been, nevertheless, the most accessible game in town.

A second major argument continues the attack on what the authors feel are the myths of home ownership. Through climometric analysis they pursue the hypothesis that home ownership impeded wealth accumulation and occupational mobility for both men and their sons. Lacking important social variables for their cases – especially age and family size – their arguments are less conclusive than the reality that is their stated result. A third significant claim represents an attempt to revise a conclusion posited in another Boston-based study, Sam Bass Warner Jr's *Streetcar Suburbs*, the classic account of suburbanization. Focusing on building contractors, Warner portrayed the building of suburbia as a weave of small patterns, an achievement of thousands of actors. *Shaky Palaces* shifts the focus and emphasizes the paramountcy of a few major land developers. Unfortunately, it fails to consider the actual degree of concentration in the land-development process at any particular time. Furthermore, a concentrated land market did not always translate into a small number of decision makers in dwelling construction. As we have indicated in previous chapters, the land-development process in Hamilton in the nineteenth and twentieth century has been complex. Concentration and windfall profits have been evident, but so has competition, diffusion, risk, and failure. Moreover, *Shaky Palaces* also fails to consider the adaptation of the land-development process to social change reflected in consumer demand, although its authors register the very important point that "even developers operate within a large milieu. Consumer demand is a partial limit to their power."[18] Once again we are pulled back to the central riddle. Do producers lead or follow demand?

If home ownership has constituted, in most historical eras since 1870, an unwise investment for common folk but a very rewarding one for large-scale land developers, why have North Americans committed themselves to this dream? *Shaky Palaces* suggests an intriguing answer that presents its authors' fourth major argument. Aware of the new labour history, which accents the achievements of workers through struggle, Edel, Sklar, and Luria steer around the manipulated environment argument of Harvey and Castells. "A capitalist class may dominate the investments society makes, command the economy, and exercise considerable influence over the thoughts of others, but it does not do so in a vacuum. Workers, too, develop ideas and organizational power. There are two sides to a class conflict."[19] Thus enters the Marxist dialectic. Workers' pressure for better housing mounted influentially in the 1880s and 1890s. Machine politicians, the single-tax movement led by Henry George, and urban socialists demanded better living conditions. According to *Shaky Palaces*, while they did not necessarily support home ownership as the answer, they did not resist the notion. Business leaders, meanwhile, began to direct technological and capital-raising efforts toward suburbanization. To the authors of *Shaky Palaces* a clash had produced "the suburban compromise."[20] There may be something to the claim, but a more complete and subtle historical account might show that some of the radical impetus of the 1880s truly did support home ownership or at least the broader concept that land ownership bestowed freedom and the opportunity for virtuous productive labour. The Georgeite movement especially spread and contributed to these values, as we explain in this chapter. Furthermore, the labour versus capital dialectic poses a narrow social dynamic. Agrarians and an emerging middle class had come to see the freehold plot as part of a North American promise of bounty and liberty. In other words, the revisionist arguments of *Shaky Palaces* derive from such a determined hostility toward suburban home ownership that they overlook the possibilities that this residential form was strongly desired, that it was not accepted merely as a compromise but truly yearned for because of motives based on rational economic considerations and profound cultural or psychological drives. Status considerations or self-definition, for example, have too firm a place in the literature to have been ignored for self-consciously revisionist reasons. In the few examples of testimony by common men, the mix of practical economic motives, sheer speculation, and self-esteem are quite apparent.

THE SUBURBAN TRACT AND THE
DEVELOPER'S IMAGE MAKING

Before assembling the evidence for the spontaneity of the will to possess, particularly as it arose and flourished in its heyday from the 1820s to the 1890s, the marketing of city and suburban lots must be considered. The aim here is not to recapitulate our account of the business practices, legal and technological framework for the property industry, and potential for profits and losses. Nor is the subject to be power, in the sense that land developers have manipulated the political and cultural life of the city. Instead, the following discussion probes the conduct that enabled developers to harmonize their activities and their products with the wider world of commerce and industry. The essential activities of land developers melded with other socio-economic trends, making for a culturally unobtrusive activity that, ironically, was highly visible. Interesting in itself, the cultural invisibility of land development or its chameleon-like facility for blending with the way things were done in other areas of society has a bearing on the argument that the North American taste for property ownership was primarily a creation of capitalists employing advertising and related manipulations. Historically, the eruption of image making in the property industry occurred in the twentieth century and now has surely saturated the process of suburb creation. While promotion of a demand for owner-occupied houses was not historically the leading force in the creation of that demand, it has assumed great significance – or at least great visibility –since World War II. Today, image making is an intrinsic part of the process, influencing the actual design of suburbs. It is no longer an afterthought, as it was in the early twentieth century. The dovetailing of suburban planning with symbolic theme concepts in survey names and street names and with other details of image building have doubtless persuaded casual and serious observers to see in blatant promotional ventures a potent constant factor in the makeup of demand for property ownership. Among other things, the following discussion demonstrates a more evolutionary process in the history of image making. What Victorian land developers basically produced were surveyed lots in a fairly crude state; what developers do today includes built-in representations of a style of life. There is a history of demand stimulation, but it was not always the forceful ingredient that strikes commentators on contemporary urban living.

In Canada in the early nineteenth century, the little image making that accompanied land development bound the developer to the site. An image was not being fashioned for the buyer, but rather for the developer. This occurred because of the wider business context wherein family networks and partnerships gave business a personalized character focusing on the merchant instead of the client. Family name, a reputation for prudence, and references figured as elements in credit worthiness. Firms bore the names of senior partners, and recast these names when new partners entered. Artisans displayed their names on their shop fronts and, along with merchants, put their names in the boldest print in early newspaper advertisements. An individual's name was the gist of public notice. Similarly, with respect to land development, the owner of an entire townsite or of a subdivision invariably attached his name to the endeavour. The imprinting extended to naming streets after parents, children, and wives. Parcels of land were not yet commodities to be packaged for quick sale, but remained linked to a family, which sometimes retained lots or developed rental properties. Besides doing very little by way of promoting any image about the land, the early developers added little to the property itself. Individual idiosyncracies appeared when developers varied the dimensions of the nearly universally applied grid. Some estates generated elongated blocks, others more squared blocks; some blocks had lots facing two streets, others four, and a few three. Streets from one landowner's survey did not necessarily flow into another's estate. With the prideful, individualistic, and name-conscious property owner at the centre of development, lot and street arrangements presented modest diversity. In recent times variety, more deliberately planned and keyed into advertising, has been aimed at the buyer.

By the mid-nineteenth century, property developers had begun to promote land sales with straightforward advertisements and popular social events meant to draw a crowd to an auction of lots, but there was still no promotional dimension that could be linked to attempts to manipulate taste. Around the turn of the century, major image-related changes occurred on several fronts. Personalized land development faded and, just as family names and partnerships ceded prominence to corporate entities in other areas of business, syndicates and impersonal labels began to crop up in the property industry. Image making also started to appear in advertising, but for a while the subdivisions expressed nothing in themselves that supported, however tenuously, the images expressed in hyperbolic advertisements. Like the manufacture of long runs of standardized stampings or castings, land developers generally created large subdivisions of uniform lots. In the United States the typical subdivision before 1930 consisted of a gridiron of streets producing blocks of about 600 by 250 feet.[21] A virtually identical size characterized the industrial-era block in Hamilton where 600 by 200 became the developers' canon. Hamilton's lots were 30 by 100 feet. Blueprinted subdivision maps with precisely indicated surveyors' monuments fixed by longitude and latitude, uniform block characteristics, bland street names or numbered streets, and continuous streets running from one survey through another suggested the volume batches of industrial products turned out by calibrated machine tools. The factory-produced hardware of city utilities literally reinforced the straight lines of the easily surveyed and simply described numbered lots of a gridiron subdivision. Cast iron pipes, concrete conduits, iron and steel rails, and the overhead wires for telephones and electrical power accented the prevailing linear bias.[22]

Putting in little of their personality or name, a few developers of industrial-city suburbs originated some of the first advertised images, most of which studiously denied commonplace linearity, and almost all adopted names touting a park or garden: Bedford Park, Centennial Park, Dominion Park, Hamilton Park, Hyde Park Survey, Pelham Park, Sunshine Park, Wellington Park, Mount Hamilton Gardens, Roxborough Gardens, and Tuxedo Garden. Undoubtedly a concurrent and enthusiastic parks movement throughout urban North America helped inspire the enthusiasm for such names. Moreover, the garden label had just been popularized through North American interpretations of Ebenezer Howard's *Garden Cities of Tomorrow* (1898).[23] For the first time, in Hamilton at least, an advertising campaign replete with illustrations idealizing land had appeared. Nothing in the surveys themselves suggested a park or garden, for the land was usually bereft of developer-furnished landscaping or greenery. The garden or park aspect existed in the limited potential of the private yard. Gardening appealed to the self-improvement aspirations of modest-income families, all the more so when the cities were attracting many former rural residents. It must be recalled too that the variance between garden image and industrial reality may not have seemed outlandishly deceitful to a working family contemplating a move from a rented row house or cottage in the older blocks of the inner city.

Industrial subdivisions proliferated, but even before World War I there were a handful of better designed surveys for the middle class. Stuart Blumin's masterful attempt to establish a methodology for the study of the American middle class in the nineteenth century emphasized patterns of consumption, including residential space. Managers, salesmen, and senior clerks, for example, were striving on every conceivable path to distance themselves from labourers. Middle-class surveys, therefore, were frequently set off from others by a park, major street, or geographical feature like a ravine or peninsular location such as Winnipeg's Armstrong Point. Given the monotonous grid of the working-class tracts, hallmarks of distinction could be minimal: slightly larger lots, boulevards, a curving street. A few better-groomed suburbs benefited from the talents of landscape architects or representatives of the newly established planning profession. Hamilton's Westdale, designed by the New York planner Robert Anderson Pope, was representative of the genre.[24] The advent of self-conscious professions from roughly 1880 to 1920 and the emergence of management hierarchies in joint-stock corporations created a demand for fine tuning in the consumption of status objects. Among white-collar workers, men bore titles, sat behind desks of different scale, occupied offices providing distinct degrees of privacy, and aspired to promotion. The institutionalized signs of rank at work carried over into residential space. Westdale, for example, contained a set of surveys with varying lot sizes and design features. In other words, while advertising certainly had begun to play a role in fostering images in conjunction with parcels of land, the socio-economic context of the era also stimulated consumption of specialized commodities like lots in planned subdivisions.

Planning and advertising, especially for middle-class surveys, drew together because former means of achieving a social-spatial segregation could no longer be practiced readily. Well-situated land on heights had often been occupied by the urban elite early in a city's development. Limited in their natural supply of such land, particularly in relation to the growth of the middle class, developers responded with aid from advertising agencies and planners. A built environment supported by slick art and copy in advertisements and brochures replaced purely natural distinctions. Statistics on the composition of the Canadian workforce show that the number of owners and managers, a flawed but available measure of the size of the middle class, grew briskly between 1901 and 1911, modestly in the 1920s, and quickly again from 1951 to 1961. The expansion of a body of consumers interested in distinctions challenged developers to create refinement, despite the restricted supply of distinguished sites. During the 1950s and 1960s developers also sought land-use designs that conveyed modernity, because city dwellers had had to make do with existing shelter during the depression and the war. In many fields of consumer design – automobiles present an obvious example – manufacturers looked for ways to break with nearly twenty years of drabness. As we noted in chapter 3, subdividers discovered winding street layouts and cul-de-sacs, creating asymmetrical patterns that provided home buyers with moderately individualized sites. Interestingly, the advertisements in Hamilton now stressed the total package: neighbourhood, lot, and house. During the first stages in the introduction of advertising, only the raw lot in the subdivision was promoted.

All Canadian cities evolved a band of the new alignments – the spaghetti configuration – around their outskirts. The depth of the band indicates the city's degree of growth since 1950. To the west of the City of Toronto, the borough of Etobicoke consists almost entirely of the winding forms, as does Mississauga and Brampton. The growth of Calgary since 1950 can be witnessed in the abundance of new design surveys on all but its western side. In contrast, the Halifax region indicates its slow growth by the confinement of its curvilinear streets to Herring Cove Road, Beachwood Park, and Dartmouth. The greater Montreal area had had less abundant new design areas than Toronto, but the suburban community of the City of Laval and the closer suburb of St Leonard reflected the prevalence of the form. The new street arrangement affected every Canadian city, although street was too common and bland a term for promotion, which aimed at presenting a picture of betterment. The roads in the newly opened residential tracts were named, in order of descending street length: drives, crescents, places, and courts. Plebeian streets, avenues, and roads passed through new areas, but they ran as extensions of older developments or of pre-existing traffic arteries. Past and present alone did not make for the mix of linear and curved, of streets and drives. The new arrangement separated local traffic from high speed traffic.[25]

Beyond the benefits of reducing traffic and providing moderately individualized lots, the new suburbs provided images often sustained through the crafting of names. Advertising and suburban creation had drawn together. In Hamilton and apparently elsewhere in Canada, the suffixes glen or wood turned up in many new street names; Arcadian terms formed the most popular category of winding street names. Within that group, quite a few names had associations with golf, a sport long employed in advertisements for many products in an effort to project privilege, recreation, style, and manicured landscaping. British names – especially with elite associations – formed the second largest category in Hamilton's new suburbs. References to sunshine and warm climes, conceivably representing the Canadian middle-class yearning for travel to escape winter, were plentiful if place names in Florida and Italy are included. Although the batch of sunny names was fanciful, a number of factors precluded a critical dissonance between the Arcadian images and reality. Nearby parks, conservation areas, golf courses, and a few scenic outlooks allowed imagery a basis in fact. Commercial nurseries supported the motif. The deflection of heavy traffic helped too. The shortness of the many drives, crescents, places, and courts established an abundance of distinctive addresses. The combination of low house numbers and an elegant sounding street name broadcast exclusiveness – 18 Camille Court, 6 Bentley Place, and 9 Fairway Drive, for example. The obvious promotional exertions of at least the last forty years and the traces of advertising reaching back to 1900 make the notion that demand for ownership of lots or single-family detached houses comes from manipulation by developers understandable and compelling, except for two observations. First, a powerful popular demand for land ownership and house ownership preceded the emergence of sophisticated promotion. Second, the promotional exercises, whether just raw advertising or a polished combination of planning and advertising, probably had to accord with pre-established cultural tendencies. It is conceivable that developers have reacted to consumers' taste at least as much as they have led it.

INDEPENDENCE AND A WILL TO POSSESS

First-hand accounts of nineteenth-century life, especially those written by common men, are to be found in considerable number in the United Kingdom, in somewhat lesser abundance in the United States, and rarely in Canada.[26] Three rare pre-1860 Canadian affirmations hint at what a more exhaustive and systematic review of immigrant letters could disclose about newcomers' motives and zeal in securing land. The first two, from the 1830s, came from radicals who took to speculative buying. Speculation, in fact, was considered a noble occupation in the mid-nineteenth century. The US census of 1850 listed 363 people who called themselves speculators, a figure that had increased to 1,982 by 1860.[27] This association of radical interests and a plunging into real estate does not seem coincidental or paradoxical. Both radical attitudes and property ownership expressed the pursuit of freedom from the decaying remnants of feudal bonds.

John McAdam, a Scottish shoemaker and agitator for the first English Reform Act who later returned to the United Kingdom and became a Chartist, travelled through Upper Canada in 1833. Finding the property in Hamilton too expensive, he pressed on to London where he practiced his craft. He also purchased real estate in Port Talbot. Writing to his mother in October 1833, he reported on why he had secured a half-acre lot. "I will be able to realise a good sum, as when there is a house put on, which can be got for 1½ or

2 hundred dollars, it will be worth 500 or 600 dollars at least, as they have been sold for more in worse situations."[28] Carpenter George Martin – from the tone of his letters home, a Chartist sympathizer – arrived in Hamilton in 1837. He wrote home boasting to his father in 1837: "I hope to be able to pay by September [50?] towards my Lot, which if I can manage will save one 3 Dollars of interest for the next year ... I should like to pay it before if I can, for to get my deed for it in my own hands, and when I have done that I think I shall be happy, when I can call it my own ... I was offered 600 dollars for house & lot a little while ago, but I did not want to sell. Besides ... if I keep it for 2 or 3 years longer it will be worth $1000 or more."[29] Such visions were spurred on by the passage of the Great Western Bill early in 1837. Few could resist the temptations of the land market. According to Adam Hope, "schemers, planners, and jobbers are in ecstasies." With uncanny prescience, Hope warned his father that "speculation in town lots is rife [but] some will burn their fingers for all this some day."[30] George Martin should have sold. The real-estate market collapsed shortly after he wrote the letter. For him, the *Shaky Palaces*' thesis of a poor return on real-estate investment proved catastrophically true. However, that hindsight judgment fails to tell us anything about the rational and psychological processes that directed George Martin's decision to purchase a lot and build a house. Buying a lot was essentially a new-world opportunity that allowed common men entrance into a form of bourse. There were no alternative common investments and buying into a rising market is a familiar and rational enough decision. Besides these points explaining the reasonableness of George's purchase, his letter anticipates fulfillment in getting a deed into his hands.

It took James Thompson, a baker from Scotland who arrived in Canada in 1844, ten years to save a down payment for his dream – a farm. Shortly after his purchase of a farm in 1854, his letters to kin radiated self-satisfaction. Several letters enclosed his sketches of the farm and improvements. Of his father-in-law, he wrote "Marys [sic] Father commenced the world a poor man. He has been industrious and persevering. He is now the owner of 200 acres of land."[31] Land in Canada and the United States, he may have reflected, passed to men because of their own effort and not as a result of the caprice of birth. Optimistic immigrants like George Martin and James Thompson might have misjudged the market for land or, in Thompson's case the quality of the soil, but for a while they could feel more the masters of their own fate. We suggest that this attitude, whose roots will be examined further later in this chapter, accounts for the yearning to acquire and improve property noted in several studies of urban and rural land transactions in the nineteenth century. Charlotte Erickson, whose collection of letters written by English immigrants to families back home is a wonderful source for delving into the impressions and hopes of newcomers, made the following observation: "The fulfilment of a dream of homeownership was clearly one thing that held migrants in the United States."[32] Erickson's commentary underscores the appeal of independence from rural tenancy, but the supporting letters also make clear that the dubious profitability of real-estate investment demonstrated by *Shaky Palaces* was brought home to many immigrants who lost when gambling on property appreciation. Let us return now to theoretical statements and more recent testimonies that stress psychological rather than material gain.

When discussing lots and houses as status symbols or as part of their owners' self-definition, Thorstein Veblen's *Theory of the Leisure Class* must be acknowledged. Published in 1899, this classic retains its appeal, mainly because of Veblen's acerbic and modern prose. His reputation as a pioneer social scientist remains in spite of a complete absence of empirical inquiry. Direct at times, ironic at others, and always able to make an unexpected observation appear simultaneously outrageous and true, Veblen was a genius of style and insight, but not a father of any social-science method. His theory rests on nothing more than *a priori* claims about human nature referenced to generalizations about race and the history of mankind: "With the exception of the instinct of self-preservation, the propensity for emulation is probably the strongest and most alert and persistent of the economic motives proper. In an industrial community this propensity for emulation expresses itself in pecuniary emulation; and this, so far as regards the Western civilised communities of the present, is virtually equivalent to saying that it expresses itself in some form of conspicuous waste."[33] For Veblen, taste in shelter or real estate could be embedded in a race memory. A lawn would have greater beauty "to the eye of the dolicho-blond than to most other varieties of men." This racial type could "readily find pleasure in contemplating a well-preserved pasture or grazing land."[34] Since Veblen did not equate the leisure class with anything below the truly rich – the middle class he called "the lower or doubtful leisure class" – he had little to say about a commonplace like home ownership and much to discuss concerning clothes, pets, sports, and higher learning.

The step from a critical analysis of conspicuous consumption by the rich to an application of the concept to other classes was taken in several empirically based works on American communities.

Middletown by Robert and Helen Lynd repays many visits. Very often what they found concerning housing in Muncie, Indiana in their two studies (1929 and 1937) accords with Hamilton trends. From their field work, the Lynds concluded that "there is a deep-rooted sentiment in Middletown that homeownership is a mark of independence, of respectability, of belonging."[35] Apparently the psychic satisfaction of owning was profound enough to propel home buying through mortgage finance at a cost that the Lynds approximated as being 50 per cent higher than the rent for a comparable house. Unfortunately, they did not spell out the basis of comparability, which is important because our Hamilton data suggest that owner-occupied dwellings are seldom comparable to rented dwellings. But the Lynds were the first social scientists to see something more to housing than shelter, although Veblen had come close. Another classic community study glanced on the topic of house ownership. W. Lloyd Warner's *Yankee City* research, begun in 1930, sought to define classes by collecting and analysing attitudes about stratification. For the upper-upper class, Warner observed a role for houses. "New people" could acquire, after several generations, membership in the upper-upper class by establishing themselves in fine old houses. The houses embodied the heritage to elevate the occupant. Only in an old and historically conscious community could this mechanism work.[36] Arguably, then, the history of mid-America, not that of the east coast, holds the more pertinent meaning of the house. Art Gallaher studied a Missouri farming community from 1940 to 1955, a setting with a shorter settlement history and less pretention than Warner's Newburyport. Gallaher's Plainvillers believed "that every family should own a home and that preferably a man oughta have the money ready when he buys or builds," although mortgage finance was acceptable.[37]

Overall, the empirical work on American stratification and culture has confirmed that a house brought membership in a particular social world. Through social status issues, home ownership had slipped into the topics now covered in community studies, although researchers had not set out principally to probe home ownership. More recently, a few investigators have taken up, for specific analysis, the topic of the home as a feature of identity. This has come about not because of any inexorable path of refinement in the social sciences, but because rising construction and land costs, increasing indebtedness among buyers, new shelter needs (for example to house single-parent families), and the politics of zoning have awakened an interest in the apparent resilience of the owner-occupied, single-family detached house.

Everything in Its Place: Social Order and Land Use in America by Constance Perin brought contemporary social issues and anthropological methods into a stimulating union. Few of the concepts discussed in this chapter were missed by Perin and her insights have informed our own wrestling with the rise of home ownership. She can be credited with seeing the need for a new methodological foray into the topic of housing. Perin maintained that economists had managed to convey the impression that economic constructs have a specific, phenomenal, and measurable existence. They also had succeeded in putting housing onto their plate. Cultural constructs, she hoped, could attain a comparable stature, because policy making and intelligent citizenship needed more than one set of theories. Perin had a prescriptive intent. She argued that shelter ideals had ossified while work relations and social interaction had proceeded toward toleration of non-conformity. She wished to know what had obstructed a closer matching of social advance with shelter forms. If segregation by class and race had collapsed at work, why could it not break down in residential realms? The authors of *Shaky Palaces* had a similar purpose and they emphasized what struck them as a misguided attempt to defend conformity in order to protect property value. They had used research tools from economics, but Perin had been formulating anthropological concepts. With these she aimed to take apart and appraise a social convention, something that people took for granted.

Owner-occupied houses were signs, but of what? Literary sources as disparate in time and culture as Samuel Taylor Coleridge's *On the Constitution of the Church and State* (1839) and Mordecai Richler's *The Apprenticeship of Duddy Kravitz* (1959) provide clues about the longing for property and concepts of individual worth. Coleridge wrote that "the notion of superior dignity will always be attached in the minds of men to that kind of property with which they have most associated the idea of permanence: and the land is the synonyme of country."[38] Duddy's grandfather, speaking over a century later in Montreal, made a similar observation more concisely. "A man without land is nobody. Remember that, Duddel." Duddy, seven at the time, recalls the maxim when, as an adult, he is possessed by the vision of buying and developing lake-front property.[39] Belonging and status were also the glancing summaries of ownership's meaning offered by the Lynds and others. Perin pressed for expansion and exactitude, using the principle that "in American society the form of tenure – whether a household owns or rents its place of residence – is read as a primary social sign, used in categorizing and evaluating people, in much the same way that race, income, occupation, and education are."[40] The categories owner and renter were real and symbolic. Their symbolic meaning had to be read and then explained. Basically, Perin proposed that American culture had its own ladder of life, a sequence of events to be lived out in a proper order. At each rung there had to be or should have been — lest the non-conformist label outlaw the violator — a matching of age, appropriate marital status, income, ages and number of children, school years completed, leisure tastes, housing tenure form, and housing type. One moved along the ladder's rungs in correct sequence. At the suitable point in life, one nested into an owner-occupied dwelling. Family and citizenship sanctified this move. In chapter 7, the elegantly simple metaphor of a ladder, the concept of stages in a life-cycle, instructs our analysis of home ownership in the city of Hamilton from the mid-nineteenth century to the mid-twentieth century. Yet we are ever mindful that these stages will not be passed in the same sequence or even totally encountered by all households.

Perin cited the Lynds. On Robert Lynd's return to Middletown in 1935, he drew up a rough list of things that Middletown was for and against. "Homeownership is a good thing for the family and also makes for good citizenship."[41] The people Perin interviewed formulated a similar equation. Tenants, they implied or expressed directly, were less-involved citizens. Furthermore, Perin introduced the idea that home ownership possessed a sacredness. Perhaps impishly, she commented that Sunday, the Lord's Day, was often reserved for newspaper advertising of homes directed at families, while Saturday ads specialized in apartments, directed largely at single people. She also planted the thought that Saturn's Day in ancient Rome was a time of unrestrained license, an interesting and extreme attempt to draw together cultural connections.[42] Her sharp eye resembled that of Castells as he appraised the symbolic functions of the modern American suburb. The important differences were that her interviews offered documentary substance for main points and that she employed history to explain why or how the conventions may have come into being. The colonial and early federal attacks on feudal vestiges and the later flights from feudal remnants undertaken by nineteenth-century immigrants to North America identified property ownership with freedom from customary restrictions. That is not all. Leasing and renting were, despite such powerful historical events, to remain a part of the American scene. The law continued to perpetuate a secondary status for tenants, manifested in landlords' minimal obligations and tenants' maximum liabilities.

The nineteenth-century history of North American land policy and property law indicates a partial revolution, one carried further than in Europe but with the emotional and legal stigma of tenancy enduring. North American jurisdictions had had to revise property laws, but not so totally as to extend to tenants the freedom from the owner's right of entry to inspect the premises. Perhaps, as an advocate of more flexible housing arrangements, Perin had a bias that led her to assign to the law a hefty portion of the blame for separating society into two groups. Another element in the established set of community authorities, the bank, received similar treatment from Perin. Bankers issue the credit rating that acts as "a major threshold of American social personhood crucial in the correct traversal of the ladder of life."[43] The buyer of a house signals that the bank has bestowed personhood by securing and then improving property. The well-mown lawn is not the racial atavism alleged by Veblen; it is not the bridge between alienated man and remote nature cited by Castells; it signals that its occupier has received the bank's approval. When Veblen and the Lynds wrote, the institutional mortgage was not the pervasive instrument that it had become in Perin's time. The concept of the bank's approval conveying stature is not a farfetched hypothesis. We have not tested this specific claim, but in chapter 7 we demonstrate empirically the soundness of Perin's observations on the association between home ownership and a ladder of life as well as on home ownership and what it meant to immigrants seeking respect in a new society. How owner-occupancy became an ideal and how tenancy declined as consumers and investors were affected by their cultural circumstances occupy the remainder of our discussion of home ownership in this chapter.

The history of land tenure, though involving unique legislative details in many individual American states and Canadian provinces, involved a sequence of struggles, of greater or lesser friction, to extirpate lingering aristocratic and feudal characteristics stemming from assorted colonial beginnings. By the mid-nineteenth century, certainly by the time of the American (1862) and Canadian (1872) homestead acts, the domain of real property had been altered greatly in favour of individuals who lived and worked on the land. Quit-rents had been eliminated, primogeniture and entail essentially abolished, and large private estates with a tenant class had become exceedingly rare. The modernization occurring across one hundred years, beginning with the American revolution, had minimized the marks of feudal custom on real property law. In the English colonies and in New France, land holding had essentially subscribed to the formula *nulle terre sans seigneur*. The corporate colonies of Massachusetts, Rhode Island, and Connecticut were somewhat exceptional because the Puritans opposed a distinction between lord and tenant.

Nevertheless, the New England corporations actually held their lands from the king by a socage tenure. The type of tenure in the English colonies was styled free and common socage or tenure in fee-simple or freehold, the terms of which were fealty and a fixed rent. Fealty was the bond between lord and tenant; it survived in the oath of allegiance to the Crown. Rent, usually called quit-rent, was the bond between the lord and the land, the symbol of territorial ownership. The term quit-rent derived from the fact that such a rent represented a commutation in money of certain feudal obligations, such as personal service. The payment in no way hampered the occupant's control of the land, for although actual ownership remained elsewhere the freeholder was free to alienate or bequeath the real estate providing the terms of tenure were respected. The lands were thus "free" because they were alienable and heritable. However, freehold remained contingent. Legally speaking, ownership was held elsewhere. The quit-rents in colonial times often had a real money value and were not nominal. Moreover, the existence of legal forms does not give an account of social and economic relations as practiced. Therefore, free and common socage or tenure in fee simple contained important principles about the security and freedom of tenure, but only for those able to take advantage of ownership.[44] The history of land tenure from the American revolution until the mid-nineteenth century included attacks on remaining feudal traits in freehold and on the rare instances of agrarian landlordship.

In the Puritan colonies, there were no great individual proprietors. Towns receiving grants of land held the land in free and common socage and distributed plots to individuals who held them in free and common socage: "to be held forever in fee without any incumbrance whatever." In Virginia, by way of contrast, some lords proprietor held land in free and common socage and leased their lands to farmers. Similarly, the old Dutch estates of New York functioned as manors with tenant farmers. By the time of the revolution, freehold tenure was the rule for land holding in the first instance; but, in some places, below the direct holder of the freehold, tenants farmed large estates and their contractual relationships with the landlord varied greatly. Coincident with the overthrow of royal authority during the American revolution, one feature of the traditional freehold system was eliminated. The quit-rents collapsed. These replacements for feudal dues were quickly deemed incompatible with land tenure for an independent people. The British government acknowledged the failure of quit-rents after the revolution when it gave up these charges in Canada. In North America, the freehold system had evolved to the state where even the fixed-rent dimension had vanished. During the decade after the revolution, many states also abolished primogeniture and entail, the devices used to hold together large landed estates when the landlord had more than one son. Perhaps these actions were symbolic, for primogeniture and entail were on their way out in any case, but as Richard B. Morris wrote in *The American Revolution Reconsidered*, their abolition had "importance as a symbolic blow against a class-structured society."[45] Henceforth, too, proprietors of great landed estates in the United States and British North America experienced difficulties that, in one way or another, challenged their domains.

For financial reasons, the Calvert proprietor of Maryland already had begun to sell off manors as early as 1765. A few Virginia proprietors lost their estates for supporting the Crown during the revolution. The Penn family lost most of its lands when in 1779 the Pennsylvania assembly vested the estates of the proprietors in the commonwealth. The proprietors were allowed to retain title to their smaller private estates. In the tenancy counties of revolutionary New York, manor-rich families like the Livingstons fought two concurrent wars, one against the British and another against their own tenants. The landlords' loyalty to the revolutionary cause saved their estates. Lease terms on these surviving New York estates varied considerably from manor to manor. Some leases ran for a set period of years; others bound the land to a farmer's family for two lifetimes and then the land reverted to the landlord.

Crises gradually wore down the anachronistic estates. When one of the heirs of Rensselaerswyck tried to collect back rents in 1839, he ignited an anti-rent agitation that peaked in the mid-1840s and flared up intermittently until well after the American civil war. The protest provoked more than a conflict in a tiny community. First, the area was extensive – in 1848 an estimated 1.8 million acres in New York were under lease. The Van Rensselaer estate alone contained over 3,000 farms and 436,000 acres. Second, the conflict received publicity. The rent issue became a major topic of debates in the assembly.[46] The New York incidents dramatized the evils of land monopoly and aroused Americans to the importance of accessible land for the common man. The spectre of the landlord or great landowner attained an accepted place in American popular culture, often in the guise of the cattle baron who menaced the farmers, and the events in New York provided a backdrop for a novel and film with one of the better personifications of aristocratic decadence in America – Vincent Price portrayed the vile Hudson-Valley manor owner of *Dragonwyck*, while a crowd of vengeful commonfolk assembled as the instrument of justice.

Freehold tenure served as the basis of both modern English and American real property law, but America, with its lack of a titled aristocracy, its repudiation of quit-rents, its abolition of primogeniture and entail, and its signal clashes between landlords and tenants, which were resolved in favour of the latter, created a different legal and mental framework for conveying and thinking about land. Of course, legislators definitely did not pursue an equality in distribution of land. Notwithstanding a persistent concentration of real property, access appeared comparatively open and with that access came notions of independence. Moreover, that comparative openness had required enough political struggle and direct action to sanctify small proprietorship as a national good bound closely to the mythology of the revolution. In matters of real property law and market practices, the British North American colonies that became Canada had more in common with the United States than with the United Kingdom by the mid-nineteenth century. In Quebec the British inherited the French seigneurial tenure system by which seigneurs held land from the Crown. A set of rights and dues bound seigneur and tenants. In 1854 the government of Canada eliminated the two-hundred-year-old practices of seigneurial tenure. The legislation ending the system provided for the conversion of ancient rents in kind or service into a low cash rent to be paid annually or commuted into a lump sum. The movement to abolish the seigneurial system was almost certainly made possible by a bourgeois leadership that alleged that progress would attend the opening up of the land market. Whether the leaders had widespread support of both the urban working class and rural tenants is not documented, although it seems that the latter were sceptical.[47] In the tiny colony of Prince Edward Island, the modernization of tenure came slightly later than in Quebec, but here the popular support seems unequivocal. By a lottery in 1767, the British government had distributed the island's sixty-four townships to proprietors who paid quit-rents to the Crown and leased lands to settlers. After decades of agitation, the provincial assembly in 1875 passed a compulsory Land Purchase Act whereby the leases were eliminated.[48]

The first significant break from a land system based on landlords and tenants in British North America had already come with the creation of the colony of Upper Canada in 1791. Putting aside the rights of the region's aboriginal people, Upper Canada presented a virgin landscape onto which colonial authorities overlaid a grid survey of land to produce roughly uniform parcels of real estate, extended freehold tenure, and dispensed with quit-rents. A 1795 act established a system for the registration of property instruments. A vigorous land market quickly developed. To help guarantee that purchasers held a strong title, Canada adopted mandatory land registration in 1847 for Upper Canada, as noted in our account of land development in chapter 1. Within a few years, the merchants and agriculturalists of Upper Canada openly coveted the lands held by the Hudson's Bay Company. Their social vision of a nation of small proprietors eventually migrated westward following the purchase of the bulk of the Hudson's Bay Company's domain in the North American interior. Settlement excited a volatile prairie land market and overran the small hunting cultures of Amerindians and Métis. Individual small proprietors free to dispose of land occupied the Dominion just as they had occupied the Republic. Tenancy remained, especially in the cities, but the form of real-estate law and the bourse-like character of land sales indicate that the continent accommodated the glimmers of independence that common folk – native-born and immigrant alike – came to associate with owning land.

An aspiration for independence, the assertion of individualism against the pull of collective obligations, and the centrality of land-based relationships to each of these traits of modern western behaviour have surfaced as major concerns in a debate among English historians about the origins of English individualism. Alan Macfarlane, who precipitated the controversy by arguing that as early as the thirteenth century England was mainly peopled by rampant individualists, has been criticized effectively for having misinterpreted and ignored communal bars against an individual's freedom to alienate land or otherwise engage in land transactions. Despite Macfarlane's eccentric argument, the important and obvious limitations on property rights of English villagers in the pre-modern era checked individualism.[49] Puritan communalism or proprietory colonies established bridgeheads for English restraints on individualism, although, as we have seen, these restraints tumbled down in the hundred years from the 1770s to the 1870s. What was it about freehold tenure without a quit-rent obligation that tied into a spirit of independence?

Spun of myth and fact, several bonds composed the connection between land ownership and independence. The holding of land in free and common socage implied the retention of all produce. Ownership by freehold meant the independence to devise and to alienate, and consequently fed a trust in the worth of improving one's land. It also meant living without the practices of deference toward landlords. In his account of the Irish immigration to Canada, Bruce Elliott illustrated the perceived contrast of the old and new worlds with an extract from Francis Evans's *The Emigrants' Directory*. Evans led his readers to believe that the farmer in Canada was "the lord and master of his own estate."[50] Let us quickly add that, in reality, land ownership did not bestow freedom from obligations. Mortgages introduced elements of deference between mortgagor and mortgagee. Painting the house before the mortgage had to be renewed was a common practice illustrative of the subtle ties between owner and mortgagee. Even so, the mortgage gave telling rights to the debtor. Before the courts, as we demonstrate in chapter 5, the two parties had to prove fair treatment of each other or face penalties. Moreover, the passionate agitation that had compelled statutory action to facilitate freehold tenure and a land market reminds us that contemporaries feared immediate restraints and charges based on experiences and current history, not future limitations implied in an altered state of property relationships. Contemporary aspirations and anxieties and not prescient forecast is what should be expected of our forefathers. The mortgage was an obligation, but also an instrument of independence.

One ancient and explicit restraint had applied to non-freeholders. In England, the colonies, and in the new republic until the 1830s, the statutes defining an electorate tended to restrict the franchise on the basis of a relationship to real property. The quick application of universal white-manhood suffrage in the United States may have made the ownership of a lot or a dwelling less instrumental in fulfilling a common man's self-image as independent agent, since American citizenship and residence in a state for a requisite period often sufficed as qualifications for voting in state and federal elections. In England and Canada assorted property qualifications restricted the electorate throughout most of the nineteenth century. George Martin, the Hamilton carpenter and erstwhile English Chartist who in 1837 had looked toward the appreciation of his property, stated the following about his house: "if I can but get it my own, and then I have a vote to send a member to Parliament."[51] This goal of the parliamentary vote through home ownership was soon to become an element of the Chartist program of Feargus O'Connor. His National Land Company, established in 1845, endeavoured to settle English factory workers on small holdings. The experiment collapsed with the death of Chartism, having produced only 250 cottages. The scheme, however, had exemplified an ambition apparent in Martin's letter.[52] Dorothy Thompson has proposed that the Chartist land plan held the movement together during the 1840s. Furthermore, she has indicated that the Chartist advocacy of small landholdings was "not confined to the utopian part of the working-class movement," but was discussed widely by philanthropists, political econ-

omists, and liberals.[53] Even if most jurisdictions in the United States
had instituted universal white-manhood suffrage by the 1830s, it can
be suggested that lore about land and access to full citizenship priv-
ileges had built up under British regimes of concentrated land holdings and restricted franchises. The "invisible immigrants" from the
United Kingdom remembered the basis of old country economic
and political power.

In Europe, the feudal threads had not been readily severed.
Lengthy and bitterly contested land reformation efforts confronted
feudal remnants in the form of a politically privileged and frequently
titled land-owning class. In North America by the day of the Homestead Acts, the association of land and independence had progressed
from the plane of advances against feudal legacies to a new level of
concern. Land reformation advocates in America had somewhat different objects to attack.

In the eastern seaboard cities, investment in commercial and domestic property and geographic constraints had raised the cost of
proprietorship. In New York, as Elizabeth Blackmar has documented, the appearance of tenant houses "opened new fields of real
estate investment that contributed to the closing down of earlier
institutions of proprietory independence."[54] For New Yorkers,
Blackmar has argued, the solution to the crisis posed by rising
landlordship in a republic was found in the myth of the home as
sanctuary. Perhaps the idea of dwelling apart from the site of work
cushioned the reality of tenancy. However, there remained the remedy of westward migration to rural and urban places with cheaper
land. But even the west seemed ready to fall into the hands of
landlords after mid-century. An anxiety about access to land in North
America sprang from the two contradictory policies pursued by governments in both the United States and Canada. With the 1862
Homestead Act, Congress gave settlers free land. It also would enact
measures granting vast tracts to railways. The Homestead Act also
precipitated the Kansas fever. Land-policy historian Paul Gates attributed the boom in the population of Kansas between 1865 and
1880 partly to the flight of tenant farmers from the corn belt areas
of Illinois and Iowa. Tenancy's unpopularity was borne out by this
episode. The advertising of the Chicago, Burlington, and Quincy
Railroad played upon the mounting rejection of tenancy: "Life is
too short to be wasted on a rental farm."[55] The homestead measures
and even the land promotion ventures of railways intensified the
public awareness of access to freehold land. The railway's participation in fostering hope for independence was paradoxical. Railway
land grants, branded by a few critics as a new feudalism, kept the

ideal of democratic access to land prominent as a public question because the grants seemed to threaten a reversion to great concentrations of land, no longer held for rent but now for speculative sales. In California, the grants provoked the crusade of the century's greatest land reform advocate and the continent's most internationally renowned social critic, Henry George, who also amplified the urge to own real estate.

In comprehending the cultural significance of ownership of real estate, the role of Henry George as sounding board for the democratic values of America and their apotheosis in land is of central importance. A sounding board, which picks up and amplifies vibrations, provides the perfect metaphor for the vociferous George who, in the 1880s, lectured widely in the United States, the United Kingdom, and Canada. Between 1878, the date of its first publication, and 1920 *Progress and Poverty*, his masterpiece, went through dozens of printings. Over forty in English are listed in the Library of Congress. He inspired a Georgeite movement that channelled the continent's reform energies in ways that set some of them at variance with the European left, for George essentially accepted the slogan of Pierre-Joseph Proudhon – "property is theft" – and then proceeded in a haze of rhetoric to defend what amounted to a revised concept of privately held real estate. For a century now, commentators have scrutinized, assailed, or defended his economics, and debated the practicality of his single tax remedy. Most recently, John Thomas, an intellectual historian, put George in relation to his times, presenting the inveterate lecture-hall campaigner and would-be economist as an American democrat inspired by natural law, the independence rhetoric of the founding fathers of the republic, and evangelical Christianity.[56] As an evangelical nationalist proudly defensive of the moral mission entrusted to America – the mission of freedom – George witnessed events that he believed jeopardized the providential course. The Civil War levied a bloody setback, but for George the original qualities of the country were embattled during peace and prosperity too. The prospect of a nation of small landed producers, a vision of the founding fathers, faced elimination by economic concentration, particularly by the consolidation of land in the hands of what George believed to be non-productive speculators who supposedly withheld land from productive use in order to drive up rental payments and the cost of those lands that they sold.

Winning plaudits as an orator, first in San Francisco and then nationally and internationally, George adapted his concern about property concentration to the idiom of particular audiences or special occasions. At a Fourth of July celebration, he extolled republican democracy and cited the fall of Rome as a warning against concentrated wealth. To religious assemblies, his sermons linked Biblical texts to demonstrate the God-given right to land. "Thou shalt not steal." Among the Knights of Labor in New York, he renounced the evils of child labour and tenement dwellings. George scooped up many commonplace sentiments; he also added a voice to the denunciation of social injustices, and evolved a not wholly original reform measure – a single weighty tax imposed purely on land, not on land and improvements, that would force equitable land redistribution. His nostalgic view of the past, his evangelical nationalism, his attacks on privilege, and his single tax scheme all supported his concept of getting the land into the constructive hands of farmers, artisans, and labourers. To fight poverty, he espoused the ideal that "each family have its healthful home, set in its garden."[57]

The history of California gave George's democratic idealism a particular twist. In 1871 he published *Our Land and Land Policy*. His obsession with land, possibly American culture's obsession with land, took reference points from the liberal democratic experiment of the United States, Biblical allusions that fitted the new nation's hopes and resources, and a pragmatism that centred on breaking Malthusian pessimism through a policy of helping true producers. Small proprietors related to all three cultural precepts. Only the vigorous and vigilant farmer or artisan could keep alive life, liberty, and the pursuit of happiness. The United States was God's country, and the Bible placed man in a garden, which was the true focus of civilization. Every man should sit by his own vine and fig tree, with none to vex him and make him afraid. "The ill-paid, overworked mechanic of the city could [with a rational settlement policy] find a home on the soil, where he would not have to abandon all the comforts of civilization."[58] Land redistribution also had a "practical conclusion." The worker on land was a better worker if he were also a land owner. In *Progress and Poverty*, George elaborated on the escape from the Malthusian trap. If China, India, or Ireland could not escape famine, blame had to be placed on the land-tenure system. "The tenant dared not accumulate capital, even if he could get it, for fear the landlord would demand it in the rent."[59] The victories won for proprietorship by mid-century throughout the United States and British North America should have been known to George, but as a prophet he dealt in lamentations.

As a California journalist, George had come to fear a resurgence of feudalism. Threats to the common man arose from new variations on old-world themes. Forecasting social retrogression, he understandably drew upon historical and contemporary examples, fusing them into a danger that readers and audiences could easily understand. So he saw railroad companies as establishing a landed aristocracy in America – "The landed aristocracy of California" consisted primarily of members of the Central-Southern Pacific Railroad Corporation. They and urban land speculators, he claimed, *"rent* their land; they will not sell it."[60] This blight of tenancy allegedly threatened American democracy and the heart of a producer economy. And George perceived other evils preparing a retrograde path for America. An opponent of Chinese immigration, George felt that the arrival of Orientals heralded the introduction of serfdom. Free men could not earn a living in a California of the corporations, consequently a labour system tantamount to slavery had begun to be introduced. "Clear out the land-grabber and the Chinaman must go. Root the white race in the soil and all the millions of Asia cannot dispossess it."[61] Although George hinted that he sensed a wider basis of power than land – foreseeing the day of the titanic corporation – he never moved on to evaluate it. His first visit to Ireland and England in 1881–2 and several subsequent voyages to a kingdom in the midst of controversies over land issues and poverty latched him firmly onto the land question. Looking into the British land system after his California experience could have detracted from his understanding of the real property business in much of North America. He misunderstood the dynamic in urban land development that had progressed even as he wrote and lectured. Landed wealth in North America did not necessarily seek to retain estates with rental dwellings for revenue. Some estate managers did, others did not, and some surely mixed estate management strategies. Many propertied interests worked to liquidate their land and to place profits in new investments involving less exacting attention and less friction. George failed to apply himself to political economy and to serious investigation of the processes he attacked. His speeches and, not coincidentally, his popularity originated from reiterating certain public opinions that he first encountered as a California journalist.

If George had a nose for public sentiment and the truism, he was no indolent cracker-barrel sage. He ultimately confronted a few classics of political economy and he avidly collected aphorisms of ancient wisdom. To Biblical references he would add a truth from a Stoic, the wisdom of the Brahmins, and a passage from the Indian myth, the *Ramayana*. George cultivated the eclecticism of a Victorian gadfly. Above all other strengths as a crusader, he exercised an extraordinary power of language. A poet of petty proprietorship, he eulogized land in ways that could easily have contributed to his cool reception by English socialists. His ultimate message amounted to a flat rejection of socialism. By 1881, after the publication of *Progress and Poverty* in 1878, he had enough of a following to launch a lecture tour in the northeast. In June it took him to Montreal, Ottawa, and Toronto.[62] Presently the book made George a transatlantic celebrity, and ultimately a renowned figure throughout the English-speaking world. In 1890 he undertook a tour of New Zealand and Australia. Part jeremiad, part watered down-political economy, George's ticket to fame, the much-reprinted *Progress and Poverty*, touched the nerves of aspiration and anxiety. Let us consider its argument first, then probe the reasons for its tremendous appeal, and finally consider its crucial bridging of radicalism and the will to possess.

George had to justify his earlier claims that land owners siphoned off the product of labour. The proof anyone could see – where land was expensive, squalor was great. Ownership of land had been, and would remain, the foundation of the aristocracy. The stronger and more cunning people would acquire a superior share as long as land could be alienated from the protection of common use. Having moved to the brink of land nationalization, George jumped back. He would leave title intact. "Let the individuals who now hold it still retain, if they want to, possession of what they are pleased to call *their* land. Let them continue to call it *their* land. Let them buy and sell, and bequeath and devise it. We may safely leave them the shell, if we take the kernel. *It is not necessary to confiscate land; it is only necessary to confiscate rent.*"[63] The single tax would accomplish that, and, in George's estimation, would preclude the assembly of large estates for rental or speculative gains. The single-tax idea outlived George, though not without controversy. The debates over the specific remedy had narrowed the broader meaning of George's crusade for nearly a century until John Thomas's *Alternative America* located him in evangelical nationalism. Thomas's intellectual history evaded the centrality of the materialistic thrust, the producer emphasis, and the equation of land holding with human fulfillment. Inadvertently perhaps, but quite emphatically and persuasively, George was, to return to our metaphor, a sounding board for the celebration of owner-occupancy.

George's stress on rent as a source of pauperization led to his characterization of a tenant as "an abject slave, who, at the nod of a human being like himself, might at any time be driven from his miserable mud cabin, a houseless, homeless, starving wanderer, forbidden even to pluck the spontaneous fruits of the earth, or to trap a wild hare to satisfy his hunger."[64] The disturbingly evocative imagery, plausible as an exposure of the Irish land issue, would have touched the emotions of many North Americans of recent arrival. It rang true to immigrants who, personally or from family lore, knew of the travails of tenancy. If, for George, tenants lacked hope, possibility of gain from energy and thrift, and self-respect, then landlords wore the black mantles of villains who subjugated by resorting to agents and, in crises, by appealing to the armed force of the state. The moral drama amplified by George's crusade had its basis in major indigenous rural confrontations within recent memory. The tenancy relationship bristled with unhappy memories for those who had been tenants and even for landlords. Tainted as it was by recent history, tenancy withered under George's attacks. His poetic exultation of land ownership, in spite of the very logic of some of his arguments that brought him so close to land nationalization, made the ethos of suburban home ownership irresistible. Before advertising copy might be thought to have seduced the public into a taste for the suburban lot, George had rhapsodized about the virtues of land in the hands of the common man. The following passage, one of George's most lyrical and revealing, speaks volumes for the sources of the home ownership demand in American society.

The great cause of inequality in the distribution of wealth is inequality in the ownership of land. The ownership of land is the great fundamental fact which ultimately determines the social, the political, and consequently the intellectual and moral conditions of a people. And it must be so. For land is the habitation of man, the storehouse upon which he must draw for all his needs, the material to which his labor must be applied for the supply of all his desires; for even the products of the sea cannot be taken, the light of the sun enjoyed, or any of the forces of nature utilized, without the use of land or its products. On the land we are born, from it we live, to it we return again – children of the soil as truly as is the blade of grass or the flower in the field. Take away from man all that belongs to land, and he is but a disembodied spirit.[65]

The ethos of property ownership that George, as sounding board, took in and then intensified by text and platform had numerous origins, none entirely distinct from others. For example, the eighteenth- and nineteenth-century struggles against quit-rents and tenancy in North America should not be sequestered from the wider challenges to propertied wealth that arose in the name of liberty across Europe and in the United Kingdom. The circumstances of discontent and conflict in Ireland alone would have surfaced as a significant cultural force in the United States and British North

America. A likely urge to possess land as the source of security, touchstone of liberty, and route to prosperity accompanied many farmer and artisan groups immigrating in the nineteenth century. Many more declarations about these hopes are likely to be uncovered, beyond the few that we have cited. However, another probable psychological source for the ownership impulse deserves mention. The grassroots and Georgeite attitudes that expressed a mythology of landed independence flourishing in the abundance of the North American continent had largely sprung from notions about the value of rural land and, although the urban populations of America surely possessed plentiful rural origins, a more exclusively urban dynamic pulls the discussion back to its principal topic – urban housing.

The Georgeite movement and the increases in home ownership in many North American cities and towns roughly coincided with the restructuring of productive relations in urban centres. The widening gap between the work and the social and creative nature of the worker is a well-known hypothesis supported by a shelf-full of studies about industrializing communities or individual industries. A major issue tackled by labour, social, and even intellectual historians concerns the reconciliation or, more correctly, the synthesis between, on the one hand, the ideal of artisan independence and the work ethic and, on the other, the invasion of machines and scientific management. Reconciliation may apply to some features of the synthesis, but conflict was very much a part of the historical record. Amid the ever-increasing monotony of tasks in the factory system, where was the space for the mind or for individual creativity? As labour historians have argued, the struggle to retain a work-related independence on the shop floor and the maintenance of traditional leisure activities, as well as the creation of new ones, provided a portion of space for the mind and creativity. Fraternal associations also kept alive human relations and self-esteem. During the 1880s the Knights of Labor in the United States and Canada united the struggle with associational life. But where else could a worker find space for independent accomplishments, satisfaction in work, and a belief in control of financial rewards for personal work?

The home and its modest plot of land could have raised at least the promise of a domain where labour in construction, improving, do-it-yourself projects, and gardening afforded close and continuing compensatory opportunities. Sociologist Jane Synge, who interviewed many Hamilton senior citizens in the 1970s, observed that working-class respondents, when reporting on the work their fathers had done around the house, emphasized the garden. "My father used to plant all our vegetables in the yard. It was really good."

"Father did the vegetables." "He worked outside. We had a vast vegetable garden."[66] Gardens formed a significant part of the North American city. As a truism, suburbs take land out of commercial agricultural production. On the other side of the ledger, their lots have frequently supported labour-intensive non-commercial food production. A real-property survey conducted in sixty-four United States cities during 1933 and 1934 included reporting on summer vegetable gardens. There was great regional variation. New England and South Western cities had the fewest, while the Great-Lakes and mid-Western cities had the most. In Albuquerque the percentage of all dwelling units with a vegetable garden was merely 2.7; the highest percentage was 56.1, reported in Springfield, Missouri. Among the Great-Lakes and mid-Western centres most resembling Hamilton in size, age of city, and economic base, about a quarter of all dwelling units had gardens.[67]

Conceivably the depression had compelled some of the urban gardening, but even so, the garden's potential for assisting with the family budget must have been part of common wisdom long before this particular collapse. The single-family dwelling with its lot had a material advantage and a role to play in family strategy. It also possessed a psychological appeal connected with agriculture, as John Stilgoe concluded in *Common Landscape of America*, presenting a miniature farm, a compressed pastoral setting that expressed a traditionally loved landscape.[68] The imagery of near congruence between beloved farm and suburban plot is not the sole basis for the parallel. To Stilgoe's astute discovery of a connection between an old and a new landscape, we can add the linkages of work. The pioneers built, improved, and nurtured. Thanks to the mass marketing of building or gardening materials and journals, the factory labourer or office worker could share these creative experiences on a small scale. One did not have to own a farm or spend years as an apprentice joiner to enjoy traditionally ennobled work.

An explicit connection between the international Georgite movement, the labour movement, and the locale for much of our quantitative research existed from January 1883 to December 1886 in the form of a Hamilton-based labour and reform newspaper. This journal, serving southern Ontario, including Toronto, and known first as the *Labor Union* and subsequently as the *Palladium of Labor*, urged its readers to familiarize themselves with Henry George's writing. In its 20 January 1883 issue, *Labor Union* announced that "Henry George's book is now to be had for 20 cents complete. Buy it, read it, get its truth by heart, and then lend it to your neighbors."[69] The *Labor Union's* successor, the *Palladium of Labor*, enthused over George during his August 1884 visit to Hamilton and for two months published a column essentially about George entitled "Love of the Land." The journal stated that "the only solution of the social problem is the restoration to the people of their natural rights in the soil."[70] It followed the land struggles in Ireland, Italy, and even Texas, where farmers clashed with ranchers, and gave George's lecture tours enthusiastic coverage.

The reform hero, whose pictures were being sold at the *Palladium* office for parlour decorations, delivered a florid oration that elevated the significance of land to the very essence of material existence. "Land is the element upon which alone we can draw our subsistence. Deprive man of it, and what have you left him? He is nothing but a disembodied spirit. We come from the land, and we must return to it; even our flesh and blood must become a part of it. Give to one man the control of the land on which the others must live, and he is their master, even to life and death."[71] At last the great crusader had arrived in Hamilton to address the annual demonstration of the Knights of Labor at the city's Crystal Palace. Punctuated by applause, George's address struck at inequality with arguments appealing to common sense. George's lectures had so many aspects that various groups and interests could cull what they wanted. The Knights stressed his attacks on inequality, but passages like the one above could lead in a different direction. In fact, on 4 September 1886, Patterson Brothers, an innovative and opportunistic firm of builders declared a commonplace that many others had and would use to sell houses, but they integrated the idea and their merchandise into the Georgite excitement by bold advertising in the *Palladium*:

The True Solution to Land Question
Be Your own Landlord
Buy a house when it costs you
No more than rent.[72]

Evidence of links between radical causes and the will to possess vanished during the decline of worker's movements in the 1890s, only to reappear in 1912. Responding to high shelter costs forced up by the city's industrial boom, industrialists, politicians, and labour leaders discussed a limited-dividend housing corporation. Alan Studholme, arguably Ontario's most respected and successful labour politician of the time, endorsed the scheme, but not unequivocally. The moderate and popular labourite said he "would prefer a proposition by which the workingman would secure his home instead of becoming a perpetual tenant."[73] Studholme, whose original trade as a stove mounter placed him in the ranks of skilled workers, had been a member of a Hamilton local of the Knights of Labor and of its political association. What he said about housing a quarter century later held a flashback to the radicalism of Henry George. It is impossible to say how much support Studholme had for his vision of a city of worker proprietors. Since he was a moderate reformer who came to political maturity in the 1880s, Studholme's outlook is not surprising and it probably did not damage him politically in this city of homes. First elected to represent Hamilton East in the Ontario legislature in 1906, he died in office in 1919. His funeral was reported as one of the city's largest. Studholme knew his enfranchised constituents.

A paradox of mass culture and industrialism was that they threatened craft-based independence while they provided material for access to a heritage of independence. The heritage was certainly "bowdlerized," compressed, cheapened, and laden with constraints and costs. Nevertheless, as we show in chapter 5, the machine-tool revolution and other industrial innovations held out opportunities for common folk to raise, purchase, or improve their homes according to styles of the day. The suburban lot bestowed a sense of independence on its owner because of a web of cultural values that equated land with freedom from subordination. It also secured an unrestricted workplace for the self-reliant at a moment when the factory and office minimized the opportunities for the pleasures of tinkering, and agricultural mechanization and urban growth narrowed the numbers that could participate in real farm cultivation. The alienation accompanying industrial work and the potential loss of a feeling of independence facing the rural migrant to the city might have been mitigated by the suburban lot. The claim is, therefore, that in advance of any advertising to create a taste for home buying there were profound sources for that longing. Radicalism had helped glorify the goal of land owning. Industrialism made the lot with house a refuge for self-expression. It is conceivable that this historical background helps to account for the recent findings of sociologist Lois A. Vitt. Using United States survey data from 1972, 1985, and 1989, she has discovered that home ownership has brought many owners "non-financial feelings of satisfaction with family life, friendships – even general happiness." It is significant too that home-owning working-class people do not feel "more middle class" on account of their houses; they may not generalize the benefits they find in home owning to altering their place in the class structure.[74]

Such notions have proved to be enduring. By the end of World War II, real-estate analysts had exploited these undercurrents to expand the advantages of ownership into long lists. Writing about the meaning of housing in America, geographer John S. Adams summarized many sentiments into the pithy affirmation that "a house is a structure; a home is an experience."[75] The will to possess persists, but the reality of land ownership has become more elusive in North America's urban centres. In the late 1980s, observers were bemoaning the lack of affordable homes in Southern Ontario. Too few renters, perhaps only 3 per cent, were financially capable of entering the ranks of home owners.[76] A Gallup poll taken during the summer of 1988 revealed a great deal of pessimism concerning the possibilities for home ownership in Canada. Nationally, only half of the Canadian public believed that this traditional dream of young couples was within reach. In Ontario the figure was just 38 per cent.[77] But this cultural perception is a new development whose meaning has yet to crystallize. Let us now consider how the will to own prompted changes in the shelter industry, influencing the history of contracting, building technology, and house design.

During the boom of the early 1900s land developers minimized their
outlays, but on Barnesdale Avenue the developer provided a boulevard, one of several in the city in that period.

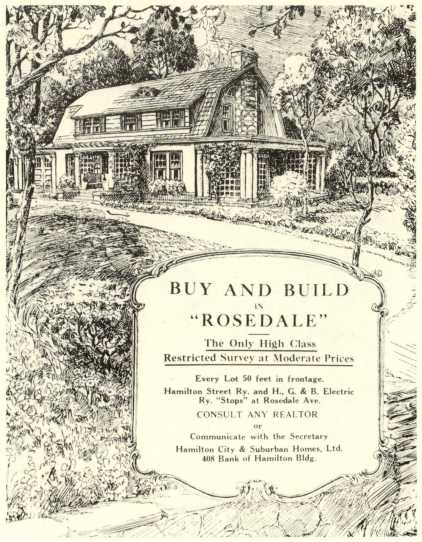

BUY AND BUILD
IN
"ROSEDALE"

The Only High Class
Restricted Survey at Moderate Prices

Every Lot 50 feet in frontage.
Hamilton Street Ry. and H., G. & B. Electric
Ry. "Stops" at Rosedale Ave.
CONSULT ANY REALTOR
or
Communicate with the Secretary
Hamilton City & Suburban Homes, Ltd.
408 Bank of Hamilton Bldg.

During the 1920s one device used by developers to promote their surveys as distinctive was the "restriction." This often meant racial or religious discrimination and was practised in several of Hamilton's subdivisions. The advertisement for Rosedale is distinctive only because its depiction of bucolic settings is more lavish and detailed than those mendaciously portraying park and garden settings for the working class around 1910.

The mountain became the haven for the city's better-paid white- and blue-collar employees. The frame houses were constructed just after the end of World War II. The small brick houses were constructed in surveys developed by Grisenthwaite Construction, one of the city's most prominent land developers during the 1950s. DeSantis was one of Hamilton's largest developers in the 1980s.

The long interludes between initial survey and eventual construction as well as the absence of controls on building were features of Victorian land development that led to a juxtaposition of housing styles.
Top Ray Street south of Hunter. *Bottom* West side of McNab Street at Ferrie Street.

The celebration of civic centennials included statements about housing. In 1913 the house built in a day proclaimed the achievements of the building trades and the progress of the city; in 1946 a float celebrated a home of one's own.

The formal attire of these masons posing at the house built in a day
attests to craft pride, although the uniformity of the bricks and lumber
studs exemplifies the major technological innovations that had challenged the standing of carpenters and masons.

This frame house was one of several built by contractor William Wray for real-estate agent William Moore around 1908. They were based on plans from a pattern book.

A major builder of prefabricated dwellings, Halliday Homes had originated in 1910, flourished in the 1920s, survived the depression, and experienced enormous growth during the post-war building boom. There was no time for posing on this busy site.

5 Material Culture and the North American House: The Era of the Common Man, 1870s–1980s

The history of housing for the workers in urban North America is at once a chronicle of great continuity, many subtle changes, and several upheavals. This complex pacing along several tracks demands an appreciation of tradition, entrepreneurship, and technology. It also requires a preliminary sense of the scope and idiosyncrasies of a major activity. Residential building has been one of the most important sectors of the North American economy, accounting for between 50 and 65 per cent of total construction in most years during the nineteenth and twentieth centuries. In 1987 residential construction accounted for just over half of the $24 billion spent on building in Canada. The importance of residential construction to both stimulation of the economy and job creation is widely recognized by governments in North America.[1] If logging and finance were added to construction trades, it would have required thirty to fifty specialists to complete a house by 1920.[2] This significant manufacturing activity has displayed unique traits distinguishing it from other consumer-durables industries. First, it has experienced economic cycles of twenty years duration.[3] Second, the logistics of moving building materials from source to scattered subdivisions as well as variations in local land-use regulations and urban services have helped to check corporate consolidation until very recently. Currently the housing industry is witnessing instances of rationalized activity by builders of packaged suburbs as well as through a series of corporate takeovers.[4] Tentative vertical integration actually began about one hundred years ago, but knowledgable analysts insist that dispersal rather than corporate consolidation may still

Table 5.1
The Expansion of Home Ownership in Hamilton, 1851–1931

Year	Percentage of Home Ownership, Actual Count of Assessment Rolls or by Census Enumeration	Percentage of Home Ownership in the Sample Blocks[1]
1852	26.5	33.2
1856		25.3
1861	25.6	26.0
1864		21.0
1866		26.6
1871	23.7	38.4 (1872)
1876		43.3[2]
1881	30.5	37.1
1886		37.2
1891	34.5	34.4
1896		33.4
1901	33.3	35.2 (1899)
1906		38.1
1911	51.4	42.5
1916		42.4
1921	50.4	51.5
1926		48.1
1931	50.9	43.7

Sources: Hamilton Assessment Rolls and, beginning in 1901, *Census of Canada*
[1] The sample blocks were drawn from blocks outside the central business district. This means that in the early years the per cent of tenants was understated because of the absence of tenants in rooms above commercial properties. In annexed areas, commercial areas were retained, as they usually occupied only one face of a block.
[2] The 1870s marked a recovery in the local economy as its industries expanded.

provide the most efficient mode for management in the industry.[5] Routine home construction has shown an obstinant tendency toward local organization. Housing construction, therefore, has often beckoned as an industry with relative ease of ingress for new operators familiar only with their locale. These dimensions of the shelter industry bear witness to continuity or slowness of change in urban history.

The themes of long duration offer one chronological agenda, but other features show shorter-term swings. For all the unique and nearly static qualities that the housing industry has displayed, it figured prominently in a major cultural transformation. The years from 1880 to 1918, according to a recent cultural history by Stephen Kern, form "part of a general cultural reorientation that was essentially pluralistic and democratic."[6] The evolution of the single-family detached dwelling in many modest architectural guises accompanied

Table 5.2
Hamilton Home Ownership Levels Compared for the Original City and
Annexed Areas, 1899–1966

	Original City			Annexed Areas		
Year	% of City Area	Number of Sample Cases	% of Home Owners in Sample	% of City Area	Number of Sample Cases	% of Home Owners in Sample
1901	76.7	1169	34.8	23.3	38	47.4
1906	65.1	1260	37.1	34.9	85	51.8
1911	50.2	1306	39.0	49.8	382	56.3
1916	42.8	1557	37.7	'57.2	752	52.3
1921	38.1	1552	41.6	61.9	933	70.3
1926	36.5	1680	37.6	63.5	1243	62.4
1931	31.5	1889	35.2	68.5	1494	54.7
1936	31.5	1940	34.0	68.5	1593	45.6
1941	31.5	1756	26.4	68.5	2024	43.4
1956	16.0	2426	38.7	84.0	3680	70.4
1966	8.9	1665[1]	4.9	91.1	4221	76.1

Sources: Hamilton Assessment Rolls; John C. Weaver, *Hamilton: An Illustrated History*
[1]Massive urban renewal and the clearing of old housing stock for future apartment building
construction reduced the number of sample households.

the popular press, the cinema, the telephone, and mass transit in a
great social and cultural reorientation. From the 1870s to 1900s
many North American families benefitted from cheap new housing,
as tenants and as home-owners. Indices of urban residential construction plus calculations for the mean ages of cities' residential
structures betoken extraordinary waves of shelter construction in
the 1880s and early 1900s.[7]

Growing home ownership also characterized these years (see
table 5.1). Various series of estimates for annual wages in the United
States suggest a partial explanation for this significant social fact
because they indicate general improvement with rare reversals. Important influences abetting home ownership came from the housing
industry itself. In younger cities, not hemmed in by port facilities
and water barriers, the availability of suburban land and streetcars
assisted home ownership. For example, the home-ownership levels
in Hamilton differed widely between the old city as defined in 1846
and the annexed tracts that began to be added in 1891, doubling
the area of the city between 1901 and 1916 (see table 5.2). Annexation provided new land for residential use and, as stressed in chapter 2, the city quickly extended urban services through direct
expenditure and pressures on the streetcar company.

Mortgage finance also may have been an influence. The golden age of housing for the common folk, particularly the late nineteenth century, was characterized by remarkably stable and even declining interest rates. Blended payments and longer amortization, new to the era, spread the burden of high initial payments of conventional mortgage financing, but at first applied only to new housing.[8] Technological and architectural shifts also nurtured the North American pattern of shelter – the balloon-frame structures, the prefabrication of housing components, the modest single-family detached houses, the use of new materials, the provision of new utilities, and the goal of home ownership. From 1870 to 1900 housing-construction costs fell or remained deflated (see table 5.3). The technological and materials considerations and their influence on residential architecture and housing costs form a major part of the current discussion.

Unfortunately, the statistical evidence for the continent's urban-housing trends is fragmentary. The United States census began to record home ownership in 1890 while the Canadian census started to do so only in 1901, but studies of individual cities support an impression of significant change. Research on Detroit and Pittsburgh indicates a substantial level of home ownership among working-class ethnic communities at the turn of the century.[9] One mechanism for the attainment of the goal appears to have been the practice of family members and friends working together to construct each other's dwelling. The technological innovations and mass dissemination of construction information initiated during the golden age made this feasible. The assessment rolls of Hamilton revealed that the percentage of dwellings that were owner-occupied rose as follows: 26 per cent in 1861; 35 per cent in 1891; 51 per cent in 1911. Hamilton contractors erected many worker's dwellings applying the technological advances that had reduced construction costs and made do-it-yourself building possible.[10]

After 1900 inflationary surges thrust housing construction costs to a peak in the early 1920s (see table 5.3). One outcome of this arresting jolt to a happy course was philanthropic and political discussion of a housing crisis and the inception of housing reforms. Another was an apartment construction wave that broke upon the urban landscape to house young middle-class families. There were also intense campaigns to promote small homes on suburban lots. The memory of earlier advances and the industry's promotional efforts kept alive the prospect of home ownership and help to explain why, when real costs climbed, home ownership remained embedded in the continent's popular culture. Both popular and technical journals urged people from all classes to aspire to home ownership. They

Table 5.3
Home Construction Cost Trends at Five-Year Intervals, 1870–1920

Year	Cost per Square Foot for Inexpensive Housing	% Change	Cost per Square Foot for Expensive Housing	% Change	Construction Cost Index (1913 = 100)	% Change
1870	1.20		2.50		95.3	
1875	1.20		2.00	− 20	82.0	− 13
1880	1.00	− 17	1.75	− 13	73.2	− 11
1885	1.00	0	1.50	− 14	73.1	0
1890	1.00	0	1.30	− 15	73.3	0
1895	1.00	0	1.30	0	69.8	− 4
1900	1.00	0	1.50	+ 15	79.9	+ 14
1905	1.00	0	1.70	+ 18	90.6	+ 13
1910	1.30	+ 30	2.00	+ 18	96.3	+ 6
1915	2.10	+ 62	3.50	+ 75	100.9	+ 5
1920	5.50	+ 162	7.50	+ 114	251.3	+ 150

Sources: Assorted building trade journals. The construction cost index appeared in United States Department of Commerce, Bureau of Census, *Historical Statistics of the United States, Colonial Times to 1970*, part 2 (Washington, 1975), series N138–193.

often stressed the economic folly of rental payments, occasionally they sentimentalized the non-economic virtues of home ownership captured in the poem "Home":

I want to have a little house
   With sunlight on the floor,
A chimney with a rosy hearth,
   And lilacs by the door;

With windows looking east and west,
   And a crooked apple tree,
And room beside the garden fence
   For hollyhocks to be.

Oh, all my life I've wandered round,
   But the heart is quick at knowing
Its own roof and its own blush
   And its own boughs blowing.

And when I find that little house –
   At noon or dusk or dawn –
I'll walk right in and light the fire
   And put the kettle on![11]

The years from 1870 to 1900 initiated possibilities for home ownership that persisted in spite of later inflationary surges in construction costs. Home-ownership ambitions continued to be realized
whenever deflation, government programs, and more economical
blueprints rekindled consumer interest. Consequently in the United
States 36.9 per cent of non-farm dwellings were owner-occupied in
1890, 46 per cent in 1930, 62 per cent in 1970, with modest gains
in the 1970s. Similar gains were experienced in most major Canadian
cities but there home-ownership levels peaked at 56.9 per cent in
1961, falling to 53 per cent in 1976.[12] Nevertheless, the dream of
home ownership remains resilient enough to influence both family
strategies and government policies.

NORTH AMERICAN DESIGN AND
BUILDING PRACTICES

The ownership impulse and housing design history transcended the
international boundary. Pattern books popularized identical plans
at all points of the compass with only minor adaptations for the
Pacific coast. In the South, builders dispensed with basements and
the cost of constructing a house was 10 to 20 per cent lower than
elsewhere. Design regions tended to be international.[13] Nova Sco-
tians assumed prominence as "wood butchers" in Boston.[14] American plans circulated in Canadian periodicals while a few Canadian
plans appeared in American trade journals. The California bungalow moved up the Pacific coast to Vancouver and eastward into
Chicago during the early 1900s.[15] Chicago served as the great international clearing centre for ideas in materials and housing design.
The mid-American metropolis had figured prominently in the introduction of the balloon frame in the 1830s and it grew into a vital
marketplace for lumber, millwork, hardware, and machine tools.

The career of Fred T. Hodgson (1838–1919) illustrates both the
Chicago influence and the international connections. A carpenter
from Collingwood, Ontario, Hodgson edited (1878 to 1919) the
American Builder, a Chicago-based journal. He wrote, compiled, edited, or revised over one hundred and forty books dealing with home
carpentry. Hodgson helped popularize the California bungalow and
occasionally contributed to the New York-based *Carpentry and Building* and to the *Canadian Architect and Builder*.[16] The vast continental
public reached by Hodgson and others like him had ample access
to the styles initiated in other regions. Just as importantly, the publication reflected or even catalysed the interest in the private home.
Nevertheless, a few cities developed distinctive shelter traits: New

York at the turn of the century had become a metropolis of tenements and apartment houses; Baltimore, Washington, and Philadelphia had considerable areas of terrace-style dwellings; and Montreal had a substantial number of rental duplex and row houses.[17] However, such urban structures were, as Robert Barrows argued, exceptional.[18] In any event the prevailing construction techniques were generally applied to row housing as well as to the single-family detached house.

During most of the nineteenth and twentieth centuries, residences have been constructed on site, according to the balloon-frame method, and through the auspices of some variation of the contracting system. Probably invented in the 1830s, the balloon frame came to dominate home-construction technology, replacing the old method employing cumbersome girts and posts. The technique expressed the more general and already evident American genius for cutting corners. Writing home from London, Upper Canada, in 1833, Scottish shoemaker John McAdam described wonderfully the slapdash approach to manufacturing as to "Yankee the job." The phrase summed up the assorted innovations that together formed the balloon frame. In dwellings with the balloon frame, light studs, joists, roof rafters, and purloins are joined by simple nailing. The mass production of slit nails in the 1830s and the use of even cheaper wire nails by the 1880s facilitated the technique. According to architectural historian Sigfried Giedion, the invention of the balloon frame "practically converted building in wood from a complicated craft, practiced by skilled labor, into an industry."[19] George E. Woodward, author of *Woodward's Country Homes*, calculated in 1865 that the new frame cost 40 per cent less than the mortice-and-tenon one.[20] Without such a labour-saving technique it is doubtful that decent housing could have arisen as rapidly and as cheaply as it did in the urban expansions of 1850 to 1910. This seems especially true of new Great-Lakes and western communities, which lacked the extensive and more diversified stock of housing found in the older and larger eastern-seaboard cities.[21]

Normally carpenters who erected the simplified frame also functioned as co-ordinators for other trades that completed a dwelling. Such carpenters have formed a subset of the construction business. Construction contractors specializing in railways, commercial or industrial structures, and public works usually possessed some plant facilities, equipment, and a small permanent staff that expanded during seasonal or cyclical upturns. Usually residential contractors had none of the above, although at the peak of building cycles "master builders" in most cities have expanded boldly into building supplies and land development; and a few have also launched large-scale enterprises erecting entire streets of dwellings.[22] Regardless of scale, the contractor is best understood in relation to a system of coordinating and remunerating tasks.

Contracting, and hence contractors, represented a relatively new system at mid-century, but by the 1880s the system was entrenched. If working for an investor, contractors made an estimate for the entire cost of a job and submitted bids, a process that became more precise as time passed. This arrangement generally intensified a search for labour-saving practices. It usually generated a sequence of bidding since contractors subcontracted much of their work. Contractors inspired materials suppliers to innovate in order to meet the contractors' and subcontractors' perpetual quest for savings so that they might tender successful bids or otherwise undercut their many rivals. Contracting also contributed to a destabilized labour situation by provoking the construction unions to action when new materials or unskilled labour were introduced.[23] Ruthless competition meant that the typically small contractors were quite vulnerable to labour union pressure during construction booms, but there was more to the story of contracting than rivalry and ceaseless innovation. Limits were recognized. A double bind of bidding competition and labour action was met by the formation of builders' exchanges, a movement that waxed and waned with the business cycle from 1890 to 1920. Probably too, family and ethnic ties functioned as informal checks against the system's competitive momentum, tempering a raw pursuit of the lowest bid with a preference for working within a familiar network.

Despite attempts to curb competition, residential contracting approximated pure free enterprise because of its essentially competitive nature and because it allowed easy entry to hopeful builders and evidently passed on savings as well as shoddy work to buyers. "Anybody," complained the *American Builder* in 1874, "however ignorant or however capable can run up houses for sale – it is no special business and the cheaper and more 'scamped' the work is, the better it pays."[24] In every city, a set of contractors possessing little capital was the rule. Contractor Michael Pigott, a Hamilton carpenter who launched a major Canadian construction firm, put his finger on both the remarkable openness of the residential-housing industry and the hazards confronting casual newcomers when he said that "the step from a journeyman to a contractor is the simple matter of getting a small job which a man may figure up after supper in the evening. He will probably be a contractor next morning, oftener through guessing than any real knowledge of the value of the work he has taken."[25] To minimize risks carpenter-contractors could refer to estimators' manuals. Trade journals featured columns discussing estimating and bookkeeping practices, and a number of companies advertised and sold formal systems to facilitate the estimating process. Stubbornly small scale, contracting remained for most, however, a business of informed guesswork. In most instances, contractors pursued innovation, but did so primarily within the limits of well-understood practices and materials. They could "figure up" estimates where the supplies and equipment were within a woodworking tradition. Locally based, under-capitalized, and hidebound in some regards, contractors were apparently unable to launch and sustain radical departures in home-construction methods. There would be no true paradigm shift in home construction after the balloon-frame revolution, only isolated radical experiments and a significant process of exceedingly gradual innovation.

In the late nineteenth century, contractors co-ordinated a half-dozen or so basic subcontractors: excavators, masons, tinsmiths, plasterers, glaziers, bricklayers, and painters. Contractors usually handled the carpentry, certainly the balloon frame. The range of hardware installed in houses increased from 1890 to 1920, adding to the number of subcontractors: public health authorities across the continent forced sewer connections, making plumbing a necessity by the 1890s; both hot water and hot air central heating had become more sophisticated in the 1880s, and in 1886 the *National Builder* forecast the widespread use of electric lighting; in 1906 electricians, contractors, and architects established standardized wiring symbols; and by 1922, builders had come to recommend at least one electrical outlet in each room. However, other changes in the make-up of subcontractors stemmed from costing considerations – time and labour.[26]

The nearly contemporary careers of three builders in our subject city are instructive. Based on quite different source materials, the profiles of Charles Emery, William Moore, and Thomas Casey provide glances into an extremely fluid business in the early twentieth century.[27] Charles Emery's father, Charles Sr, had emigrated from England to Canada when he was only 18. For a while Charles Sr worked a farm near Ingersoll, Ontario, on the basis of shares with its owner. Charles Jr was born in 1890 and in 1893 the family moved to Chicago, where a cemetery hired Charles Sr as a gardener. Subsequently, he worked for the Asbestos Fire Proofing Co., the Pullman Car Co., and as a salesman for Bullong's Pickles before settling into a job as a steamfitter for the Deering Harvester Co. In 1904 this firm transferred him to Hamilton, where "many US people were ...

working." Charles claimed to have read "all of Alger's books, Henty's books, as well as Cooper's. I really believed in them and I think that I did get great inspiration from them. They seemed to give me much ambition and desire to do things." Charles also took an interest in the flourishing state of Chicago construction. "I rushed home from school each day so I could help the carpenters who were building a row of houses on our street, so I learned a lot about building." Doubtless the mix of books that romanticized "pluck and luck" and of urban surroundings that exuded exertion helped make the future contractor.

As a teenager in the flourishing industrial city of Hamilton, Emery held an astonishing variety of jobs. He worked at a newspaper printing press, did odd chores for a railway, sweated as a furnace hand at a tool company, stoked on a freight boat, and laboured on assorted carpentry and plumbing undertakings. For a while industrial employers blacklisted him for his participation in a strike at International Harvester around 1907. He had also gone to western Canada to help with the wheat harvest around 1910. In 1910 when he was about twenty Emery acted as a contractor of sorts in his spare time. He built a house on a spare lot owned by his father and he finished off his in-laws' house. These undertakings, he recalled, gave him experience and "confidence to get into the building business." As he put it, in classic terms, "if I could make money for someone else I could make more for myself." In 1911 he took up lathing with Hank Potter, a southpaw from Kankakee, Illinois. They could work efficiently – Potter took the left cuts and Emery the right. Hamilton rode the crest of a building boom and soon Emery employed nineteen men. His wife kept the books. Business was slack across the border in Buffalo and he recalled "a lot of Buffalo men working in Hamilton. They would commute each weekend." However, the 1913 recession struck hard in Hamilton and the Emery family, both generations, returned to farming near Ingersoll, where he dabbled in the lathing business. War reinvigorated the industrial economy of Hamilton, and Emery returned in 1915 and worked at International Harvester. Fellow workers elected him to the industrial council or company union.

After the war, Emery worked as a construction foreman for his brother, who had put up capital for a contracting business, but the arrangement broke up in 1924. Charles left the city to work as a lather in Detroit and Buffalo during 1924. When construction fell off in Buffalo, the union would not allow him to work there. Returning to Hamilton, he went to work as a carpenter-contractor, building homes and apartments. The latter were a leading compo-

Table 5.4
Contractors Listed in Hamilton City Directories at Five-Year Intervals, 1866–1936

Year	Number[1]	Number of Contractors per 1,000 People
1866	25	1.25
1871 (1871–72)	31	1.15
1876 (1875–76)	34	1.04
1881 (1880–81)	20	0.55
1886 (1887–88)	16	0.38
1891 (1890–91)	9	0.18
1896 (1895–96)	6	0.12
1901	75	1.42
1906	59	0.90
1911	186	1.05
1916	134	1.28
1921	110	0.96
1926	273	2.12
1931	184	1.18
1936	138	0.90
All Years		0.93

[1]Contractors include the following cited headings: builders, contractors, contractors and builders, contractors (building), and contractors (carpenters). Subcontractors were excluded.

[2]The directories were not always clear about the process of selection for the business directory pages. In 1891 the directory stated it only listed businesses that subscribed to the directory. This selection device – actually a payment for inclusion – is not likely to have altered the trends dramatically. An active contractor in good times would want to announce his business. In poor times, it was common to withdraw into the wage labour force or pure carpentry. Consequently, there would be no wish to subscribe or advertise.

nent in an extraordinary building boom and the ranks of contractors swelled to an unprecedented scale (see table 5.4). As Emery himself volunteered: "Hamilton was a very competitive city to work in at that time so one had to figure very close." Actually, one apartment-building merchant builder who had given Charles a contract went into bankruptcy on the eve of the great depression. Emery left contracting for good during the 1930s and became a high-school shop instructor.

Emery's memoir, prepared for a family member long before our research began, spontaneously demonstrates propositions about contracting in the early twentieth century. First, it is significant that he made no mention of formal training or an apprenticeship. Tucked into an extensive and disorderly account of many jobs he noted working for carpenters and a plasterer. The message is one of great informality and simplicity in embarking on the contracting business.

Second, aside from the possiblity of entry without long formal tutelage, contracting – indeed the building trades generally – had another alluring feature. For a few factory labourers, the only way to beat the industrial wage labour system was to get out of it. Emery's checkered employment record, his reflection about having been ambitious, and his participation in a labour dispute and in labour organizations hint at a desire to secure something better than wage labour. Third, the contractor required little capital. Emery got out of wage labour with very little backing or savings. Fourth, family networks provided a little work, a little capital, partnerships, and headaches. He remained testy about who among his brothers did what to whom and who had the best eye for building. In a modestly capitalized business, kinship ties often provided the corporate glue as well as the solvent for dissolution. Fifth, all of the above meant that in very good times contracting became extremely competitive. The unconsolidated structure of the industry in Hamilton from 1866 to 1936 produced a nearly standard ratio of one contractor for every 1,000 inhabitants, except that in depressions it fell remarkably and during boom periods it rose by 50 to 100 per cent (see table 5.4). Finally, for Hamilton at least, the labour market occasionally proved inviting to Americans. Hamilton carpenters, meanwhile, were not shy about working in Detroit and Buffalo. This exchange provides small additional confirmation of the international character of material culture and housing.

Emery's memoir lacks a rendering of accounts. How profitable could contracting be? How were home-construction jobs financed? How were deals struck and business connections executed? Quite impermanent and, in an era before income tax, sales tax, and government incentive to the shelter industry, having little need of records, contractors left faint tracks. One detailed rendering of some of their operations, including costs and fees, can be provided. From 1907 to 1909 a cautious family firm of Hamilton property managers and mortgage brokers sized up the city's great growth and decided to take a plunge as a merchant builder. The merchant builders should not be confused with the building contractors. The latter, men like Emery, tried to make a living by selling simple skills and labour; they worked on site and also managed the team of subcontractors who finished the house. The former tried to profit from bringing land, capital, and contractors together. On occasion, the merchant builder might manage subcontracting.

William Ghent Moore, son of the firm's founder, William Pitt Moore, was in an excellent position to understand the interplay of the business cycle, demography, and neighbourhood quality because the family business since 1858 had dealt with rent collection, insurance, and mortgages. The firm's excellent records are used in our discussion of mortgages in chapter 6. The family had avoided risky real-estate purchases, but in 1907 the younger Moore realized that a boom was underway and that the south-east side of the city, though relatively close to the exploding industrial district, was far enough away from the noise, fumes, and unsightly factories in the north-east to have some appeal to potential home buyers. Thus in 1907 and 1909 he became a merchant builder in the south-east, contracting out for the construction of fourteen single-family homes. His contractor, William Wray, committed no funds, not even for the short-term purchase of materials. Wray arranged for the subcontracting: excavation (by manual labour), masonry work, carpentry, tinsmithing, plastering, painting, and glazing. When he had approved of the work, he forwarded the subcontractor's request for payment to Moore. Contractor Wray did not have to display great skill in carpentry. Doors, sashes, mouldings, baseboards, newel posts, banisters, and other wood components were being turned out in city mills. Designs came from plans in an uncited pattern book. Horse-drawn drays hauled the supplies to the construction sites. A punch clock at the building supply dealer's sheds recorded the departure times on the delivery tickets. When the materials arrived, the contractor signed the ticket, which then became the basis for billings mailed to Moore by the dealer.

Moore, a businessman who kept accounts, retained copies of supply tickets and itemized other expenses. Since we discuss the critical importance of lumber later, it should be noted here that, excluding the total wage bill for all trades, lumber was the largest item, 21 to 27 per cent of all costs, compared to 23 to 29 per cent for labour. The contractor's fees varied in proportion to the scale of the dwelling. For supervising the construction and finishing of eleven houses in 1908 and 1909, Wray received $1,000. The houses went up in under two months and quite likely Wray contracted for other jobs during the roughly six-month building season. At that time, carpenters and joiners in the Hamilton area received an hourly wage of 40¢ to 50¢ and worked eight hour days for six days a week. In other words, for the building season, a carpenter might have earned $500 to $600. The incentive to have broken away into contracting is obvious. The hitch came in securing lines of credit.

Thomas Casey, interviewed in 1976 about his career as a contractor in the 1920s, explained how a contractor "building on spec" got access to land and capital. When Wray worked for Moore, he accepted a flat rate; there was no risk and no possibility of reaping any windfall from a rising market. "Building on spec" was a different operation, requiring the contractor to scramble in order to build up the greatest amount of leverage with a trifling amount of capital. The apparently relaxed credit arrangements that opened in the house and the apartment construction sectors in the 1920s may have been the important factor in expanding the number of Hamilton contractors well beyond historical levels (see table 5.4). It is certain that the suddenly overbuilt housing market contributed greatly to both the local misery of the Great Depression and to the arguments against social housing. Casey's activities chart the adjustments to unprecedented boom and drastic collapse commonly made by contractors.

During the frantic 1920s there was a great assortment of contractors in Hamilton. Small ones like Emery scraped along while others, such as the firm of Armstrong Brothers, seemed to have an integrated operation that included a real-estate agent and an architectural department. In 1930 the Armstrongs claimed to have built some 600 dwellings during their history. Casey's practices suggest, in contrast, techniques employed by contractors with few assets save work and ambition. Raised in County Cork, Ireland, Casey had laboured in Liverpool before emigrating to Canada in 1914. Between 1914 and 1920 he laboured on construction in Toronto and Hamilton, and for a while, he was a hand in Pittsburgh steel mills. In 1919 he had built himself a worker's cottage in Hamilton. He continued to erect very modest frame houses in the early 1920s when the soaring costs of building materials restricted the scale of workers' dwellings. By the prosperous mid and late 1920s, Casey was building more fashionable brick homes for the middle class. During these years Casey selected a parcel of adjacent lots in a survey and signed an agreement to purchase with the subdivider. He could not afford clear title. Next he installed a number of basements simultaneously and started to erect the frame for one dwelling. Other work he subcontracted to what he called "the seven trades": masons, lathers, plasterers, electricians, plumbers, roofers, and tinsmiths. By staggering construction along a row of basements, Casey provided a sequence of work for these subcontractors. More importantly, he did not exceed his credit limits and the stages in home construction were paced by his line of credit.

The half-dozen major office buildings in downtown Hamilton housed a large and yet almost invisible capital market. Banks and trust companies provided building loans and at least thirty non-institutional mortgage brokers advertised in real-estate columns during the 1920s. In some instances, barristers functioned as brokers, handling estate funds or working for clients with capital. Casey pre-

ferred borrowing through the barristers, because institutional lenders demanded quarterly instalment payments that included principal
as well as interest while private lenders accepted payments of interest
and a lump sum for principal when a house was sold. For the contractor, loans came in "three draws". When the roof went on, the
lender's agent drove out to inspect the dwelling. If he approved,
Casey could make his initial draught. When the white finishing plaster dried, he was entitled to a second, and when the interior trim
and fixtures were in place, he could make a final draught. Each
draught helped to finance subcontracting work on adjacent houses.
Optimism and loans carried Casey into his busiest season, 1930, when
he employed eight carpenters and had at least fifteen homes under
construction. From 1933 to 1939 he built one house a year. During
the early 1930s he had to let several of his newly completed homes
to tenants and he even exchanged one middle-class house for a cheaper one in the industrial area, believing it would sell more readily.

From the 1930s to the present the fragmentation of the housing
industry and the centrality of men such as Emery, Moore, and Casey
have bothered analysts charged with recommending improvements
for shelter to task forces and governments. According to advocates
of modernization, house building has been frozen in a handicraft
stage of industrial development and contracting has lacked strong
management dedicated to technical advancement. These characterizations, as we shall demonstrate, have ignored noteworthy changes
and have downplayed the role of consumer interest in traces of
craftsmanship and in traditional materials or familiar forms. The
complaints have also lacked an appreciation for wider economic and
social considerations. The failure of the home-construction industry
to consolidate swiftly into large corporate entities with streamlined
assembly procedures resembling factory production was not an act
of perversity from a backward economic sector. The uncertainties
and sharp collapses in the shelter-building industry may well have
discouraged heavy capitalization among most contractors. Unlike
new consumer-durables manufacturers whose products (household
appliances and automobiles) had no long-established market histories, the housing and the real-property markets had histories of risk
and also of sudden, almost infectious, demand. The relatively small-
scale and open contracting system reflected the fickleness of capital
and demand in a peculiar area of productive activity. Nevertheless,
the contracting system cannot escape from some criticism, for the
image of competition which benefits the consumer is offset by a
recognition of efforts to control the friction of competition. Before
considering the attempted checks against competition, let us consider

more completely the course of innovations in the critical period for the expansion of home ownership.

LABOUR-SAVING INNOVATIONS

The dynamics of the relationships between merchant builders, contractors, subcontractors, and unions produced too many permutations for simple summary, especially when the variable of the construction cycle is introduced. Contractors bid close to cost. Lacking capital, the numerous small contractors relied upon credit from lenders or materials suppliers. Furthermore, in Canada and the North-east and North-central United States, they were trapped by seasonal conditions. With a small profit margin, supplier credit, and climatic constraints, contractors pushed for speedy completions. "Throughputs" of 40 to 60 days were usual at the turn of the century. Because timing was absolutely critical to survival, the construction unions could count on pressing residential builders more effectively than large-scale contractors. Unions counted on small builders to be the first to crack under the threat of a strike or during strikes. Understandably, contractors warmed to new practices and materials that they could afford and understand and that supplanted or challenged organized labour. By 1920 the few more heavily capitalized contractors or those integrated with materials suppliers could operate with half of the labour force of small contractors who functioned without benefit of mortar mixers, concrete mixers, metal concrete forms, metal scaffolds, floor surfacers, and woodworking machinery.

Efforts to cut labour's leverage extended to all manner of innovations beneath the house, in the erection of the exterior, and in the finishing of the interior. Two drastic efficiency approaches were stillborn, their failure demonstrating a guiding circumspection shared by labour, consumers, and poorly capitalized builders. By World War I, contractors could build a house in one day, to the chagrin of the building trades unions whose leaders perceived threats to jobs and diminished quality in housing as possible consequences if these holiday or fund-raising stunts coalesced into routine practice. Piecemeal changes were difficult for labour to resist; a full restructuring was not. The "house-in-a-day" experiments went forward because the trade unions accepted a once-in-a-lifetime display of supreme exertion. And there the matter nearly died, suggesting what was feasible but impossible. During World War I, however, speed-up practices were introduced on some large-scale war-industry housing projects.[28]

Concrete rode a brief wave of popularity as the miracle material of the 1890s and early 1900s. The short-lived (1907–09) Minneapolis-based *Journal of Modern Construction* placed most of its residential emphasis on the use of concrete materials of various kinds. Concrete blocks challenged stone masons, for until the 1890s North-American home foundations consisted of rubble stone. The labour-intensive construction of rubble foundations provided costly basements. For a short time, when the cost of lumber escalated and when bricklayers successfully pushed up the hourly rate, concrete blocks were conceived of being an eventual replacement for traditional materials. In 1905 the *Canadian Architect and Builder* reported over two hundred requests for plans that incorporated concrete blocks for walls as well as for foundations.[29]

Experiments with poured concrete were more radical, for it was not a domain of craft unions. In 1904 the *Canadian Architect and Builder* predicted that the house of the future would be poured into forms and the average buyer would select a house from standard designs. Thomas Edison repeated the forecast in 1911 claiming "the immediate advent of the concrete house poured complete in one operation, with all wiring, plumbing, casing and fixtures installed."[30] In fact, a patented system of reusable pressed-steel forms had been employed during 1913 in the Chicago suburb of High Lake, at Virginia Highlands near Washington, DC, and at West Nanticoke, Pennsylvania.[31] In 1915 "a genius builder," the leading contractor of Berlin, Ontario, realized Edison's prophecy. Caspar Braun applied his factory construction techniques to a scheme for mass-producing houses. A mixer platform and distribution apparatus moved down a street filling house moulds on either side. Significantly, Braun, the factory builder, was not typical of house builders. His scheme developed outside of carpentry and required large capital outlays.[32] Consumer resistance to the alien appearance thwarted full-blown success for homes of concrete blocks or poured concrete. This unheralded episode in North American material culture demonstrates the tension in the housing industry during boom periods. Innovation altered a few ingredients, but the conservative influence of both small carpenter contractors and public taste sustained the momentum of established materials and building practices.

Instead of overwhelming tradition by sweeping frontal assaults, improved efficiency came in ceaseless small steps. Most happened in the materials supply sector where, because of scale and rationalization, there could be increased capitalization. The investment risks and discovery of new bottlenecks after innovation produced the piecemeal pace of change. Brickmaking is a case in point. Brickmaking machines had been tried in the United States as early as the 1790s. At first, the crude shaping implements were hand operated, but machines using horse power seem to have come into limited use in the 1830s. By 1840 a steam-powered machine had been introduced into Philadelphia, only to be destroyed and thrown into the Schuylkill River, "such was the prejudice and antipathy of the old-time brick-yard employees to any methods or improvements that appeared an innovation to their old-time way."[33] Retrospective accounts agreed, however, that innovations truly spread during the 1870s and 1880s. It was safe to state in 1889 that hand-made bricks "no longer maintain the ascendency over machinery."[34] In that year, some machines could form 50,000 bricks in a day, but cost confined their use to "large capitalists."[35] Investment in one of the several technological routes available for replacing handmade bricks promised labour savings. A strike of brick-manufacturing labourers at Chicago in 1882–3 was credited with accelerating the introduction of brick forming machinery.[36] Machinery remedied one bottleneck, but it engendered another. Kilns interrupted the production flow, but to maintain supply flows by building up inventory entailed risk. Investing in more kilns alleviated the constraint, possibly too late and to the detriment of the short-term profits that were sought by small local brickyards. As a result, contractors using brick confronted crippling delays during the transitions from slump to boom that accompanied the building cycle.

For contractors, the identical sizes of machine-made bricks offered the advantage of demanding "less skilled labour in building and no extra time in preparation."[37] Handmade bricks had to be broken or trimmed to fit. The savings made possible by uniform bricks were universal. In the mid-1870s English experiments with small concrete blocks (1" x 2" x 9") as a substitute for bricks achieved a "saving of over 200 per cent in labor."[38] Eventually, the leaps in capitalization embraced the entire process of brick manufacture. To guarantee swift throughputs after the shaping of the raw brick, special drying sheds and efficient kilns fueled by natural and coal gas were constructed. In 1897 W.A. Eudaly's Cincinnati-based firm designed motion-efficient plants with improved kilns, constructed the facilities, and maintained experts to run the operation while training the owner's crews.[39] Greater capacity and lower labour costs required substantial capitalization, so brick costs fell only slightly.

In North American bricklaying everything was done to reduce labour and to expedite the work. The construction of load-bearing walls was symptomatic of American haste. English bricklayers still carefully bonded bricks while Americans dumped mortar and scrap between two curtain walls. In the 1890s there were English experiments with lock-joint bricks that required no binding course or wall ties, thereby offering a labour saving of an estimated 20 per cent.[40] Yet American bricklaying remained the more slapdash business. Contractors paid premiums to a pace-setting hod carrier. If others dropped out of line they were discharged. Frederick W. Taylor's disciple in the field of motion study, Frank B. Gilbreth, undertook an investigation of bricklaying in 1909 and devised an adjustable scaffold for piling up bricks that allegedly tripled worker output. As has often been remarked, a scarcity of skilled labour on this side of the Atlantic encouraged invention and a substitution of impatient assembly for manual art. Such practices apparently did not sit well with the public, so in 1925 the Common Brick Manufacturers' Association of America founded a Cleveland-based monthly, *Building Economy*, to challenge the critics of brick homes. The positions taken in this journal reflected many of the tensions evident within the traditional approaches to home construction. In particular, the attitudes of the lumber manufactures toward fire-limits laws in cities were ridiculed and the relative cheapness of the brick home was touted. Toward the end of the decade the brick manufacturers, along with other building suppliers, newspapers, and realtors banded together under the Home Owners Institute, Inc. of New York to produce Master Model Homes that were certified as meeting specified standards of quality in both finish and materials.[41]

Inside houses, traditional lathing and plastering practices were challenged by innovations. For a high quality finish, subcontractors put on three coats of plaster: scratch, brown, and putty. Scratch was the base that held the other coats onto wood laths. Wire and metal mesh began to replace the latter in the 1890s and two coats were judged to be adequate. A special union of metal lath workers was formed in 1901. Pressed-metal centre pieces for ceilings replaced ornamental plaster work. "Compo-Board a Substitute for Lath and Plaster" made its debut before 1910, backed by advertising calculated to appeal to cost-trimming contractors. Sheets in several standard sizes could "be put on by any carpenter." Beaver Lumber, an early chain of international building-materials suppliers, introduced its highly successful Beaverboard around 1910 and maintained a department of design and decoration to promote the new material.[42] In the 1890s Augustine Sackett was perfecting Gypsum Drywall, and in 1909 the US Gypsum Company acquired Sackett's plants. The firm's engineers commenced a continuous research effort to streamline production and to facilitate the use of drywall by builders. It has been estimated that contractors could shave off two weeks from the time needed to complete houses by avoiding lathing and plastering. Thus an earlier cash flow, added to the labour savings, caused a widespread switch to the drywall system by the 1940s.[43]

Of all materials, wood reigned supreme in North America. When English stonemason James B. Turnbull arrived in Toronto in 1872 he found no work in his craft. It was early in the building season, "and besides, there was very little of it [stonework] used; all the houses in course of erection were being built of wood."[44] Wood materials remained the largest single cost factor in house construction – about 50 per cent in 1880 and 40 to 45 per cent in 1920. The abundance of the North American forests provided a cheap stock of wood for urban shelter. Pliable and durable, wood was a mainstay for the balloon frame, doors, sashes, trim, and interior finishing. True, fire-zone by-laws in the late nineteenth century often forbade the use of wood on the exteriors of central-city dwellings, but in the suburbs frame houses were abundant. Lumber was harvested in the same wasteful and labour intensive way decade after decade, but the characteristic absence of technological improvements prevailing in the forest harvest of wood upstream contrasted with technological refinement as the product moved downstream. Everything about the conversion of wood into either rough frame pieces or sophisticated embellishments changed constantly in the second half of the century. All of the previously discussed innovations with materials pale in comparison to the cascade of woodworking advances. Like the spread of the balloon frame, they amounted to a revolution.

The North American machine-tool industry, intrinsic to the Yankee reputation for ingenuity, revolutionized the home-construction industry with an extensive assembly of shaping tools powered by steam and later by electricity. With a will to master details and to recommend improvements, trade journals revelled in outlining the merits of new pieces of machinery from precision adjustment screws to cutting blades. The impressive technological onslaught on the skills of traditional carpentry and joining began in earnest during the 1860s in urban lumber yards, and at the Centennial Exposition in Philadelphia in 1876 many complex machines were proudly displayed. Certainly by the early 1870s carpenters and joiners had experienced both the joys and the challenges associated with using machined wood pieces. Optimists who admired machined perfection in moulded and joined pieces welcomed the range of exquisitely fashioned wood forms. The better world of producers, artisans proud of their design talents, was to be powered into realization by machinery that would provide splendid materials for their capable hands. In a wishful display of enthusiasm, the *American Builder* fore-

Table 5.5
Hand and Machine Labour Costs in the Woodworking Industry – 1858
and 1896 Compared

	Labour Cost	
Item	By Hand	By Machine
1,000 ft. 41/2″ yellow pine ceiling	$14.00 (1859)	$0.44 (1896)
1,000 ft. 3″ oak flooring	$21.00 (1858)	$0.54 (1896)
1,000 ft. 41/2″ yellow pine flooring	$12.00 (1859)	$0.36 (1896)
50 pair yellow pine sashes	$60.00 (1858)	$9.00 (1896)
10 sets stair risers and treads	$26.00 (1858)	$2.00 (1896)

Source: Robert A. Christie, *Empire in Wood: A History of the Carpenters' Union* (Ithaca: Cornell University Press, 1956), 26

cast the demise of capitalism. "The reign of mere capital, if we are not much mistaken has passed the zenith of its power."[45] The innocent and wonderful psychology of hoping to realize abundance through honest labour and the shrewd inventive nature of the mechanic subsided as evidence of threats to old trades accumulated. What had begun with M. J. Brunel's invention of a circular saw for use by the British Admiralty Board in Portsmouth in 1804 had climaxed in the 1880s with the likes of the aptly named "universal wood worker" that could plane, bore, rout, mould, match, and mortise.[46] The multiple functions describe many of the manual skills formerly required on a construction site and increasingly relegated to the lumber mill. The machine tool came as the first and most potent breakthrough in a highly competitive industry's search for labour-saving practices (see table 5.5).

Machines not only performed numerous manual tasks but did so with greater accuracy and lower costs. Planing mills and mortising machines in the 1870s simply produced "better doors than hand made."[47] In the estimation of the *American Builder* in 1874, "the past few years have been wonderfully prolific in producing labour-saving machinery. Especially in the advancement noticeable in the construction of wood-working tools and machinery, by use of which the cost of production is materially lessened, the quality of manufacturers greatly improved, and the manual labour of the work man largely diminished."[48] In 1887 the same journal summarized the effect on construction sites with an arresting anecdote about an aging pauper who no longer could find shavings for his stove when he searched for them along streets with construction. "We don't have any shavings in the houses now; they are all made at the mill and you will have to go there for them. I don't believe that the carpenters now a-days make more than a barrel of shavings in building a house. Modern houses are put up pretty much as Solomon's temple was, the parts are brought together all prepared and fitted, and it is short and easy work to put them together."[49] The article compared the new construction process to the contemporary introduction of ready-made clothing. Many decades later, another analogy came to mind. During a 1939 Michigan conference on housing, one delegate proposed that "in a certain bracket in the lower cost field, we could use the idea of a basic plan like an automobile chassis."[50] Home construction has often been seen to be changing from a process of on-site artisan fabrication to one of assembly, following the lead of some new industrial sector. At the other extreme, terms that were inappropriate to the modern building site lingered, for even as old practices vanished carpenters were written of fraternally as "brother chips." The saying "a workman is known by his chips" ceased to apply and a small box of tools replaced his chest or two of equipment. An old-time carpenter laid out twice to five times as much as his less-skilled successor.[51]

Machine tools certainly pushed the general practices of the industry onto a new plateau of activity. The sheer scale of trade in wooden articles of machine manufacture confirms that assertion and, coincidently, the significance of the housing industry. During the urban construction boom before World War I, Sears, Roebuck had expanded its mail order business to embrace house components. In 1911 the firm's sash department had 100,000 in stock and produced 5,000 per day; there were 75,000 doors on hand and 1,500 were manufactured per day. The company had a supply of 1,250 carloads of lumber.[52] From 1908 to 1940 Sears had shipped by rail about 100,000 houses (in 450 styles) within the United States. Despite such a remarkable rationalization of production and distribution, the benefits of a decentralized corps of building trades remained sufficient in conventional housing markets to preclude a truly drastic reorganization of residential construction. Apparently, few contractors or developers ordered the Sears units. The largest single sale was of 100 houses to a Massapequa, New York, developer in 1931.[53] Carpenter-contractors were fortunate in that their knowledge of local laws, civic politics, attributes of specific building sites, markets for dwellings, and credit arrangements helped to protect them from the widespread marketing of prefabricated dwellings.

Wholly prefabricated dwellings of English manufacture had been tested in the 1820s and 1830s in the arid regions of India and Australia. Very simple assembly skills were required so that the lack of skilled labour in the colonies would not preclude home-country shelter forms. Their component sections were hand crafted. By the 1840s the English prefabricated-housing industry had switched to iron beams and corrugated iron sheets.[54] In North America, prefabricated frame structures – incorporating the balloon frame – appeared by the 1850s, possibly earlier. The *New York Tribune* of 18 January 1855 mentioned that "the western prairies are dotted with houses which have been shipped here all made, and the various pieces numbered."[55] The small scale and rudimentary form resembled the initial English models. In 1861 Bostonians D. N. Skillings and D. B. Flint advertised a patented construction system requiring just two or three men "without mechanical knowledge or experience in building."[56] They could erect a shell in less than three hours. Suppliers aimed at sales to railway construction firms, to the military, to plantations, and to regions where lumber was scarce. Canadian manufacturers earned an envied reputation for prefabricated houses in the 1870s and 1880s. A Quebec firm in 1883 built them for the Canadian Pacific Railway and allegedly exported some to Panama to house crews engaged to survey canal routes.[57] In the same year, the Truaxes' Planing Mills at Walkerton, Ontario, had shipped entire streets of dwellings to Brandon and Winnipeg, Manitoba. In one season, this small-town mill had sent 219 car loads of "knock down house material" to the Canadian west.[58] Early in the twentieth century, British Columbia firms shipped homes to prairie cities such as Calgary, and during the same years of urban boom, the Harris Brothers of Chicago produced a line of five- to eight-room houses.[59]

One Canadian prefabricated housing firm proved extremely tenacious. Coincidentally, it settled in Hamilton. Halliday Homes originated around 1910 to serve the surge in demand for homes in the rapidly expanding cities on the prairies. Halliday drew upon eastern Canadian forest resources and concentrated its milling, manufacturing, and mail order facilities in Hamilton. The fortunes of the company indicate the character of the market that it would secure from time to time. Halliday flourished in periods of very strong housing demand when escalating supply costs and delays from conventional suppliers offset some of the bias against prefabricated homes. It expanded during the 1920s – a 1925 catalogue illustrated over 70 plans, ranging from three to eight rooms. Many plans bore the name of a Hamilton neighbourhood and one of the largest houses was the "Hamilton." During the 1930s Halliday struggled as a builders' supply outlet. Only in the post-war boom did it return to its former specialties, issuing annual catalogues of homes, and adding garages and summer cottages.[60]

Only on the treeless frontiers of the Empire or America, places that lacked the infrastructure of trades and suppliers, or during extraordinary periods of urban growth, could flimsy shells become a feature of the housing stock. The ideal of a rationalized housing industry capable of establishing prefabricated homes as a substantial share of new dwellings would endure to the present in the visions of a few housing reformers and entrepreneurs. Traditional builders, however, have usually mocked prefabrication.[61] The appearance of short-lived producers of shells could not lure the industry away from a traditional reliance on on-site skills. The pattern books of contractors presented more style choices than the catalogues of prefabricated housing mills. Even on the new settlement frontiers, a desire for style, requirements of durability, development of local materials, and formation of pools of skilled trades outweighed the initial necessity of importing prefabricated units. Contrary to analogies, dwellings were not quite like ready-made garments. In spite of improvements in quality and the use of sophisticated design technology, only 6 to 7 per cent of the dwellings put up in Canada in the late 1980s were prefabricated or, to use the industry's preferred terminology, "manufactured" or "pre-engineered." In conformity with the historical pattern, one of the most recent radical innovations in construction occurred just as soaring prices peaked. It will be interesting to see whether hinged "stressed skin" panels pioneered by Dutch engineers and Canadian manufacturers will survive a recession in the early 1990s or will have to be reborn in a later boom.[62]

Technological innovation checked by both tradition and the limits inherent in the housing durable itself effected a compromise mediated by carpenter-contractors. Mass production of materials was fused to local assembly and perhaps some use of local materials. The writings of Hodgson, "the most prominent authority on building construction in the country,"[63] exemplified the synthesis of old and new by his mingling of a revolutionary mass-market approach to carpentry with an apparently sincere yearning for the old ways. In the 1870s and 1880s he and others despaired of a new generation of apprentice carpenters and joiners. Men who knew nothing of the adze – as old as the trade itself! Men who allegedly squandered money on fancy canes and big cigars! These apprentices took no pride in collecting fine tools – "chisels handleless, their saws worse than briar roots."[64] Hodgson romanticized the apprentice system and the artisan's supposedly inherent passion for working geometric puzzles, building up fine tool chests, and reading an interminable stream of epistles on saw sharpening, stair construction, and so forth. Others shared his perception. *Carpentry and Building* in 1880 professed that "the model foreman" had to "make up his mind to be virtually a student for life."[65] Others proposed that well-educated foremen should arrange each week for several hours of lectures for

their workmen. But would mechanical invention encourage or dis-
courage these lofty aspirations?

Whether one was an optimist, a pessimist, or a realist who spotted
a dilemma, the role of the machine was a much-debated issue in the
1870s and 1880s. *The American Builder*, serving both machine-tool
advertisers and craft readers, had split vision. It reported breath-
lessly about inventions only to draw back to wonder, as it did in 1874,
if there was a net advance. The machine "is simply marvellous. But
it can also, – and here it is that the artist has occasion, now a-days,
to tremble a little – not only make bricks, and saw and plane timber,
but it moulds them into 'gracious forms' ... But we may ask, is this
all gain?"[66]

The on-site assembly of machined pieces opened up construction
to unskilled workers. By the late nineteenth century, nativism occasionally crept into fears about unskilled "hammer and saw men"
or "green hands" taking the places of native-born artisans.[67] Ironically, given his craft idealism, Hodgson's books assisted in weakening
apprenticeship skills by making carpentry a more accessible vocation
and avocation. In 1907 he alleged that *The Steel Square and Its Uses*
and *Practical Carpentry* alone had sold "a million or more copies since
they were first issued" thirty years earlier.[68] Distribution of many of
his titles by Sears, Roebuck further intimates the remarkable mass
market associated with wood construction. The timing of the publication of Hodgson's numerous how-to-do-it books – 25 prior to
1901, 102 from 1901 to 1919, and 13 posthumously – suggests the
important part played by the printed word in educating the expanding labour pool. New carpentry journals and the increasing size
and circulation of the established ones support the same inference
(see table 5.6). The accessibility of manuals, magazines, and ready-
made components helps to explain how it was that "Building a home
was a neighborhood business, a community affair, since workers
could not rely on the formal homebuilding industry to build houses
for them."[69] Seeking improvement through access to information
and mass-produced materials had a mighty influence on urban
North Americans.

Machined products moved jobs away from the construction site
and concentrated them in wood-finishing mills. During the 1870s
and 1880s building materials producers assembled the array of new
machines under one roof and in 1879 the *American Builder* presented
a series on how to evaluate, tender for, and care for a complete
mill.[70] An 1882 article in the same journal illustrated a model layout
of equipment that allowed for the efficient progression of pieces
from one stand to another. At one end, a boiler and engine room

Table 5.6
The Circulation of Carpentry and Housing Journals in Selected Years,
1882–1925[1]

Year	National Builder (Chicago)	Carpentry and Building (New York)	Keith's Magazine (Minneapolis)	Bungalow (Seattle)	Small Home (Minneapolis)	Canadian Architect and Builder (Toronto)
1882	–	3,500	–	–	–	
1887	–	decline	–	–	–	
1890	2,000	20,000	–	–	–	1,500
1893	8,000	20,000	–	–	–	1,500
1897	6,500	not listed	–	–	–	1,550
1901	7,500	not listed	7,500	–	–	1,500
1904	9,200	13,500	8,000	–	–	1,750
1907	12,000	13,500	9,000	–	–	2,800
1912	29,000	13,783	16,000	–	–	2,800
1915	24,000	19,486	11,000	41,856	–	5,700
1920	40,955	10,567	12,400	–	–	–
1922	37,994	10,567	12,400	–	–	–
1925	60,676	66,065	12,600	–	5,000	–

Sources: For 1882 and 1887, Edwin Aldin and Brothers, *American Newspaper Catalogue* (Cincinnati);
for other years, Lord and Thomas, *Pocket Directory of the American Press* (Chicago)
[1] The titles of the journals changed slightly and there were mergers.

fuelled by scrap and shavings powered a central shaft that ran the
length of the work area; belts from the power shaft ran a planer,
three circular saws, a mortising mill, two matchers, and a surfacer.[71]
Not only had a technological leap been made, but so had a rudimentary study of motion and space that foreshadowed Taylorism.

Carpenters were not passive witnesses to the pressures of the contracting system or, later, to the onslaught of the machine-tool revolution. Altogether carpenters made three attempts to form a
national union in the United States: in 1835, 1853, and 1865.[72] The
depression of 1873–77 ended, for a time, these efforts, which had
included openings to Canadian locals.[73] In early campaigns, the
unions worked to protect craftsmen from employers who tried to
force long hours – 12 to 15 hours per day in summer months.[74]
The terms of the struggle for control of the workplace changed as
the balloon frame and wood-working machinery deprived outside
carpenters and joiners of some of their work.[75] Having lost some
control over the trade at the point of construction, the carpenters'
and joiners' craft unions endeavoured to organize wood-working
shops, beginning in 1881 when the United Brotherhood of Carpenters and Joiners organized some of the men who operated mill ma-

Table 5.7
Price Indices of Lumber, All Building Materials, and an Index of Wage Rates for
Ten-Year Intervals, 1850–1920 (1850 = 100)

Year	Annual Average Price of Lumber in the United States	Annual Average Price of all Building Materials	Index of Wages
1850	100	100	100
1860	107	98	109
1870	222	145	195
1880	197	128	154
1890	203	121	172
1900	241	119	186
1910	347	143	239
1920	1156	387	575

Source: Joseph Zaremba, Economics of the American Lumber Industry (New York: Robert Speller and Sons, 1963), 72–4

chinery. After 1890 other machine operators were signed up by the Wood Workers' International Union, fostering a jurisdictional clash that continued until 1912.[76] However, it was not just labour that attempted to control the disruptive impact of technological innovatons fanned by the contracting system. The agencies of capital also began to rein in competition and thereby affected the cost and form of urban shelter. By the early twentieth century, all parties involved in residential construction struggled to restrain the genies of free enterprise set loose during the previous half century.

DISTRIBUTIONAL COALITIONS

Housing costs held fast or even declined between 1870 and 1900. Mortgage interest rates, steady or even slipping marginally lower for thirty years, had begun to edge up again around 1905.[77] A post-1900 rise in building costs coincided with a major construction boom spurred by immigration and a second industrial revolution of electrically powered factories. Therefore, mortgage interest and demand factors must be acknowledged in any account of the end to deflation. Wage settlements do not appear to have been a potent stimulus to inflation (see table 5.7). There were, however, drastic supply-side transformations that can be summarized partly under the heading of an end to competition. Lumber prices were particularly sensitive to combined demand and supply changes.

Demand for lumber is inelastic.[78] Rises in the price of lumber are not ordinarily accompanied by a drastic reduction in the quantity demanded. New materials had proved unsuitable as substitutes for lumber – consumer taste and familiarity, weight, ease of working, and strength fixed it as a preferred material. To some degree its use in dwellings was reduced by eliminating porches, bay windows, mouldings, and other embellishments. However, the impetus to seek out and test substitutes tended to occur in booms, when bottlenecks and rising prices had provoked contractors. Even so, the length of time required to develop, test, and market a new product tended to delay introduction until the building cycle started a downward correction. A prolonged slump could destroy the infrastructure associated with innovation. The limits to lumber's inelasticity have been tested time and again so that gradual elimination has occurred, but between 1900 and 1920 substitution had commenced in only a minor way. Consequently, lumber led the way in an escalation of housing-construction costs.

Three price factors came to bear on the increases in the cost of lumber; economists cannot disentangle their relative influence. Suffice it to say that two factors – diminished supplies in mid-America and historically low productivity in lumber harvesting – assisted the third factor, the formation of producer cartels. Real supply bottlenecks strengthened the rationales for price increases that were coordinated by trade associations. In the sequence of their ever more distant timber rights, two Hamilton-based lumber dealers charted the forests' retreat. The Michael Brennan lumber firm operated a mill that supplied local builders through its linkages to lumber camps. In 1903 it controlled pine timber limits of 100,000 acres, and the 400 men in its mills turned out 200,000 feet of lumber a day. The business first bought timber lands in Simcoe County in 1880 at a site only 100 miles from Hamilton. Soon it had moved into neighbouring Muskoka. Eventually, it had timber rights on the Spanish River and in the Algoma District. Essentially, the second firm, the Flatt brothers of Hamilton, exported to England. The family secured Muskoka timber rights in the 1880s, but in the 1890s had camps in Michigan and Wisconsin. During a peak year, their dozen or so camps cut 400 miles of road and cleared off 20,000 acres. If other cities and towns of the heartland had such active timber harvesters, it is easy to understand why the readily accessible forest reserves in the Mississippi Valley and near the rivers of the Great Lakes had all but vanished by around 1900.[79]

Large operators like the Weyerhaueser company expanded into the south and onto the Pacific coast when pine in mid-America became scarce and when builders resisted its replacement by other woods. Taste and experience inhibited substitution. Greater distance meant higher transportation costs and not until the 1920s, when logging trucks were introduced into the forests, were there countervailing gains in productivity. In fact, the lumber industry has been one of the least successful in raising productivity. One economist has estimated that the most optimistic rate of annual increase would be 1.1 per cent per man-hour from 1899 to 1954.[80] In the midst of a construction boom in 1910, lumber manufacturers began a reform that confirmed their awareness of the productivity problem. Instead of producing lumber only in lengths of even feet, which generated a 2 per cent waste, henceforth the industry also produced odd lengths.[81]

Like most major North American industrialists, the lumbermen embarked on reorganizations to reduce competition. The initial moves took place at the local level in the 1870s, as producers in the then timber-rich valleys supplying Chicago and the Mississippi Valley markets set up pooling operations to allocate log quotas and prevent ruinous oversupply. The Chicago Lumberman's Exchange had been created in 1869 and achieved little until it began to compile statistics on deliveries in 1874. That same year an ineffectual National Association of Lumbermen tried to control production to stem the decline of prices during a construction collapse. Having acquired experience during preliminary rounds, the industry worked to establish standards for grading lumber before attempting to structure prices by issuing price lists during the 1890s. Much of the restructuring pertained to the pine industry, but even the smaller hardwood industry had begun to issue bi-weekly reports on market conditions and prices by 1891. After 1900 a few of the associations attempted to restrict production.[82]

Other materials suppliers pursued a parallel course, initially establishing uniform shapes, designs, and quality standards. Codification enhanced efforts to regulate prices. A National Brick Manufacturers' Association had been organized around 1885 and during the construction slump of the 1890s, when brick manufacturers responded to oversupply with price cutting, a few formed local combines – bearing deceptive labels as exchanges or clubs – in city after city: Pittsburgh, Washington, DC, New York, New Haven, Cincinnati, Toronto, Louisville, Salt Lake City, and Seattle (to name a few cited in trade journals). They had the avowed purpose of controlling standards, but occasionally, more plausible admissions – agreements on price schedules – slipped into reports.[83] In 1904 the manufacturers of builders' hardware in the United States were reputed to have come to an understanding whereby all designs had been classified and priced. "It is said that the better goods are to be advanced from 25 to 50 per cent, at once, and that a 10 per cent advance on all goods is slated for the near future."[84] These supplier organizations provided a mere taste of the collusive, restrictive, and oligopolistic arrangements for which the building-materials industry would become notorious. Throughout the history of American anti-trust action to 1940, a quarter of the cases dealt with building-materials or building trades.[85] Concurrent with the formation of manufacturers' associations, the lumber industry evolved toward cor-porate consolidation, a transformation accompanying the shift from a dependence on the north-eastern and Great-Lakes forests to an increasing reliance on the Pacific coast. Due to policies that assisted railroad construction, large tracts of forested land in the North-west came into the hands of the Northern Pacific Railroad and thus under the control of the Weyerhaueser interests. In the newly developed area, therefore, corporate concentration was greater than in the other regions.[86]

Contractors formed their own trade associations, known as build-ers' exchanges, which ultimately gained strength in the early 1900s. For example, in Canada, Toronto contractors had formed an as-sociation as early as 1867 and Montreal organized in 1899. By 1908 the Montreal exchange maintained an office with a secretary and a library. Contractors and subcontractors left bids and messages at the office. The exchange had also begun to compile coded records – "a sort of Berthillon card system" – on construction workers, and mem-bers of the exchange received the master code required to read the reports.[87] In 1899 the omnipresent Fred T. Hodgson supplied an-swers to "Builders' Exchange, Why?" in the *Canadian Architect and Builder*. Clinging to craft ideas, Hodgson hoped that exchanges would host lectures, stock books, and uplift the building trades. But he went slightly further. "By rubbing against each other in the Ex-change, many of the sharp edges of business rivalry get worn off, and many misunderstandings are explained and rectified and jeal-ousies allayed, and the dignity of the trades upheld."[88] Hodgson remained vague and keen about labour's dignity, but the secretary of the Montreal Builders' Exchange furnished a far more acute declaration on the issue by noting that combines were illegal but exchanges could impress upon members a common-sense attitude. "There is enough to go around without sacrificing legitimate profit."[89] The first annual meeting of a nascent Canada-wide build-ers' exchange gathered in Montreal early in 1907 and a stronger and more comprehensive organization, the Canadian Construction Association, was formed in 1920. National organization in the United States first occurred in 1886, but the original National Association of Builders faltered during the recession of the 1890s. There was a successor, for a Chicago convention with representatives from 34 cities assembled in 1919 to establish a National Association of Building Trades Employers. The picture of an end to competition is even more entangled, since there were some arrangements between labour unions and builders' exchanges whereby union men refused to work for independent contractors. Economist Mancur Olson, in his ambitious explanation of "stagflation," has arraigned such "distributive coalitions" for arresting economic growth and afflicting relatively powerless consumers. In the housing industry, this process had begun in North America at the turn of the century.[90]

THE SMALL HOUSE AND BUNGALOW

The phases in the development of the common house – the first launching cheap construction and the second driving up costs – inspired corresponding architectural considerations. By popular demand the inexpensive house became a concern of serious architectural discussion, then, pressured by cost factors, it posed the challenge of combining shrinkage, new technological systems, and appealing styles. Both experiences created a store of knowledge about design and construction that would inform the builders of modest suburban residences to the present day. Within the balloon-frame and machine-tool revolutions, it is probable that the only dramatic architectural innovation was the bungalow. The inimitable Fred T. Hodgson helped introduce the style and touted its importance. "It is not too much to say that these bungalows are on the whole the best type of cheap house which has been erected in large numbers in this country since the New England farmhouse went out of fashion."[91]

Before the bungalow attained prominence, there had been a gradual preparatory phase from the 1870s to the 1890s when the inexpensive dwelling first assumed near-legitimacy as an architectural form. At mid-century, writings about house design had concerned translation of class-bound cultural ideals into tangible forms for the benefit of an affluent minority. At their most lofty, reflections on the house incorporated transcendentalist convictions comparable to those promoted by landscape architect and park planner Frederick Law Olmsted. Ideally, picturesque trim imitated the lines and variety of a natural environment. Zebulon Baker, author of *The Cottage Builder's Manual* (1856), mused that children raised in a bucolic setting "are more likely to retain the integrity and virtue of childhood in mature years; their aims in life will be true and higher; and patriotism almost alone here fostered and encouraged."[92] His manual, however, offered few plan details for attaining these goals. *Woodward's National Architect* (1869), a more practical treatise, depicted a range of house styles. Even George Woodward's most inexpensive dwellings, such as a $2,000 Gothic cottage, had lavish decorations bearing erudite labels: frame brackets on bales, step butresses, plinths, balustrades above bay windows, and fine plaster work.[93] Belief that good health derived from space and ventilation influenced the addition of a large verandah and bay windows. Formality required the separation of floor area into public, private, and functional space. Parlours and halls provided the public area, while libraries, dining rooms, and bedrooms offered varying degrees of privacy. The division of function meant that pantries as well as kitchens were commonplace in plans for the houses of the middle class. In sum, the laying out of a house by an architect or a draughtsman builder posed challenges within a set of cultural precepts. To assist with the tasks, pattern books and manuals presented artisan builders or would-be architects with plans, problems in geometry and carpentry, and sometimes philosophical ruminations. Early pattern-book compilers, however, paid little attention to housing for common labourers. During the 1860s and 1870s journals serving builders and architects shared the cultural values of an elite. That changed.

Against the backdrop of idealistic reflection on middle-class homes, individuals, investors, and builders had nonetheless combined to erect abundant plain and cheap dwellings throughout urban North America. Dwellings for the working class, therefore, had attained forms independent of serious discussion on architecture let alone comment on the small home's role in a democratic society. Until the 1870s there appear to have been no sustained discussions in the trade papers and manuals about the need for carpenters and architects to consider planning workers' homes. Gradually, however, the slighted topic became a matter of serious study, and eventually an architectural industry centred on the small home.

There are several reasons for initial interest in the subject. In 1880 the *Builder and Wood Worker* suggested that Prince Albert's 1851 offer of a £500 prize for the best design of a worker's home at the London World's Fair had prompted some early interest in the subject.[94] Inquiries from the readers of building journals had an influence on topics covered, despite an ingrained contempt for cheap dwellings. On the one hand, the Chicago editors of the *Architect and Builder* wavered when the subject first appeared on their pages in the early 1870s. They disapproved of artless dwellings put up by speculative builders who allegedly debased the crafts. On the other hand, tenements threatened an even greater evil. Cheap dwellings at least kept alive the ideal of home ownership among the working class. In a list of "Suggestions to Artisans" prepared in 1873, the journal recommended securing land immediately. "Do not be afraid to go in debt for land."[95] After many issues had appeared with plates and plans for fine town houses, churches, and interior decorations, the *American Builder* inserted a page of plans for worker's cottages in 1874 and reproduced them in 1882.[96]

The New York-based *Carpentry and Building* devoted greater attention to small frame homes. Still, in an 1880 article describing plans submitted by a Toronto builder, the journal let slip a hint of residual snobbery. "As a general rule, small houses are not considered worth careful attention upon the part of builders and architects. The drawings prepared for them are frequently deficient in finish and lacking detail. All cheap houses must be open to criticism."[97] Swallowing its pride, the same journal set its thirteenth design competition (1884) for a frame house costing about $800, and the number of submissions surprised the editors.[98] *Carpentry and Building* repeated the competition, for a $1,000 house, in 1893.[99] Throughout the 1890s, the journal published letters from readers requesting plans for inexpensive houses. The ground swell of popular demand broke through prejudice so completely that the magazine announced a special end-of-century contest (1898) "relating to competitions in low cost dwellings calling for designs of frame houses costing $750, $1,000, and $1,500."[100] Meanwhile the *Canadian Architect and Builder*, in its first year of operation (1888), published a "design for cheap cottages" in response to popular demand "plainly evidenced by correspondence received at our office."[101]

Architectural ideals had shifted in a direction favouring all manner of economically conceived structures where form followed function. In an 1881 article on advice to "American architectural students," the *Builder and Wood Worker* stated as its third principle that "as construction necessarily implies a purpose, utility must have precedence of decoration."[102] The lure of profit, too, ultimately worked its magic. By 1900 some architects and draughtsmen began to depend on bulk sales of standard plans as opposed to individual commissions. Minneapolis architect W. J. Keith, appropriately ensconced in the Lumber Exchange Building, published a series of plans of inexpensive homes in the *Ladies' Home Journal* in 1897 and 1898. In his 1898 pattern book, *The Building of It*, he advertised that he had prepared eighty plans for cottages each costing less than $1,500. Fred T. Hodgson assembled his first collection of cottages and bungalows, *Hodgson's Low Cost American Homes*, in 1904.[103] *The Radford*

Review, a Chicago-based monthly launched in 1898 and "devoted to the interests of home-builders and home-lovers," caught in its motto the new spirit at work on the fringes of the architectural profession: "Give the People What They Want."[104]

Until about 1900 plans for inexpensive homes seemed to be scaled-down versions of more grandiose inventions. The cheapest dwellings had the outward form of abridged rectangular Gothic cottages stripped of decoration. Somewhat larger dwellings were tall and angular, some looking like transplanted farm houses and others like scaled-down versions of the Queen Anne style. In general, small front porches replaced spacious verandahs, eaves were narrowed, and bay windows were withdrawn. Inside, vestibules and entrance halls disappeared, and often entrance to rooms was direct as halls were eliminated. Kitchens encompassed the functions of separate cooking areas, pantries, and dining rooms. In two-storey houses, load-bearing walls were aligned above one another. When plumbing became essential, architects and builders placed facilities in line to save material. Economization demanded no particular genius, but a true challenge lay in rendering austerity in pleasing forms. An escalating cost of home construction failed to stifle what Gwendolyn Wright has described as "a preoccupation with the private dwelling."[105] Rising expenses at the turn of the century were met by shaving down the scale of dwellings and minimizing decorations. A crude estimate of declining floor space based on a limited sample of published plans supports this interpretation (see table 5.8). By the early 1900s it was rare to have single-purpose rooms such as libraries, pantries, sewing rooms, and spare bedrooms.[106]

Conceived around 1900, the 1,000 to 1,200 square-foot bungalow offered the best blend of economy and credible artistry. First stirring extraordinary interest on the west coast, it then spread to mid-America. *The Bungalow Book* compiled by Henry L. Wilson had gone through four printings in 1907 and 1908; the fourth produced 21,000 copies.[107] Bungalow designers and boosters such as Wilson, who also edited *Bungalow Magazine* (Los Angeles), were conscious of the style's successful combination. Wilson's portrait in the journal carried the salutation "Yours for Artistic Homes."[108] But, as he expressed it in "The Architect as Evangelist," the practical considerations had priority. "For, as Christ recognized, it is of first importance to care for the material needs of mankind before presenting spiritual need."[109] Evangelists for the bungalow combined the practical and the aesthetic by exposing structural form with ceiling beams, reducing the pitch of the roof and lowering ceilings, extending the eaves, and reincorporating the verandah. The picturesque Gothic

Table 5.8
Estimated Range of Interior Area of Inexpensive and Expensive Houses by
Ten-Year Intervals, 1870–1920[1]

Year	Range of Square Feet for Inexpensive Houses	Range of Square Feet for Expensive Houses
1870	600–1,700	1,800–2,700
1880	700–1,700	not available
1890	600–1,800	1,350–2,800
1900	1,000–1,500	1,000–1,900
1910	800–1,100	1,000–1,900
1920	650–1,100	900–1,900

Sources: Approximately 100 house plans appearing in trade journals and pattern books. One-of-a-kind dwellings were not considered; houses in the lower and middle range of construction costs were selected.

[1] Inexpensive houses had a simple floor plan, small front porch, no basement or partial basement, and few frills such as bay windows. Expensive but relatively common houses tended to have larger porches, fancy mill work, full basement, and frills like elaborate dormers or bay windows.

cottage aimed to uplift the civic virtue of the republic through association with nature. The bungalow "nested in a bit of shrubbery" aimed to be "chummy."[110]

The term "bungalow" was often used indiscriminately for any small house. At the end of World War I, the true bungalow figured as merely one arrangement for a small dwelling. The proven success of the bungalow encouraged architects to experiment with answers to the challenge of providing attractive but inexpensive homes. Furthermore, wartime and immediate post-war action by governments presented the small house as a buttress for social stability and inspired practical measures in the construction of cheap housing. The small house was being promoted as never before – small had become beautiful. In 1921 the small-house plans submitted to a United States national architectural competition included exteriors influenced by Colonial, Tudor, and Spanish American styles.[111] The Small Home, which had begun to appear shortly after the demise of both Bungalow Magazines around 1919, contained plans of homes described as Dutch colonial, Tudor, and modern, as well as cottage and bungalow.[112] If the bungalow radiated a chummy aura, the colonial staked its appeal on "the homelike effect, the dignity, and the convenience."[113] The range of moods and styles associated with small dwellings had been expanded without completely sacrificing the fundamentals of economical shelter. Yet there were boundaries of taste, reinforced by the conservative lending institutions and zoning by-laws, that militated against the incorporation into the mainstream of modestly priced dwellings homes designed by modernists such as Frank Lloyd Wright and Richard Neutra. Their use of flat roofs, slab bases, concrete, steel, and glass received guarded acceptance only in the 1940s and 1950s.[114]

The *Small Home* of the early 1920s epitomized the corporate embrace of inexpensive and architecturally safe homes. The march of consolidation in the lumber industry and in contracting found a counterpart in architecture: during the 1870s and 1880s a few daring builders had submitted plans of inexpensive dwellings to the critical snobbery of trade journals, for pride or for prizes; during the 1890s and early 1900s hustling architects or compilers marketed so-called bungalow plans; and in the 1920s, the American Institute of Architects launched the Architects' Small House Service Bureau of the United States, Incorporated, with the blessing of the Department of Commerce.[115] This operation, which began with thirteen branches in 1921 and gradually added more, epitomized the evolution of residential architecture from craft to corporate business. Based in Minneapolis, more than coincidentally the home of the Weyerhaueser concerns, the bureau prepared and sold quality plans for homes of up to six rooms.[116] As a nation-wide business enterprise, the bureau sought publicity. The *Small Home* boosted the bureau, and the magazine's editor, Maurice I. Flagg, lectured on the radio about "Lessening Home Building Costs." Both *Collier's* and *Good Housekeeping* also gave the bureau coverage.[117] After its first ten months (May 1921 to March 1922), the bureau boasted that its plans had reached "all parts of the United States – El Paso, Texas, New Orleans, New York, Salt Lake City, Winnipeg, Canada, Boston, Massachusetts, Chicago, Illinois – and we might enumerate at length until nearly every state in the nation is listed."[118]

With the attainment of a basic state of development for the small house with modern services in the 1920s, the evolution of the single-family detached dwelling had reached a plateau. External features and dimensions would fluctuate, but not the underlying technology or business of contracting. Let us consider what would and would not change. For the bourgeoisie, especially in the upper income brackets, there would always be a shifting historicism in architectural styles. Tudor revival flourished in the prosperous later 1920s; Georgian symmetry reappeared in the 1980s. Historicism had regional variations: the Cape Cod designs retained great popularity throughout the eastern United States; Spanish mission architecture remained a prominent idiom in the American south-west; imitations of the manor houses of New France appeared in the more expensive sub-

Table 5.9
An Estimation of the Character and Scale of Hamilton's Housing Stock,
1876–1936

| | *Estimated Percentage of* | | |
	Poor Houses of No More Than 6 Rooms	*Modest Houses of 6 Rooms*	*Better Houses of More Than 6 Rooms*
1876	42.7	30.6	26.7
1886	46.5	30.1	23.4
1896	56.6	17.3	26.1
1906	48.0	16.9	35.1
1916	39.5	30.7	29.8
1926	21.2	19.6	59.2
1936	18.6	22.8	58.6

Source: Hamilton Assessment Rolls

urbs of Quebec. Also, from the 1920s to the 1980s, the scale of new homes increased or declined in concert with economic circumstances – the trimming exercises from the 1890s to the early 1920s did not constitute an irreversible trend. In Hamilton the construction boom of the 1920s changed the city's housing stock by adding an appreciable number of larger homes.

Building permits do not survive for this period, but by using the assessed values in the exploratory manner described in chapter 10, it was possible to detect something of the probable trend toward building larger homes (see table 5.9). Assessed values reflected many things – market conditions, scale, and state of repair – and fine distinctions between poor and modest are a matter of educated guesses. The trend in the better houses column is too robust to be dismissed. Confirmation of a dramatic shift in the city's housing stock comes from a search for the city blocks that supplied the bulk of the higher value assessments. The contributing blocks were in the white-collar suburb of Westdale and in the south-east, around Gage Park, where clerical and managerial employees of the city's industries had begun to locate (see chapter 2). These areas have remained enclaves of larger houses. There can be no question that for Hamilton and likely elsewhere the houses built in the late 1920s, while squarely along the evolutionary path of housing construction, were different. They were generally larger than houses had been for decades. The cycle of variations in scale continued after the hiatus of 1930–45. In the post-war decades, trends in housing size can be based on an examination of CMHC publications. In 1949 CMHC published three booklets of standard plans. Out of 75 plans, 57 (76 per cent) presented dwellings with under 1,200 square feet of living area. The mean for all plans was 995 square feet – snug bungalows predominated. By the late 1950s floor area was expanding and one-and-one-half-storey houses were out of fashion, replaced by split-level dwellings. In a CMHC booklet of 79 plans, published in 1971, only 20 (25 per cent) described houses of less than 1,200 square feet. The mean had climbed to 1,376 square feet. In the United States the median-priced new house of 1965 was 1,495 square feet. By 1979 it had increased 10 per cent to 1,645 square feet. Such postwar increases in scale complicate comparisons of house prices over time. Furthermore, new houses in recent years have been outfitted with assorted amenities: energy-efficient hardware, air-conditioning, *en-suite* bathrooms, built-in appliances, and fireplaces. North Americans are likely spending more income on shelter in the 1980s than in the 1950s, but they are buying a lot more house.[119] Obviously, demographic and socio-economic conditions can transfigure the new house, making its scale an indicator of short-term economic settings. Nevertheless, the house has been a superficially mutable artifact. Underneath the surface, little would change.

By the early 1920s detached housing for the working class had acquired refinement. Its proliferation was a corporate objective that was increasingly believed to be indispensable for social stability in North America. It had reached a level of sophistication in its manufacture that, for all the bottlenecks and traditional constraints, was remarkably modern, so much so that Ned Eichler has proudly, if incorrectly, credited the industry as he had known it in the 1940s and 1950s with initiating a series of cost-saving innovations, including eliminating the basement, reducing the number of exterior wall breaks, lining up bearing walls, aligning wall lengths with stud separation, standardizing window and door sizes, bringing plumbing together, and eliminating hips in roofs.[120] As for construction practices, the large merchant builders avoided choking the construction site with chaotic activity. The modernizing builders separated field operations from supply; in some cases a separate "supply compound was set up on or near the site."[121]

The design and construction practices described by Eichler were anything but new. Unintentionally, Eichler bridged past and present; he confirmed the persistence of old challenges and the revival of responses witnessed in earlier boom periods. The practices he described had been commonplace in the late nineteenth century. Moreover, a 1918 article in the *National Builder*, "Systematizing the

Building of Workingmen's Homes," outlined precisely the same re-organized site that Eichler had attributed to the legendary mass builder of the late 1940s, Bill Levitt of Levittown.[122] Eichler's scepticism about "experimentation with more dramatic changes in technology or construction method" and his criticism of the "relative inefficiency in management" of national companies reaffirm deep structural features of the construction industry.[123] Despite the research on new building technology since 1945, consumer taste and the high prices of innovative materials have prevented any major restructuring, though as a recent report by CMHC has documented, house-building technology has not remained static during the postwar period. New machinery and materials are now being employed on Canadian and presumably American building sites (see table 5.10). Resistance to "manufactured housing" may derive from middle-class consumers who continue to look for well-known signs of domestic comfort, prestige, and traditional materials. Yet in the critical estimations of Gwendolyn Wright and Dolores Hayden, written in the wake of profound shifts in family composition, the resulting product – the single-family detached dwelling – is a shelter dinosaur.[124] Their contemporary concerns ratify the themes of nearly imperceptible changes in a hidebound industry and of the burden of housing lore generated in the formative years 1870–1920. Recent demographic changes, particularly those relating to smaller family size and to the increasing number of households headed by women employed in lower-income female job ghettos, may well re-awaken North American interest in smaller dwellings.

Other observers have added credence to the sense of long-term patterns. According to lumber-industry economists, the residential-construction industry's constant bumping against inelastic demand has led to a reduction in the use of lumber, particularly in non-structural features such as millwork.[125] Houses for the working class continue their evolution. Nevertheless, they do so within a shaping and confining set of material and cultural factors. In a way, the history of North American housing counterbalances the epochal periodization of urban history as proposed in Sam Bass Warner Jr's "scaffolding" with its commercial, industrial, metropolitan, and corporate phases. Changes in the residential construction industry have come gradually, have run into fundamental given conditions – the construction cycle, the non-portable character of the product, the balloon frame, the pressures of the contracting system, and the inertia of consumer taste and contractors' habits – and have included dead-ends like poured concrete as well as lessons forgotten only to be rediscovered during a later construction boom. One recurrent

Table 5.10
Single-Detached House-Building Technology: Examples of Evolution in Processes

Process	Practice		
	1946	1966	1986
Excavation	Bulldozer	Backhoe	No change
Basement	Concrete block and site-mixed concrete used with site-built board form-work; boards then re-used as wall and roof sheathing	Transit-mixed concrete used with prefabricated form-work	Little change but some spread of preserved wood foundations
Wall framing	Platform frame; some stationary assembly-line processes; little use of power equipment or piece-work sub-trades	Precut studs; tilt up, stationary assembly line with sequencing of piece-work produced by sub-trades	Little change
Roof	Laid out and erected by skilled tradesmen	Introduction of engineered, prefabricated roof trusses into general use	Little change
Wall and roof sheathing	Boards	Plywood sheets	Waterboard sheets
Siding	Wood clapboard, brick	Precoated aluminium and hard board introduced	Introduction of vinyl siding
Plumbing and heating	Site-fitted and installed	Prefabricated chimneys; some ductwork sub-assemblies	All-plastic plumbing; chimneys and flues prefabricated
Interiors	Wet-finished with plaster, cured, and brush painted	Dry-finished with drywall and roller painted	Little change
Windows/cabinetry/doors	Fabricated on site	Prefabricated windows, cabinetry, and countertops	Introduction of prehung doors and prefabricated stair units

Source: CMHC, *Housing in Canada 1945 to 1986: An Overview and Lessons Learned* (Ottawa: CMHC, 1987), 23

"novelty" has been prefabricated housing. It has waxed and waned, engaging nineteenth-century manufacturers and the modernist genius of Walter Gropius.[126] The packaged house together with mobile homes have accounted for about 20 per cent of new dwellings in the United States from the mid-1970s to the mid-1980s, and up to 14 per cent of new Canadian dwellings. In the remote areas of Canada, such as the Arctic or the west coast of Newfoundland, pre-fabricated houses abound. For over a century, they have threatened, but never quite managed, to make a major breakthrough to home buyers and mortgage lenders. For almost as long, carpenters' representatives have appraised industrially produced housing as a threat. A builder cited in Tracy Kidder's *House* spoke for generations of the continent's carpenters. "Increasingly, the organization of a house building site resembles that of an assembly line ... Out on the job, young carpenters don't learn how to build houses but to hang prefabricated doors or lay floors or to put up cabinets."[127]

A further long-term theme is implied in this chapter and in the preceding discussion of the will to possess. Several recent accounts of the obsession with the house in the American past treat it as an alienating emblem. This sentiment is too narrowly conceived. Thus, the full record of accomplishment in housing cannot be found in the intellectual history of Jan Cohn's *The Palace or the Poorhouse: The American House as a Cultural Symbol*, for it does not set out to encompass cost trends, the operations of the housing industry, and technological considerations.[128] Similarly Wright's *Building the American Dream* cites home ownership data selectively to leave an impression of the failure of the single-family detached house. Both seem intent on exposing the emblem in terms of its legacy to current housing problems. A historical account, however, requires completeness of reporting to include contrary evidence. It also obliges us to understand the past in terms of its own standards and problems. The originator of North American housing history, Sam Bass Warner Jr, has articulated a more rounded sense of achievement and failure in *Streetcar Suburbs* and, oddly enough, in his chronicle of failure, *The Urban Wilderness*: "For housing built after about 1880 a new minimum standard prevailed. In most cities the standard manifested itself in miles upon miles of small wooden freestanding houses. The cumulative effect of forty years of such construction was to free middle-class and typical working-class Americans from the dangers of alley housing, boardinghouses, and jerry built conversions typical of the high land values of the big city of the early nineteenth century."[129]

Urban America glimmered with faint but real achievements that contrasted with the hopeless prospects that drove immigrants to opportunity. In his discussion of landlord and tenant relations in British cities from 1838 to 1918, David Englander charted assorted means of protesting against and coping with rents, rates, and want of basic amenities. Englander is surely correct in assuming that few slum dwellers tried to escape and fewer succeeded.[130] However, the strong emigration flow, whatever its precise composition, must include as part of its background the British housing crisis that began in the 1880s and that contrasted with the housing optimism across the Atlantic. Racism, abject poverty, and inequality tarnished the record, but the North American urban masses, including newcomers, pursued "the dream" and, by and large, got industry, finance, architects, and governments to comply. Even if acquiescence was grudging, self-interested, contingent, and ultimately vitiated by "distributional coalitions," it represented something novel in the history of urban dwellers. A home that one could improve, even build, on freehold land made "the new world" a meaningful term. The burden of this perspective is to see today's detached homes through a century of antecedents as an embodiment of an often ingenious North American urban culture. Many of the roots of this ingenuity evolved between 1870 and 1920 – the golden age of housing.

6 Crafting Home Finance: Mortgages as Artifacts of Law, Business, and the Interventionist State, 1790s–1980s

The mortgage, like the balloon-frame house and its site on a suburban survey, has had a crucial role in creating the state of contemporary North American housing. It has mobilized resources for house buying and for the building of rental units. It has figured in popular culture, appearing as an instrument for threatening the virtue of heroines in silent films. In the 1946 Frank Capra classic *It's a Wonderful Life*, mortgages held by the kindly little building and loan office run by James Stewart's decent relatives rescued the folk of Bedford Falls from another fate worse than death: tenancy. By the late 1980s the image of the lending institutions, particularly in the United States, had been tarnished by white-collar criminals. The mortgage, then, is familiar, mythic, and newsworthy. It is also ancient and poorly understood. Like the house, its commonplace quality masks an absorbing history. From the standpoint of legal history, the mortgage is a very old and much tinkered with instrument. In modern history a host of court and legislative measures has continued an ancient struggle between mortgagee and mortgagor interests. Through legislation states and provinces have blazed their distinctive trails with respect to the rights of those who hold the mortgages (mortgagees) and those who must meet the terms of repayment (mortgagors). In essence, there has been a century-and-a-half tug-of-war over matters such as default. The following account concentrates on the Ontario jurisdiction, but its provincial attributes do not obscure more general principles.

The mortgage is a legal instrument with a protean history in common law, equity, and statutes. It has long figured as an exceedingly important asset in the portfolios of individuals and later financial institutions. Furthermore, the mortgage as a security has peculiar attributes, since it is one of the least liquid of all of capitalism's paper assets. This feature has helped to determine its evolution from an instrument passing between individuals to an instrument of concern to governments and major financial institutions. Again like the house, today's mortgage is a product of a basic traditional framework and of pressures for improvement. Like house building, mortgage shaping has undergone a ceaseless process of adjusting and redesigning the basic item to ensure that home buying – that salient economic activity in North America – remains robust, and it has required a team of skilled personnel. In fairly modern times this team has expanded from lawyers, judges, and brokers to include appraisers, institutional loan officers, and civil servants. The mortgage is embedded in the contemporary legal and socio-economic life of urban North America. But that is not all. Culturally, along with the will to possess, the mortgage has engaged the hopes, anxieties, and status attitudes of North Americans. The debt has opened doors and troubled those who entered. Jim, the contractor in Tracy Kidder's *House*, bought an old home and owed money. "It was a mortgage, *that most respectable form of indebtedness*, but it bothered him."[1]

THE MORTGAGE IN LAW

Early common law had held that a mortgage brought about a conveyance of an interest in land to secure a debt. The transfer of the interest was accompanied with a stipulation that if the debt were repaid by an appointed day the conveyance became void and the interest in the land reverted to the mortgagor. The mortgagee came to own the interest absolutely if the mortgagor failed to pay off the debt by the date specified. The mortgagee was still entitled to the payment of the debt and could sue for it while occupying the mortgagor's former land. This commanding position of the lender in common law produced injustices. Equity law, a separate path of recourse, allowed remedies for borrowers. Equity, a system of jurisprudence serving to remedy the limitations and the inflexibility of the common law, evolved from the petitions of subjects to the Crown. These were directed through the Chancellor, hence the development of an equitable jurisdiction known as a Court of Chancery. A decision in chancery in England could restrain mortgagees from suing for these debts unless they agreed to reconvey the land to the mortgagors on payment of the debt. Equity thus extended an opportunity for the mortgagor to redeem the property after defaulting.

Gradually, in England, chancery courts treated the mortgagor's right of redemption ever more liberally. Whereas common law had favoured the mortgagee, equity favoured the mortgagor, although not invariably. Under equity, mortgagors could appear and endeavour to force the mortgagee, who may have occupied the land, to reconvey it under tender of satisfaction of the debt. By the nineteenth century, the English dual law system had settled into conventions about the rights of mortgagees and mortgagors. Mortgagors were granted the right of redemption of their equity in the land, and this right was accepted as usually extending six months from the date that the court heard the mortgagee's petition to foreclose.[2]

North American colonies adopted English law at separate times. Canadian jurisdictions, for example, adopted English law as it existed at the time when each obtained its first legislative body. For Nova Scotia, that was 1756, and for Alberta and Saskatchewan it was 1870. English law had evolved between those dates, so that the provinces had slightly different features in their mortgage law. All jurisdictions also made statutory amendments resulting in further variety. Yet the study of one North American jurisdiction indicates the interests that met and occasionally clashed when a mortgage agreement was made. The statutory history of the mortgage in Ontario begins in 1795 when an act created a public registry system and provided examples of documents that might be registered: Bargain and Sale, Mortgage, and Certificate of Redemption.[3] The Certificate of Redemption confirmed the discharge of a mortgage. In common law, the mortgage was a conveyance of an interest in real property. The object was not to effect the sale of that real property, but to provide security for the payment of a debt. If the debt was paid on time, the mortgage would be redeemed, thus reversing the transfer of the interest in the property. If the borrower fell behind in making payments, a suit could be initiated to bar and foreclose this right. Known as strict foreclosure, the suit effectively conveyed the property to the mortgagee with a loss of the mortgagor's equity in the property. In England the remedy through equity law had been to grant the mortgagor a period of grace for collecting funds to pay off the debt after the expiration of the date stated in the mortgage. In 1792 Upper Canada adopted the laws of England; however, no steps were taken for over forty years to establish a Court of Chancery. The 1795 Registry Act only fleshed out the common law. Equity remained in limbo.

Failure to establish a Court of Chancery promptly was no trifle. Upper Canada's law officers, especially Attorney-General John Beverley Robinson, were hostile to a system of justice that they knew little about and that had developed to mitigate common law. Robinson recalled in 1828 that he might have discussed the issue of an equitable jurisdiction with colonial-office authorities when in England in 1822. Lieutenant-Governor Sir Peregrine Maitland had instructed him to do so again in 1825. Robinson returned to Canada with the news that to secure a good barrister to put an equitable jurisdiction in place would cost £2,000 per annum. The reform majority in the Assembly would not hear of it. This outcome of parsimonious dissent, for once suited the conservative Attorney-General. However, the colonial office pressed the matter and, in the summer of 1827, sent a noted equity barrister to Upper Canada. But even as John Walpole Willis crossed the Atlantic, the Crown's law officers concluded that a judge of equity should receive authority from an act of the colonial legislature and not by Crown commission. When this news reached Canada, Willis was on the Court of King's Bench; he awaited the outcome of the colonial government's deliberations.

The Assembly disliked the idea of another costly salary and some members felt equity made for long and expensive actions. Robinson, as the government's member in the Assembly, introduced a draft bill sure to draw opposition lightning. It would have made the judge's tenure a matter of executive pleasure and the reformers favoured the appointment of judges based on good behaviour. The latter made judges subject to legislative censure rather than executive recall. Before Robinson had finessed the Assembly into a predictably negative response, he had had to commit his own thoughts to paper. So too had the Solicitor-General, Henry J. Boulton, Chief Justice William Campbell, and the judges of the Court of King's Bench, Levius P. Sherwood and John Walpole Willis. Maitland had requested their opinions. Different positions were taken, but Robinson's alone suggested problems with adopting equity. He argued that equity was a complex and ancient system suitable for an advanced society but not for a colony. It should not be introduced in its entirety. He worried about how equity would affect contracts made at a time when contracts did not anticipate such a result. He cited mortgages as an example. Robinson wanted transactions kept simple – in the instance of mortgages, this meant preference for strict common law measures. Commentary on the failure to create an equitable jurisdiction in 1828 has assumed that the problem was related to the rivalry between Robinson and Willis. Some of that friction had its roots in a basic tension between two legal systems and their different principles concerning the mortgage. The upshot was that for another six years Upper Canada retained a legal system that permitted advantages to mortgagees.[4]

Common law demanded from mortgagors a strict performance of conditions. Failure to meet a payment, even by a day, allowed mortgagees to pray the court to issue a writ of *fieri facias* to seize goods; in the instance of a defaulting mortgagor, the rèal property could be seized. So long as equity could not be tried, the mortgagee had a fairly swift action. In 1846 in a major decision on mortgage law that tried to clean up retroactively the muddle of earlier decades, John Beverley Robinson claimed that when he came to the bench in 1829 the procedures governing mortgages were purely those of the common law. There had been no mortgagor interest in the property after the conveyance to the mortgagee became absolute. That is to say, the mortgagor had no equity of redemption. His allegation, though consistent with his 1828 report on the alleged mischief of allowing a full equitable jurisdiction, was, like that earlier claim, prejudicially oversimplified. The statute establishing law in Upper Canada may have included equity. Other legal minds did not share Robinson's certainty about the absence of equity and were unnerved by the fact that equity truly slumbered. Once awakened it might give mortgagors claims on their lost estates. Therefore, in a bizarre exercise, forced by the legal mess promoted by Robinson himself, Upper Canadian mortgagees used the common law writ of *fieri facias*, valid only to seize goods, to obtain a judgment against the mortgagor's equity of redemption, often selling it and the real property to a third party. Robinson described this as a nugatory act, for it transferred that which he felt did not exist in the absence of an equitable jurisdiction.[5]

Wanting no costly superstructure of courts, reformers avoided the equity issue when they had their majority from 1829 to 1834. However, in the next parliament the Legislative Council and reformers in the House of Assembly forwarded the standing of equity without having to broach the touchy court issue. This "Act to amend the Law respecting Real Property" fixed a twenty-year limit on making claims against titles and declared that the act left open recourse to equitable jurisdiction when it would be invoked at some future time. This first statutory mention of equity kept it alive. Another act in 1834, "An Act concerning the release of Mortgages" shook common law more forcefully. It effectively granted a right to mortgagors of a practice that assumed the actuality of the equity of redemption. If, within the period during which a mortgagor had a right in equity to redeem (six months, usually), the mortgagor registered a certificate of payment of the mortgage money, the certificate would act as a reconveyance of the real property and defeat the mortgagee's title as well as that of any party to whom the mortgagee might have sold the property. This measure threw into doubt the unmitigated common-law expedients formerly used to deal with defaulting debtors. Now equity stirred.

In 1846 John Beverley Robinson inadvertently displayed attitudes that help to explain why he disliked equity. His opinion in *Simpson v. Smyth* inveighed against the 1834 acts for placing mortgagees at a disadvantage. They did not deserve to be forced by law to extend a period of grace for mortgagors to redeem. Lenders, as gentlemen, would honour an effort to redeem even if the property had been seized months or even years beforehand. He cited an anecdote from the annals of Toronto real estate attesting to the willingness of holders of seized estates to accept a payment of a debt in order to redeem a property that had become worth much more than the debt. The mortgagee had behaved honourably. The reasoning was naive, if not disingenuous. Robinson, even after the fact, was celebrating a pure and inflexible body of law, although the fount of that law – England – had no such one-sided regime. However, in 1837, with equity awakened by statutes, Robinson had joined his two colleagues on the bench and the colony's Attorney-General and Solicitor-General in approving a bill to establish a Court of Chancery. The legislators took into account New York's act and the colony's legal experts preferred that state's verbal process in chancery to the English written process. An equitable jurisdiction, a Court of Chancery, actually proved not to be simply an agency to secure the six-month period for the equity of redemption. Chancery decisions sometimes proffered occasions for relief to mortgagees. Anticipating that a Court of Chancery would not necessarily work to the common residents' advantage, reformer Dr John Rolph had opposed the 1837 bill that would establish the court. He preferred remedies by statute and cited mortgages as an example of where legislation could redress problems.[6]

The preamble to the 1837 act creating a Court of Chancery confessed that mortgages were in a mess and proposed that an equitable jurisdiction would enable mortgagees to foreclose and mortgagors to exercise their equity of redemption. The court would resolve confusion. Of course, complexity, if not confusion, was the very substance of the relationship of mortgagee and mortgagor and what a court would accomplish would be the partial removal of controversy and friction from politics. Whereas a radical majority might openly swing the law by statute onto the debtor's side, as would occur in some jurisdictions in the United States, a court could less visibly establish the rules that kept the instrument viable. For example, chancery decisions on foreclosure, reported from the 1850s, recognized the mortgagee's interest in a mortgaged property to the very important extent of entertaining reductions to the usual six-month period for redemption ordinarily recognized in English equity law. If the mortgagee could demonstrate that the mortgaged premises were likely to deteriorate in value, courts could order an immediate sale. Chancery judges operated here on the principle that there had to be fairness and justice to the creditor as well as to the debtor. This open-ended rule depicted, as lucidly as any statement on mortgages, the considerable scope for disputes between mortgagee and mortgagor.

In addition, Canadian legislators frequently introduced nuances into mortgage law. A reform administration in 1849 sought to bring Canadian (Canada West) practices into line with those of England and to regulate a measure of potential relief to mortgagees and occasionally to mortgagors. Under existing practices into the 1840s, once a mortgagee had disposed of a property after foreclosure, he forfeited any right to demand repayment of a debt. Short of cash himself, the mortgagee might succeed in foreclosing, but have to sell the real estate at a loss, thereby forfeiting any recourse for collecting the outstanding portion of the debt through court action. The situation of the mortgagor was no better. The right of redemption created some relief, because it permitted a variable period for scrambling to collect money to prevent foreclosure and thereby to protect equity – the labour or material put into a property by way of improvement. It is impossible to discern how often a desperate mortgagor used the six months to try to raise enough money to redeem; it is equally impossible to know how the struggles were engineered and how often they succeeded. The campaign of Catherine Hill of Elora has survived, recorded on a rough and deferential note folded inside a mortgage handled by Toronto mortgage and land agent Weymouth George Schreiber. She and her husband Samuel had taken out a small mortgage in February 1870. It was due the next year. Samuel, a carpenter, fell to his death from the railway bridge at Caledonia during its construction. His widow had an infant and six other children to raise and a house to save. Five months after the due date of the mortgage, she wrote asking for the exact sum owing, for she hoped to "Raise the Money in some way to Redeem the place." In fact, the spirited widow was "going to ask the people of Elora for a subscription to help me pay what is owing on the land." She succeeded in saving her unfinished house and members of the family resided on the property until 1911.[7]

Foreclosure after the six months meant a loss of mortgagor's equity. Into the unhappy situation attending absolute foreclosure, the idea of a sale by the court came as an invention to give both parties openings to additional remedies. An 1850 Chancery Court ruling confirmed a practice that may have been used for several years. After taking into account English conduct, the judge averred that "in this country it is competent to a mortgagee, in every case, to pray a sale, instead of a foreclosure."[8] An 1849 act set out to confirm and control something that seems to have been introduced nearly concurrently through Chancery. The act, however, reformed the process by making the county courts competent to administer sales. A sale of the mortgaged land by the court gave the proceeds to the mortgagee; if the sum were smaller than the total of the loan, accrued interest, and court costs, he retained the right to seek judgment against the mortgagor for the deficiency. To avoid low bids, mortgagors occasionally employed "puffers" to try to drive up the bidding. If the sale produced a sum larger than debts and costs, the surplus would be returned to the mortgagor. This added dimension to mortgage laws spread by statutes across North America. In 1953 only Connecticut and Vermont among the states had not introduced foreclosure by sale. Interestingly, during the late nineteenth century when Ontario loan companies increased greatly in number and in assets, they may have steered clear of operating in Quebec because of a unique provision in the Quebec law on court or sheriffs' sales. Such sales in Quebec cancelled all liens, barring the mortgagee from further action to recover the balance of a debt. This nipped the Quebec loan business in the bud, for it threw "on the mortgagee the obligation of watching constantly lest his security be sold without his knowledge for less than sufficient to cover the mortgage."[9] If this plausible deterrent did apply, then it serves as an example of how the mortgage laws might affect access to home ownership, thereby joining a list of causes cited for the persistently lower levels of home ownership in urban Francophone Quebec.

The courts in Ontario not only issued the orders of foreclosure or ordered and supervised sales, but they also have had to act continually as umpires when the exact meaning of fairness to debtor or creditor was being disputed. Whether mortgagees moved for sales within the rules established by the courts or for court-supervised auction sales, they had to tread very carefully. The courts recognized that mortgagees were trustees of a special type – surely not acting on behalf of the mortgagors, for they had their own interests to protect. Nevertheless, they were never at liberty to look after their interests alone. They could bid at the auctions, but usually this required the leave of the court. In a multitude of cases during the late nineteenth century, the courts reiterated that a mortgagee could never fraudulently, wilfully, or recklessly sacrifice the property of the mortgagor. To wrongly do so risked a decision to pay damages for the loss. An action against the Hamilton Provident and Loan Company in 1898 illustrates how a mortgagee could damage the mortgagor. Under its powers of sale, the loan company advertised a Manitoba farm property for sale but placed the notice in a newspaper seventy miles away from the real estate and obscured the presence of improvements. The Hamilton firm had to pay damages, because it had not adopted such sale practices as would have been followed by a prudent person to get the best price possible. The sale mechanism could not do the impossible – it could not, in most cases, resolve a default to the complete satisfaction of both creditor and debtor. Litigation, therefore, established a guide for conduct that once more accented fairness, but in this area a fairness to the mortgagor.[10]

Eventually some mortgages in Ontario gave the mortgagee a contractual power of sale that could be exercised without direct court action or court supervision. In these mortgages, the mortgagor has agreed to grant a power of sale to the mortgagee. The Ontario legislature has required the mortgagee to provide advance notice to the mortgagor about the date of the sale. Over the last hundred years, the legislature has narrowed the time between default and notice of sale, and between notice of sale and the sale itself. In 1879 the sale could not take place less than three months after the service of notice by the mortgagee or less than six calendar months after the date of default. In 1970 the sale could not take place less than fifteen days after default or until thirty-five days after the notice of sale had been given.[11] Court decisions and distinctive legislation by Canada's provinces and American states have created very different processes throughout the continent. State laws, according to a 1938 review, varied greatly. Due to distinct redemption practices, foreclosure in Texas took a few days, while it required eighteen months in Illinois. In most cases, the debtor was entitled to possession of the property during the period of redemption. The costs of foreclosure varied. Texas, apparently a mortgagee's haven, had a quite inexpensive process, but in New York and Chicago, foreclosure was costly. Precise features evolved through social and political processes that remain unchronicled, but in every North American jurisdiction creditor and debtor positions surely tangled and then were resolved by different formulas. Participants in these many and presumably continuing legislative disputes would have recognized at least a common legal language and familiar basic issues in dispute had they been transported to another jurisdiction. But the different formulas were more than just variations on a common theme, they meant something to lenders and seem to have influenced their practices. It is just possible that foreclosure laws favouring mortgagors had the ironic effect of damaging small borrowers' interests by promoting highly conservative and discriminatory lending practices.[12]

A commonplace agreement whose bewildering legal possibilities call for the public's most frequent contact with lawyers, the mortgage has not achieved its complexity through machinations to render the law arcane. Lawbooks describe a knotty situation because precedents and statutes demarcate a shifting front in the tug of war between the interests of borrowers and those of lenders. Courts and governments under various pressures cobbled together an instrument that could be trusted by borrowers and lenders. Though not giving perfectly satisfactory conditions to either party, the evolving law of the mortgage successively narrowed risks and defined rights so that both parties might trust in the mechanism while remaining sceptical about each other. Rare forceful legislative acts, like provincial debt moratoria during the 1930s, have jeopardized the position of mortgagees and threatened the crucial balance sustaining one of Canadian capitalism's central instruments. The legal history of the mortgage suggests how a court-managed conflict had developed and sustained an absolutely vital credit device by keeping creditors and debtors engaged. Less visible than the balloon frame or the suburban lot, but still an expression of specialized crafting and experiment, the mortgage document embodies an essential artifact of city-building in a capitalist economy. Like the frame house and the lot, the mortgage has been the product of an industry with a mix of small individual producers, and, increasingly, large institutional producers. Like architectural styles or suburban layouts, the mortgage has displayed regional variations, but beneath the surface appearance of differences, differences which incidentally help define the autonomy of provincial bar associations, the mortgage is a standard feature of North American housing. Lawyers working in one jurisdiction based on English law might not know the mortgage law of another jurisdiction, but they would recognize the language, principles, and parties.

As we turn to the large and small producers of mortgages – the mortgagees – several questions have to be kept in mind. How have mortgagees raised capital in the amounts required for a vigorous shelter-construction industry in a settlement colony? This question is all the more interesting when it is recalled that Canada has been capital-dependent throughout much of its history. Or has it? How have amortization periods and repayment schedules been amended in order to liberalize access to lenders' funds? How were property valuations and borrowers' equity dealt with to achieve an expansion of consumers' credit and to sustain the lenders who naturally wanted protection against default? In sum, how has capital mobilization worked in the mortgage market in this corner of North America? An insight into the market's developing practices is provided through an account of the institutional lenders, case studies of two mortgage brokers, and a consideration of the role of the federal government from 1919 to the 1980s.

THE INSTITUTIONAL MORTGAGE LENDERS, 1846–1985

Institutional lenders, until the post-World War II decades, accounted for considerably less than half of the capital for mortgages in Hamilton and probably in most of urban North America. However, they are examined first and at length, because they have left a trail in statute books and public records, and because today they provide the bulk of financing for new houses. Of the four most common types of mortgage-lending institutions – loan companies, insurance companies, chartered banks, and trust companies – the first two have a continuous history of involvement with mortgages in Ontario, beginning around the mid-nineteenth century. Trust companies generally were not incorporated in the province until 1882. In the early 1920s loan companies began to move into the trust business because the trusts and insurance firms had started to act more aggressively in the mortgage market. Meanwhile, the federal government had banned banks from dealing with real-estate mortgages from 1872 to 1954.[13]

Most Ontario loan companies originated as building societies, which in turn had grown out of several forms of eighteenth-century English benefit societies. Like the friendly societies that aimed at helping workers in times of misfortune, the building societies collected regular contributions from members, but like the Tontines that collected sums for annuities, they demanded such large instalments that they could hardly claim the social conscience of the friendly societies. Membership in the building societies, as the historian of the English movement put it, was "likely to go to smaller trades and artisans."[14] The earliest known society existed in Birmingham as early as 1775. At the end of the Napoleonic wars, the number of societies expanded rapidly. The individual societies differed only in small details. Essentially, they accumulated deposits from a fixed number of members. The deposits represented payments for a subscribed number of shares in the society, and the societies' rules specified fines for failure to pay subscriptions on time. As capital accumulated, the society purchased land and materials, with the object of contracting for the building of a house or houses. The houses might then be let out to tenants until rents, fines, and share instalments had generated enough capital to purchase or build a house for each member of the society. Another practice called for members to draw lots or roll dice for the first house or for loans from the accumulated funds. Whatever the protocols for the sequence of who got access to what and when, the early societies typically dissolved when all members had a house. Consequently, these earliest building societies, and some later ones that retained these features, were known as "terminating societies."[15]

At approximately the same time that building societies appeared in Canada, in the 1840s, English societies had been innovating with the basic form, creating financial institutions out of the clubs, which had sought merely to furnish houses to a score or so of members. The early societies had borrowed money to speed up the granting of advances to members. Increasingly, the English societies took on the appearance of banks with small depositors. To expand into this service, the societies in question had to become permanent rather than terminating. Permanence, investors, and depositors had transformed the English building societies from co-operative efforts to resolve economic problems into capitalist lending enterprises bereft of mutuality.[16] This was not always the case, but the burden of changes sustains the generalization. Canadian building societies progressed along a similar path with an interesting exception that points to the differences between the metropolis floating in assets and the colony shy of capital: English societies had flourishing deposit branches by the late nineteenth century while Canadian societies raised capital by issuing debentures and chasing down investors in the money markets of the United Kingdom.

The first chartered building society in Canada, in Montreal, secured its act of incorporation in 1845. In Ontario, the Lambton Loan and Investment Society possibly commenced without a charter in 1844. A flurry of enthusiasm for building societies, including 1846 petitions from Toronto and London, motivated the Legislature of Canada to pass an act to allow the formation of societies in Canada West.[17] Within three years of the passage of this general act for Canada West, New Brunswick and Nova Scotia had followed with similar legislation. In the United States, the first building society had been founded in 1831 in a community that became part of Philadelphia.[18] The stirrings of the new financial institutions caused a substantial amount of legislation – by one count, from 1846 to 1893 legislative authorities in Canada had passed forty acts.[19] Much of the legislation permitted changes to the lending and borrowing powers of the societies, and judging by the statutes, the societies' character swiftly shifted away from mutuality and toward business enterprises. However, even as late as the 1890 a few of the Ontario societies bobbed behind the capitalist wake, remaining mutual in inspiration and modest in scale. Permissive legislation, in contrast, inked maps for the future at the behest of businessmen charting the conversion of building societies into loan companies. During several inquiries into the Canadian mortgage market during the 1960s, question were raised about the country's lack of building societies. Perhaps misunderstanding that, except for a different label and possibly a somewhat less local character, loan companies were more akin to than different from modern British building societies, the 1969 Task Force on Housing suggested that Canada encourage their formation. Canadian building societies evolved differently from their British counterparts, but their reinvention was unnecessary.[20]

A prototype for today's mortgage lending institutions originated several years before the first wave of building-society measures. In 1842 six Kingston notables, including John A. Macdonald, successfully petitioned for a private act: "An Act for Incorporating and Granting Certain Powers to the Upper Canada Trust and Loan Company."[21] With over one-hundred sections, it was one of the longer private measures of its time. As well as bestowing powers to raise capital by stock subscription and to lend money, it empowered the company to lease and sell lands. Truly a corporation having much in common with today's trust and loan companies, the Kingston enterprise nevertheless did not serve precisely as a model for the majority of loan company charters. Instead, institutional mortgage lending evolved most frequently out of the 1846 general act for building societies. However, the Upper Canada Trust and Loan Company indirectly accomplished the removal of a significant obstruction that had been impeding the flow of loan capital from the United Kingdom into Canada. At first the company limped along. Eight years after its incorporation, its stock was not taken up adequately, because, like most transatlantic and North American jurisdictions of the early nineteenth century, Canada retained a usury law.[22] Since 1811 this law forbade charging more than 6 per cent interest on loans. Any greater charge was legally void and the lender liable to a fine of treble the sum that had been lent. Some debtors had been known to inform in order to evade repayment. Thus foreign investors shied away from the trust and loan company's stock. Since Canadians with capital circumvented the usury law by a variety of expedients best managed in face-to-face dealings between parties who knew one other, Canadian capitalists also steered clear of the company's stock.[23]

In 1850, following a petition from Oliver Mowat, future Premier of Ontario, and the shareholders of the Kingston trust and loan company, the legislature granted the company the right to lend at 8 per cent interest and to act as a mortgage broker between British investors and Canadian borrowers. Occasionally when the colony itself wished to raise funds abroad for several public works, it had also made exceptions to its own usury law. The hypocrisies had to be confronted.[24] The Board of Trade of the City of Hamilton petitioned for a repeal of the usury law in October 1852, adding pressure for an action that was producing one of the more remarkably ingenuous debates on capitalism in Canadian history. The discussion had begun shortly before the presentation of Hamilton's petition. On 30 September 1852 George Brown moved second reading of a bill to abolish the penalty for usury. Eventually, at least twenty members of the Assembly spoke and, in a nearly straight sectional vote between the abolitionists from Canada West and the defenders from Canada East, the bill passed this reading by the narrowest of margins, 31 to 30.[25] The debate is fascinating because the members obviously believed that they were disputing the meaning of money and credit in a contest to determine how their society would develop. In the end, by one vote, a relatively new conception of economic relations – a vigorously capitalistic one – triumphed.

The debate marks a revolution in the development of finance capitalism in Canada: not only were legal interest rates to be raised, but the Crown's protection of consumers – prominent in the statute books though not seriously enforced – was minimized in exchange for a vision of better living through freer and hence broader credit. The defenders of the usury law regarded its maintenance as a defence for the country's farmers. When it was pointed out that the law failed here and encouraged the invention of extortionary dodges, they demanded rigorous enforcement.[26] One speaker proposed that since the state created money it had the right to control its price. This thoughtful proposition came in response to George Brown's opening address in which he called money "a merchantable commodity."[27] Brown and Francis Hincks carried the attack on the usury law: Brown painting the idyllic picture of progress and Hincks rising to sketch in prosaically the practical considerations. Brown's vivid prose and his genius for placing the gloss of natural right and true progress on measures sponsoring growth shone brilliantly as he illustrated the community benefits soon to adorn Canadian society once politicians dissolved a lingering medieval restrictions on the free movement of goods. In an expression of the liberalism of the age that had nurtured the will to possess, Brown argued that as land now traded easily, so too should money.[28]

One of Brown's examples of how unfettered lending could speed industrial and commercial headway shows him, and presumably other urban men of affairs, as well attuned to the very innovations in the building trades that shortly would render "brother chips" archaic. By artificially depressing interest rates, Brown insisted, the colony had repelled outside investors and kept capital in excessively short supply. Lenders resident in the colony and wise in the ways of subverting the usury law then extracted a heavy price by discounting. How did this affect industry? "The carpenter has a number of men at work with hand-tools; by the introduction of machinery he might borrow the purchase money, and repay it from the savings between hand-labour and machine-labour – but the law will not permit it."[29] Repeatedly, the advocates of abolition attacked the law's alleged adverse effects on the flow of British capital, sometimes citing how the exemption made for the Kingston-based trust and loan company had cleared the way for it to prosper. Then, too, once that exemption had been made, the law had to be made consistent.[30] The issue fanned tempers, and accusations slipped out that left a record of how the mortgage market functioned under the conditions of a fixed interest rate. Thus, if Brown, Hincks, and Macdonald expressed the attitudes and techniques of capitalist innovations in the age of bourgeois liberalism, Thomas Clark Street embodied old colonial wealth that had found the tight money regime profitable. The great Niagara-district money lender – singled out as the only creditor in the Assembly – had purchased mortgages at a 20 to 30 per cent discount. When a member of the assembly referred to this practice as charging a 20 to 30 per cent interest rate, Street fervently denied this slur on his character, but he did not deny that he dealt in mortgages by discounting at these rates. The former conduct would have been a clear violation of the law, the latter merely subverted its spirit. Street squirmed and hoped the debate would move on. According to several speakers, other money lenders ignored legal niceties altogether and charged usurious rates.[31]

During this significant debate, William Lyon Mackenzie coined a few choice epithets. Ever the acid-tongued critic of the corporate creations of the legislators, he went after "the two great guns" – Hincks and Macdonald – who had put through the 1850 exemption for the trust and loan company, saying that "they were the legalized usurers in the country." Holding office as Inspector General of Canada West, and thus the administrator of the colony's finances, Hincks was labelled "the usurer general" by Mackenzie.[32] On the vote, Mackenzie, defender of agrarians, sided with Street, defender of tight money. Another rare English-Canadian supporter of the usury law, Henry Smith from Frontenac County, made sharp verbal sallies against the bourgeois political centre, appealing for protection of farmers, and relating a most interesting anecdote to illustrate how the exemption of the trust and loan company had failed to make money less expensive. When John Scott, a stockman and a constituent, wished to raise £100, Smith had suggested a visit to the trust and loan company in nearby Kingston. Scott got his money, but he had had to pay a legal fee of £7 10s to John A. Macdonald, plus an appraisal fee, and an interest deduction to the company. Scott returned home with £87 16s 3d and a £100 debt. Hincks conferred with Macdonald, and rising later to defend the maligned member from Kingston, he indicated that Scott, an intelligent livestock dealer, had made money with the loan. Indeed, he had returned to borrow an additional £50.[33] Scott might have a claim to be the prototypical modern debtor or mortgagor. Promised cheap money, he had to pay hidden costs, but swallowed the hook. The lure of credit even attracted him again, as it would millions of subsequent borrowers.

Eliminating the usury law freed the wheels of credit and where they encountered obstacles in later years governments allowed additional revisions to the corporate machinery. The legislation of later decades resembled that of the 1840s and 1850s in that acts permitted finance capitalism ever-widening scope. Starting with a modest housing program in 1919 and building to a series of housing acts from 1935 to 1954, the Government of Canada actually began to do more than enable lending institutions to amplify their leverage. It engaged in credit-enhancing schemes that are explained later in this chapter. The essential point is that the devices for collecting and pumping capital through the mortgage market have never had a fixed form for long. Like the search for house-construction short cuts or new ways to design and promote suburban lots, the pursuit of credit enlargement with minimal investor risk has been a constant in the shelter industry. On the one hand, the early ceaseless quest for novel credit practices that would generate more volume and greater profits denotes change. On the other hand, as the 1853 debate laid bare, the more things change, the more they remain the same.

A usury law, loans in small figures, and 6 per cent interest rates appear antique, yet today's consumer groups and representatives for lending institutions might identify counterparts and familiar arguments in the debate. Macdonald, Brown, and Hincks had no exact conceptions about the future they were promoting. While they likely believed sincerely that they worked for progress, they had not the slightest conception of themselves as progenitors of corporations with billions of dollars in assets. In fact, the Kingston trust and loan company charter and the elimination of penalties for usury did not remake finance capitalism quickly. New barriers to greater capital mobilization were encountered: the colonial administration feared that institutional failures would damage its own reputation, and different international conditions threw up barriers and called for distinct remedies. So gradualism should be added to the themes of innovation and of continuity. Some things, like the usury law, experienced a sudden statutory change, but other facets of borrowing and lending have an eternal quality centring on disputed concepts of fairness. The operation of the mortgage market has always depended on the appearance of fairness and justice to creditor and debtor. Meanwhile, the precise legal and institutional procedures have been recast in related steps often spaced many years apart. Perhaps most astonishing, all these steps furthered the growth of institutions whose share of the mortgage market may have remained smaller than that serviced by individuals until as late as the 1960s. The greatest continuity of all was the predominance of the private mortgage.

The trust and loan company, a novel venture whose travails had forced a significant statutory change in lending, succeeded in its objective of getting shares subscribed in the United Kingdom. It outgrew Kingston and moved its main business office to Toronto in 1872, where it was renamed the Trust and Loan Company of Canada. It continued to have an instrumental role in mortgage-lending history, for as the plaintiff in many suits it helped establish precedents that shaped Canadian mortgage law. By the late nineteenth century, its British directors had plunged heavily into western Canadian mortgages. Then in 1910, from its head office at 7 Great Winchester St in London, they announced its withdrawal from the Ontario market. From its inception until 1910, Ontario's first trust and loan company had gone its own way and very few other companies in the mortgage market resembled it.[34] Most operated out of Ontario head offices, and held considerably less than half of its assets. Almost all had secured incorporation by the provisions of a general act and not by a private act, because it was remarkably simple to register a building society – that is, a loan company – under the terms of an 1846 act whose basic procedures stood until 1897.

The 1846 act specified that a building society, consisting of twenty or more people, must register its intent to function as a building society. This registration had only to be deposited with the Clerk of the Peace of the District or later the County. The clerk received a fee of 2s 6p. In England, after the passage of an 1836 act, societies had to register with a central office, usually in London. Here their by-laws came under critical if inconsistent judgment.[35] Members in a registered Canadian society could purchase shares, but no share was to have a value of more than £100 and monthly payments on subscribed shares were not to exceed 20s per month. The act protected the rules of the society against legal process.[36] This act envisioned the formation of modest mutual societies consisting of individuals with income security. Amendments soon brought a succession of crucial changes, first in the 1850s and next in the 1870s. Powers were enhanced in 1850 so that societies could expel members for arrears and could sell a mortgaged property to recover the debts. The concept of membership was modified to allow corporate bodies and co-partners to hold shares in any society incorporated under the act.[37] This measure indicated that society shares were sometimes regarded as investments and not just as devices for enforced savings. It remained easy to secure incorporation of what were, in fact, financial institutions. A group of Toronto investors, for example, filed with the Clerk of the Peace for the United Counties of York and Peel on 1 March 1855. Their Canada Permanent Building and Savings Society, renamed Canada Permanent Loan and Savings Company was, by measure of assets, the largest loan company in late nineteenth-century Ontario. It remained a financial services giant in the late 1980s and later merged to become Canada Trustco.[38] The passing of a chain of related and permissive acts enabled this and other companies to become substantial operators.

In 1859 the legislature of Canada granted major new powers to building societies, while retaining the uncomplicated registration procedures. The act stated that the legislators, understanding that a few building societies had established themselves on a permanent basis, wished to bring them under the protection of the 1846 act, clarifying their legal status. Furthermore, it empowered the societies to borrow money up to three-quarters of the amount of capital actually paid in on unadvanced shares and invested in real securities and to possess a business office and hold real estate to the annual value of $6,000. This act confirmed building societies as potential commercial loan companies. Subsequent federal and provincial acts added to the borrowing powers so that societies' managers could procure greater leverage from their assets.[39]

After Confederation, the Dominion of Canada and the province of Ontario shared authority for regulating building societies. From

1859 to 1874 legislators were rarely called upon by the budding financial service companies. The statutory conditions under which they operated remained basically the same. The financial panic of the late 1850s, the Civil War in the United States, and tensions between the republic and the United Kingdom had dampened British interest in Canadian investments. Frances Kestemann, a shrewd widow whose assets helped to back the brokerage business of her Toronto nephew Weymouth George Schreiber, wrote in 1860 that she had been astonished to have read in the *Times* that two new Canadian loan companies were peddling their debentures in England. She implied little hope. Canadian mortgages brought in 6 to 7 per cent interest, but recent bad news from Canada, she wrote, "make us wish all our money safely back in England's Funds at 3 per cent."[40] By the 1870s this cautiousness in England had yielded to new enthusiasm. The pressure of savings at home pushed money toward colonial and American agents and institutional leaders. The federal Parliament, undoubtedly at the behest of loan companies, passed an act affecting the management of permanent building societies carrying on business in Ontario. While the act's terms demonstrated eagerness to assist institutional lenders in raising English capital, they also showed prudence. Legislators, with one financial panic just behind them, now strove to balance the urge to permit considerable credit leverage on a limited capital base against a recognition that an institutional failure could retard the continuous campaign to induce British investment in government as well as private securities. Therefore, under the terms of 1874 legislation, building societies with more than $200,000 capital could lend "to any person or persons or body corporate at such rates of interest as may be agreed upon without requiring any of such borrowers to become subscribers to stock or members of such society."[41] In essence, the societies could cease being societies, an option taken by many, who confirmed the new status of their corporations by renaming them. They shed the garb of mutuality and put on the vestments of solid lenders. Another extremely important point for the capital-raising side of the business, the eligible companies could issue debentures and receive money on deposit: They no longer had to depend on capital raised by share subscriptions and loans against a proportion of that capital. As a precaution, Ottawa enabled only larger societies to tap into the new capital source.

The sweeping transformation of names may have been calculated to entice British capitalists who, it was hoped, would snap up Canadian debentures. Getting funds across the Atlantic and into Ontario mortgages was the express purpose of the transmuted societies as they grew in the 1870s and 1880s. Indeed, their success in gaining British investors' confidence, which had been shaken in the 1850s and 1860s, was so complete that within a few years they were "champing at the bit" to obtain authority for working up greater leverage from their assets so as to borrow larger amounts. The Parliament of Canada forthrightly declared this purpose in an 1882 act designed to enable the building societies to raise more funds "from beyond the limits of the Province."[42] An earlier government act setting out to accomplish this in 1877 was introduced by Andrew Trew Wood, who represented Hamilton and who counted among his diverse interests (centred on wealth accumulated as a leading hardware wholesaler) a directorship in the Hamilton Provident and Loan Society. Both his supporting remarks and those of a sole critic demonstrated that there was a great distance between the lending societies of the 1840s and those of the 1870s. The act was framed to help the societies plunge into the British money market. D'Alton McCarthy wondered who would benefit, alleging that the societies deceived borrowers and entangled them in heavier obligations than those implied in company circulars. In effect, he charged that societies used false or misleading information in advertising interest rates. He cited an example that resembled the case of John Scott nearly a quarter century earlier. McCarthy's criticisms stirred no great fuss, but the modern concerns of consumers' groups are clearly foreshadowed in his objections. One of his remarks could pass as twentieth-century naïve commentary. "The profits [of the loan societies] were so large that the salaries of the managers of some of these societies were larger than the salary of the Premier of Canada."[43]

Meanwhile, the Ontario legislature had been passing a batch of its own statutes. Jurisdictional matters were to be resolved in 1914 by allowing the loan and trust companies to choose between federal and provincial incorporation and regulation. However, Ontario's legislation in the 1870s largely duplicated the federal measures. This repetition of effort was one of the era's many examples of the tangled affairs produced by the combined activities of a new federal state and new agencies of capitalism. It also underlined the intention of both levels of government to facilitate the pursuit of British funds. An 1877 act enabled any British building and loan society to apply for a licence from the Provincial Secretary and, if licensed, to transact mortgage lending. That same year and again in the next, the province expanded the crucial debenture issuing capacity of the loan companies.[44] By the late 1870s many of the building societies could appropriately be called loan companies. An Ontario act of 1877 permitted societies to change their names and many of them adopted

explicit titles. The many companies operating under charters required special acts to alter their names. The tone of the changes
deserves comment. The London Freehold and Leasehold Land Benefit Building Society – a descriptive building-society label if there
ever was one – was merged with other societies to form the Agricultural Savings and Loan Company. The Canadian Permanent
Building and Savings Society – another revealing name – had been
founded as a permanent building society in 1855, but in the mid-
1870s it became the Canada Permanent Loan and Savings Company,
and in 1894 the Happy Home Building Society became the Sun
Savings and Loan Company. Two alterations were common. First,
the term "permanent building" was either wholly replaced by the
word "loan" or spliced to become "permanent loan". "Permanent"
had had a particular meaning when taken up by building societies,
but it was translated suddenly into an adjective designed to inspire
confidence among investors. Second, the once-significant noun "society" was displaced by "company". Perhaps to worldly financiers and
the British investors of the era, "society" evoked familiar images of
a casual club of artisans and small merchants gathering monthly at
tavern, home, or meeting hall to co-ordinate mutually beneficial
objectives. Overall the new names flatly declared business objectives.

True building societies remained on stage, though much in the
background. In 1897 the first public report on building societies,
loan companies, and trust companies made this clear. The activities
of the institutions were partly disclosed by their names, but the report
also published balance sheets and information about interest rates
and ratios of property values to mortgages issued.[45] The following
were almost certainly true building societies: The Frontenac Loan
and Investment Society, The Ontario Building and Savings Society,
The Home Building and Savings Association of Ottawa, The Peoples'
Building and Loan Association of London, The Peterborough Workingmen's Building and Savings Society, and The Sons of England
Building, Loan and Savings Association.[46] These and others were
small and local, and retained a fraternal element in their names.
The 1897 report affords a good glimpse into the structure of the
societies and companies, in relation to their comparative standing as
suppliers of mortgage funds. Of the 96 firms reporting, 10 per cent
held about 40 per cent of the group's assets (see table 6.1). The
institutional loan business had become concentrated. Minor, ephem-
eral, hopeful, and mutualistic building societies likely had a negligible impact on capital mobilization for housing in the late nineteenth
century. In truth, until the 1897 public report, the activities of the
societies and of the loan companies left scant traces. Newspapers

Table 6.1
Summary Characteristics of Loan and Trust Companies in Ontario, 1897

Dates of Incorporation			Head Office Location in Ontario		
	n	%		n	%
Before 1860	6	6	Toronto	43	45
1860–1869	5	5	London	9	9
1870–1879	33	35	St Thomas	5	5
1880–1889	23	24	Ottawa	4	4
1890–1897	29	30	Hamilton	3	3
			Other	32	34
Totals	96	100		96	100

Concentration of Capital

10% of trust and loan companies held 40% of assets

Sources of Capital	$1,000	%
Shareholders' Equity	59,504	39
Deposits	17,928	12
Debentures Payable in Canada	11,538	7
Debentures Payable Elsewhere	43,583	28
Other (Loans etc.)	21,337	14
Totals	153,890	100

Mortgages as Percentage of Assets

Loan, Trust, and Landed Loan Companies = 60%

Interest Rates

Average Rate Earned on Realty Loans	= 5.9%
Average Paid on Deposits	= 3.5%
Average Paid on Debentures	= 4.1%

Source: Ontario, Sessional Papers (no. 36), *Loan Corporations Statements ... for the Year Ending 31 December 1897* (Toronto: Warwick Bro's and Rutter, 1898)

occasionally published the societies' annual statements or announced meetings, and one sequence of newspaper items on a mutualistic building society underscores their modest scale.

An account of the Hamilton Homestead, Loan and Savings Society returns the study to its subject city and establishes another connection between workers' causes in the late nineteenth century, the land-reform crusade of Henry George, and the popular will to possess. The Hamilton Homestead, Loan and Savings Society (HHLSS) published its first annual report in January 1884, placing its audited statement and a description of its activities in the Hamilton-based *Palladium of Labor*, a journal of radical commentary covering labour news in southern Ontario. The HHLSS did not raise capital by borrowing through debentures hawked in England. In one year it had raised nearly $11,000 solely from the dues of an unspecified number of members. Possibly several hundred had subscribed, because the principal was considerable when it is recalled that a subscriber's dues would have been limited to only a few dollars per month. Admittance fees had brought in almost $300 and fines another $45.67. In its first year of operation, the society had made eighteen loans, but the report did not explain how the society allotted these to its members. Given the potentially destructive effect of disclosing that new members would likely have to wait in a queue, bid for early loan rights, or accept a position by lottery, the omission is understandable.

The mean mortgage loan extended by the society – $775.92 – was roughly half that made by one of the city's mortgage brokers during the same period, suggesting that members took out loans for materials and some subcontracting while putting in their own "sweat equity," or that they purchased very small houses. Of the eighteen loans, fifteen were mortgages on an owner-occupied dwelling. The remaining mortgages were taken out on rental properties. With its report, the HHLSS printed a breathlessly confusing and optimistic statement about how the society operated. It did not mention that members would have to wait for enough capital to accumulate before they could borrow. New memberships, pursued in this prospectus, would have lessened the wait for established members. So, in a sense, the report was self-serving and suspect. However, its choice of appeal has interesting implications. The fact that a small building society would choose to play up the idea of a worker's desire for independence through home ownership suggests once again that ownership sentiments were too varied in origin and too diffuse in expression to have been foisted on the public. The building society's pitch, though tainted, was scarcely a part of an advertising blitz to dupe workers into home ownership. It had a grassroots appeal and showed naïveté as well as calculation: "Each month's interest and dues are carried to the credit of the borrower and reduces his liability by just that amount, so that it may be said he saves his rent and lives rent free, as the rent he should as a tenant pay a landlord, he goes to make a home, and at the end of the year the borrower has that much saved."[47]

This mutual society broadcast the theory of ownership by instalment as a wise investment. A letter to the editor, published in a later issue, praised the HHLSS and asked: "Why are the working classes (of Pennsylvania) so well off?"[48] The answer was home ownership and the subscription to shares "in one or more of the Homestead Societies." The obvious promotion of the HHLSS among the *Palladium's* readership paralleled the Knights of Labor and *Palladium* build-up for the August 1884 visit of Henry George.

When the Royal Commission on the Relations of Labor and Capital in Canada held hearings in Hamilton in January 1888, the secretary of a Hamilton Homestead Loan and Building Society appeared and reported on the inner workings of a building society. It is likely that this society was the same as the HHLSS. The secretary stated that the society made loans almost exclusively to "the working classes" or to mechanics. While his report further supports the ideas about a demand for home ownership among working families spelled out in chapter 4, the details on the society's operations give a rare glimpse into how true building societies functioned in Canada. The Hamilton organization gave labourers an opportunity to save and to share in the profits of a financial enterprise. Participants subscribed shares and the minimum payment was a dollar a month. A full share cost $200, but subscribers likely could sign on for half shares. In 1888, 649½ shares had been subscribed for by mechanics or labourers, 33 by doctors and lawyers, and 130½ by women "who earned their bread as serving girls and servants." The women may have been accumulating dowries. When the monthly payments built up capital to a certain point, the right to a loan was made by auction. Let us assume that a mechanic had subscribed for five shares. Paying at $10 a month, his shares would have been paid up in eight years, the limit set by this society. Now in eight years, the shares, with a par value of $1,000, would belong to the mechanic, but he would want capital sooner. Consequently, he could attend the auction and bid with his shares by discounting them. He would agree to a $1,000 mortgage debt and receive a discounted sum. For example, he might bid $666.69 and, providing no others bid lower, get that sum. He would continue to pay on his shares until they had been paid up. At that point, he would turn over his shares to the society and liquidate his $1,000 mortgage. At the discount rate he had bid, he was paying roughly 4½ per cent interest. In effect, the share subscription and bidding by discount produced a mortgage repayment scheme that blended principal and interest at a time when mortgages typically required a principal payment at the end of the term. Here, then, was a novel approach that would catch on gradually among commercial lenders. The society's property committee inspected the mortgagor's property before agreeing to a mortgage loan. It also made loans to men who planned only to live in the dwellings they built or bought.[49]

If the HHLSS was set against commercial practices and followed a prudent but mutualistic course, another Hamilton building society exemplified the path of development that led to the creation of very large and quite commercial enterprises. Although it was called a building society because it only lent on the security of real estate, the Hamilton Provident and Loan Society was founded in 1871 by prominent manufacturers and merchants, not mechanics looking for ways to secure home ownership. At first, their share subscriptions provided the capital. In 1871 the society had invested just over $34,000 in mortgages. Five years later, with funds increased largely by savings deposits, it had nearly $700,000 invested in mortgages. A surge of debenture sales boosted investments to over a million dollars by 1878. The Hamilton capitalists benefitted greatly from the legislative changes, pressed by their colleague, Andrew Trew Wood, in the federal Parliament. They were local representatives of a major phase in the growth of Ontario lending institutions.[50]

For shrewd facilitators of capital movements across the Atlantic, the 1870s and 1880s were palmy days. As never before and only rarely since, would-be financiers set up new loan companies. Significantly, the stream of mortgage capital from the imperial centre coincided with the dissemination of plans for and discussions about houses for the working class. There was a conjuncture of technology, communication, and capital mobilization. Canada and Ontario had encouraged this remarkable period of sustained expansion in the number of financial service companies with legislation that responded to the needs of growing loan companies. Rounding out a felicitous convergence, the United Kingdom continued to swell with pools of savings poised to flow toward the higher interest rates of that new and exciting enterprise – Canada. Reports for all of Canada show less than twenty thousand dollars of British funds invested in trust and loan companies in 1874, the year in which the debenture route to raising funds had been approved by Ottawa. When the federal parliamentarians had liberalized the process further in 1877, debentures sold in the United Kingdom had raised just under four million dollars. In a spurt of investment, the sum had risen to about twenty three million in 1879, and it attained its historic peak of forty-nine million in 1893.[51] At that time, roughly 80 per cent of Ontario loan company debentures were payable outside Canada. According to an analyst of the time, these debentures were "mostly held by the landed gentry in England and Scotland, especially Scotland."[52]

The British funds that streamed to Canada by way of loan-company debenture transactions mostly poured into Ontario, although some Ontario firms moved money into western Canada. It is impossible to pin down precisely how much of the money lent out by loan companies and secured by Ontario mortgages originated in the United Kingdom at any one time. It must have been considerable during most of the nineteenth century, judging from the circumstances attending the nurturing of the Trust and Loan Company of Upper Canada and from the overtures implied in the legislation of the 1850s and 1870s. Assuming that British investors held half of shareholders' equity and around 80 per cent of the debentures, then in 1897 half of the capital in Ontario loan companies had originated in the United Kingdom. British debenture holdings would fall and so probably would British shareholders' equity. By the 1920s, if not much earlier, Canadians supplied the major portion of Ontario loan company capital (see table 6.2). Therefore, the trust and loan companies of the 1920s, having raised hundreds of millions of dollars in Canada, demonstrated how unlike their ancestors they had become. British capital put life into the Trust and Loan Company of Upper Canada during the 1840s and 1850s. It speaks to the maturity of Ontario's loan companies that the 1910 withdrawal of the oldest firm from new Ontario business was recorded by footnotes in government reports.[53]

The relative fading of British capital was one development in the early twentieth century, consolidation was another. The Canada Permanent Loan and Savings Company, reorganized as The Canada Permanent and Western Canada Mortgage Company in 1899 and renamed Canada Permanent Mortgage Company in 1903 had been the largest Ontario loan company. It had merged several substantial competitors in 1899: Western Canada Loan and Savings Company; Freehold Loan and Savings Company; and The London and Ontario Investment Company. Canada Permanent took over The Oxford Permanent Loan and Savings Society (1918), The London and Canadian Loan and Agency Company (1921), The Royal Loan and Savings Company (1927), The British Columbia Permanent Loan Company (1927), and The Canada Landed and National Investment Company (1927). The Huron and Erie Loan and Savings Company, another major firm, likewise absorbed several loan companies. Toward the end of the decades of great expansion, in 1897, ninety-six firms had registered in Ontario. By 1928 the number had dropped to fifty-six, including the asset-rich Montreal Trust Company, which had begun to operate in Ontario in 1909. Ontario's trust and loan companies had become fewer and wealthier (see table 6.2). The depression and the wartime curtailment of the mortgage business stilled the formation and reconstruction of loan and trust companies during the 1930s and 1940s. During the 1940s investment funds moved away from deposits and debentures and into government

Table 6.2
Summary Characteristics of Loan and Trust Companies in Ontario, 1928

Dates of Incorporation			Head Office Locations		
	n	%		n	%
Before 1860	2	4	Toronto	20	38
1860–1869	1	2	London	8	15
1870–1879	8	15	Ottawa	2	4
1880–1889	9	17	Hamilton	1	2
1890–1899	14	27	Sarnia	3	6
1900–1909	6	12	Other	18	35
1910–1919	7	13			
1920–1929	5	10			
Totals	52	100		52	100

Concentration of Capital

4% of loan companies held 50% of loan companies' assets
4% of Ontario-based trust companies held 43% of trust companies' assets

Sources of Capital for Loan Companies	$1,000	%
Shareholders' Equity	78,716	34
Deposits	35,612	16
Debentures Payable in Canada	60,477	27
Debentures Payable Elsewhere	48,822	22
Other	2,428	1
Totals	226,055	100

Mortgages as Percentage of Assets

Loan Companies	79%
Trust Company Funds	38%
Trust Company Guaranteed Funds	52%

Interest Rates

	Loan Companies	Trust Companies
Received on Mortgages	6.8	6.3
Paid on Deposits	3.6	3.7
Paid on Debentures	5.5	5.4

Percentage of Mortgage Investments Located in Ontario

Loan Companies	50%
Trust Company Funds	46%
Trust Company Guaranteed Funds	71%

Source: Ontario, *Loan and Trust Corporations' Statements ... for the Year Ended 31st December 1928* (Toronto: Printer to the King's Most Excellent Majesty, 1929)

bonds. Recovery for the trust and loan companies came slowly. Only by the 1960s had the companies regained the share of financial business that they had held in the 1920s, once more achieving assets valued at approximately 20 per cent of those in the hands of the chartered banks.

The 1960s and 1970s resembled the expansive years of the late nineteenth century as loan firms were launched and out-of-province companies moved into the booming Ontario real-estate markets (see table 6.3). In other developments, loan companies merged with trust companies or otherwise acquired trust powers. By the 1980s the largest trust companies – National Victoria and Grey Trust, Royal Trust, Canada Trustco, Guaranty Trust, Montreal Trust, and Central Trust – had combined assets of nearly $70 billion. Several handled real-estate sales, managed trust accounts, traded securities, functioned as banks, and energetically participated in the mortgage market. Canada Trustco, the result of a merger of Canada Trust and Canada Permanent, was the largest with assets of $24 billion in 1987. In the Toronto region alone it operated seventy-eight financial branches and twenty-eight realty offices. In the hands of unscrupulous manipulators, a few small trust and loan firms, such as Security Trust, had engaged in the inflation of real-estate appraisals and subsequently brought their companies crashing down in 1982. Headline news in the 1980s, the trust and loan establishments had evolved to resemble American savings and loan companies. From new suburban branches they competed with the federally-chartered banks for deposits and dealt directly with consumers who sought mortgage finance, frequently for older housing stock. Throughout their histories the loan companies had concentrated on mortgage lending to the exclusion of all other investments except those required for equity and for establishing funds for further mortgage investment. As a result, they have had about 70 per cent of their assets in mortgages whereas trust companies have maintained broader portfolios with only 45 to 60 per cent of their assets in mortgages.[54]

Less visible than the publicity-prone and well-promoted trust and loan companies of recent times, insurance companies have also staked out a position in the mortgage market. Known for their policy sales, solidly discreet architecture, and corporate images of studied caution, insurance companies functioned as important institutional lenders with singular hallmarks. Because they lacked a network of neighbourhood branches, their mortgage lending tended to be of the kinds that demanded minimal paperwork and personnel. Not meeting directly with borrowers, they cut their risks by concentrating

Table 6.3
Summary Characteristics of Loan and Trust Companies in Ontario, 1985

Dates of Incorporation			*Head Office Locations*		
	n	%		n	%
Before 1900	9	11	Toronto	38	45
1900–1929	10	12	Montreal	15	18
1930–1939	0	0	Calgary	4	5
1940–1949	0	0	London	3	4
1950–1959	0	0	Halifax	4	5
1960–1969	18	21	Hamilton	2	2
1970–1979	36	43	Other	18	21
1980–1985	11	13			
Totals	84	100		84	100

Concentration of Capital

10% of loan companies held 48% of loan companies' assets
10% of trust companies held 52% of trust companies' assets

Sources of Capital for Loan Companies	$1,000	%
Shareholders' Equity	2,309,236	4
Deposits (Demand)	6,842,907	13
Deposits and Debentures (Term)	43,600,685	83
Totals	52,752,828	100

Mortgages as Percentage of Assets

Loan Companies	77.4%
Trust Company Funds	33.5%
Trust Company Guaranteed Funds	63.1%

Interest Rates

	Average Earned on Mortgages	Average Paid on Shares	Average Paid on Deposits
Loan Companies	12.1%	7.7%	11.6%
Trust Company Funds	8.5	11.6	–
Trust Company Guaranteed Funds	12.4	–	10.7

Source: Ontario, Ministry of Financial Institutions, *Report of the Registrar, Business of 1985, Loan and Trust Corporations, 89th Report*

on new housing where the structure alone rather than the mort-gagor's credit worthiness weighed in the decision to lend. Sometimes they held down overhead by preferring to deal with clients who had large needs, like apartment-building developers or commercial and industrial borrowers. Also, while the trust and loan companies have preferred to lend for five-year terms, the insurance lenders wrote or bought mortgages with twenty- or thirty-year terms. Although not visible in the current day-to-day mortgage market for home buyers, insurance companies historically have participated in new suburban undertakings and played an important role in the federal government's intervention into the mortgage market immediately after World War II.

Insurance companies carrying marine and fire risk had entered British North America routinely during the early nineteenth cen-tury. Dealing with risk, the fire insurance companies acquired knowl-edge about industries, building materials, and fire equipment. By the later years of the century, their research and detailed fire in-surance atlases influenced urban planning and house-building prac-tices.[55] Life assurance companies developed in Canada at mid-century and they also accumulated important skills, pioneering in actuarial studies and hence the development of epidemiology and demography. In terms of their public and declared purposes, there-fore, insurance companies were to have a tremendous influence on Canadian life. Their impact extended, through a relatively unap-parent dimension of their business, to home ownership. Insurance firms were simply wonderful instruments for accumulating capital. A former insurance-company mortgage specialist and CMHC official, David Mansur, admittedly speaking about a specific period, bluntly expressed a more universal truth: "The life companies seemed stuffed full of money."[56] Premiums rolled in on a regular basis and, for the most part, the normal intervals between payouts meant that the firms were not compelled to maintain large liquid reserves. In short, insurance companies could pour regularly collected revenues into long-term investments. Like their corporate cousins, the loan and trust companies, indigenous insurance companies cropped up in Ontario just before 1850. Mutual fire insurance companies had begun to appear in the 1830s, and by 1897 the province had ninety mutual fire insurance operations.[57] Most were minuscule and many, but not all, were township-based. Legislation in 1859 had permitted the directors of mutual insurance companies to invest funds in mort-gages on real estate, bank stock, or shares in building societies. The techniques for accumulating and recycling funds were from the start to form a fabric of interconnected interests.

A group of Hamiltonians had founded the country's first life insurance company in 1846, securing a charter as the Canada Life
Assurance Company in 1849. The directors received authority to
invest in mortgages.[58] Across a mere two decades of boom and bust
for the city and the colony, the Hamilton firm had succeeded in
amassing assets of $1 million.[59] In 1871 mortgages accounted for
nearly 30 per cent of the investments. With typical residential mortgages for that period being worth $500 to $1,000, it is conceivable
that Canada Life could have held 300 to 600 residential mortgages.
Regrettably, it is impossible to estimate the impact on Hamilton, for
the company spread out its investments geographically and may well
have invested in mortgages on industrial and commercial properties.
Nevertheless, it is worthwhile to speculate on how Hamilton's housing market and building trades were affected by the presence of
locally-based financial institutions during the late nineteenth century: Canada Life Assurance (1846–1899); Landed Banking and
Loan Company (founded 1876); and The Hamilton Provident and
Loan Society (founded 1871). As we shall see, until the 1920s the
institutional lenders carried only a small fraction of Ontario and
Hamilton mortgages, although they facilitated the property industry's activities in other ways. Investments in civic bonds financed
utilities, and loans to "agents" on the security of stocks, bonds, or
out-of-town properties enabled the agents to invest in mortgages in
their own names, using loan company funds. Loan company credit
also enabled well-placed individuals to invest in mortgages as as-
signees or to lend to builders and building supply manufacturers,
who were then able to take back mortgages on properties they sold.
Once capital had been accumulated by a loan company, there were
many legal instruments and personal channels for recycling the
funds into house building and house buying. In a reconstruction of
how the activities of the Landed Banking and Loan Company in the
1880s meshed with the private recycling of capital by John M. Gibson,
the city's representative in the provincial cabinet, Gibson's biogra-
pher, Carolyn Gray, has uncovered the threads of deals involving
individuals and the backing of the loan company. To see the loan
companies and the insurance companies as participating only in the
mortgage market as mortgage holders offers a narrow view of their
city-building roles. They appear to have assisted many who then
took on mortgages in their own names. It is little wonder that towns
across Ontario had spawned so many building societies and insurance ventures.[60]

By the late nineteenth century, the variety of insurance companies
operating in Canada was extensive and several firms had far more

mortgage business than others. The federal-provincial jurisdictional division affected the industry, just as it had the loan companies. Small mutual companies and friendly societies registered under the laws of Ontario and rarely invested in mortgages, preferring to place funds in bonds, stocks, and debentures. This practice illustrates the general North American rule that only the larger insurance firms, with enough staff to manage appraisals, have taken an interest in mortgages, although another trickle of insurance funds flowed into loan companies through the purchase of loan company debentures. The federal government registered the British and American insurance companies that operated throughout the country, and issued federal charters to insurance companies. Among the federally supervised companies, some giants in the industry had taken up very large positions in the institutional sector of the mortgage market. Companies based in the United States rarely placed funds in Canadian mortgage-backed loans because state laws sometimes forbade investments beyond their boundaries. The Republic, like the Dominion, was struggling to accumulate capital for internal development. British firms behaved differently since, as we have mentioned, the United Kingdom was exporting capital. Two British fire and marine companies put considerable sums into Canadian mortgages. The North British Company, the largest player, had put in over $2 million according to its report for the calendar year 1897. The second largest sum of mortgage investments originated with the Liverpool and London and Globe Company. Life insurance companies were the major players, having taken the largest share of the insurance companies' mortgage investments.[61]

Canadian life insurance firms were very significant lenders. Placing about a quarter of their assets into mortgages, they had nearly $16 million invested in the area in 1897. Canada Life and Sun Life had grown into the industry's titans. British companies had channelled roughly a third of their Canadian assets into mortgages, and altogether their mortgage investments in the Dominion were less than half that of the Canadian firms. Like their fire insurance compatriots, United States life firms seldom invested in Canadian real estate. Finally, there was a small class of so-called assessment system life insurers, essentially lodges, of which the International Order of Foresters was the largest. Lodges placed about half of their investments into mortgages.[62] As a rule, the insurance companies rarely put more than half of their assets into mortgages whereas the Ontario loan companies historically have carried 60 to 80 per cent of their assets as mortgages. The 1897 investment practices, as revealed in reports to federal and provincial authorities, appear to have been part of a continental historical pattern and not exceptional. A study of twenty-nine of the larger United States insurance companies disclosed that, with one notable exception, the percentage of their assets in mortgages from 1870 to 1950 ranged between 25 and 40 per cent. A 1964 Royal Commission on Banking and Finance reported 40 per cent as the normal condition in Canada.[63]

During World War II, the insurance companies in Canada and the United States drastically deviated from tradition in assembling their investment portfolios. Of course, they had allowed their mortgage portfolios to decline in the 1930s, for they avoided new risks. In the 1940s, they soaked up war bonds. Their mortgage assets declined to 10 per cent or less. One consequence was that, after the conflict, the insurance firms wished to liquidate vast holdings of low-yielding bonds and increase their mortgage assets.[64] Undoubtedly, they had a vested interest in seeing a post-war shelter construction boom. The larger companies, like Sun Life of Canada, worked with the government in Ottawa to plan for a privately-financed but government-guaranteed mortgage lending scheme that would convert company assets quickly into mortgages.[65] In Canada the liquidation of government bonds and the reciprocal vigorous moves into the mortgage markets had restored the normal balance in insurance company portfolios by 1952. Once settled happily into their pre-war pattern, the insurance companies slowed their mortgage-lending activities. In scanning the land registry records for new suburbs in Hamilton from 1948 to 1952, the surge of insurance funds stands out. The names that figured prominently were the Equitable Insurance Company, London Life, Imperial Life, Sun Life, and Mutual Life (see table 6.4). Almost as perceptibly, they scaled down after 1952. Yet home ownership demands remained strong, forcing the federal government itself to resort briefly to massive direct lending. The Ministry of Finance disapproved of this lending and discovered an expedient that brought a third type of financial institution into mortgage lending.[66]

Against the backdrop of the real-estate boom and collapse of 1857 to 1865, the federal government in 1871 introduced a bank act that banned chartered banks from lending on the security of real estate. The United States bank act of 1864 had done the same. Not only had the market collapse demonstrated that real estate was notoriously difficult to appraise, but financial wisdom in both countries now held that the banks' liabilities – their holdings of short-term deposits – were incompatible with longer-term investments like mortgages, which were felt to be insufficiently liquid to meet a stream of panic withdrawals by depositors. By the mid-twentieth century,

Table 6.4
Several Attributes of the Hamilton Mortgage Market, 1870-1955

Years	Area of the City and Character of the Area	Number of Mortgages in Sample	% of Mortgages in Private Hands	Mean Value of Mortgages	% of Mortgage Value to Sale Price
1847-81	City wide	1,700	93	–	–
1870-1905	City wide	110	99	$ 540	64
1870-90	North End: Small frame dwellings and brick row houses	30	100	470	50
1870-90	Centre City: Mixed housing	20	90	620	70
1920-30	Barton Roxborough: Houses of skilled labourers	20	80	2630	–
1910-30	Cumberland: Poorer brick houses with tenants	40	100	3000	–
1920-40	Westdale: High quality new middle-class area	40	45[1]	3300	–
1948-52	Centre Mountain: Modest new brick houses owned by skilled labourers	30	25[2]	3700	–

Source: Wentworth County Registry office records
[1]Given its tone and new dwellings, Westdale was popular with the loan and trust companies (Huron and Erie, Canada Permanent) and the insurance companies (Equitable Life, London Life, Mutual Life, Sun Life).
[2]The mortgages here often indicated a CMHC guarantee. The prominence of insurance company mortgagees illustrates their post-war move to adjust investment portfolios. The following institutional lenders were prominent: Canada Permanent, Dominion Life, Equitable Life, Huron and Erie, Imperial Life, London Life, Mutual Life, Stelco Credit Union, Sun Life, and Western Life.

Canadian bankers no longer worried much about this liquidity issue, although as a business convenience they wanted to see the establishment of a formal mortgage market wherein they could trade mortgages, expanding their available cash for new investment opportunities should these appear just when they had a bounty of mortgages. Essentially, Canadian bankers believed that their liabilities had come to include a hard core of long-term depositors and that these now permitted the banks to venture safely into mortgages. In the United States, restrictions on national banks lending on real estate had begun to be lifted as early as 1913. While liquidity was not a serious problem, Canadian banks continued into the 1970s to advocate measures or to create institutions that could improve mortgage liquidity. Also by mid-century, another former source of bankers' caution had diminished.

Bankers historically had worried about the costs and losses of foreclosure. They were not as free as the trust and loan companies in charging interest rates high enough to include a margin to offset risks. However, the adoption of blended-payment amortization schedules that included a portion representing taxes became widespread among the savings and loan banks of the United States in the immediate post-war years. Their experiments were watched in Canada. Lending officers believed that the new mortgage-repayment scheme gave the mortgagee closer supervision and better security. The mortgagor's equity also increased with each monthly PIT (principal, interest, and taxes) payment, making default a greater loss to mortgagors than under other repayment schemes. Furthermore, the blended payments had built-in liquidity, for the principal was gradually being repaid. Still the Canadian banking fraternity held back. It did not insist on access to mortgages, remaining slightly sceptical about the security of blended payments, wary of bad publicity that might accompany the eviction of mortgagors, and concerned about the overhead costs of training and hiring appraisal officers. Also, by the early 1950s, they must have recognized the strength of their position. Bankers were used to the long game – politicians needed immediate achievements. The government was quite desperate to mobilize the banks' resources and to make use of their 4,000 or so branches to promote home buying. Therefore, in late 1953 and early 1954, the banks and the federal government honed an act that dispensed with the risk of default by insuring lenders to 98 per cent of the mortgage and interest. Home buyers actually paid a one-time insurance premium, which was then blended into the monthly repayments. Typically, the cost was two hundred dollars and it worked out to an added half per cent on the monthly instalments. CMHC, the federal housing agency, would handle the appraisals of properties. Banks, however, could not evade the burden of foreclosure. They had to turn over empty premises to CMHC in order to secure the 98 per cent refund. The other terms of the program will be discussed when looking at the use of tax revenues for housing measures later in this chapter. Here we focus on the banks' mortgage business.[67]

Bankers held most of the trump cards in 1953–4. They controlled the money and they had the time. To government officials, they grumbled softly about having been dragged into the mortgage market, but in almost the next breath, they boasted about how they had over achieved, exceeding government estimates on the volume of initial lending. From March to the end of 1954, the banks had lent 60 per cent more money than projected and enough mortgages for 17,000 housing units.[68] As always, it seems, the capacity to lend would eventually bump into a business reluctance to go further, a capital shortage, or a legislative barrier. In 1959 two of those conditions temporarily took the banks out of the mortgage business. The 1853 abolition of the usury law had not been absolute – the chartered banks had a 6 per cent lending rate fixed by federal law and in 1959 the National Housing Act interest rate crept above that. Bankers thought, or at least claimed, that further lending by their institutions would have been illegal, so they stopped taking on new mortgages. The banks had also become "fully loaned up" in a period of tight money.[69] By the early 1960s, however, the bankers had abandoned all pretence of aversion to mortgage lending. Loan and trust companies, hard hit by the depression and the wartime controls on construction, had resumed expansion and had poached on the banks' deposit business. Having begun to learn about mortgage operations in the 1950s, the bankers now had the appetite to go head-to-head against the booming "near-banks."[70]

To convince the government to permit the banks to raise their interest rates above 6 per cent, the bankers hinted that their institutions would then move into the older-house mortgage market, supposedly bringing efficiency and competition to an area dominated by private lenders and the loan and trust companies. With assurances from the bankers that their institutions would not be "inners and outers" in the mortgage market, the federal government clarified the matter of the interest-rate ceiling in 1967 and welcomed banks into the mortgage business. Thus, by the 1970s and 1980s, the mortgage market in Ontario appeared to have become the domain of institutional lenders, although the scale of the private lending market remained a matter of guesswork. In Hamilton, the institutional lenders, commencing in the 1960s, worked closely with contractors so that new-home buyers normally assumed a mortgage from a bank or from a bank's hostage mortgage company. As subdivisions matured and houses were resold, private mortgages or mortgages financed by credit unions began to appear. Teachers and steelworkers were especially well-served by credit unions, a fact that reinforces the observations about the occupational mix of home owners made in chapter 7 (see table 6.5). Not only had the mortgage

Table 6.5
The Transition to Institutional Mortgage Lending in Hamilton, 1956-1987

Area of the City and Character of the Area	Number of Mortgages in Sample	% of Mortgages in Private Hands	Mean Value of Mortgages[3]	% of Mortgage Value to Sale Price
NATHANIEL HUGHSON SURVEY: NORTH-END LOW INCOME				
1956-65	25	68	4450	–
1966-75	30	72	8150	–
1976-85[1]	40	17	26,100	–
TUXEDO PARK: MIDDLE-INCOME AREA NEAR A PARK				
1956-65	20	44	9030	–
1966-75	30	30	18,750	–
1976-85	30	20	34,650	–
MOUNTAIN VIEW: WEST MOUNTAIN NEW HOUSES				
1969-75[2]	40	10	19,180	–
1976-87	40	24	45,140	–
UPPER PARADISE: EAST MOUNTAIN NEW HOUSES				
1972-75[2]	60	7	21,620	–
1976-87	80	13	40,270	–
KING'S FOREST: EAST END NEAR A PARK				
1973-75[2]	40	10	28,000	–
1976-87	35	9	38,650	–
LINCOLN ESTATES: BEYOND CITY'S EASTERN BOUNDARY				
1973-75[2]	20	4	33,160	–
1976-87	20	4	58,680	–
ALL SAMPLE SURVEYS: TRANSACTIONS FOR 1986-87 WHERE SALE PRICES ARE QUOTED	11	9	85,000	69

Sources: Until the late 1960s new surveys continued to be registered by the land registry system whose abstract books are slightly more complicated than the land titles system introduced in the 1960s. The separate offices that maintain the abstract books and property instruments of the two systems share space and are under the direction of the Ontario Ministry of Consumer and Corporate Affairs.

[1] The streets sampled were the object of urban renewal and the new house construction of the 1970s. Many of the houses were owned by Portuguese immigrants.

[2] These dates represented the initial building phase when institutional lending through the contractors was pronounced. In later years, resales and second mortgages introduced non-institutional lenders.

[3] The average value of mortgages is lower than the average debt burden of households because second mortgages were treated individually and not as part of a household's mortgage total indebtedness. Second mortgages were for small sums and often represented a purchaser's desire not to consolidate and lose a vendor's older mortgage with good terms.

market been taken over by the institutional lenders, but during 1986 and 1987 the federal government announced measures to permit the integration of financial services. Trust, loan, and insurance companies were granted full consumer lending powers. Institutional lenders were growing, overlapping, and coming together with official sanction.[71]

Beginning with the major revisions to the Bank Act in 1967, the federal government initiated a sequence of permissive legislation calculated to stimulate the cycling of capital within the mortgage market. The 1967 act allowed banks to charge more than 6 per cent interest and admitted them to the conventional mortgage market with the stipulation that conventional loans not exceed 75 per cent of the appraised property value. In 1970 the banks were permitted to lend more than 75 per cent, provided the excess was covered by mortgage insurance. This measure stimulated a modest expansion of the private mortgage insurance business that had begun in Canada only in 1963. Several consortia of financial service companies assembled the mortgage insurance companies, producing a new network of associations in what was already an institutional mortgage market of interlocked complexity. Banks and trust companies had their captive mortgage insurance firms, and some trust companies also handled the investment of mortgage funds for life insurance companies, including American ones. To promote the collection of capital and to encourage households to plan for the consumption of owner-occupied dwellings, in 1973 the federal government announced a tax incentive, the Registered Homeownership Savings Plan (RHOSP), which it dropped in 1985.[72]

Measured against all previous decades, the 1970s contained a super-abundance of government measures to stir the mortgage market into greater activity. Arguably, no Canadian industry received as much attention as shelter construction and its ancillaries. During the early 1970s two new types of mortgage institutions were approved. Ontario permitted the operation of Real Estate Investment Trusts (REIT), which issued trust units comparable to common shares. REITs could borrow money to reinvest in mortgages, but had to borrow from financial institutions and not directly from the public. After all, the point was not to compete with existing operations but to draw new investors into mortgage lending. This goal also accounted for the creation of mortgage investment companies (MIC), a corporate form allowed by 1973 federal legislation. These enterprises could not accept deposits, but they could issue debentures. They seemed to have been set up for use by pension funds. Both the REITs and MICs could invest only in revenue-producing properties such as shopping plazas, apartment buildings, and nursing homes. In essence, using its powers to regulate and to permit, the federal government managed to entice private-sector institutions to increase their expertise in mortgage lending, to develop high-ratio mortgages, and to secure more capital for the mortgage market.[73]

A century earlier, the pursuit of capital and of mortgagors had inspired the drafting of candid corporate names and the erasure of those burdened with vestiges of tractable mutuality. At that point in the progression of Canadian capital markets and consumer credit, the loan and trust companies required a strong public image – the name and the architecture had to attract both investors and borrowers. For new loan and trust firms incorporated during the 1960s and 1970s, reaching out to the public remained a crucial exercise calling for an impressive and memorable name, highly visible locations, and promotional campaigns. In contrast, the new breeds of financial services operated without public dimensions – they required no presence. Loan and trust companies, chartered banks, and insurance companies have co-operated through the medium of newly formed service firms that they jointly financed. Novel entities catering only to the corporate requirements of the mortgage market or circumscribed by legislative restrictions on their scope for borrowing and lending, the service firm and new vehicles for mortgage investment operated away from public view. In anonymous offices, contrasting with the public banking halls and the confidence-promoting edifices of the principal capital accumulators, these enterprises adopted suitably bland and unrevealing titles. For example, in 1973 a group of major capital collectors including the Canadian Imperial Bank of Commerce, Canada Trust, Montreal Trust, and Imperial Life Assurance founded Insmor Mortgage Insurance Company, a distinctly anonymous corporation. In the same year, the Bank of Montreal and Royal Trust established a REIT to which they affixed their monograms: BM-RT Realty Investments.[74] The highly interdependent institutional mortgage market of recent years insists on a great deal of information about its borrowers. The key accumulators of money have posed visibly enough to attract investors and borrowers, but they keep their captive service companies off stage. Therefore, in working to perfect the mortgage market from their side, institutional lenders have evolved approaches and designed faces to meet the public from calculated stances while insisting on guileless disclosures from clients. The mortgage, always a document encapsulating a power relationship, has sponsored the rise of institutions that state this relationship in manifold ways apparent to borrowers or understood through semiotics. A social trade-off is implied. The social power of institutional lenders – their power to probe into private lives and yet to insulate themselves – may well be the price for spreading access to the popularly rooted social asset of home ownership. Ironically, institutional aggrandizement has accompanied a particular expression of individualism – the will to possess.

The frenzy of legislative and administrative activity to revamp the mortgage market in the 1960s and 1970s occurred for assorted political reasons. Governments took pride in the abundance of new housing stock in Canada and in its spacious and sanitary quality. They foresaw the pressure of increasing household formation, both family and non-family, especially with the maturity of the baby boomers in the 1970s. Given the cultural importance of the single-family detached dwelling across most of the country, new shelter meant houses requiring consumer mortgage financing. As an industry with few imports and considerable local impact, housing construction was always eyed as a politically sound means to stimulate employment. Moreover, for many years, CMHC's position as a direct mortgage lender and as a mortgage insurer had been criticized for diverting the agency's attention from needed social housing. Its corporate name was held by some wags to declare this preference. CMHC put the mortgage before the housing.[75] A private institutional-mortgage market rich with networks for mobilizing capital, with resources and confidence to meet demands from the house-loving middle-class and better-paid working-class families, was a political requisite that would give CMHC elbow-room for greater effort on the social-housing front. The story of federal and provincial action in the mortgage market and in the provision of social housing has been deferred until the end of the chapter because, as active agents, governments came late to the housing scene, but social housing did occupy much of CMHC's attention during the 1960s. The character and volume of social housing would change, but it remained visible. During the 1970s and 1980s, in concert with other measures calculated to foster a particular vision of national political culture, the federal authorities advertised a bilingual presence: Central Mortgage and Housing became Canada Mortgage and Housing; a few low-income housing projects, many senior citizens' housing complexes, and some co-operative housing ventures announced their federal and provincial sponsorship with requisite emblems in both official languages. In contrast to the conspicuousness of the new housing agencies and the physical prominence of institutional lenders, the invisible stalwarts of the mortgage market continued to ply their trade, although with declining influence. Virtually undiscernible, individuals historically had constituted a massive source for home-buying loans.

THE UNINCORPORATED MORTGAGE
BROKER, 1850s–1980s

Samples of mortgages taken from the Wentworth County land registry records (see tables 6.4 and 6.5) confirm what all inquirers into mortgage banking have guessed. Individual or unincorporated agents and private deals between vendors and buyers could have accounted for half of mortgage lending at least until the collapse of the securities markets in 1929. As late as 1962, when CMHC officials testified before the Royal Commission on Banking and Finance, it was suggested that private mortgage funds might constitute 60 per cent of the residential market. Ontario data on private sector mortgages began to be collected in 1969, and in that year almost half of the province's mortgages came from that part of the market. By 1973 it had dropped to below a third. The transition in the United States seems to have started earlier and progressed further, because in 1946 individuals were estimated to hold 26.8 per cent of outstanding mortgages and by 1960 their share had plummeted to 7.2 per cent. In both countries, however, the mortgage market after the great depression had become increasingly the place of institutional lenders and of governments. An arresting way of expressing this development is to report on the dollar value of institutional mortgage lending. In Canada it had been less than $1 billion in 1945, but by 1983 it was almost $100 billion.[76]

The private mortgage broker, not involved in the politics of loan company incorporation and not subject to routine inspection, stood outside the limelight of financial developments during the nineteenth century. However, two rare and complementary sets of records, both covering the second half of that century, offer glimpses into the non-institutional features of the market. The records of Toronto broker Weymouth George Schreiber, limited to incoming letters and several crude transaction journals, largely dwell on the years from 1856 to 1880, difficult times for the Canadian economy. Letters to Schreiber illuminate how capital had been secured for investments and how mortgagors contested with mortgagees for relaxed terms. In 1871 Schreiber received a letter from the Hamilton real estate and mortgage firm of Moore and Davis. Schreiber represented the vendor of a parcel of land just beyond Hamilton's city limits. Moore and Davis negotiated for a buyer. The exchange also appears in a massive collection of letterbooks and mortgage ledgers retained by the Hamilton business. From this material, a detailed picture of the mortgage market from the late 1850s to 1919 can be fashioned. The characteristics of this market and the conduct of the non-institutional lenders have much in common with what American historian Allan Bogue described in his account of farm mortgages in the mid-west in the later nineteenth century and what Canadian historian David Gagan portrayed in his study of Peel County at mid-century.[77]

Neither of the Canadian enterprises specialized. Schreiber operated an insurance agency, had a hand in managing the Denison estate in Toronto, worked as a freelance city assessor, and speculated in urban and rural properties. Born in 1826, Schreiber had twelve siblings. His father, a Church of England cleric, had brought the family out to Canada, but in 1850 they all returned home. Weymouth came back to settle in Toronto in 1856, shortly after his marriage to Harriet de Lisle of Jersey.[78] A charmer who convinced his family and in-laws to back him in setting up an agency for English investments in Canada, young Weymouth exemplifies the personal link in the era's characteristic transfer of British funds to colonial railway and city building. In the months just following his marriage, his connections plied him with capital. Canada's prospects still sparkled in the summer of 1856. Thus Weymouth's father, seeing in him the family's hope, entrusted him with the worldly interest of a large clan. The de Lisles promptly chipped in £3,500 for investment, and another familiar party, possibly related, placed £5,000 in Weymouth's hands. Cryptic account books for 1856–7 indicate that he was probably acting as broker for eighteen individuals. Credits deposited in their names totalled nearly $8,700. Assuming that most of this consisted of interest at the rate of 6 per cent, the payments represented a principle sum of almost $145,000.[79] Agents operating in Canada were not the only ones to place British capital in mortgages. Bogue found that an extremely ambitious Kansas broker, Jabez Bunting Watkins, drew 28 per cent of his investors from the United Kingdom. Almost half of Watkins's investors were women. Schreiber and the firm of Moore and Davis likewise handled the estates of widows.

Unfortunately for his father and other clients, George Schreiber's investments coincided with a financial panic. For years he received their tolerant appeals to bear down on delinquent accounts. A shrewd wealthy aunt, a widow who resided in London and on the Isle of Wight and held several Canadian mortgages, made a series of fascinating comments on the capital markets and on the way an agent should conduct business. Writing in 1860, early in the prolonged financial slump, she noted that the 6 to 7 per cent returns on Canadian mortgages were excellent, but the bad news from Canada had made "us wish all our money safely back in England's funds at 3 per cent."[80] Seventeen years later, Weymouth still handled delinquent accounts for her and she fretted about small sums. Denying that she was "very avaricious", she complained that "it is so hard to carry all these debits in my old head."[81] Besides, she worried about a son who had failed to get a commission in the Horse Guards and had gone to New Zealand. In the hope of straightening up her Canadian affairs in 1877, she issued a frank appraisal and firm instructions to Weymouth:

I fancy you are somewhat too easy and debtors take advantage of it. I think it would be well in future, in all cases of mortgage, to foreclose immediately the law permits. With my present increased expenses every pound is of consequence. I do not at all like the system adopted by Mr. Dwyer of working out his land, and being nearly two years in arrears of Interest on his mortgage. Under the circumstances Mr. Dwyer should have given you a Bill of Sale over his crops as further security when asking for extra extensions of time. In future, by acting strictly up to the letter of the law, we may avoid similar risks. And should the defaulters think or say you are acting hardly – *lay all the blame on me*, and excuse yourself as being simply an Agent and bound to obey instructions. I hope Bigelow, Jagger, and Scott could pay you as expected.[82]

The problem with foreclosure and sale was that, unless the mortgage had been taken up or bought at a discount, the depreciation in real estate values could leave a loss, and pursuing the balance in court incurred legal expenses. Knowing this, delinquent mortgagors pleaded for time – sometimes noting improvements to the property as a sign of commitment and as added collateral. Weymouth tried to keep the accounts active. Eventually, toward the end of the century, he invested to avoid the friction of a mortgage broker by purchasing loan company shares, including those of the Canada Landed Credit Company launched in 1860 by his brother-in-law George William Allen.[83] From middleman for English capital to coupon clipper, Schreiber's evolution prefigured a tendency in the mortgage market that became more generalized long after his death. Lending would become ever more institutionally concentrated.

William Pitt Moore and John Gage Davis opened their agency in 1858, and for the first few years the business dealt with more than property. Like Schreiber, the firm sought bonds and stocks for correspondents, and handled insurance. However, property was the major if not the sole concern of the partners. When acting as mortgage broker, the firm sometimes worked for several interlocking commissions. Moore and Davis charged clients a fee of 2 per cent on any mortgage funds that it obtained. Such funds might have helped close a property sale for which the firm also charged a commission. Often Moore and Davis sold the insurance policy that protected the mortgagee against loss by fire.[84] Beginning in the 1860s, the firm handled estate funds for trustees and widows. Six major estates seem to have constituted the bulk of the firm's mortgage business during the 1870s and 1880s. Counting renewals, 706 mortgages with a total value of $1,073,268 originated from the six sources between 1860 and 1909. The most common values were $1,000 (n = 48), $2,000 (n = 38), $1,500 (n = 33), $500 (n = 32), and $600 (n = 25), while the average value across the 49 years was $1,560. These mortgages financed the purchase of small dwellings for owner occupants, revenue properties, and building lots.

Moore and Davis did not arrange mortgages with a blended payment scheme. One of Hamilton's property specialists, the large-scale and forward-looking Patterson Brothers who built large numbers of workers' houses, had initiated a blended scheme in the 1880s, but they were concerned with volume sales and assigned mortgages. They were not investors' agents. The latter, represented by Moore and Davis, moved cautiously and arranged short-term loans whose principal was to be paid off at the end of the term. Of the 706 mortgages arranged by Moore and Davis, 221 (31.9 per cent) ran for three years; 204 (28.9 per cent) for five years; and 102 (14.4 per cent) for two years, with an average term of 3.6 years. The contemporary American agencies studied by Bogue preferred five-year terms. High down payments and a close credit check by the firm ensured that mortgagors were excellent risks, who evidently endeavoured to meet the principal due at the end of these fairly short terms, because 461 mortgages (56.3 per cent) were not renewed. Only twice did Moore and Davis draw up mortgages with interest due monthly. The half-yearly payment, occurring 619 times (87.7 per cent), predominated. In sum, the mortgages resembled those noted for Boston by Sam Bass Warner Jr in his *Streetcar Suburbs*.

The rate of interest arranged by the firm declined steadily from the aftermath of panic in the early 1860s to the depression of the 1890s. During the 1870s interest rates averaged just over 7.5 per cent, falling to an average of below 5 per cent in the early 1900s. The firm's papers only vaguely mention the availability of money, but it is probable that the flow of low-interest British funds, and moderate rather than extraordinary growth in Canada, kept the rates sliding downward. The financial crisis in the late 1850s, which made British investors wary of Canadian risks, and the accompanying high interest rates of 10 or more per cent could have depressed home buying, whereas the relatively lower rates at the end of the century could have encouraged buying. Moore and Davis, inciden-

Table 6.6
Examples of Mortgage Interest Rates Charged in Ontario and
in the American Mid-West, 1850-1930

	Rates Charged by Moore and Davis	Range of Rates among Largest Loan Companies in Ontario	Rates Charged by John and Ira Davenport in Illinois	Average Mortgage Interest Rate for Previous Ten Years in Peel Country, Ontario
1850				7.0(1840s)
1860	10.0%			8.3(1850s)
1865	10.0			
1870	7.8		10.0%	7.4(1860s)
1875	8.0		10.0	
1880	7.4		8.0	
1885	6.7		7.0	
1890	6.1		6.0	
1895	6.0	5.9-6.0%	6.5	
1900	5.3	5.2-6.1		
1905	5.0	5.4-6.1		
1910	6.0	5.9-6.5		
1915		7.0-7.4		
1920		7.3-7.4		
1925		7.1-7.7		
1930		6.9-7.1		

Sources: For the Moore and Davis rates see Michael Doucet and John C. Weaver, "The North American Shelter Business: A Study of a Canadian Real Estate and Property Management Agency, 1860-1920", Business History Review 58 (summer 1984): 246. The loan company interest rates were taken from the annual reports made by loan corporations to the Ontario government. The title for these changed, but they exist as a complete series from the report for 1897 to that for the current year. The Davenport rates were cited in Allan G. Bogue, Money at Interest: The Farm Mortgage on the Middle Border (New York: Russell and Russell, 1968), 13. The Peel County rates appear in David Gagan, Hopeful Travellers: Families, Land, and Social Change in Mid-Victorian Peel County, Canada West (Toronto: University of Toronto Press, 1981), 47.

tally, appear to have arranged rates within the range of those charged by the building and loan societies and by the agents whom Bogue had examined. Rates were falling everywhere in late nineteenth century North America (see table 6.6).

The lenders behind Moore and Davis's six principal sources of mortgage funds had curtailed their investments by the early twentieth century, but the firm continued to find clients with funds from similar sources. No ledgers survive for these later mortgages, but the firm's letterbooks indicate the continuity. In 1913, for example, the firm handled about $100,000 in funds for mortgages. About $40,000 originated in the estate of a retail merchant, and Mrs Mary

Hopkins placed $26,300. The three-dozen mortgages in operation in 1913 had an average value of just over $3,000 or nearly double the average of those handled by the firm in the late nineteenth century. Presumably some of the increase derived from the post-1900 inflation that afflicted Canadian city-building into the 1920s. All of the 1913 mortgages were on a short-term basis and mortgagors paid interest semi-annually. A report on the retailer's estate described an assortment of mortgages as well as the firm's responsibilities with respect to the loans. Interest was overdue on two of eleven mortgages and it was the firm's duty to pursue these accounts. Each mortgagor had slightly different terms for repayment of principal. One privileged borrower was permitted to pay all or any part at any time. Another was to pay $100 or its multiple annually. Mortgagors worked out individual arrangements. This flexibility may have been a feature of private lending that kept it a popular source of mortgage funds for the borrowing public.

Several further insights into mortgage brokerage (in the mid-twentieth century) came from discussions with a lawyer whose experience with the business began in the 1920s. D'Arcy Lee had worked as a law clerk for a Hamilton firm with an extensive mortgage business. The firm had acted in the 1920s as a favoured conveyor of insurance company funds into the hands of local builders and home buyers. In the twenties' boom, this law firm had a truly land-office business. On some days contractors lined up for their cheques from the insurance companies and on others homebuyers poured in to pay their interest. Lee opened his own office in Dundas, a town just outside of Hamilton, in 1935, a terrible time for the mortgage business. Mortgagors came to his office hoping that he could persuade mortgagees to extend relief. Many did so, because foreclosure was costly in time and money, an empty house was a poor asset in hard times. Foreclosure was practised most commonly when a family had simply abandoned its house and the mortgagee sought to recover title. Court-managed auctions of foreclosed properties were common, but in difficult times there were no bidders. In the case of an over-extended landlord, holding perhaps twenty or thirty houses, Lee recommended that he give quit-deeds to the mortgagees rather than declare bankruptcy. In effect, this office's first experiences with the mortgage market amounted to exercises in damage control. However, after the war, the brokerage of mortgages revived.

The mortgagors often came from Hamilton, while the mortgagees often came from the surrounding countryside. "The farmers with the worst clothes and dirtiest boots were the fellows with the money." Occasionally, the trustees of an estate came in, having cashed in low interest bonds and seeking higher interest mortgages. Widows investing in mortgages normally did not take an interest in the mortgaged property, yet one of Lee's most capable appraisers was a widowed investor. Lee never lent more than half the value of what he felt a house was worth. To increase the mortgagor's equity in the house, he inserted in every one of his mortgages a clause allowing the repayment in $100 units of any sum of principal on the interest dates. In return for placing the funds, sending out notices, and collecting the payments, the law office received a half per cent added to the interest charges. The brokerage side of the law practice flourished for about twenty years.[85] Eventually, during the thirty or forty years after the 1935 Dominion Housing Act, the non-institutional lender was hived off into specializing in older house sales or second mortgages. Whatever details might have been negotiated between private lenders and mortgagors, the low down-payment mortgage or the lower-interest insured mortgage came to consumers from institutional lenders prodded by the federal government. Since non-institutional brokers and vendors were not among the government's approved lenders, the small brokers were often relegated to higher risk older homes and second mortgages.

THE INTERVENTIONIST STATE AND THE MORTGAGE MARKET, 1919–1980s

By the mid or late nineteenth century, lenders and borrowers trusted mortgages because courts and legislative bodies had established a dense legal fabric around them. The formation of financial guarantees and market mechanisms in which lenders could place confidence progressed more slowly. For many investors, the prospect of clipping coupons seemed preferable to a mortgage investment that might turn sour and demand unpleasant confrontations and legal actions. Better to spread out risk and leave conflict to experts, better to invest in a loan and trust company. As David Mansur recollected before The Royal Commission on Banking and Finance (1963), in the 1920s "bond boys found out that they had an instrument that could be sold."[86] American historian Robert Fishman, writing specifically about Los Angeles in the 1920s, proposed a related concept that dealt with the other side of the business of accumulating funds. He felt that investors pushed the mortgage market and that drove the real estate market. Or, as he put it, "the salesman selling lots or the builder selling houses was simply one element in a system the ultimate purpose of which was to sell money."[87] The will to possess surely nudged consumers toward mortgage finance, but financial institutions were increasingly becoming active city-builders putting funds into specific urban developments. Furthermore, after the real estate boom of the 1920s they were to influence state formation in order to increase the security of their businesses.

The great depression made the bonds and debentures of loan companies insecure as real estate prices caved in. The mortgage-investment crisis developed when mortgagors found their equity in a house had been whittled to nothing because their mortgage debt exceeded the market value of their real estate. Quite a few, faced with this loss as well as reduced income, defaulted on payments, leaving institutions with the Hobson's choice that had faced brokers like Schreiber in the 1860s. They could foreclose and accumulate inventories of shelter. In many parts of the continent the depreciated real estate market meant that they could not sell and expect to recover the money owing, and renting such properties to tenants incurred overhead costs. Alternatively, mortgagees could leave the mortgagor undisturbed, cajoling payments by threat. In effect, lending institutions had spread out the investors' risks, but they had not yet drawn in the state to reduce those risks. On the contrary, many North American jurisdictions during the 1930s passed laws protecting defaulting mortgagors. As late as 1962 mortgage experts in Canada intimated that Alberta legislation protecting the borrower and Quebec court processes that – as in the nineteenth century – took time made them unattractive jurisdictions for lenders.[88]

Ontario brought in a mortgagors' and purchasers' relief act in 1932 and superseded it in 1933 with another that authorized the court to suspend or delay legal remedies against debtors during a period of extreme financial distress. The court could grant time to the debtor in respect of payments due, and the court's ruling could be appealed. The measure applied beneficially to small proprietors: owner-occupied one and two family dwellings; any small retail business or petty trade that might also contain one or two self-contained apartments in one of which the retailer resided; farm land upon which the mortgagor resided.[89] Understandably, institutional lenders stopped placing their funds and, when called upon by the federal government during the 1930s and 1940s to comment on the shelter situation, lobbied for government intervention to minimize risks in the mortgage market. A string of housing acts, amendments, and administrative adjustments accorded institutional lenders the option of protected investments. Between the first housing act in 1935 and a number of social housing measures in the 1960s, the symbiosis of government and lending institutions largely functioned to the exclusion of low-income housing. In the first stages of growth, the

Canadian interventionist state adopted a housing policy whose changing features operated primarily to improve the mortgage market in the estimation of institutional lenders.

The Government of Canada first intervened directly in the mortgage market in 1919. Its hastily conceived program demonstrated the hazards of the higher ratio or low down-payment loans that it boldly introduced, contrary to the wisdom of conventional lenders. In later schemes, federal governments consistently worked to fashion mortgage-lending practices characterized by high ratios. Having a sense of the public's goals, they realized that the will to possess had to be served. The first effort, the 1919 operation, earned no lasting credit for government intervention. The riddle of how to channel more loans out of the financial institutions and into the hands of consumers through high ratio mortgages remained. Indeed, the plan's failings became part of the mortgage-lending fraternity's lore against easy terms.[90]

In late 1918 at a federal-provincial conference held to consider the transition to a peacetime society, the acting prime minister, Sir Thomas White, agreed to a national housing program. The federal treasury would make available $25 million at 5 per cent, a rate below market interest.[91] A sequence of filter-down lending would follow. The provinces could claim a sum of the federal money in proportion to their populations. However, the operation of the scheme rested with municipal housing commissions. They borrowed from the province's allocation and then worked like conventional mortgage lenders except that they could lend only for the construction of relatively cheap dwellings. Local housing commissions lent to contractors who built approved small houses. When the contractors sold the units, the municipality became the mortgagee. The mortgages were amortized for 20 years at 5 per cent. The buyer's equity – normally 40 per cent or more in the conventional mortgage market – was cut to 10 per cent.[92] This expedient for establishing high ratio mortgages proved ill-timed. Wartime inflation and postwar shortages had boosted building costs so that buyers from 1919 to about 1922 had purchased over-valued shelter. The private sector soon reduced the housing crisis with suburban developments and apartment buildings. House prices declined and the cheap units insisted upon by the government scheme dropped sufficiently in value to wipe out some purchasers' equity. Compounding the lenders' risk, the buyers were forced to make up for the low down payments with relatively high monthly instalments. Buyers had little to lose when they lapsed into arrears. Therefore, municipalities had to foreclose on an abnormally high proportion of mortgages, acquiring a stock of dwellings that were difficult to market. Hamilton's Housing Commission, formed to administer the scheme, made 122 loans and acquired a few other creditors when it annexed parts of neighbouring townships. The default rate by 1940 was about 70 per cent and in 1946, after six years of selling effort, the city still held 39 houses given up by defaulters in 1940.[93]

The blunders had been many. First, in some cities, the plan had stirred opposition because it competed with the private sector (Toronto), would not accommodate local housing forms (the Montreal duplex), or involved favouritism (Quebec City). Future schemes endeavoured to work more closely with lending and realty interests on the basis that risk and bad publicity could and should be shunted onto the private sector. Second, housing appraisal and design review had failed, for though it was not conducted recklessly, architects and housing authorities had had to work within impossible circumstances: inflation and a tight budget on each unit. The plan probably could not have done much to ensure that sponsored units would retain a high market value. Subsequent federal housing measures either charged lending institutions with making appraisals or, as with the creation of CMHC in 1945, enforced a building code, disseminated plans, inspected sites, and tried not to saturate local markets. The challenge of ensuring that mortgaged properties held their value, so that buyers retained enough equity to keep up payments or, failing that, that foreclosure brought a decent return to mortgagees, was intrinsic to achieving sound high ratio lending. Finally, the 1919 venture underscored the standard economic problem of foreclosure, namely how to extract money owing from a piece of property in a weak market. Municipal governments were stuck with the debts and sometimes the depreciated real estate. If high ratio lending were to be encouraged, it would have to be on properties fashioned to hold their value, otherwise mortgages would have to be guaranteed against default. Both tasks were part of CMHC's assignment.

In a way, the failings of the 1919 scheme prefigured the mortgage lending crisis of the 1930s and the lessons learned from both would influence the directions taken by private and government experts when attempting to restore home-buyer credit. Whether for apartment building ventures or houses, mortgages negotiated during the real estate boom of the 1920s had been secured on what proved to be inflated appraisals. Consequently, in addition to the measures taken by the United States and the Canadian governments to protect lenders, the lenders themselves reconsidered their practices. From the 1930s, they finely tuned property appraisal, preferring always a new house in a new area in a good city. Thereby, within and among cities, advantages were loaded onto the advantaged. Also, the mortgage lending fraternity henceforth paid more systematic attention to the borrower's education, training, age, health, and other personal traits affecting employability.[94]

In an account of how the mortgage market has matured, the many thoughtful calls for public housing cannot receive much attention. Ultimately, federal, provincial, and municipal governments would work on a variety of methods to have more citizens better housed: some public housing would be built; a few co-operative ventures would be supported; and substantial rent subsidy programs would be negotiated. However, the high ratio mortgage had favoured status. It harmonized with one of finance capitalism's huge investment markets and, unlike a direct public-housing program, it would not undermine the business of realtors by producing a stock of housing outside the purview of private enterprise. Moreover, the high ratio mortgage, if perfected, would institute a short cut to the ever-popular goal of house buying. As advocates of public housing forecast, tinkering with the mortgage market failed to help the tenants of sub-standard shelter in the midst of a depression. In fact, the first serious discussions of urban housing since the 1919 plan occurred in the early 1930s and recommended federal financing for locally administered public-housing projects. Despite these entreaties, which were reiterated in the recommendations of a 1935 parliamentary committee on housing, the Dominion Housing Act (DHA) of the same year addressed a collapse in the construction industry and the deterioration of the country's housing stock with innovations in mortgage lending.[95]

The deputy minister of finance, W. C. Clark, had intervened in the legislative discussions of 1935 to prepare a mortgage measure. An economist and a former vice-president of an American mortgage investment company, Clark had an affinity for the financial community. He also believed that the Canadian exchequer carried high existing financial burdens. Spurning the parliamentary committee, Clark drafted a bill meant to coax out private capital and to revive consumer interest in house buying.[96] The DHA created a mortgage with three inducements: it had a twenty-year amortization period, a 5 per cent interest rate when the market was 5⅔ per cent, and a high ratio, for it covered up to 80 per cent of the construction costs or appraised value, which ever was less. Lending institutions administered the program and put in their standard share of the mortgage capital, namely 50 to 60 per cent of a property's value. They also received the market interest rate on their portion of the money lent. The federal government provided the funds for up to 20 per cent of the mortgage and charged an interest rate low enough to pull down the overall rate to 5 percent. As a further incentive to the lending institutions to place high-ratio mortgages, the government committed to pay a portion of the lender's loss when, after foreclosure, a house sold for less than the lender's share of the loan. Depending on the size of the loan, the government would have made up 70 to 80 per cent of the company's loss.[97] In spite of the steps taken to use public funds to launch sound high-ratio mortgages, the DHA failed. However, the joint lending approach dominated peacetime housing measures until 1954.

Loan and insurance companies, claiming that even slight losses were unacceptable, dragged their feet with the DHA in high default areas of the country, such as the Prairies. Lenders also reacted conservatively to low-cost market housing, as illustrated by the case of Hamilton's Halliday Company. Having survived during the depression by concentrating on home-repair items, this once major producer of prefabricated houses geared up for sales of low-cost house kits on the assumption that the DHA would stimulate mortgage lending. Buyers made deposits and Halliday's agents went to work negotiating loans for these customers at lending institutions. The company had no problem obtaining mortgages for buyers of more expensive units, but applications for houses and lots costing under $4,000 were often rejected by a number of life insurance and loan companies.

The 1938 National Housing Act (NHA) attempted to deal with the persistent reticence of lenders badly burned by the financial crisis. The new act guaranteed all of a lenders' losses on joint loans for homes not exceeding $4,000 in value. It also created a class of joint government and lending institution mortgages that required only 10 per cent owner's equity in the property. Still the institutions balked at investing in mortgages on low-cost houses in western Canada. Lenders concentrated their high ratio mortgages in what they believed were good risk centres – suburban Toronto, Hamilton, and Vancouver received 45 per cent of such loans during the first year.[98] These urban and regional preferences of major lenders were acknowledged as late as the early 1960s and may persist today as one obstacle to a truly national mortgage market.[99] In any event, the regional biases were not solved by the 1944 NHA, which perpetuated the joint loans and the guarantee against loss. Legislation in 1945 created Central Mortgage and Housing Corporation (CMHC) to administer the NHA and to dispose of the Crown's housing assets assembled during the war. With insurance companies anxious to convert government bonds into mortgages during the post-war years, the joint lending scheme at last worked well to finance a house-construction boom. It was noted earlier that by 1953 the insurance companies, having rebalanced their investment portfolios, eased away from the mortgage market and left CMHC high and dry to lend directly in a very active housing market. From a tentative joint lender, the government had graduated into a mortgage banker tapping public revenue.

The 1954 NHA drew the banks into the mortgage market and terminated the joint lending scheme. CMHC remained charged principally with the task of increasing the quantity of new housing and of getting the dwellings into the hands of buyers who could make only small down payments. Such buyers, as the Hamilton data demonstrate, included a great many young household heads. A historically unique shelter episode had begun – young people were no longer barred from home buying. To attain the goal of an extensive high-ratio mortgage market, CMHC had two instruments. First, it had the authority to lend directly. Within limits, it could draw upon public funds to act as a lender of last resort. Second, to encourage the loan and insurance companies – and now the banks – to issue high-ratio mortgages, CMHC administered a mortgage loan guarantee or insurance program. The precise rules for insurance eligibility would change many times. For example, the down-payment minimum would be lowered and the value of eligible houses would be raised. Interest rates on insured mortgages were set at about 2 $\frac{1}{4}$ per cent above the rate paid on twenty-year-term Canada savings bonds, although the operating rate was set by Order-in-Council. When interest rates rose rapidly in the later 1960s, the occasional adjustments by Order-in-Council were not sufficient to keep the rate evenly competitive. Finally, in 1969, the statutory maximum rate was removed.[100]

Regardless of the shifting terms of NHA mortgages, the lending process for new houses began when a building contractor approached one of the participating institutions and applied for a construction loan, filling out an application for participation in the NHA scheme. The builder had to adhere to a national building code worked out by the National Research Council. CMHC officials at regional offices had to approve the plans and inspect several stages of construction. Approval at each stage freed a portion of the construction loan for the builder. When houses were finished, the builder ascertained the credit worthiness of buyers and forwarded agreements of sale to the financial institution that had provided the building loan.[101] When the transactions closed, the mortgagee added a sum of 2 per cent to the mortgage debt. This sum represented the mortgage insurance premium that went to CMHC. Mortgagors paid off the premium as part of the blended monthly instalments. CMHC managed the premiums as a fund out of which mortgagees were paid when mortgagors defaulted. For many years, the fund suffered few serious runs. In 1962 a CMHC official reported that defaults had amounted to around ½ per cent of the mortgages covered by the fund. Nevertheless, CMHC may have taken on more risk than the administrators of the Federal Housing Act insurance plan in the United States, a scheme instituted in 1934. The Americans enforced a harder line on mortgagor eligibility and charged a higher premium than the Canadians did. CMHC allowed regional development needs and social objectives to enter its decisions to insure. Elliot Lake, an Ontario mining town where no reasonable insurer would have acted, was an early case. More serious than losses there in the early 1960s were the demands flooding into the fund during the collapse of the Alberta real estate market in the early 1980s. Notwithstanding the important difference between the Canadian and American insturance operations, they both had rearranged the historical relationship between age and house buying, by facilitating the acceptance of the low down-payment mortgage by lending institutions. Concurrently, because the insurance plans dealt only with approved lenders, they strengthened the position of institutional lenders.[102]

During periods when the insurance ploy failed to draw the major lenders into supplying high-ratio and lower-than-conventional interest mortgages, CMHC had to operate as a direct lender, drawing on federal revenues up to a ceiling. Its first major intervention had occurred before the insurance scheme, in 1953, when the insurance companies had eased out of the market and before the banks were enticed to lend. Under the insurance measure, the banks slowed down their rate of participation in late 1957 and CMHC filled the void with vigorous direct lending. At the end of 1959 CMHC had attained its statutory lending limit, and in the mid-1960s the ceiling was lifted. Direct lending realized its greatest volume in 1967.[103] Thereafter, permissive legislation attracted banks back into the market, and CMHC began to direct its attention toward a greater range of housing issues. Many of the corporation's officials, though proud of their role in improving enormously the country's housing stock, were distressed by the lack of important social-housing measures. The political mood of the late 1960s encouraged public discussions leading to a variety of new programs. CMHC laboured at the same time to improve the functioning of the mortgage market by encouraging a secondary market. Such a market, if truly effective, was seen as a boon to consumers and investors.[104]

A place to trade in mortgages, to make them a more liquid asset, was thought to be a way to reduce interest rates and to attract further investment capital. The reasoning derived from distinctions between a bond and a mortgage. Managers of investment funds accumulated bonds because they constituted a readily saleable item. They believed that by sales and purchases they could increase the yields of their investment portfolios. In effect, they were willing to sacrifice interest points in return for holding easily traded paper. Mortgages lacked the critical element of ready transfer. The United States government, aiming to stimulate the mortgage market during the great depression, employed two devices to promote a secondary market. The federal mortgage insurance scheme imposed criteria on the borrower, lender, and property. Therefore, investors anywhere in the country could purchase such a mortgage and supposedly feel confident about its basic quality. As well, the government chartered the Federal National Mortgage Association (Fanny Mae) in 1938 to purchase insured mortgages from the primary lenders. Comparable steps in Canada came between twenty-five and forty years later.[105]

Investment lore had maintained that lenders needed to evaluate each mortgage in terms of the property and the credit worthiness of the mortgagor. Beginning in 1961, at a time when it had accumulated mortgages through its direct lending, CMHC attempted to erode the practice of individual appraisal as a step toward fostering a secondary market in Canada. The corporation sent out packets of mortgages to managers of investment funds and called for bids. Mortgages in the packets were identified only by street; the street addresses and mortgagors' names were excluded. The exercise failed to alter habits. A lesson of the 1930s had become ingrained and lenders continued to protect themselves not only by valuation but also by credit analysis. CMHC certainly did not address liquidity head on, for investment funds and small financial institutions, like credit unions or pension funds, required a place to sell mortgages. CMHC only sold. No private-sector institutions had worked to form a secondary market for mortgages. After all, lenders remained wary of the idea that mortgages were all alike. Politically, the federal government had a need for a secondary mortgage market because it continued to require new sources of mortgage money in a society where, as CMHC President Stewart Bates said, "there is a tremendous fetish for home ownership."[106] The task of providing the secondary market, therefore, fell to the federal government. The call for this market surfaced as one of the recommendations in the *Report of the Federal Task Force on Housing and Urban Development* (1969).[107] An act in 1973 created the Federal Mortgage Exchange Corporation to improve mortgage liquidity and to make mortgages more appealing to investment managers of credit unions and pension funds.[108]

A variety of private and public institutions has evolved and virtually replaced the unincorporated brokers, establishing a crazy-quilt mortgage market in both Canada and the United States. Nevertheless, it has serviced the will to possess remarkably well. However, the roller-coaster movement of interest rates and housing starts during the 1980s has forced rapid innovations and reappraisals of mortgage details among housing experts. Mortgage finance is not the sole target of revisionist thought, for energy costs have inspired new housing standards and the rental market has had its share of crises. The rapidity and complexity of change in the mortgage market have focused attention on housing issues, bringing them to the doorsteps of millions of households and would-be home buyers. Historically speaking, the so-called revolution in real estate finance may not be particularly profound, although any change may strike contemporaries as revolutionary or innovative. There was nothing unique about soaring interest rates – these had plagued North America in the 1860s. There was nothing unusual about seeking new sources of funds – the search for pools of mortgage capital has gone on without cease. Pension funds are only the latest in a sequence: private estates (c. 1850), loan company debenture sales, insurance premiums, and tax revenue (c. 1950). Even the limited experimentation with alternative mortgage instruments – the variable rate mortgages and renegotiated rate mortgages – continue a tradition of redesigning: short-term mortgages with semi-annual payments (c. 1850), blended monthly payments (c. 1920), and the high-ratio mortgage (c. 1935). The basic instrument and its problems remain visible across jurisdictional boundaries and over time.[109] This analysis is not meant to deny the uniqueness of historical crises and the particular responses to them, but it suggests that the law, habits of property appraisal, and caution have been parts of the process. The elements of continuity have also made North American social housing develop with mortgage-market attributes and connections. The mortgage haunts the history of North American shelter.

THE INTERVENTIONIST STATE, SOCIAL HOUSING, AND MORTGAGE FINANCE, 1935–1980s

The attachment to home buying in a democracy has assured a drive by the interventionist state to expand the mortgage market. For many reasons, including the prevalence of home ownership as measured through households' life cycles, the dedication of governments to social housing has been problematic and irregular, although not without accomplishments. Social housing means dwellings whose primary function is to provide people who might not afford adequate housing in market forms with decent shelter. There are two pure and direct forms of social housing and two marginal forms. The former consist of shelter units owned and operated by governments as well as units owned and operated by non-profit organizations. Marginal forms of social housing include limited dividend corporations and profit-making enterprises which mix rent-subsidized units with market units. Only the limited dividend corporations have failed historically to deliver needed low-cost housing. Favoured frequently by governments reluctant to undercut the private market, limited dividend housing legislation really substituted promising words for deeds.[110] Co-operative housing ventures, which have supplied non-market housing in Nova Scotia and Quebec, have not necessarily been examples of social housing, because social housing must take its definition from its recipients.[111] Social-housing tenants include welfare recipients, the working poor, and the elderly with reduced income. A slender case could be made for university students as consumers of social housing and the Ontario government treated them as such by creating an Ontario Student Housing Corporation in 1966. Of the true consumers of real social-housing units, the elderly or seniors have provoked little controversy among politicians and building managers. Moreover, moving them out of conventional housing has the probable result of increasing housing stock.

Social housing in Canada progressed dramatically in the 1960s. Social housing had come to stay, and in principle, it had support. Controversies on important specifics would still flare up often, especially for units directed toward the needs of welfare recipients and the working poor. Home buyers, as future home sellers guarding their equity, made location a nettlesome political issue. Politically easy remedies that placed units in low-value areas or on raw urban fringes raised the legitimate ire of tenants and social critics. Even the 1970s approach of mixing subsidized tenants with market tenants in low-rise units ran into locational controversies. Fiscal concerns, including the annual charges for substantial rent subsidies, joined the backlash over location in causing a slow-down in social-housing construction by the late 1970s.[112] This thumbnail sketch expounds on the differences between social housing, which has been controversial, and mortgage-market measures, which have been far less contentious to the majority.

The limited-dividend housing corporation, that marginal form of social housing with least claim to the label, had an unsuccessful Canadian trial in the Toronto Housing Corporation, which had built 242 units by the end of World War I. As leaders of the city's organized labour movement had forecast, only the better-paid could afford the rents.[113] Almost concurrently, the Halifax Relief Commission built rental units to replace some of the shelter obliterated by the 1917 munitions explosion. The Halifax project demonstrated the capacity of government to achieve something admirable in the housing sector by direct action.[114] During the 1930s a substantial alliance of groups advocating social housing would come away from Ottawa with an insubstantial and frustratingly complicated limited-dividend scheme in the 1938 NHA.[115] As housing historian John Bacher and planning analyst David Hulchanski have demonstrated, the Deputy Minister of Finance, W. C. Clark, turned the 1935 DHA and the 1938 NHA into mortgage-market acts and made the latter's limited-dividend section unworkable.[116] The housing emergency generated by wartime industries in the early 1940s created an opportunity for the building of a stock of housing with post-war social potential. Once more, Clark deflected the direct approach to social housing.

By early 1941 most Canadian industrial centres reported no vacant dwellings. Hamilton, one of those centres most affected by wartime demands, expected that war work would bring an influx of 20,000 new residents before the end of the year.[117] An Order-in-Council created a federal housing corporation under the supervision of the Minister of Munitions and Supply. The President of Wartime Housing Limited (WHL), the vigorous and innovative Hamilton contractor Joseph Pigott, was well-qualified for the national task and he had a social vision. Pigott had made the family firm into a major builder of commercial and public-works structures. He had little in common with the typical residential contractor. He had expressed strong views since the 1920s on the need to train skilled building tradesmen and to retain their skills with public works during slumps. His business life by no means clashed with forwarding social housing.[118] The Canadian Construction Association, consisting of similar large-scale builders, had lobbied for social housing in the 1930s.

Pigott's WHL achieved much across Canada, but he favoured Hamilton with a social housing experiment that created a flap, not only because it would benefit his city. WHL had to serve the war industries, but in Halifax Pigott had negotiated a project open to servicemen's families and hardship cases. For Hamilton, he arranged a similar deal, but for more houses and on somewhat different terms. He proposed lending the city enough money to build 300 permanent houses. The city would let these to low-income families and pay back the federal loan over 30 years at 3 per cent interest. Of course, Toronto, Windsor, and Ottawa building and reality interests demanded a repeal of the agreement, which gave Hamilton a bit of

relief from wartime sacrifices. The Dominion Mortgage and Investments Association observed critically that Pigott had undercut
current mortgage rates including that of the joint lending plan of
the 1938 NHA. Pigott had his wings clipped: the federal government
prohibited WHL from negotiating the construction of permanent
housing; and WHL plans came under tighter formal scrutiny. The
taint of favouritism, the lobby against social housing, and wartime
demands on materials and labour crushed Pigott's experiment.[119]

Embracing the intellectuals of the League for Social Reconstruction, the Cooperative Commonwealth Federation party, social workers, a number of civic politicians, large-scale contractors, and
representation from organized labour, Canada's social housing advocates had their position capably aired once more during the war.
The federal cabinet had formed a committee on demobilization and
reconstruction in December 1939. Early in 1941, it had appointed
an advisory committee on post-war planning. This group started to
consider housing and, by January 1943, admitting the topic's complexity, had the cabinet establish a special housing sub-committee of
the advisory committee. Named for its chairman, economist C. A.
Curtis, the Curtis sub-committee included several champions of social housing. Released in March 1944, the Curtis report amassed
abundant data on housing conditions.[120] Some of its findings for
Hamilton will inform the later discussion of housing quality. The
Curtis sub-committee endorsed home ownership and recommended
the provision of mortgage insurance, which the United States had
introduced in 1934.[121] The report maintained that stimulating home
buying through the mortgage market could not supply all housing
needs. The report recommended subsidized low-rent housing administered by local housing authorities. Federal low-interest loans,
it proposed, were to finance the construction of the low-rent units
and federal subsidies were to secure affordable rents. The United
States Housing Authority, created in 1937, had already operated an
identical program and had overcome opposition from realtors who
were experiencing difficulty in avoiding losses on their own property.
An identical contest arose in Canada, but it ended differently. There
was spirited opposition to the Curtis report from those with interests
in real estate, house building, or mortgage investment.[122]

Ultimately, the social housing scheme ran afoul of W.C. Clark
and the federal cabinet. Family allowances, favoured by Clark, replaced subsidized housing. The "baby bonus" promised a popular
direct transfer from government to families and aid for household
budgets supposedly sufficient to preclude the need for municipal
social-housing authorities. The 1944 NHA steered clear of social

housing, issuing no initiatives in the area of subsidized housing and giving limited-dividend housing an undue reprise. No limited-dividend units were built until 1949. Albert Rose, author of a survey history of Canadian housing policy, was whistling in the dark when he wrote: "There was never in Ottawa the strong anti-public housing lobbies which were evident in Washington from 1933 onwards."[123]

In the post-war years, urban housing crises erupted across the country. A moribund house-construction industry could not gear up quickly enough. During the emergency, many of WHL's temporary houses were moved to new sites, placed on permanent foundations, and given proper furnaces. Joseph Pigott had been vindicated. In his hometown, trucks hauled approximately 450 WHL units up Hamilton's mountain and to a survey administered by the city's housing authority. Some were knocked down in 1975, but a few stand today.[124] Stop-gap measures and the housing boom of the 1950s did not answer social-housing needs. Depending on the definition of social housing, Canada had added only 10,000 to 12,000 new social-housing units between 1945 and 1960.[125] Twenty years after Ottawa turned its back on the Curtis report, the federal government amended the NHA to help the municipal and provincial authorities with their social-housing shortages. CMHC could make loans to provincial and municipal housing corporations for up to 90 per cent of the cost of acquiring and servicing land and for up to 90 per cent of the cost of a project as determined by CMHC, and it could contribute to the operating losses of housing established for the benefit of low-income tenants. Public-housing production peaked in 1970 at 20,000 units or 16 per cent of all dwelling units funded under the NHA during that year. Projects built by the Ontario government were usually large high-rise developments.[126]

A swift public reaction against the towering and concentrated social-housing developments made an impression upon the federal task force on housing and urban development. Its 1969 report noted tenants' complaints about stigmatization, inadequate recreational facilities, and vandalism.[127] A shift in conventional wisdom about social housing occurred in North America during the 1970s as a response to widely publicized shortcomings in the larger-scale high-rise developments. To achieve low-rise dispersal, federal legislation and CMHC implementation of the NHA encouraged several forms of social housing. Co-operative ventures were assisted, provided that they included units for low income tenants. In addition, private investors who built rental housing with social-housing units integrated into market units were eligible for government loans.[128] In 1988 the Hamilton-Wentworth Housing Authority administered about half of the city's social housing, but the authority had not been growing in recent years because of a shift in policy that favoured private non-profit developments. Decentralization and diffusion of social housing was designed to achieve a political purpose in addition to a sociological one. Responsibility for locating, building, and managing social housing could not be fixed on a central political authority.

Throughout the many adjustments in social-housing programs, CMHC retained the word "mortgage" in its title. The designation described the significance of CMHC's work in the private mortgage market. The primacy of the mortgage in the agency's name also squared with the Canadian approach to all forms of social housing, past and present, and with American housing programs, although there the social component pioneered in the 1930s faded just as that in Canada accelerated. Lending and shelter, historically associated with and inextricably tied to the investment industry of small and large operators, were bound together so profoundly across North America that the loan nexus influenced social housing. The negotiated loan has been central to every federal initiative in low-income housing, including Canada's original great undertaking (WHL). Even that which seems unconventional – social housing – has been arranged using contracts and terminology familiar to the financing of market-rental housing and house buying.[129]

In the United States and in Canada, civil servants, the investment lobby, and the financial press have attuned the politicians to the mortgage market and its practices. Interest rates, amortization periods, and appraisal apply to both private and public sectors. Federal governments in both countries have employed public revenues like the premiums of insurance firms, the trust funds of trust companies, the deposits held by banks, and the capital of lending institutions. State formation in North America has often entailed a meshing with the private sector, a formation of government agencies along private corporate lines, and the exchange of personnel. In the historical literature, transportation projects have provided examples of a mixed economy rooted in nineteenth-century public works and government bonuses to railways and industries. Shelter joined the mixed economy much later, but has had much in common with the forerunners of state intervention. Just as aid to other economic sectors was to stimulate free enterprise, so too was aid to housing, whether market housing or, in the main, social housing. Government intervention in housing expressed the economic and political importance of construction and capital recycling industries. The building of most of North America in the nineteenth and twentieth century, a dynamic process engaging immigrants and credit ties to Europe, accented construction, borrowing, and the government's helping hand. In a general sense, the drawing of housing into the mixed economies of Canada and the United States accorded with the sweep of settlement history. Governments facilitated the occupation of the crabgrass frontier as they had that of agrarian and resource frontiers and for much the same reason – the will to possess.

7 The Social Contours of Home Ownership in a Middle-Rank North American City, 1856–1981

A positive relationship between age and home ownership is a conventional finding in housing research in North America, but that basic conclusion is surrounded by complexities. For example, does ownership rise, peak at middle age, and then level off? Does it fall after middle age, or does it continue to increase? A quest for explanations for the age and ownership connection has occupied writers on urban questions and policy. To these concerns, we add the consideration of changes over time. Economists who have studied the demand for home ownership have dealt overwhelmingly with the eternal present, using single cross sections and perhaps slipping in an aside about the past for interest or for impressionistic contrast. One significant exception, based on a sample of urban industrial workers in the United States for 1889–90, found that ownership was "an increasing function of family income and age of family head. This is a persistent pattern in the twentieth century – older persons have had more time to acquire wealth and higher income permits higher rates of wealth accumulation."[1] Still, this unusual study of the past was not a part of a systematic time-series analysis. In Canada, one of the most thorough examinations of the demand for home ownership was undertaken by Marion Steele. Her work, which employed 1971 data, is a splendid basis for introducing the themes that must be treated in an analysis of age and home ownership, but her monograph considered a particular cross section that represents an unusual pattern in the broader sweep of housing history.

Both our study and Marion Steele's have employed the concept of household heads. Our choice of this term was forced by the sources at hand. Assessment rolls principally treat units of shelter for households and household heads. Steele selected the household head data from the 1971 census on the rational basis that a consumer's housing choice may be ordered into a hierarchy of decisions. "The first of these is whether or not to occupy a separate dwelling unit. This is equivalent to the decision whether or not to head a separate household."[2] In other words, the concept of household that was forced onto our data base by the records at hand has a sound theoretical grounding. The concept's imperfections present problems for the historian concerned with total history rather than for the economist inquiring into the demand for housing. There are several consequences for social history arising from the study of household heads. For example, home ownership rates have to be regarded warily as indirect indicators of social gains for occupational or ethnic groups. What the data often lose are the young people who may have had to defer household formation due to bad times, the elderly who have had to give up independent households, boarders, and the homeless of diverse origin. Household head data are biased socially toward the great generality of lives lived as household heads. The bias has to be kept in mind. For example, during the great depression of the 1930s, the real impact of hard times that we will see levied on the young may be understated by looking purely at household heads and thereby missing the thwarting of household formation by unemployment or under-employment.

Steele's 1971 data catches Canadian housing in a period that has no parallel. Her analysis emphasized the existence of a housing-consumption curve, which shows that, while housing ownership essentially increased with age, ownership peaked at 45–54. Indeed, our assessment data for 1966 and our 1981 census sample neatly conform to the 1971 Canada-wide pattern reported by Steele. The drop of ownership for older household heads, noted Steele, was not evident in data for the United States, leading her to conjecture that an American capital gains tax inhibited the liquidation of the house as asset. This would not explain the monotonic relationships between age groups and ownership predominating in the Hamilton cross sections from 1856 to 1956. Historically the relationship of age to ownership for household heads has been closer to a rising curve than the arched curve found in 1966, 1971 (Steele), and 1981. If greater proportions of seniors in these three recent cross sections have been selling out more than ever before, the composition of the rental market in the city might help to explain why. Thus, in subsequent chapters, we call attention to the swelling share of the rental stock made up of apartments and to the composition of apartment dwellers.

How could, and why would, home ownership among household heads, until recently, increase with age, usually beyond middle age? The answer to the "how could" question concerns income and home finance. Young families historically had been restrained by lower incomes and by an inability to accumulate large down payments quickly. However, the radical adjustments to the mortgage market noted in the last chapter have altered historical experience. Since the end of World War II, down payments have declined as a proportion of prices. Along with a more generally relaxed attitude to credit worthiness, reduced down payments fostered by mortgage insurance have helped families to acquire houses earlier in their life cycles. A study of Canadian consumer credit, based on 1970s data, indicated that the average mortgage debt for home buyers under 35 exceeded their average income. The reverse was true for other age groups.[3] The surge of home ownership for household heads aged 25–34, a remarkable event evident in and after the 1956 cross section, is impressive and confirms that a significant change had occurred after the war. Shifting financial-service practices largely account for younger families recently getting into their own homes. The answer to the "why would" question is much trickier and finally elusive. Using aggregate data for all of Canada, economist John Miron arrived at a comparable conclusion when looking at post-war household formation and shelter. He could not attribute relative weight to the many factors influencing household formation because of an unquantifiable variable: the desire for privacy. Privacy, the will to possess, and status, all based in culture and psychology, are associated with the stages in the individual and family life cycle.[4]

Housing analyst and anthropologist, Constance Perin, addressing the character of contemporary society in the United States, alleged that Americans had a culturally rooted conception of what constituted the normal life course. Housing types played an essential part in the definition of the expected course – dwellings functioned as markers. Perin developed the metaphor of the ladder of life (see figure 7.1). Households scaled this ladder by moving from tenancy to home ownership. Failure to follow the prescribed course courted suspicion or stigmatization. People who did not move out of tenancy at the correct point could be labelled transients, drifters, rootless, strangers, or failures. Perin's interviews with people in the housing industry convinced her that American culture sanctioned a sequence of shelter forms appropriate for different phases in the life cycle of

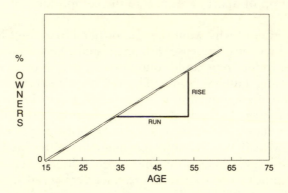

1. Normal Times

2. Boom Times
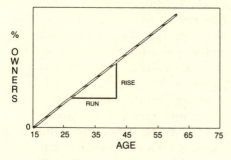

3. Bust Times
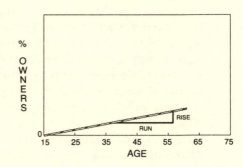

Figure 7.1
The ladders of life

the traditional family. Moving up was part of growing up. The relatively affluent young couple in Kidder's *House* epitomized the norm. "For eight years, Judith and Jonathan shared a duplex with another young couple, but both families grew too large for the place, and reluctantly they all agreed that *the time for moving on had come*."[5] When Perin discussed credit, she did not do so from the conventional position of credit as just a facilitator for acquiring a consumer durable. For her, home ownership was the visible manifestation of credit worthiness, which itself was a potent subliminal or even overt objective. To Perin the anthropologist, the owned house signified that capitalist society's chief locus of power – the bank – had conferred full tribal membership. From what we have reported on mortgage lending, the concept is not far-fetched. In any event, her unique approach to home ownership promotes two thoughts. First, there is the ladder of life notion, which we are about to describe in connection with the Hamilton household-head data for over one hundred years. Second, there are no simple conclusions and even fewer simple explanations.

The second conceptual point hardly needs illustration. By swinging the discussion back to the bedrock of material conditions and away from Perin's bold psychological or anthropological exploration and by interjecting an important element of probability, complexity is all too apparent. The case for the complicating probability is presented in the chapters on rental housing and the apartment building. Quite likely most rental stock in all eras has been smaller in living area than the living areas available in the stock of owner occupied dwellings. Certainly this condition typified the compressed apartment units in new apartment towers. Here, then, enters the *prima facie* case for assuming a connection between the apartment, especially the zealous construction of apartment towers in Hamilton and elsewhere in the 1960s and 1970s, and the peaking of home ownership at age 45–54 in the 1966, 1971 (Steele), and 1981 (census) cross sections. Seniors with an empty nest could liquidate the housing asset and secure smaller, more appropriate dwellings in the new apartment or, after 1967, condominium towers. A reciprocal effect may then have affected household heads in the family formation age groups. The dwindling number of rental houses – a rental market trait illustrated in chapter 9 – and the increase in smaller apartment units could have helped to deflect many growing families toward the only abundant form of larger accommodation – the single-family detached dwelling for sale in the city or suburb. Houses vacated by seniors would have joined the pool of available houses. In sum, one can imagine a round similar to musical chairs. Nevertheless, the tidiness of a formula of moves and of demand for home ownership incorporating household size and house size cannot satisfy all situations.

Apparently, the growth of household size does not in itself explain all shifts from tenancy to ownership. Consider the currency in the late 1980s of the term "Dinks" to describe an alleged wave of special home buyers across North America. Does the term, standing for the "double income, no kids" subspecies of yuppies, suggest a return to psychological motives for ownership, prudent planning for a future family, an investment fad, or all of these? And, though widely reported in the media, exactly how big a factor in the housing market were the Dinks? Housing-consumption puzzles, though easily aired, cannot be neatly sorted and resolved. Possibly survey questionnaires and follow-up studies by social scientists might elucidate contemporary events. Historical patterns may never be convincingly explained. We can only advance some of the findings of a comprehensive look at home ownership and draw qualified conclusions.

HOME OWNERSHIP AND THE LADDER OF LIFE

Perin's ladder of life is a tempting metaphor. We can picture a situation where the climb up the ladder of life is long and difficult, where the slope is steep because adverse economic conditions have crippled the ability of households to accumulate down payments and establish credit; or the ladder could commence at a high point and have a very modest rise. To ascertain a ladder's slope there are two necessary measurements – the elevation being climbed and the horizontal distance that needs to be covered. To develop the analogy, let us assume that the elevation is measured in terms of a population's attainment of home ownership and the horizontal distance is measured by units of age. The natural hypothesis is that as household heads age and thus cover ground they would experience a rising percentage of home ownership, advancing up the ladder of life from tenancy to home ownership. This climb, a struggle and a progression, was accepted as part of life until the widespread introduction of low down payments in the 1950s. Previously, the experience of George Harper, a compositor in a Hamilton printing shop in the 1880s, was probably typical of skilled workers' progression. Harper testified before the Royal Commission on the Relations of Labor and Capital in January 1889.

Q.   What I want to find out is, whether they [workers] purchased this property from the amount of wages they earned by their trade, and if so, how long it took them to do it?
A.   To answer that I would have to take individual cases.
Q.   Do you know of any individual cases?
A.   I could refer you to my own case. Of course, I have been working at the business for twelve or thirteen years, and I own the house I live in. I suppose it will take a man about ten years to secure a house of his own, and perhaps longer.[6]

Assuming that Harper dated the period of saving to purchase a house from the time of a man's marriage and settling into a trade, he was placing the age of home buying in the mid to late thirties. The Hamilton data largely confirm this relationship between age and ownership. Moreover, three patterns of ladder arrangements emerge from the data.

To compress the statistics, age has been presented in five groups, each according roughly to stages in family and employment history. The first (25–34) constitutes the family-formation stage. Young married couples have typically characterized the population in this group. Personal assets and individual earnings tend to be minimal compared to later age cohorts. The second group (35–44) theoretically finds the family close to its maximum earnings and at its maximum size. The third (45–54) brings the family closer to or into its maximum earnings. In the fourth group (55–64), the need for a house due to perceived requirements for rearing a family had passed, and it is plausible to assume diminished income and greater likelihood of illness and layoffs. Finally, households heads aged over 64 would have faced retirement, illness, and possibly the loss of a mate. For household heads in this group the house serves not as a nest but as an asset for conversion into money or, as sometimes happened, flats for shelter and income. The age groups, therefore, imply shelter needs and asset accumulation that constitute a plausible five-step ladder of life.

Before plotting the characteristics of the ladders for the sample years from 1866 to 1981, consider the data in table 6.1. Even a quick scanning confirms the value of the ladder analogy. Youth invariably stood more precariously on a lower rung of home ownership than seniors, with the middle-aged groups attaining the intermediate rungs. In most of the sample years, the over-64 age group had a higher percentage of homeowners than any other age group, a phenomenon attributable to the use of household data. This underscores one of the probable areas of bias associated with the use of household data. Presumably, before the days of pensions, nursing homes, seniors' housing, and numerous apartments, many seniors had to surrender household autonomy and move in with family. Those who could have clung to independence likely possessed the superior resources of their age group and continued to reside in an owner-occupied house.

It is apparent that the sample years presented distinct home-ownership opportunities and limitations. Not all ladders were set at the same angle, as two extremely different years demonstrate. In the 1899 sample, the youth group had a miserably low degree of home ownership. The other age groups also had relatively low ownership percentages. The depression of the 1890s had pulled down almost everyone on the ladder of life. A remarkably different situation ruled in 1966, when the youth group's home-ownership percentage was more than double that of its 1899 counterpart. All other age groups in 1966 also attained extremely high home-ownership percentages (see table 7.1). Descriptive accounts of the relationship between age

Table 7.1
Home-Ownership Percentages by Age Cohorts, 1872-1981

Year[1]	Age Groups[2]					Description of the Economic Times as Illustrated by Ownership[3]
	25-34	35-44	45-54	55-64	over 64[4]	
1866	11.0	22.5	42.5	34.4	33.3	Collapse
1872	31.0	40.2	50.0*	44.2	72.8	Moderately Good
1876	35.3	47.4	43.4	56.0*	54.5*	Youth Opportunity
1881	25.5	41.7	47.7	55.1*	61.1*	Moderately Good
1886	25.9	37.8	47.7	44.4	64.5*	Moderately Good
1891	26.3	36.7	43.6	47.8	50.0*	Moderately Good
1896	20.0	28.5	46.7	48.7	47.3	Collapse
1899	19.6	29.5	41.5	57.9*	48.6	Collapse
1906	17.7	37.6	46.0	57.4*	60.0*	Moderately Good
1911	29.6	41.3	51.2	53.3*	61.3*	Moderately Good
1916	25.5	38.9	52.1*	55.7*	74.8*	Moderately Good
1921	36.4	52.9*	57.3*	68.9*	65.7*	Youth Opportunity
1926	30.0	49.1	53.2*	66.0*	71.5*	Moderately Good
1931	23.6	44.4	52.5*	55.7*	66.5*	Moderately Good
1936	13.7	33.3	49.8	52.6*	56.9*	Collapse
1941	13.7	27.4	44.6	50.2*	51.6*	Collapse
1956	46.3	57.9*	63.9*	64.9*	64.8*	Youth Opportunity
1966	48.7	70.8*	75.0*	72.7*	70.1*	Youth Opportunity
1981	45.2	65.4	72.4*	68.5*	55.9*	Youth Opportunity

Sources: Hamilton Assessment Rolls and 1981 Census of Canada

*Indicates home ownership of 50 percent or better.

[1] The samples for 1856, 1858, and 1864 were not considered because of low numbers in some of the groups.

[2] A few cases had ages below 25, but the small numbers precluded establishing a group.

[3] The criteria for moderately good years were: a level of ownership of 20 to 33 per cent for the 25-34 year group and a positive slope in the linear regression. For the youth opportunity years, the criteria were: over 33 per cent for the 25-34 year group and a relatively flat slope by linear regression. The criteria for economic collapse were: under 20 per cent for the 25-34 year group and lower ownership for most other age groups than the levels in the previous several cross sections.

[4] The numbers were small in the over-64 cohort until the 1886 sample.

and home ownership for each year would tell of slightly different ladders of life for each cross section. Since data can be tedious we have sought to collate the information, and thus to simplify the relationship between age and home ownership for the cross sections. It is reasonable to expect that booms and depressions, tight credit times and easy credit times, would set up different ladders of life. It is possible to assemble clusters of similar ladders, but inevitably such generalizations slough off details.

Three idealized ladders of life stand out from table 7.1. The most frequently occurring one, characterizing a particular age and ownership connection in about half the sample years, can be claimed to represent moderately good economic times. On this ladder, the youth group of household heads possessed a moderate degree of home ownership and homeownership increased substantially in each subsequent group. Since the great depression, this version of the ladder of life has vanished. Conceivably, the age and ownership association could be reverting to this form, which prevailed for many decades. The 1981 data disclose slippage in home ownership, which could indicate a looming crisis because, if a pronounced reversion occurs, or is occurring, the shelter industry and government will have to be exceedingly inventive in devising new ways of keeping ownership accessible to the young or in shifting tastes toward a variety of rental solutions and thereby confronting the great North American myth of proprietorship realized by unusually large numbers of young families after World War II. This scenario of a crisis in expectations remains hypothetical.

Several trends cloud a rose-tinted forecast. Since 1966 home ownership has been declining not only in Hamilton but in virtually all Canadian cities outside of Quebec. Data on household incomes and house prices depict an increasingly large gap.[7] Therefore, it seems certain that, in relation to the past, the 1956, 1966, and 1981 cross sections produced unusual ladders. These three cross sections and those for 1876 and 1921 produced an age and ownership relationship representing youth opportunity, when an unusually high percentage of young and middle-aged household heads owned homes. Several discrete explanations account for the striking bias toward youth in 1876 and 1921. The ownership situation in 1876 related to the 1873 economic panic and a local economic recovery from the calamitous depression of the late 1850s, which had continued to afflict Hamilton into the mid-1860s. Growth in Hamilton's manufacturing sector – foundries, sewing machine manufactories, farm implements producers, a ready-made clothing establishment, and boot and shoe producers – had overtaken the collapse of the city's commercial fortunes, which had plunged the city into demographic, real estate, and fiscal crises by 1864. Moreover, the real estate collapse kept down land prices into the early 1870s.

The buoyancy of ownership in 1921 was an unexpected finding given the post-war inflation, the soaring of construction costs throughout North America, and the recession of the early 1920s (see chapter 5). A technical explanation underlines the importance of becoming thoroughly familiar with the source materials before drawing conclusions. The 1921 cross section captured the post-war construction of very modest homes on cheap lots in areas recently annexed by the city. The percentage of household heads who lived in their own homes in the original city was just under 42 per cent. In annexed areas, home ownership had shot up to 70 per cent, demonstrating the boost that peripheral land gave to the attainment of the bungalow and garden. A drop in home ownership in the annexed areas appeared in the 1926 cross section. It probably had derived from the construction of commercial structures with flats and the erection of apartment buildings. In effect, urbanization had followed swiftly upon suburbanization and deflated home ownership, but in 1921 unusually pure single-family detached dwelling areas had made their mark on our data.

The extremely high ownership percentages for household heads aged 25–34 in 1956, 1966, and 1981 – indeed the high percentages for all age groups at these cross sections – have their origins in the juxtaposition of fifteen years of depression and wartime controls with another fifteen or more years of industrial prosperity. The former conditions established a pent-up demand for consumer goods including new housing stock, while the latter made the objectives of a consumer-oriented society feasible, especially with the gains made for workers by the organized labour movement. It is impossible to weigh these factors or the many others involved. Certainly a relaxation of credit was important and has been discussed in chapter 6. Down-payment requirements were lowered and a new source of mortgage capital was tapped when banks were allowed to operate in the mortgage market. Behind the multitude of housing-ownership programs lay a governmental attitude that was radical, in the light of historical experience. The state was catering to a nineteenth-century conception of independence, and effectively conveying this ideal of independence by proprietorship to an age group that would come to regard as a right what had once been a goal attained by saving.

Though young families lacked savings for a down payment, they had achieved government support for access to home ownership. When house prices climbed in the 1960s, the *Report of the Federal Task Force on Housing and Urban Development* asserted the claims of young families who could carry monthly payments. "Their problem is to get that mortgage and that house now when their children are young and their need for family accommodation is the greatest."[8] James Hatch, in a treatise on the Canadian mortgage market published by the Government of Ontario in 1975, indicated that the 25 to 29-year-olds were the prime group for house buying. In the very short term of the post-war decades, this may have been true, but government action had made it so. Moreover, the National Housing Act, by confining itself to mortgages for new housing, had skewed the market. It induced "young people with $500 to go into a new house rather than an old house."[9] The era of demographic politics had commenced. Interestingly enough, the young beneficiaries of the post-war high-ratio mortgage soon will be the recipients of geriatic entitlements including social housing. Another demographic consideration, namely new Canadians, also may have affected home ownership. Conceivably but unverifiably, post-war immigrants to Canada brought with them an especially strong desire to own real property. The immigrant factor introduces a host of cultural hypotheses outlined later in this chapter.

A final set of ladders appeared frequently. It reflected the consequences of dismal circumstances, expressing very low home-ownership percentages for youths. The improvement with age was modest. The effects of the depression of the 1890s were conveyed in the 1899 data, but 1906 a prosperous year, stands forth as an anomaly whose peculiarly low youth performance is dealt with in later discussions about occupation and ownership. Like the 1921 cross section, it requires a special explanation. A recession from 1913 to 1915 and the enlistment of young males for the war dragged down the home-ownership percentages for the two youngest groups in the 1916 sample. Data for 1936 and 1941 capture the consequences of the great depression and its shattering of young people's aspirations as measured by the standard of home ownership – the continental culture's gauge of success, independence, and credit worthiness.

Let us summarize the findings. First, as expected, home ownership has been affected strongly by age. This fact warns against treating a city's level of home ownership as a meaningful social indicator. Unless the life cycle is featured in home-ownership discussions, they will have limited value. Second, given the North American preference for home ownership, the degree of attainment of this principal consumer goal by age groups does offer a means for ascertaining the city's economic performance. For any sample year, the degree of home-ownership realized among the youngest age groups and the overall type of ladder of life displayed can serve as useful supplements to the conventional knowledge about a city's depressions or booms. Finally, the long time series accents the aberrantly high level of home ownership among younger heads of households during the last thirty to forty years and places current discussions about its durability into historical context.

The foregoing cross-sectional analysis of age groups and home ownership has limitations. It completely lacks a dynamic expression of how people fared over their lifetimes. Instead, the cross sections look at age groups frozen in one instant of time, or rather over the several weeks of a civic assessment. The lack of detailed accounts of common lives bars access to either the material or psychological impact of the economic times in which, for example, young people came of age and started up households. A cross section presents their situation at one point only. A massive expenditure of resources would be necessary to collect a sample of people who could be studied from cradle to grave. That research would be costly even if it merely dealt with the austere basics of social history: moves, jobs held, and an indirect measure of economic stability, such as home ownership. Charting these basics would demand an extraordinary research effort. Undoubtedly, even if available, the data alone would generate a bloodless account that would still fall short of a satisfying portrait of crisis and routine in common lives. Therefore, the further manipulation of our existing data base to achieve a little better understanding of how economic conditions and the will to possess real property interacted across the lifetimes of household heads will come as a short cut with two strikes against it – it cannot penetrate directly even into the basic life course of a sample population; neither can it express intimately how lives were affected by assorted economic times.

The constricted short cut consists of considering the lifetime experience of generational groups. A group selected for its age range in a particular year can be followed through subsequent decades, a process known as cohort analysis. As an expedient measure, it has its share of problems. A generational group, selected and traced for several decades, will not consist of the same people. Not all members of the selected group in a particular year will have stayed on in our sample blocks or even have continued to reside in Hamilton, and a few most certainly will have died, so there is attrition in the sample. It is possible too that a cohort actually might increase in size as a consequence of migration and immigration. The theoretical justification for looking at cohorts is that the experiences of the newcomers and of the persisters would not have been greatly different from those who had left. The economic and family lives of people in a generational group presumably would have been much the same whether they stayed in Hamilton or not, whether they came from abroad or were born in the city. The assumption of interchangeability, however, must arouse some scepticism. For example, it is demonstrated later in this chapter that there are grounds for be-

Table 7.2
The Progress of Home Ownership for Cohorts Selected at Ten-Year Intervals

Year	Ownership of Group 25-34 in the Year Cited	Ownership of the Cohort as it matured and became Age Groups		
		35-44	45-54	55-64
1866*	11.0	47.4	47.7	48.7
1876	35.3	37.8	46.7	57.4
1886	25.9	28.5	46.0	55.7
1896*	20.0	37.6	52.1	66.0
1906	17.7	38.9	53.2	52.6
1916	25.5	49.1	19.8	n.a.
1926	30.0	33.3	n.a.	64.9
1936*	13.7	n.a.	63.9	72.7

Source: Hamilton Assessment Rolls
*Years of economic collapse

lieving that some ethnic groups have had a far greater desire to hold real property than others. The changing ethnic mix of a cohort is an alteration that challenges the concept of generational conformity.

The many shortcomings of cohort analysis as a historical method call for restraint when interpreting data. Nevertheless, while cohort analysis does not treat the same body of people over time, the same city or almost the same city provides the context for home ownership. Even if a cohort were to be swollen by an influx of decidedly more property-conscious people and the cohort thereby experienced a distorted and soaring level of home-ownership as it aged, this in itself implies something about the capacity of the city's employment opportunities, wages, and property industry to deliver the goods. The cohort data actually suggest that this is the case. Consider the youthful age groups that experienced depressions as they embarked on family formation (see table 7.2). The household heads who were in the 25–34 age group during the years confirmed as ones of economic collapse eventually realized levels of home ownership quite as good as or even higher than cohorts that commenced on the sounder footings of prosperous years. That observation, resounding with Candide-like cheer, clearly should not divert attention from the agonies of the moment, from the cyclical unemployment that upset young lives. Yet the capacity of a society to forget or set aside the bad experiences and, in twentieth-century Canada, to mostly eschew radical politics likely have some connections with the memory-erasing and caution-inducing facts and lore about economic recovery. To assert that society is heterogeneous is commonplace, nevertheless this assertion should be examined from several perspectives. Age is one, as this section has emphasized. Age highlights the existence of many concurrent life experiences and of associated generational lore about the good times and the bad. That a fair number of men in the early years of the depression of the 1930s could recall both the experience of prior economic upsets and the subsequent acquisition of property was a fragile brace for the *status quo*. But it was a credible brace all the same.

Such men may also have suspected that, however well they and their co-workers of the same generation had done over the long run, others had done somewhat better. The facts of home ownership would have supported such suspicions. The white-collar occupations certainly appear to have recovered more wholly from the debilitating effects of depressions on youthful household heads than the blue-collar occupations. At least, this was the situation within the cohorts experiencing the collapse of the 1860s and the slump of the 1890s. The observation about inequality of ownership levels among occupational groups is not true for the cohort that had been 25–34 in 1936, the cohort originating in the privation of the great depression (see tables 7.3 and 7.4). For essentially all occupations, this cohort's astonishing achievement of home ownership in recent times is re-examined in this chapter, in the sections dealing with occupation and ethnicity.

OCCUPATION, SOCIAL STRUCTURE, AND HOME OWNERSHIP

Class and occupational groups are different but related. Home ownership is not a precise reflection of employment security or of credit worthiness, but the concepts are related. These two propositions guide the analysis of home ownership and occupation. First we must deal with class. Class, particularly among structuralist theorists indebted to Marx, is defined by the social relations to the means of production. Occupations, individually, concern technical features of production and this makes it problematic to cluster occupations together in order to generate a class for empirical inquiry. Occupations tell little about of the ownership of the means of production or the purchase or sale of wage labour. Is the carpenter whose name has been drawn from a census enumeration sheet, assessment roll, or city directory an employer or does he sell his labour? If a contractor, what is the scale of the operation? But establishing the social relations of individuals to some degree is not impossible and in many, if not most, instances the occupations are less ambiguous than the case of carpenters. For practical purposes, the empirical investigation of society must revolve around occupations and clusters of occupations because the occupation variable is common to many routinely generated sources and, with caution and explication, may be related to class.

The effort to define class by people's relationship to the private ownership of capital and their position in the labour market produces a concept of class frequently called the "class-in-itself." While the class-in-itself cannot satisfactorily be constructed from occupation alone, there are some students of class who criticize any association of class and occupation. They find the structuralist approach, which presents some prospect of connecting class and occupation, incomplete. For them, a class becomes a class through self-awareness or a consciousness of its distinctiveness, which it achieves through the formulation of a distinctive culture, through political and economic organization to promote its interests, and ultimately through expressions of antagonism. This formulation of class often carries the label "class-for-itself." Hamilton's classes have been studied by structuralists and by a neo-Marxist interested in cultural expressions of class. Michael Katz and his colleagues concluded in *The Social Organization of Early Industrial Capitalism* that early-industrial Hamilton was characterized by two classes. Katz used both theoretical arguments and statistical methods (in the analysis of variance) to support this conclusion. Employing quite different sources and a slightly more impressionistic method, we concluded that the two class model was a useful concept when looking at the roots of Hamilton society in the 1830s. Labour historian Bryan Palmer, author of *A Culture in Conflict*, has decried the bloodless quality of structuralist class analysis – his working class becomes manifest and alive through its ideas on political economy, its recreations, and its political and economic action. Interestingly, Stuart Blumin has challenged American social historians to do for the middle class what the labour historians have done for the working class by considering the cultural context. The class and culture or class-for-itself literature possesses a rich humanistic texture. It has inspired and will continue to inspire splendid community-based studies. More than that, the class-for-itself outlook makes the use of class problematic for social scientists and historians working with quantitative data. The problems of associating occupation and class, challenging enough when dealing with the class-in-itself theories, become virtually insurmountable when seen through the eyes of the class and culture analysts.[10]

A brief survey of different schools of thought about class and occupation should alert readers to the problems inherent in our discussion. However, caveats aside, we must somehow compress

Table 7.3
The Home-Ownership Progress of Household Heads by Age Cohorts for Cohorts Launched in Depressions: Blue-Collar Groups

THE DEPRESSION OF THE 1860s

Cohort (Year)	Skilled and Semi-Skilled		Labourers		Bootmakers		Moulders		Carpenters	
	% owners	(n)	% owners	(n)	% owners	(n)	% owners	(n)	% owners	(n)
25-34 (1866)	10.8	(28)	8.0	(35)	16.7	(6)	0.0	(27)	33.3	(3)
35-44 (1876)	54.0	(100)	42.0	(31)	0.0	(7)	50.0	(8)	75.0	(16)
45-54 (1886)	61.7	(60)	35.2	(37)	0.0	(0)	83.0	(6)	16.7	(6)
55-64 (1896)	50.9	(55)	52.0	(25)	0.0	(0)	0.0	(1)	16.7	(6)

THE DEPRESSION OF THE 1890s

Cohort (Year)	Skilled and Semi-Skilled		Labourers		Machinists		Moulders		Carpenters	
	% owners	(n)	% owners	(n)	% owners	(n)	% owners	(n)	% owners	(n)
25-34 (1896)	29.7	(144)	10.0	(50)	0.0	(4)	16.7	(6)	21.4	(14)
35-44 (1906)	36.0	(189)	30.2	(53)	28.8	(14)	21.0	(19)	42.9	(7)
45-54 (1916)	50.0	(244)	39.3	(61)	30.8	(26)	30.8	(28)	66.7	(12)
55-64 (1926)	51.4	(313)	48.1	(52)	53.8	(13)	77.8	(9)	50.0	(10)

THE DEPRESSION OF THE 1930s

Cohort (Year)	Skilled and Semi-Skilled		Labourers		Machinists		Moulders		Carpenters		Steelworkers	
	% owners	(n)	% owners	(n)	% owners	(n)	% owners	(n)	% owners	(n)	% owners	(n)
25-34 (1936)	12.7	(284)	7.7	(65)	30.8	(13)	40.0	(10)	20.0	(5)	22.2	(9)
35-44 (1946)	*	*	*	*	*	*	*	*	*	*	*	*
45-54 (1956)	57.0	(681)	74.7	(83)	73.5	(34)	75.9	(29)	61.9	(21)	88.4	(69)
55-64 (1966)	63.4	(519)	87.5	(56)	85.0	(20)	88.8	(18)	83.3	(18)	87.5	(56)

Source: Hamilton Assessment Rolls

*no data

Table 7.4
The Home-Ownership Progress of Household Heads by Age Cohorts for Cohorts Launched in Depressions: White-Collar Groups

THE DEPRESSION OF THE 1860s

Cohort (Year)	White-Collar % owners	(n)	Clerks % owners	(n)	Merchants (Unspecified) % owners	(n)	Grocers % owners	(n)
25-34 (1866)	0.0	(4)	0.0	(2)	0.0	(2)	25.0	(4)
35-44 (1876)	34.8	(23)	86.0	(19)	0.0	(3)	33.3	(6)
45-54 (1886)	42.9	(14)	66.6	(3)	50.0	(4)	83.0	(6)
55-64 (1896)	100.0	(11)	100.0	(2)	33.3	(3)	50.0	(4)

THE DEPRESSION OF THE 1890s

Cohort (Year)	White-Collar % owners	(n)	Clerks % owners	(n)	Merchants % owners	(n)
25-34 (1896)	21.1	(38)	15.4	(13)	75.0	(4)
35-44 (1906)	44.2	(52)	26.3	(19)	100.0	(2)
45-54 (1916)	63.0	(73)	76.9	(13)	66.7	(21)
55-64 (1926)	72.5	(51)	100.0	(9)	80.0	(15)

THE DEPRESSION OF THE 1930s

Cohort (Year)	White-Collar % owners	(n)	Clerks % owners	(n)	Merchants % owners	(n)	Salesmen % owners	(n)
25-34 (1936)	17.5	(137)	17.9	(28)	33.3	(6)	26.9	(26)
35-44 (1946)	*	*	*	*	*	*	*	*
45-54 (1956)	59.4	(217)	70.4	(27)	63.0	(27)	59.0	(39)
55-64 (1966)	65.4	(179)	85.0	(20)	64.3	(14)	75.0	(32)

Source: Hamilton Assessment Rolls

*no data

nearly 1400 occupations, in ways that will make it possible to continue out analysis of housing and social structure. First, to clear a path toward our principal topic, it is important to stress that class is not our central subject. Some of the themes in housing history certainly have a connection with class. The segregation of groups by neighbourhoods can intimate much about class, culture, and class identity as well as, assuming differences in amenities, something about antagonism. But we will discuss ownership and quality of housing before we deal with neighbourhoods. Second, to avoid the theoretical wrangle of attempting to link occupation to class, it is possible to acknowledge the problem and to investigate social structure and housing through occupational groups defined according to specific criteria and assembled with careful judgment. If not precisely classes, the groups connote stratification.

We have recorded occupation in three different ways. First, an economic or horizontal recording placed jobs partly according to the sector of the economy in which they were located: professional and rentier group, agents and merchants, services and semi-professionals, business employees, government employees, masters and manufacturers, skilled workers, transportation workers, common labour, agricultural workers, others, and none. These designations proved useful for some minor features of the inquiry, but they did not constitute a recoding according to stratification. Crude stratification results were obtained by grouping household heads into those whose occupations were usually characterized by self-employment (entrepreneurs and professionals) and the rest (workers). Far more valuable and conventional was a recoding into six groups: professionals and proprietors, white-collar employees, skilled and semi-skilled labour, labour, women, and other. Straightforward explanations account for the designations. The professionals and proprietors engage in non-manual employment. Granted the butchershop owner might cart around a quarter of beef or the grocer a box of canned soup, nevertheless the physical chores have typically been delegated to hired help. Usually professionals and proprietors performed mental work under conditions of self-employment. Corporate lawyers and company doctors, for Hamilton during our study years, were rare and not typical of their professions. Both the professionals and proprietors worked with a measure of property and autonomy, likely espousing an ideology of independence and hard work. Within the limits imposed by competition, they could set their own hours. A few exceptional individuals – very wealthy professionals with capital invested in city enterprises or proprietors who were actually major capitalists – may have slipped into the ranks of this social group.

The adjective white-collar was applicable to a modest but growing number of household heads in the samples. The men holding white-collar jobs were employees, or so the occupational labels strongly implied. Over the more than one hundred years of our study, the occupational labels appearing in the assessment rolls remained consistent until the early twentieth century. Typical white-collar positions until then included: bookkeeper, clerk, male clerk, managing clerk, market clerk, commercial agent, salesman, secretary (male), stock keeper, receiver, and warehouse keeper. In the early twentieth century, a few new occupations joined the samples: assistant manager, district manager, and typist (male). Later a statistician, efficiency expert, payroll clerk, and office manager entered the data bank. In the 1966 sample, a computer programmer was captured in the sampling process. Although the white-collar group consisted (as far as was possible to determine) of employees, a number of attributes set it apart from the skilled and semi-skilled labour group. Above all else, the workplace of the white-collar employees was generally separate from that of the manual labour groups. By the 1920s Hamilton had many modern factories that separated offices from plant facilities. Architecturally distinct office buildings among plant facilities gave a number of white-collar employees segregated and sometimes elevated stations. Physically close to the top brass – some of whom were included in our sample – the white-collar employees showed their identification with owners or operators by rejecting unions. A few clerks handling parts and invoices, working in inventory control and shipping, must have worked on shop floors. The paraphernalia of their activities – pens, pencils, stamps, clipboards, and paper – and their specific work sites still separated them from the experiences of manual labourers. The very terms white-collar and blue-collar point to "a generic disparity."[11]

Occasionally, white-collar employees of a non-managerial kind have been thought of as sliding into the proletariat. Having surveyed research on the alleged proletarianization of white-collar workers, Anthony Giddens, admittedly no Marxist, has concluded that in the United States the mechanization of the office has not proletarianized white-collar employment. A very recent analysis of the transformation of the American class structure, 1960–1980, has echoed this claim and proposed that the expansion of managerial and expert groups has dramatically reversed prolitianization.[12] For men especially, this hypothesis could very well be true. Since the white-collar workers in the assessment samples from 1856 to 1966 were overwhelmingly male household heads, they probably escaped the thrust of office-work skill reduction. Therefore, there are good reasons for separating them from manual labourers of all kinds. By 1981 this gender situation had altered and women in clerical jobs with low pay made up a substantial proportion of the white-collar household heads. The question of a merging or a separating of white-collar and labouring household heads will be re-examined in terms of their historical experience with tenancy and housing quality.

Skilled and semi-skilled labourers constituted the plurality and almost the majority of the household heads in our samples. In a few of the sample years, they accounted for half the household heads; more often their percentage of the samples ranged between 45 and 50 per cent. The existence of skill differentials has been acknowledged among manual labourers themselves and has found obvious expression for generations in the apprenticeship system and in the precise labelling of trades. Erring on the side of caution, we united semi-skilled labourers and skilled labourers in the making of this important social category. Doubtless, the recoding process scooped up a number of very marginally skilled workers. This form of inaccuracy, however, makes the group's impressive levels of home ownership all the more noteworthy. Common labourers, in contrast, proved easier to identify because in most cases the assessment rolls described them merely as labourers. Assessors could have filled in this information based on faulty reports or on laconic responses from occupants, but overall this mistake must have been infrequent. Home-ownership data and assessed values of residences confirm a divergent path for "labourers."

Having set the theoretical basis for our discussion on class and occupational stratification, we turn to consider ownership as an index of economic prosperity – income and credit. The important cultural status of home ownership and the invention of facilities, methods, and services to meet the demand have made it a reasonable indicator of economic conditions in the city. Along with assessed values of real property, no other dependent variables exist among routinely generated sources that can portray the temporal fluctuations – as well as the social inequalities – in economic performance. Given the potent demand and supply forces built up in North America around the concept of possession of land and shelter, alterations in home ownership over time or variations in home ownership among occupational groups can be read cautiously as credible indicators of achievement, misfortune, accumulation, and disparity.

The association of home ownership with the first four occupational groups is self-evident and not startling. It shows simply that across the years from 1872 to 1966 home ownership has had a direct relationship with occupational stratification (see table 7.5). Let us move beyond that generalization to the home-ownership history of the

Table 7.5
Mean[1] Percentages of Home Ownership for Major Occupational Groups
by Age Groups, 1872-1966

	25-34	35-44	45-54	55-64	Over 64
Professionals and Proprietors	36.9	46.2	59.0	61.6	68.6
White-Collar Workers	29.3	43.8	52.5	58.6	67.4
Skilled and Semi-Skilled	28.5	46.6	52.4	57.3	57.9
Common Labour	21.3	32.9	40.0	45.6	54.7

Source: Hamilton Assessment Rolls
[1]The mean was calculated by averaging the percentages for each group from each sample year.
An alternative summary statement, namely the percentage ownership for all cases from all sample
years, would have biased the percentages toward the levels attained in the large 1956 and 1966
samples.

occupational groups themselves. To begin with, better than half of
all Hamilton's professionals and proprietors – if we ignore the age
variations of the ladder of life – could expect to have owned a home.
Except for depressions, prospects improved over time. After the
crisis of the 1850s and 1860s, the homeownership percentage registered in the mid-forties; in the early 1900s, it reached the mid-
fifties; and by the early 1920s, it had climbed to the mid-sixties.
Three-quarters of the professional or proprietor household heads
in the 1966 sample had attained home ownership. The story of
improvement over the long term was shared by other occupational
groups, although the professionals and proprietors group commenced with a much higher rate of ownership than any other group
(see tables 7.6 and 7.7).

Despite the higher plateaus of home ownership experienced by
the professionals and proprietors, the unsmoothed picture includes,
even for them, chapters of worsening conditions (see table 7.5). This
picture of flux had features in common with that for other occupational groups. For example, within all occupational clusters, the
youngest age cohort (25–34) usually behaved as an especially sensitive barometer of economic declines. The depression of the 1890s
did not diminish the access of young professionals and proprietors
to home ownership with anything like the severity felt by young men
in the trades or, particularly, young common labourers. In 1896
home ownership for the youngest cohort of professionals and proprietors was 31 per cent and in 1899 it had sunk very slightly to
29.4 per cent. In contrast to that economic collapse of the 1890s,
the depression of the 1930s stands out. It struck exceedingly hard
at the young (25–34), even within the professional and proprietor

Table 7.6
Broad Trends of Home Ownership by Two Classifications of Occupations

ALL GROUPS FOR ALL HOUSEHOLDS	
Mean Percentage of Ownership for Sample Years	1856-1966 – 40.4
	1872-1966 – 42.6
	1906-1966 – 46.6
CLASSIFICATION 1	
Mean for Entrepreneurs	1856-1966 – 39.4
	1872-1966 – 48.4
	1906-1966 – 52.9
Mean for Workers	1856-1966 – 38.5
	1872-1966 – 40.3
	1906-1966 – 43.6
CLASSIFICATION 2	
Mean for Professionals and Proprietors	1856-1966 – 51.3
	1872-1966 – 53.5
	1906-1966 – 57.4
Mean for White-Collar Workers	1856-1966 – 39.7
	1872-1966 – 43.1
	1906-1966 – 49.2
Mean for Skilled and Semi-Skilled Labourers	1856-1966 – 40.9
	1872-1966 – 42.2
	1906-1966 – 46.2
Mean for Common Labourers	1856-1966 – 31.4
	1872-1966 – 34.0
	1906-1966 – 45.5

Source: Hamilton Assessment Rolls

group, who sustained these percentages: 44 (1926); 34.7 (1931); 13.5 (1936); 22.5 (1941). When employment, access to credit, and savings diminished for youthful household heads in the 1930s, it diminished across all occupational groups. Middle-aged household heads in the professional and proprietor group had difficulty retaining homes, but the effects of the depression obviously took longer to reach them than was required to devastate the comparable age groups in the remaining three employment groups. A high percentage of ownership was sustained in 1931 for professionals and proprietors in the 35–44 age cohorts: 45.8 (1926); 53.3 (1931); 35.2 (1936); 29.6 (1941). The 45–54 cohort had these percentages: 62.7 (1926); 63.3 (1931); 50.6 (1936); 44.0 (1941). Savings and established businesses likely slowed the deterioration, sparing the older professionals and proprietors from immediate impact.

Trends in white-collar home ownership stand out in interesting ways when compared with the trends for the skilled and semi-skilled labour group. The broad trends, the mean percentages for the years

Table 7.7
Trends of Home Ownership Percentages by Occupational Group, 1856-1966

	Professionals and Proprietors	White-Collar	Skilled and Semi-Skilled	Common Labour
1856	23.1	6.7	45.5	11.4
1858	22.2	18.8	45.1	19.6
1864	41.2	14.3	25.0	6.7
1872	45.8	34.3	37.6	37.5
1876	46.7	39.3	47.1	35.5
1881	42.0	24.5	40.7	39.8
1886	47.5	32.9	37.8	27.9
1891	53.6	30.4	35.0	25.7
1896	45.5	36.3	32.3	32.6
1899	53.8	42.6	31.7	16.1
1906	55.6	40.4	35.8	23.9
1911	62.8	48.1	39.7	27.0
1916	58.3	42.2	39.1	30.3
1921	60.6	59.5	50.7	40.7
1926	57.6	53.2	50.0	35.9
1931	53.8	48.0	41.7	33.5
1936	45.6	41.4	38.0	29.2
1941	40.9	37.2	34.4	28.9
1956	63.0	54.0	58.8	57.1
1966	76.1	67.6	67.7	65.0
1981	69.8	52.5	61.3	62.6
Mean[1]	51.2	39.8	42.5	32.2

Source: Hamilton Assessment Rolls and Statistics Canada, Census of Canada 1981
[1] The 1981 census used occupational classifications that could not be disaggregated; the classifications largely dealt with related work, but grouped together skill levels and also aggregated proprietors and employers. Thus there were few proprietors in the professionals and proprietors group and some inclusion of labourers among the skilled and semi-skilled and vice versa.

from 1872 to 1966, reveal little difference; the mean percentage stood at 42.2 for the skilled and semi-skilled and at 43.1 for the white-collar workers (see table 7.4). A detailed chronological breakdown modifies the impression of a parallel experience. From 1856 to 1891 skilled and semi-skilled workers consistently scored higher percentages of home ownership than white-collar workers. The average spread between the two, across these seven early cross-sections, was 10 percentage points. Subsequently, two processes characterized the white-collar occupations and point to a well-known restructuring of capitalism. First, in the 1890s and, more markedly, in the 1920s, the white-collar share of the city's work force increased. The largest jump for any two contiguous cross sections occurred between 1921, when white-collar household heads constituted 12.8 per cent of the sample, and 1926, when they accounted for 16.1 per cent. Second, for an extended period (1896–1941) white-collar employees held an ownership edge over the skilled and semi-skilled labourers. The average spread was 5 percentage points. These two findings support a hypothesis of an early twentieth-century expansion of white-collar opportunities. In more recent years the increase in low-paying white-collar work among women, the growth in the number of female household heads, and the appearance of better blue-collar wages in the city's factories have produced the reverse situation. In Hamilton, industrial expansion after World War II and the opening of new suburban tracts assisted the skilled and semi-skilled labourers to surpass the level of home ownership for the white-collar workers in the samples taken in 1956, 1966, and 1981.

Skilled and semi-skilled labourers fared better than their showing of 42.5 per cent home ownership for all sample years from 1856 to 1981 would suggest. According to the ladder of life concept, a machinist, for example, could have expected to have had a rising probability of ownership with increasing age. The 42.5 per cent mean lumps all ages together. Earlier, we cautioned against ignoring age. Now its interaction with occupation must be considered. Assuming survival, the machinist – or any other skilled or semi-skilled tradesman – could have achieved an optimum probability of ownership in the 55–64 cohort. In that advanced age group, savings and earning power had peaked. The mean percentage of ownership across all years was 58 for this cohort. It rarely fell below 50 per cent in any of the sample years. In simple terms, the city's better-trained and specialized manual workers had a slightly better than 50–50 prospect of becoming home owners over their lifetimes. This observation neither proves an equitable distribution of society's bounty nor points to a labouring population grievously squeezed out of all hope and reward. However it could help to explain the limited popular appeal of social housing in this North American city in the twentieth century. Hamilton was more than a city of homes; the majority of better-trained blue-collar household heads could have expected to nest their families in homes of their own.

In intricate ways, the fluctuations in home ownership for the skilled and semi-skilled mirrored the fortunes of an industrial economy that at different times grew, faltered, boomed, or burst. In general, employment in the metal trades opened substantially as Hamilton became a major producer of Canada's stoves, bolts and screws, assorted moulded and rolled iron fixtures, and sewing machines. Hamilton proclaimed itself the Birmingham of Canada.

Nevertheless, the industrial economy slowed in the early 1870s, had a disappointingly unsteady rate of advance in the 1880s, and suffered considerably in the general depression of the 1890s. Around 1900 the city entered a period of remarkable industrial expansion. However, the persistence of low ownership rates into the era of an astonishing industrial boom from 1900 to 1913 is a puzzle. The building of extensive new factories and the introduction of electrically powered machinery — together heralding a second industrial revolution — promoted two potentially negative consequences for skilled and semi-skilled labourers. First, the remarkable spate of factory construction, public works, and suburban building fuelled inflation. Then, too, as noted in our account of house construction, distributional coalitions had succeeded in forcing up supply costs. What the boom granted in the form of strengthened employment it may have helped to erase with mounting prices. Second, the new industrial technology and management practices of this era frequently called for less costly common labour. These two plausible explanations tie into well-known historical events, but their influence cannot be measured, and a far more direct and verifiable reason comes to mind.

The industrial boom in Hamilton attracted migrants and immigrants. Carpenter-contractor Charles Emery, it will be remembered from chapter 5, had come to Hamilton with his family from Chicago, where his father had worked for Oliver Chilled Plow. Hamilton lured many labourers when it beckoned as an attractive hive of employment. Moreover, it was often young men who joined the ranks of the skilled and semi-skilled labour group. Among that body of household heads the 25–34 age group accounted for the following percentages: 25.3 (1899); 30.8 (1906); 31.5 (1911); 29.2 (1916); and 22.3 (1921). This was the age group with the lowest home-ownership rate. The abrupt arrival of young household heads probably pulled down the mean home-ownership percentage for the entire occupational group. Further proof of this is that skilled and semi-skilled household heads over 45 had home-ownership rates of better than 50 per cent for 1906 and 1911.

During the 1920s and 1930s the home-ownership percentages for skilled and semi-skilled household heads moved in tight correspondence with the city-wide rate, which had attained a historic high of 51.5 per cent in 1921 and tumbled to a near all-time low of 35.5 per cent in 1941. The latter year's low level of ownership not only expressed the legacy of the depression, it also denoted the growth of rental accommodations for wartime workers. Easy credit, low down payments and the post-war industrial boom, including the benefits delivered by organized labour, raised home ownership in 1956, 1966, and even 1981 to remarkable highs. For the first time since the 1920s the skilled and semi-skilled labourers had realized higher home-ownership rates than the entire sample population. Appropriately enough, a 1971 book celebrating the city's achievements and quality of living was entitled *Pardon My Lunch Bucket*. The home-ownership data fortify the impressionistic evidence that paints a picture of an unabashedly blue-collar city. The city of homes had become, to a remarkable extent, a city of home owners of all occupational backgrounds.

If home ownership – home buying might be a more accurate term given the recourse to mortgages – indicates anything about the economic circumstances of households, it underscores the limited prospects of common labourers. In the samples from 1872 to 1941 inclusive, the typical rate of home ownership was roughly one-third for labourers as household heads. Besides having a crude rate of home ownership some 10 percentage points below that of skilled and semi-skilled workers, the common labourers' experience of home ownership differed slightly from that of the more skilled in specific eras. Most notably, the depression of the 1890s struck them harder. Conceivably, the distinctions between the two groups had marked significance in the 1890s. Artisans may have been in a better position regarding wages and job security than labourers were in the distressed times of the late 1890s, whereas in the depression of the 1930s and the slumps thereafter the two groups drew closer in terms of job risks and benefits. The hypothesis – and it is only that – of a narrowing of the gap in material rewards is prompted by a look at what happened to home ownership during the two depressions and into the post-World War II era. The depression of the 1890s devastated prospects of home buying for the youngest age group of labourers. The percentages for ownership plummeted: 13.2 (1891); 9.8 (1896); and 4.9 (1899). That meant a slide of 63 per cent. Skilled and semi-skilled heads of households in the same 25–34 age group experienced a decline of only 29 per cent. In the 1899 sample there is evidence that the depression afflicted the labourers of other ages quite profoundly, for the percentages of ownership by age group were as follows: 11.1 (35–44) and 17.6 (45–54). Gradually, before World War I and again in the early 1920s, common labourers recovered the lost ground of home ownership (see table 7.7).

In the great depression, common labourers' home-ownership levels relative to those of more skilled workers retained the historical spread of 10 percentage points. The home-ownership percentages of the two clusters of household heads fell in rough proportion to

one another, unlike the situation in the 1890s. Indeed, the capacity
of common labourers to hold on – relatively but not absolutely –
becomes even more noteworthy when several age groups are examined. The youngest (25–34), as we must now expect, suffered the
greatest diminution of home buying; their percentages for home
ownership declined as follows: 27.3 (1921); 16 (1926); 14 (1931);
and 8.9 (1936). Older workers, above 55, also suffered declines,
though more modest ones. Astonishingly, one of the prime working-
age groups (45–54) actually improved its home-ownership percentages during the depression: 45.1 (1921); 48.9 (1926); 46.3 (1931);
and 49.1 (1936). Credit was closed to the young, and the ability to
search for work may have diminished for the older men. So the
young and the old were, in the context of the data, the visible victims
of the depression. For those with equity in a house who were trapped
in a buyers' market for casual labour but still had the crucial capacity
to appear fit and able to an employer, the depression forced belt-
tightening struggles to defend a hard-won goal. Mute data tell us
nothing about the family strategies for retaining the house. Without
data, however, the signs of a successful defence of ownership never
would have appeared. The point is speculative, but the personal
battles to cling to a house and yard could have contributed to the
lack of strong popular support for social housing during the depression.

   The most startling evidence for the convergence of the shelter
experiences of skilled and semi-skilled labourers and of common
labourers is the progress in ownership recorded for young household heads who had launched households in the inauspicious time
of the great depression (see table 7.3). For common labourers that
cohort had 7.7 per cent ownership in 1936; ownership had rocketed
to 87.5 per cent in 1966. As usual with a potential measure of social
equality, the picture is incomplete. Something surely was happening
to the shelter situation, but the diminished stock of rental houses as
well as the poorer quality and location of shelter for common labourers call for one hand clapping. As we note in chapter 10, a poor
situation was traded for one that was – relative to other occupational
groups – little better. Common labourers may have achieved high
levels of ownership and the young unskilled victims of the depression
may have achieved something by ownership, but what had they secured in terms of rooms, amenities, and environment?

ETHNICITY AND HOME OWNERSHIP

The ties between age and ownership and between occupational
groups and ownership follow fairly obvious hypotheses. A connec-

tion between ownership and certain ethnic groups, though less self-evident, has been noted in prominent studies on immigrant communities in American cities. Such studies have employed the United-States manuscript census. The Hamilton assessment rolls, unlike the census, recorded nothing about country of origin or ethnicity. However, commencing with the assessment of 1923, the city's assessors recorded religious denominations. In prior years, assessors had recorded only whether the householder was a Protestant or a Roman Catholic, information that was used in the allocation of tax revenues for schools. While the specific denominational information apparently had no use to the municipal government, it was nevertheless recorded. Combined with surnames, the denomination presented an opportunity to fabricate an ethnicity variable. A local folk-arts council helped us to prepare a manual for establishing the probable ethnic origin of particular household heads. The manual largely relied on the tendency of surnames to have distinguishing suffixes, letter combinations, and, sometimes, prefixes. Difficult cases were traced through the folk-art council to parish priests or cultural clubs. The determination of ethnicity does not establish whether the householders were immigrants or second or third generation Canadians. However, because continental European immigrants only began to arrive in very large numbers after World War II, it is likely that in the 1956 cross section the vast majority of household heads from most ethnic groups were immigrants.

The ethnicity variable should generate higher percentages than does the Canadian census data, which until 1961 dealt primarily with country of origin. The ethnicity variable picked up some second-generation residents. The sample data in fact disclose the higher percentage, but they also parallel the growth trend of the foreign-born reported in the census. A large spread between the census-based data on German immigrants and the assessment-based data on German ethnicity captures the prominence of German immigrants in the Hamilton region in the nineteenth-century. Since their descendants were not from Germany, the census until 1961 would have underestimated German ethnicity. Neither the term immigrant nor the label ethnicity apply to the city's Jewish population. However, the comparison of the census and assessment data for this group serves as a check on the sampling procedure and the accuracy of the assessors' information. Overall, the percentages in table 7.8 confirm the value of the sample method and the creation of an ethnicity variable. The 1961 and 1971 censuses included a question about the ethnic origin of a family traced on the male side. When these returns are compared with the ethnicity breakdown provided by the coding from assessment information, the results are encouraging.

Table 7.8
Comparison of Selected Data on Country of Origin with That of Ethnicity,
for Selected Years, 1921-1971[1]

Country of Origin or Ethnicity	Year								
	1921	1926	1931	1936	1951	1956	1961	1966	1971
ANCESTOR (MALE SIDE) BRITISH (CENSUS)								62.1	57.3
Born in Canada or the UK (census)		91.1		88.4		89.8		81.3	79.6
Anglo-Celtic (assessment)		89.4	88.5	87.1		72.3		65.5	
ANCESTOR (MALE SIDE) ITALIAN (CENSUS)							6.6		11.3
Born in Italy (census)	1.7		1.6		1.8		4.8		6.1
Italian (assessment)		2.2	2.4	2.4		5.6		8.9	
ANCESTOR (MALE SIDE) EASTERN EUROPEAN (CENSUS)								6.9	6.8
Born in Russia or Poland (census)		2.0		2.9		4.5		3.7	3.3
Eastern European (assessment)		1.7	3.4	2.8		7.4		9.8	
ANCESTOR (MALE SIDE) GERMAN (CENSUS)							5.5		4.9
Born in Germany (census)	0.8		0.6		0.3		1.9		1.4
German (assessment)		2.5	2.7	2.7		4.3		5.8	
ANCESTOR JEWISH Jewish (census)	2.0	1.7			1.5		1.0		1.0
Jewish (assessment)		1.6	1.6	1.6		1.6		1.2	

Sources: Hamilton Assessment Rolls and Census of Canada
[1]Due to the selective approach and the organization of the data, the totals were not rounded out
to 100 with the addition of an "other" category.

What does the creation of a soundly reconstructed ethnicity variable permit us to see? Above all, it discloses that several of Hamilton's largest European ethnic communities had much higher
ownership levels than the Anglo-Celtic population. Hamilton's Italians particularly have had an attachment to real property (see
table 7.9). Beginning in the 1930s, they displayed a slightly higher
tendency to own their own dwellings than did the Anglo-Celtic population – an omnibus group with surnames originating in the United
Kingdom or the Republic of Ireland. After the war ownership among
Italians soared above that for the Anglo-Celtics and continued to
climb even as other ethnic groups experienced a decline in home
ownership at the 1966 and 1981 cross sections. The Italians are easily

Table 7.9
Residential Ownership by Type of Shelter and Ethnic Group, 1926-1981

Category[1]	Year					
	1926	1931	1936	1956	1966	1981[2]
ANGLO-CELTIC						
Own house	45.3	40.3	36.4	49.0	57.7	
Own apartment	0.0	0.1	0.1	0.1	0.3	
Own above shop	1.1	1.1	1.2	1.4	1.6	
Own house with flat(s)	2.3	2.2	2.0	5.7	4.1	
Total	48.7	43.7	39.7	56.2	53.7	51.3
ITALIAN						
Own house	39.7	38.3	37.8	55.0	60.0	
Own apartment	0.0	0.0	0.0	0.0	0.0	
Own above shop	1.6	3.7	4.1	3.8	3.6	
Own house with flat(s)	2.2	3.5	5.8	14.7	12.2	
Total	46.1	45.7	53.9	59.1	78.3	83.4
EASTERN EUROPEAN						
Own house	31.1	31.8	26.9	52.7	64.9	
Own apartment	0.0	0.0	0.0	1.5	0.7	
Own above shop	0.0	0.0	2.9	3.0	3.6	
Own house with flat(s)	2.2	3.5	5.8	14.7	12.2	
Total	33.3	35.3	35.6	71.9	81.4	70.3
OTHER						
Own house	45.8	26.7	35.0	43.1	50.2	
Own apartment	0.0	0.0	0.0	0.4	0.6	
Own above shop	0.0	3.3	0.0	3.9	3.2	
Own house with flat(s)	4.2	3.3	5.0	13.8	16.0	
Total	50.0	33.3	40.0	61.2	70.0	56.4

Source: Hamilton Assessment Rolls
[1]The ethnic or cultural heading labelled Eastern European embraces ethnic Poles and Ukrainians. The sub-heading of "own apartment" means that the household head was listed as a resident and an owner of an apartment building. "Own above shop" means that the household head owned a commercial property and resided in the same. "Own house with flat(s)" means the household head was listed as the owner of a house but there were other residential quarters in the house.
[2]Data furnished by Statistics Canada on the basis of a 20 per cent sample taken during the 1981 census.

defined as a national linguistic group, even if the city's Italians came from many distinct Italian regions. The designations Polish, Ukrainian, and Russian pose considerable philological challenges to historians of Europe and of immigration alike. The boundary shifts of Eastern Europe and the mingling of Slavic peoples blur some of the distinctions, but language and denomination have defined resilient cultures. In a move that would legitimately upset the members of the cultural groups themselves, we have grouped them together for the purposes of analysis, for in the data the city's Poles and Ukrainians had nearly identical home-ownership traits.

The ownership percentages for these Eastern Europeans were lower than those for the Anglo-Celtics until the 1956 cross section. They surpassed the Italians in home ownership in 1966, achieving a level of over 80 per cent. A study of Poles in Toronto, conducted in 1978 by a survey questionnaire, established that 80 per cent of household heads owned houses. In Hamilton, Polish and Ukrainian home ownership had fallen back to a lower level by 1981. However, even this reduced level was above Anglo-Celtic ownership. Furthermore, Canada-wide data present Italians and Poles as having unusually high degrees of ownership. Among the Italians and the Eastern Europeans, the home-ownership phenomenon has been observed in studies of Detroit and Pittsburgh.[13] For these cities, it has been explained in terms of the love of property inherited through the culture of peasant societies. Canadian historian Bruno Ramirez, noting that home ownership was a high priority among Italian-Canadians, reported that lyrics to a popular Italian song of the post-war years spoke of "a little house in Canada ... admired by passersby."[14] The phenomenon of unusually high home ownership has also been accounted for by the capacity of newcomers to work collectively to achieve goals, often by both working to erect dwellings and, in the United States, forming building and loan societies. The Hamilton data permit a further contribution to the meaning of high percentages of home ownership among certain ethnic groups.

Good social-history analysis resolves into few concise and resounding declarations about the way people lived. Insufficient information dogs every significant question. What follows fortifies these observations about quantification and social history, but the analytic effort should inform and open up topics for future research. None of the American studies of ethnic groups and shelter have considered the age and gender of ethnic household heads. Our data indicate that these two variables warrant attention for any analysis of home ownership. Household heads among ethnic groups such as Ukrainians, Poles, and particularly Italians have had a distinctive age pyramid through the 1950s to the 1980s. Until quite recently their percentage of the population and of the household-head segment of the population that was 65 and older was relatively small – certainly seniors were far less abundant among Italians than seniors were among the Anglo-Celtic population. The youthfulness of post-war immigration accounts for the unusual age structure. Since seniors have a higher tenancy rate than household heads in the work-force age groups, it

Table 7.10
The Indirect Evidence for a Cultural Bias toward Home Ownership
among Italian Heads of Households, 1956, 1966, 1981

	Year		
Population Group	*1956*	*1966*	*1981*
Percentage of Italian Household Heads as Home Owners	69.1	78.3	83.4
Percentage of Anglo-Celtic Household Heads as Home Owners	56.2	63.7	51.3
Percentage of Anglo-Celtic Ownership with Age Distribution adjusted to Italian Age Distribution	59.2	60.1	55.3
Percentage Italian (25-34) as Home Owners	43.9	54.2	71.4
Percentage Italian (45-54) as Home Owners	64.8	61.4	90.9
Percentage Anglo-Celtic (25-34) as Home Owners	39.0	39.1	65.3
Percentage Anglo-Celtic (45-54) as Home Owners	54.7	67.3	66.7

Source: Hamilton Assessment Rolls and 1981 *Census of Canada*

is probable that their slight representation among Italians inflated home ownership to a modest extent. The age structure for Italians, however, cannot explain their remarkable levels of home ownership across all age groups. Ultimately, ethnic culture must enter the discussion (see table 7.10). In fact, even though gender is a demographic variable, its impact on the home-ownership pattern of Italians will be seen to be based on culture.

Which aspects of ethnic culture have worked – indisputably in a blended association – to produce a noteworthy buying of houses? American works, previously mentioned, have proposed collective effort and love of land – plausible answers, though unprovable and definitely incomplete. Two more cultural considerations could have come into play: painless assimilation and an ethnic or religiously based model of the family household. North American culture has evolved a prominent defining place for home ownership and for the house as an icon of personal success, whether or not one accepts our notions about a will to possess. To become a home owner was, for immigrants and possibly the second generation, to join the mainstream without having to make the sacrifices required for full cultural assimilation. Outward conformity and good citizenship could be acquired post-haste without a sacrifice of beliefs, customs, or language. Yards could be both North American (the front lawn) and European (the summer vegetable garden with a grape arbour). In Hamilton, Italian organizations have sought to prove the exemplary Canadian citizenship of their members partly by references to their stake in the city, including their high percentage of home ownership.[15] If this sounds like an excursion into the foggy realm of psychological drives, it is. Therefore, just as we balanced such thinking with a look at consumer pragmatism early in this chapter, we will do so once more. A practical benefit of home ownership is its provision of a larger and more readily modified dwelling. To this incontestable fact can be added another – household heads in ethnic groups have had larger households than their Anglo-Celtic counterparts. Possible preferences for an extended-family household or for more children may have necessitated home buying.

There is a fourth possibility. Could Italians have been earning more than Anglo-Celtics? Custom cross tabulations performed by Statistics Canada for census data are costly, and analytic categories such as ethnicity and household head may not square with the created concepts developed for our assessment data. Nevertheless, we examined the household-income data for Hamilton compiled in the 1981 census. The data were cross tabulated for ethnic origin, by sex, and by age. The results showed that household income for all Hamilton households peaked at the 45–54 age group, a finding congruent with the peaking of home ownership. The average income for all male household heads (described by Statistics Canada as Person I) was $24,743. It was $27,327 for British male heads (our Anglo-Celtic group) and $28,070 for Italian male heads. Female Italian household heads likewise had a slightly higher mean household income than their British counterparts. Could a margin of several percentage points have had any influence? We cannot answer this, but a further potential cultural factor was detected in the same data set. Among British (Anglo-Celtic) household heads, 65.6 per cent were male as opposed to 85.9 per cent for Italians. Other major ethnic groups were not as distinctive in their male household-head percentages. The male domination of Italian household composition is significant in that the mean income for male household heads was about double that for women. In other words, Italian household heads possessed a distinctive gender profile that made their households more likely to have the resources for home ownership. Regrettably, limited resources have precluded similiar analyses of earlier censuses. Overall, however, the census material and our assessment research point rather convincingly toward a set of familial-cultural factors favouring home ownership by Italians. The patriarchal nature of the Italian family and the Roman Catholic Church's opposition to divorce have influenced this ethnic group's distinctive association with the single-family detached house. The broad hypothesis of cultural differences has the support of a recent American study of when young people expect to move out of the parental home and establish an independent household. Among traditional Roman Catholics of southern European background, the investigators discovered *a strong avoidance of non-family living.* Apartment tenancy by singles, accepted in other cultures, apparently has little support among Italians.[16]

Starting with the 1921 cross section, the variable for home ownership and tenancy was coded to include a more precise designation of rental properties. Until that point, household heads simply owned houses or rented them. Except for a few residents of rooms above shops in the central business district, the city's tenants rented row houses, semi-detached houses, or detached houses. The assessment rolls do not address the distinction between these three types of houses. However, during the 1920s, apartment-building construction and the building of numerous two-storey brick business blocks along the arteries of the new suburbs established two distinct forms of apartment dwellings. Additionally, beginning with the 1926 cross section, a third form of rental housing appeared in notable numbers. Evidently, the city's expansion in the 1920s and the existence of old housing stock prompted the establishment of some form of assessed accommodations for a second household in houses. We have referred to these as flats. These flats, extremely uncommon in new suburbs, proliferated in the residential areas just beyond the central business district. In terms of the concentric ring theory of urban land-use, the flats occupied the zone in transition. True to the theory, many vanished during the urban-renewal and apartment-building construction efforts of the late 1960s and early 1970s.

Flats increased phenomenally at mid-century. In the 1926, 1931, and 1936 cross sections, only 2 per cent of home owners had flats in their houses; in 1956 and 1966, around 7 per cent had flats, units recognized as separate dwellings by the property tax assessors. The considerable increase in flats after the war occurred across the same decades as massive immigration. The two episodes in shelter and social history were connected. Several ethnic groups had a pronounced tendency to own houses with flats, usually flats let out to household heads of the same ethnic group. Students of Canadian immigration and of Hamilton development have remarked upon the existence of village immigration or chain migration arranged by associations of individuals linked by ties of kinship or locality. Hamilton would receive substantial Italian immigration from the villages of Racalmuto and Gagliano Aterno. Unfortunately, the existence of similar patterns for the city's substantial Eastern European and Yugoslav communities has not been documented. Around the turn of the century and even in the 1920s, boarding and rooming houses sheltered many immigrants and also presented business opportunities to a few others. In the 1950s the flats seem to have performed similar functions. But what new circumstances caused the proliferation of flats rather than boarding and rooming houses? The demise of boarding houses and the advent of flats may have owed something to the city's more rigorous public health and building codes, or even to a rise in expectations about the quality of housing. At the same time, many immigrants worked in the city's construction industries, which probably made it easier for them to repair and renovate houses.

Age, occupational groups, and ethnicity can help to establish a chronology for a history of shelter. Such a chronology should serve the needs of a vital social and urban history that retains its independence from the benchmarks of national politics and the timelessness of social-science models. Discovering the nature and the pace of innovation in housing forms and of social behaviour has been our principal concern. How does this goal apply to a summary of the social contours of home ownership? There appear to have been two fairly clear speeds of change over the one hundred or more years. First, there has been a volatility in the patterns of home ownership – the ladders of life have shifted as a consequence of well-known historical movements in the business cycle and of the ultimately successful governmental promotion of high-ratio mortgages. Another relatively swift turn of events accompanied the arrival of large numbers of European immigrants after World War II. These newcomers added to the demand for home ownership with unique zeal and their treatment of housing differed from that of the Anglo-Celtic majority. The culture of urban Canada was rendered more complex, often in very visible ways. Housing offers one means through which to view the new cultures. When the enormous physical mass of shelter required by the great social mass of people is mediated by the economy and by government policies – not just in mortgage assistance but in immigration – peaks and the troughs of ownership result. Viewed this way, shelter history is sensitive to rapid socio-economic transformations. The undoubtedly better wages brought first to blue-collar workers by unionization present another instance of fairly swift change (1930s–1950s) with an impact on housing data. By the 1950s skilled and semi-skilled household heads had closed the long-standing gap between themselves and white-collar household heads in terms of home ownership.

The second speed of change is less a matter of jarring movement than of perpetuity. Ironically, the many changes in home ownership express a persistence, permitting the reassertion of the theme of continuity or of gradual change over a long time. The drive for home ownership stands as a salient attribute of North American culture, forcefully expressed in Hamilton, a prosperous industrial city. The ability to buy, not the appeal of home ownership, has waxed and waned. Over the more than one hundred years, house buying has grown within all occupational groups, exploded for several decades among young household heads, and amounted to a passion among ethnic minorities. The demand for owner-occupied shelter has rebounded from crisis after crisis. The future of home ownership is another matter and the events of the last twenty years strongly suggest a pending turning point.

Ownership taken alone offers a narrow measure of how well urban Canadians have lived. Other standards of shelter quality enter the analysis shortly. Still, given home ownership's loaded cultural meanings, its attainment by so many in an industrial city argues against general privation. On the contrary, most Hamiltonians – certainly since the 1930s – have succeeded in buying into the North American cultural mainstream. The situations in other centres may have differed, but, in determining the extent and the nature of variation, recourse to summary data that state only the percentage of owner households can only be regarded as a first and crude step. Life cycle, the mix of skill levels in the local economy, and even ethnic cultures must figure in any discussion of ownership. When this was done for Hamilton, some new problems as well as some new observations surfaced, including the simple observation that, matter how the Hamilton ownership data have been analysed over time – by age, by occupational group, by ethnicity – they have disclosed, on balance, modest improvements. Distress existed and can be sought out readily in the hardship years and in the lives of a significant minority who failed in all eras to grasp the brass ring of ownership. In recent times this group has come to include a growing number of women as household heads. Distress, defined as the failure to realize home ownership and the usually better quality of shelter that accompanied it, is one thing. Distress has certainly affected many working families in the prosperous industrial city. Wretchedness is quite another matter, meaning lives spent in cramped and unsanitary shelter. Wretchedness can be located in the attributes of shelter units, in the social distribution of neighbourhood amenities, and in the quality of environments. The daunting question of shelter quality is addressed in the concluding chapter.

8 The Rented House: Landlords and Tenants, 1830s–1960s

The city has always furnished profuse social contacts, both voluntary and obligatory. The latter include the relations between employee and employer, debtor and creditor, and tenant and landlord. These contacts involve unequal arrangements, although they have not been purely one-way streets favouring the wielders of capital at every turn. Labour historians, for example, have recorded the resistance and the achievements of workers. While much is being discovered about the dynamics of employer-employee relations in the nineteenth and twentieth centuries, little has been written about the landlord and tenant nexus in urban social history. Yet a great majority of households, perhaps almost all, would have been tenants in the course of the life cycle of their principal wage earners – a reciprocal of the age and ownership relationship noted in the previous chapter. Evidence from the study of Hamilton and from property management journals affirms that vigorous engagements between landlords and tenants occasionally affected the city's economic and social scene. The strategies of both parties altered over time and the exact features of conflict shifted, because of many of the same circumstances that caused variations in work conditions and in employer-employee relations. The undulations in the business cycle, for example, could affect the leverage of both parties. While boom periods could improve the prospects for organized labour, they strained a city's housing stock and encouraged landlords to raise rents. Recessions left working people under-employed or unemployed, but they also brought higher than normal vacancy rates and permitted tenants to

Table 8.1
Number and Percentage of All Rental Units Owned within
Occupational Groups, 1876-1966

Year	Units Owned by Professionals and Proprietors		Units Owned by White-Collar Employees		Units Owned by Skilled and Semi-skilled Labourers		Units Owned by Labour		Units Owned by Unclassified Groups (Gentlemen, Capitalists, Widows)	
	n	(%)	n	(%)	n	(%)	n	(%)	n	(%)
1876	43	(11.0)	13	(3.3)	131	(33.6)	22	(5.6)	180	(46.2)
1886	137	(23.8)	33	(5.7)	166	(28.9)	24	(4.2)	214	(37.2)
1896	175	(21.4)	52	(6.4)	177	(21.7)	21	(2.6)	390	(47.8)
1906	208	(25.0)	61	(7.3)	258	(31.0)	21	(2.5)	285	(34.2)
1916	405	(30.2)	90	(6.7)	364	(27.2)	30	(2.2)	450	(33.6)
1926	423	(27.9)	109	(7.2)	294	(19.4)	64	(4.2)	624	(41.2)
1936	628	(29.3)	187	(8.7)	428	(19.9)	53	(2.5)	844	(39.3)
1956	328	(18.2)	171	(8.3)	587	(28.6)	127	(6.4)	710	(37.5)
1966	469	(16.9)	213	(8.8)	737	(30.2)	166	(6.5)	967	(36.5)

Source: Hamilton Assessment Rolls

negotiate lower rents and improvements to dwellings. Unfortunately, good job prospects and reasonable housing costs rarely coincided.

Just as a city's complete range of manufacturing at any recent time
would contain a mix of new and old technologies, of comparatively
large and small employers, and of users of different combinations
of skilled and unskilled labour, so too have the city's landlords been
a diversified lot. Rentier incomes were not the sole prerogative of
the urban bourgeoisie. From the 1870s to the 1960s the share of
rental housing owned by the city's skilled and semi-skilled labourers
fluctuated between a third and a fifth, dipping to low points during
depressions (see table 8.1). Evidently a fair number of workers became landlords. Despite landlord diversity and the great many rental
units in the hands of blue-collar workers, it would be wrong to
dispense altogether with the notion that landlordship essentially reinforced socio-economic differences. Two occupational groups of relatively small size owned exceptionally large holdings of rental units.
The considerable slice of rental housing in the hands of one of these
two, namely the professionals and proprietors group, is a predictable
and reasonable finding. The second, the unclassified group, con-

Table 8.2
The Concentration of Rentier Activity: The Representation of Occupational
Groups in the Ownership of Rental Units, 1876-1966

Year	Professionals and Proprietors	White-Collar	Skilled and Semi-skilled	Labour	Unclassified
1876	105.8	37.1	60.3	35.0	401.7
1886	154.5	71.3	69.5	27.5	192.7
1896	158.5	58.2	52.2	17.9	329.7
1906	233.6	62.9	63.1	18.2	226.5
1916	284.9	63.2	52.1	15.6	268.8
1926	242.6	44.7	41.7	31.8	324.4
1936	281.7	47.0	44.9	17.2	370.8
1956	154.2	49.4	57.9	66.0	304.9
1966	132.0	59.1	60.6	80.2	253.5

Source: Hamilton Assessment Rolls
100 = An occupation group's percentage of the cross section's
      rental units in proportion to its percentage of households

sisted largely of widows and gentlemen with capital. The label gentleman was taken up by retired men, especially those whose capital was in rentier investments. Wealthy widows often placed some of their assets with property agents or lawyers, who in turn invested the money in mortgages or rental property.

Two solid facts support the notion of landlords or their agents as moving in a world apart from wage earners. First, blue-collar landlords, for all the evidence of their considerable presence, held a rather small number of units in relation to their own numbers in the city. The professionals and proprietors as well as the unclassified group were greatly over-represented. The features of this over-representation, as seen across nearly a century, are detailed later in this chapter. Second, family networks and land agents definitely operated as the focal points of ownership and expertise.

The viewpoint that considerable consistency lies behind surface changes is confirmed by a look at which groups have been over- or under-represented in terms of rental units owned. Table 8.2 conveys a descending order of landlordship beginning with the unclassified group, a traditional body of investors, followed by the professionals and proprietors, followed by wage earners (white-collar and blue-collar), and completed by common labour. This ranking seldom varied, and any changes have hinged on notable cultural, material, or social developments within the city. For example, a convergence for change occasioned a shift in 1956 and 1966 when common labour realized a solid presence as landlords. At a time of considerable apartment construction, a stronger showing by traditional investor groups might have been expected, but this supposition ignores the remarkable parallel development of home conversions into flats or of homes with rental units. While this unheralded phenomenon, important both to recent social history and to the history of shelter, is expanded on later in this chapter, our focus rests on the over-representation of rentier or investor groups in the rental market. One objective is to demonstrate that major landlords and agents have engaged in operations of enormous scope. The data on over-representation hint at an extensive array of social relations in the city. These relations cannot all be discovered and documented, but a considerable number can be described and analyzed. They contribute to knowledge about the conduct of urban landed capital, the history of friction in landlord-tenant relations, and the quality or character of living conditions for over a century.

LANDED WEALTH AND MAJOR LANDLORDS

In the nineteenth century, as we have detailed in chapter 1, a few major landholders in Canadian and in smaller United States cities were city builders whose investment strategies and cultural outlooks shaped streetscapes and whose rental property holdings elevated their community standing and political influence. They figured prominently among the urban bourgeoisie, but have received relatively little attention in Canadian business history and even less in American accounts. In Canada the search for nation builders among urban merchant communities long preoccupied research and generated an interesting body of historical scholarship concerned with banks and transportation projects. Conceivably the lack of heroic dimensions to a landlord's operations discouraged research on rentier capitalism. Exceptions to neglect have appeared when studies of urban reform led into descriptions of tenements and slum landlords. Even so, the great urban landlord is a phantom figure even in the writing of American urban history. However, in recent years Canadian accounts of urban society have encountered men of influence with their portfolios of real property investments.[1]

Credit for discovering and analysing the business of landlords belongs to the Montreal historians Paul-André Linteau and Jean-Claude Robert. Seeking to dispel the myth of an absence of entrepreneurial drive among French Canadians, they found ample evidence of business skills, wealth, and exploitative capacities among a handful of men who held extensive real estate in Montreal in 1825. Shortly after their discovery, Louise Dechêne published a study about the practices of Quebec City investors in a seigneury on that city's urban fringe from 1764 to the 1850s. She demonstrated the attractiveness of real property for generating revenue and providing security. Her example, involving as it did quasi-feudal obligations, appears to have been unusual. However, she provided a glimpse into the world of legal authority and tenantry that, regardless of land tenure system, probably abided elsewhere on the continent. Robert and Linteau were not so much concerned with the mechanisms of urban tenantry as they were with concentrated real-property wealth.[2]

As we now might expect from a city in the early nineteenth century, before the era of extensive innovations in the building of houses and servicing of land, about 70 per cent of Montreal's household heads were tenants. The wealthiest property owners, some 28 men or just over .5 per cent of all household heads, earned about 2 per cent of the property-based income in the city. The very richest members of the property-owning group included an over-representation of French Canadians. The man who possessed the greatest property-based income in Montreal was P. Berthelet, who had 61 tenants. "The diversity of his holdings makes him appear as [an] authentic real-estate entrepreneur. He was also involved in a business that had an indirect connection with housing. As the owner of about 300 cast-iron stoves, he carried on a lucrative business renting out the useful devices over the winter to people who lacked the means to purchase their own." Although Robert and Linteau chose to emphasize concentration, their data would have allowed them to comment on the existence of many small proprietors of rental properties. The same choice of emphasis characterized a study of housing in Kingston, Ontario, by geographers Gregory Levine, Richard Harris, and Brian Osborne. Looking at the concentration of residential property ownership in the city in 1881, they found fourteen freeholders owning 16 per cent of the total rental stock.[3] Similar levels of concentration were apparent in Hamilton in 1881. As Michael Doucet has shown, the city's 7,091 dwellings were owned by 3,266 people. At one extreme, 67 of them (2.1 per cent) controlled 1,029 dwelling units (14.4 per cent). At the other extreme, 1,954 individuals owned one house each, but few of these owners would have been landlords. Significantly, 1,245 small landlords (2–10 houses) controlled 4,110 dwelling units, accounting for 58 per cent of Hamilton's housing stock.[4] The Kingston research report was silent about the extent of operations by petty landlords. Concentration and the conduct of major landlords are important topics, but to fix on them is to truncate efforts to understand the rental housing market.

More recently, British geographer Richard Dennis probed the Toronto rental-housing market for 1885 to 1914. The shortage of business records forced him to rely on assessment rolls and there he encountered the customary landlords with extensive holdings. Adam Armstrong, for example, owned an entire street of over thirty cottages in Toronto's Cabbagetown. Interestingly, Dennis found that in his three largely working-class areas a few landlords owned many rental units and the majority owned merely one or two. In an original and insightful conclusion, he recommended a typology of landlords that would take into account their size, business interests, and likely behaviour. There were the "accidental" landlords who inherited a dwelling in which they chose not to reside. Then came the "satisficing" landlords with an interest in a safe, steady income, who did not buy or sell and who lacked financial sophistication. A third group comprised men in the building trades, who moved their properties rather than holding onto them and, although this was not one of Dennis's observations, who sometimes became landlords because they could not sell houses built on speculation. Local tradesmen and employers, like the "satisficers," were probably insensitive to market conditions and held onto a few units for a long time. Some landlords might have been philanthropists or industrialists who treated housing as an adjunct of a labour policy. Dennis found none in his sample areas and believed there were few elsewhere in Toronto. Finally, there were the large-scale professional landlords, consisting of estate agents, financiers, lawyers, holders of landed estates, property developers, and sophisticated building tradesmen. Frequently interested in many aspects of the property industry, they were, Dennis concluded, "the nearest we can find to a 'landlord class'." We will take a look at these men.[5]

In Hamilton, as we have illustrated, small proprietors of rental dwellings, not great landlords, were the norm. That conclusion would have sat well with one of Hamilton's great landlords, the Mills family, whose members never thought of themselves as typical. In their estate management practices, political prominence, and pursuit of honours and esteem, they sought eminence. Enough fragments can be assembled from the biographical riches of Canadian history to see an outline of families similar to the Mills or of individuals comparable to the leading figures in the Mills family. Historian Brian Young has recounted that politician George-Etienne Cartier invested in Montreal property and pursued "social ambitions involving a country estate, a title, a coat of arms, and frequent visits to Europe."[6]

The Mills likewise hankered for the trappings of the *grande bourgeoisie* and the aristocracy. In Toronto the proprietors of park lots developed urban land and invested in many urban-based capitalist ventures while retaining estate seats and aristocratic manners. David Gagan's description of one of these families, the Denisons, is pertinent in terms of what it says about both the signs of family social station and the basis of wealth in real estate. The Denisons "had become local magnates in their own right, always represented on city council, the senior officers of the Toronto militia, men who consorted with the elite of two continents. The Denisons' wealth was derived from the land which they farmed and then developed for urban dwellings."[7] The description almost applied to the Mills. Only their lack of military ardour separated them from the social aspirations of the Denisons.

The Denisons and Mills were manipulators of landed capital, bourgeois specialists with an intense interest in the city who, despite being bourgeois, took the British aristocracy as role models. On the surface, this colonial mimicry of the aristocracy, even the pursuit of titles or marriages into titled families, was an expression of Victorian culture in a colonial society. For the great landlords, the aristocratic model was perhaps more than a social ideal of an elite in a particular era. It concurrently presented a context for the development and management of landed wealth. This context called for a tenantry and for landlord patronage of religious and civil life. In all probability, American families comparable to the Denisons and the Mills developed landed estates with a tenantry in United States cities, but their business activities, their cultural role models, and their ideological underpinnings are as yet unknown.

THE MAINTENANCE OF AN URBAN LANDED ESTATE

Land sales, even in the nineteenth century, seem to have been the favoured way of handling the asset of urban land, but James Mills, whose land development activities were outlined in chapter 1, ploughed his land sale profits back into property development. Additionally, he operated a trade in general merchandise and accepted, in barter exchange, lumber from his farmer clients. The shelter crisis in Hamilton during the prosperous years of the 1830s encouraged Mills to use the lumber for the construction of numerous unimpressive dwellings. Eschewing the common path of land development, he launched his family's rental enterprises. James's son Samuel followed an identical path and the town clerk of the mid-1830s,

Charles Durand, recalled that during the 1830s Samuel "began to be a great man, and to own a score or two of wooden houses."[8] Both father and son accumulated buildings during the years of city expansion in the 1840s. At the time of his death in 1852, James is likely to have owned about fifty residential units. In 1861 all family members held roughly ninety-five houses and commercial properties.

What level of income did James derive from his properties? Unfortunately, the only extant rent ledgers belonging to a family member cover 1865, a depression year with deflated rents and low real estate values. There was a considerable spread in the rents collected by the trustees for the estate of James's son John Mills in 1865. A fine stone house on a park lot brought in $200 a year; a brick house brought $120. The more typical Mills property, a modest frame or rough-cast cottage, rented for $60 a year.[9] These depressed rents may well serve as a conservative estimate for those received by James in the year of his death, when there had been an economic boom in the city. If so, James would have had the colonial currency equivalent of $2,500–$3,000 in annual rental income. The family's 1861 annual rental income would have been $7,000–$10,000. Depending on the number of days worked, a typical unskilled urban labourer would have earned $250–$350 per year in the 1850s to 1870s.

One measure of the family's wealth remains to be mentioned. In 1856 the total assessed value of Hamilton's real property was $225,000 and of that the Mills had been assessed for $27,500 or 12 per cent of the city-wide figure. The expansion of the built-up areas of the city in the 1850s and 1860s, coupled with the growth of industrial assessments, subsequently lowered the Mills's relative share of assessed real property. In 1871 the family's holdings accounted for 1.4 per cent of assessed property value; but the Mills men, women, and children accounted for only approximately .1 per cent of the city's population. Their landed wealth greatly exceeded their representation among city residents.

The Mills properties were concentrated in a proper family estate (see figure 8.1, tables 8.3 and 8.4), which made up a managed neighbourhood within the city. James Mills's homestead occupied roughly one-third of the total area of St George's and St Mary's wards. According to 1851 information, about one-fifth of the platted or surveyed expanse of these wards had been subdivided by the Mills family. Running through the middle of the estate, along an east-west axis, was the ridge where the family's homes had been concentrated. In 1851 James and adult sons, John, Nelson, and William resided on the ridge within a few blocks of one another. On either side of the heights, the land sloped away. To the north, it dropped

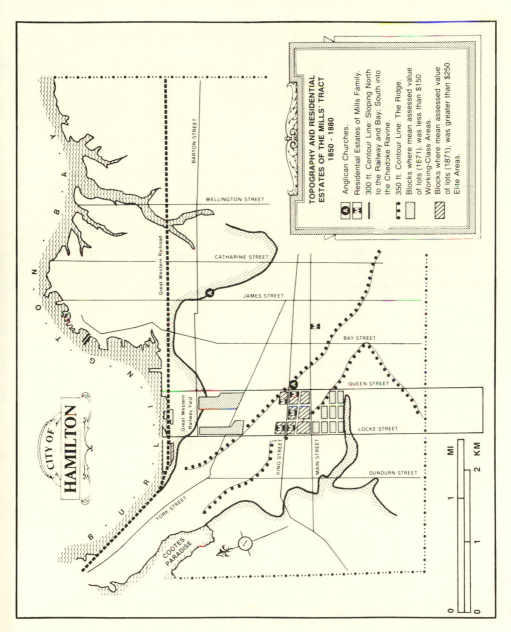

Figure 8.1
Topography and built features of the Mills's tract, 1850–1880

Table 8.3
Properties in the Original James Mills Survey Held by the Mills Family,
1861, 1871, 1881

Year	All Vacant Lots in Tract	Owned by Mills		All Rented Dwellings	Owned by Mills	
		n	(%)		n	(%)
1861	355	138	(39)	160	30	(19)
1871	247	94	(38)	288	61	(21)
1881	181	39	(22)	308	25	(08)

Source: Hamilton Assessment Rolls

down quickly and drained into the Bay at the point where the Great
Western Railway operated its shops from 1853 to 1888. To the south,
the land declined toward a ravine that severed the Mills estate and
seriously depreciated property values.

On a few of the numerous vacant lots in the 1850s and 1860s,
home owners had erected dwellings. However, home ownership in
this era extended to only a quarter of the city's households. Consequently, landlords owned most of the residences in the Mills estate;
Mills family members possessed the largest number of rental properties. The city's assessment rolls offer incomplete glimpses into the
building materials used and then only for the ten years from 1856
to 1866. For those years, the estimate of the proportion of frame
dwellings and commercial units owned by the Mills family was 70 per
cent. Roughly 25 per cent of the Mills-owned units were constructed
of brick; 5 per cent were of stone. The proportion of frame units
for dwellings ran even higher, at over 80 per cent. Class differences
were apparent in the spatial structure of the Mills estate. Both elevation and building composition accented the neighbourhood's social
hierarchy, because the brick and stone dwellings occupied by the
Mills family overlooked the frame cottages of its tenants.

From 1856 to 1896 the Mills had four types of tenants. Two identifiable groups occupied the scattered commercial blocks near the
central business district. Here the ground floors were occupied by
merchants, tailors, shoemakers, cigar makers, cobblers, and cabinet
makers. Above them resided a handful of widows, unmarried
women, and old men. Single-member households and business tenants occupied the core-area properties. On the original estate there
resided several barristers who rented elevated properties from the
family. The typical tenant, however, was a common labourer. The

Table 8.4
Concentration of Mills Family Property in St George's and St Mary's Wards

Year	Total Assessed Value of all Mills Family Property	Assessed Value of Mills Family Property in St George's and St Mary's Wards	
1861	$ 130,720	$ 99,270	(76%)
1871	$ 181,001	$ 97,741	(54%)
1881	$ 194,340	$ 86,460 (Wards 2 & 3)	(44%)
1891	$ 275,570	$114,870 (Wards 2 & 3)	(42%)
1901	$ 210,470	$ 85,980 (Wards 2 & 3)	(41%)
1921	$1,510,210	$395,350	(26%)

Source: Hamilton Assessment Rolls

assessment rolls either identified these tenants as labourers or cited no occupation. Over the years, a few exceptions appeared: several teamsters and blacksmiths dwelt in Mills cottages and in 1871 two iron puddlers employed in the railway rolling mills lived in a block near the Great Western Railway shops. The vast majority of Mills's tenants, however, were relatively poor and unskilled. In 1865 they paid a rent of $4 to $5 per month for a frame cottage; the lower charge was levied if the tenant paid the water rates. Topography, architecture, occupation, and rent defined the estate as a hierarchically ordered fragment of urban society.

The management of their estate required frequent use of the law. Through legal specialists, families like the Mills haunted the registry offices and county courthouses. Actions filed on their behalf filled the index books of judicial clerks across the colony. If Protestant Ontario read the Bible and listened to ministers on Sunday, the elite read indentures and writs and heeded its lawyers during the rest of the week. The family Bible registered births, deaths, and marriages but marriage agreements, bonds of indenture, and wills gave contractual substance to the often calculated handling of family events. The god of law as much as the Law of God ruled conduct. Proximity to the law in daily affairs made the urban elites politically active. Bonusing legislation, joint-stock company incorporation, bank and insurance legislation, and changes to the law of bankruptcy were watched with keen interest. To guide them through the law, the Mills family secured legal expertise. James Mills handled affairs in the early years. He had acquired enough practical knowledge to prepare, in the words of the family history, "deeds of land and other contracts requiring some knowledge of law."[10] His son Samuel often worked closely with lawyer George S. Tiffany and, after the death of the latter in 1854, Samuel depended on his young brother, George H. Mills. As a life member of the Legislative Council of Canada from 1849, Samuel was a lawmaker, but on mundane affairs and for news of Hamilton he relied on lawyer brother George.[11]

Relations among family members were riddled with tension, harmonized by calculated deference, bound by contractual obligations, and civilized by need. Legal agreements patched over family business relations, marriage arrangements, and even child rearing. Wills were pivotal legal instruments in estate affairs. They capsulized the strategy of the patriarch in providing for his family and forced his heirs to co-operate in an effort to rationalize the ample but fragmented estate. James Mills's will revealed his way of providing for each member of his large family in a manner that protected the young, the old, and females while also indicating how he had launched the mature sons. To the latter, he had devised numerous lots so that they might invest their birthright. Son Samuel went forth as merchant, landlord, and politician; he became one of the three wealthiest men in Hamilton. It was likely that proceeds from family land, granted to him in the 1830s, enabled him to plunge into land speculation on a grand scale. In 1838 he and Tiffany paid Nathaniel Hughson £5,000 for 50 acres on Hamilton's waterfront. In contrast, John lived on property revenues and a government patronage appointment at the customs house. The older daughters had land set aside for dowries. George and Harriet were to receive their share of the lots when they were twenty-one. In the meantime, the executors were to sell some land and convert the proceeds into an annuity of £37 10s "to be applied for the support and tuition of my son George H. Mills until he becomes of age."[12] James's wife, Christina, received the largest portion of the estate. According to the 1861 assessment roll, the assessed value of her property was $52,900 or nearly double that of Samuel, the wealthiest son, and 40 per cent of the assessed value of all Mills property in Hamilton (see table 8.5). The children borrowed from her formally.

James had died in 1852. His son John died in 1865. The latter's executors were instructed to lease out John's property to provide for the care and education of his two children until they reached twenty-one. The trustees adhered strictly to the instruction. The costs of improving the children were lodged with the trustees and recorded as journal entries against estate revenues derived from mortgages and rents. The rearing of Charles and Aurora took place in a supervised environment with aunts, uncles, and grandmother participating, but recorded business transactions often covered frills as well as essentials. Even avuncular treats like excursions to the

Table 8.5
Distribution of Properties Held by the Heirs of James Mills for 1856 and 1861

| | 1856 | | 1861 | | 1861 |
	n	%	n	%	Assessment
Christina	4[1]	–	5	–	$ 52,900
Samuel (1806-74)	68	39%	109	39%	$ 30,350
John (1809-65)	32	18%	28	10%	$ 9,700
Nelson (1819-76)	34	20%	74	26%	$ 15,870
William (1822-90)	13	8%	17	6%	$ 4,600
George (1827-1901)	16	9%	38	14%	$ 8,300
Daughters	11	6%	15	5%	$ 9,000
Total	174	100%	281	100%	$130,720

[1]Christina Mills was left choice buildings to provide her with an annuity.

circus were charged against the estate. Aurora took drawing lessons from local artist Captain John Herbert Caddy; her teeth were fixed; dresses were brought for her; she was taken to social events like concerts and weddings, and the estate also provided her with a trip to England in 1868.[13]

Senator Samuel Mills died in 1874, leaving the lion's share of his estate for the support of his wife. She was to receive an annuity of $2,000 drawn from the investment returns on Samuel's mortgages, debentures, and securities. To support the annuity, these assets must have had a liquidation value of between $30,000 and $40,000. A very fortunate skilled labourer might have earned half that much in his lifetime. Samuel also left roughly thirty houses and seven stores to four of his children; two others shared unspecified properties. By a conservative estimate based on annual assessment values, Samuel had rental revenues of $2,000 per annum. In all likelihood, his real estate holdings – lots, rural tracts, stores, and houses – were worth as much as $70,000. Total assets, therefore, approximated $100,000.

The routine recourse to legal counsel by propertied wealth was pronounced during business recessions and focused on the Court of Chancery, where mortgage foreclosures were the cases most frequently argued. The court also heard cases involving property disputes, conflicts over property titles, estate settlements, insolvency, failure to meet contractual obligations, and recovery of debts. Foreclosure was complicated. If a mortgagee, like Samuel Mills, sought to recover property and if his case against the mortgagor was upheld in Chancery, then the Court issued a final order of foreclosure. This document summoned the mortgagor to settle the debt or have the right of redemption foreclosed. If the mortgagor failed to pay the outstanding mortgage, the property was sold at public auction and the proceeds used to pay the mortgage and any judgments lodged by other creditors. The mortgagee did not assume title to the property unless he or she purchased it at auction. If the price bid at auction was less than the debt, it sometimes was in the mortgagee's interest to purchase the property. George, the family lawyer for the second generation, frequently pursued this course of action on behalf of brothers Samuel and Nelson.[14]

The recovery of property was not a blessing, for it trimmed investment income, increased the property tax expenditures, and sometimes loaded the family with a depreciating asset. In February 1860 George tried and failed to dispose of his thirty-one properties at discount prices. Five years later, claiming that his properties did not afford him "respectable competency" while he had to pay taxes upon "a large number of vacant lots," he requested Isaac Buchanan to assist him in securing government favour, saying, "I stand much in need of it." With Samuel frequently out of Hamilton to attend the meetings of the legislature, George handled the often distasteful legal work. On occasion he also pressed brother Nelson's delinquent tenants. Delinquent tenants had to be evicted and local officials were not always quick to respond. To the Sheriff of Halton County, George wrote in March 1860 "in suit Mills vs. Cargill et al., I understand that you have not dispossessed on the defendants. The writ directs you to do so and I beg that you may lose no further time in executing it."[15] Like the recovery of mortgaged property, the removal of tenants meant a loss of revenue. It demoralized landlord and tenant alike. In 1856, a boom year, the vacancy rate among the Mills properties, residential and commercial, had been 12 per cent (11 out of 95). The depression year 1861 saw the vacancy rate leap to a distressing 40 per cent (61 out of 151).

The stress of managing affairs for Samuel, the unpleasantness of the chores, and Samuel's indifference bothered George. He was also antagonized by Samuel's cheapness. Samuel, a notorious skinflint, owed George $1,417 for legal services in 1858 and 1859, but the sum remained outstanding as late as May 1860. As George recognized, he was in a miserably dependent situation. "I cannot afford to quarrel with my bread and butter," he wrote bitterly to Samuel. The tension climaxed in July 1860: "I can no longer address you as a relative. Your letters contain each an additional insult. I trust from this time forward no communication may exist between us. I totally despise the principle which appears to actuate all your conduct, and

I feel that I would be impairing my own dignity to hold further inter-course with anyone so destitute of feeling."[16] Despite this outburst, George continued to manage Samuel's affairs, but the salutation "Dear Brother" was forced and sarcastic. For his part, in 1867 Samuel pressured George to sell lots in order to pay back a loan that he had extended to George. James's legacy, the depression of 1857–1865, Samuel's political career, and George's fledgling law practice trapped the second generation in a round of squabbles. It is true, however, that the cares of the rich were quite unlike those of the city's labouring population.

The association with the law extended beyond landed wealth. Like the urban bourgeoisie of Europe and North America generally, the Mills diversified and invested in the new assets created by acts of incorporation for municipalities and companies alike. Local transportation and financial concerns were especially favoured. Landed wealth was recycled locally. Samuel's estate included municipal debentures and stocks in railways and banks. He was President of Hamilton's Gore Bank for a brief period and he was a founding director of the Bank of Hamilton in 1872. In addition, he sat on the board of the Canada Life Assurance Company and owned a substantial number of its shares. The trustees of John's estate purchased shares in the Gore Bank for Aurora's dowry. George had helped to found a local building society in the 1850s; he ran an insurance agency, and he was an investor in and a founding director of the Victoria Mutual Fire Insurance Company from 1863. In the late 60s and early 70s, he was responsible for the revival of the Hamilton and Port Dover Railway. Interestingly, while the second generation's capitalist promotions peaked during the initial stages of Hamilton's industrialization, the family did not participate in that economic transformation. The city's industrialists usually had an artisanal background, and occasionally a merchant became involved, but property wealth, as distinct from commercial wealth, stayed away from the new economic ventures of manufacturing.[17]

Within the property industry, the Mills family's third generation did not lapse into a static condition of merely collecting rents and developing land on the original family tract. By the late 1870s, given the deaths of Samuel and Nelson and the seemingly retiring ways of George, the third generation asserted itself. Charles Mills, the son of Nelson, undertook the last major subdivisions of the family's original land (on Herkimer Avenue between Queen and Locke Streets) in 1882.[18] Notwithstanding this effort, the assessed value of vacant lots in the Mills family's hands diminished in relation to that of residential and commercial properties. Along with this redistribution of wealth by property type, there was a shift in geographic concentration. In 1861, 76 per cent of the assessed value of family holdings was concentrated in St George's and St Mary's ward (later wards 2 and 3). In 1921 only 26 per cent of the assessed value was in the old tract and 70 per cent was concentrated in ward 5, which included the city's commercial core. In 1888 Nelson's other sons, Stanley, Robert, and Edwin, had established a hardware business, which they expanded into a chain of retail outlets by 1920. The brothers also established Mills Brothers Real Estate in 1913 and bought suburban land for residential development. Stanley maintained the family's involvement in financial institutions, serving as a director of the Mercantile Trust Company (first appointed in 1907) and of the Hamilton Provident and Loan Company (first appointed in 1915). His brother Charles had been elected a director of the Landed Banking and Loan Company in 1901. Unquestionably, the third generation knew well enough how the original fortune had been made.[19]

The first and second generations of the Mills family were deeply involved in the affairs of their estate which was, after all, their neighbourhood. For the urban elite of the mid-nineteenth century, the heterogeneous social character of residential areas posed problems in that it exposed them and their children to less affluent city dwellers. The city divided by large-scale socially exclusive neighbourhoods came as a post-1900 phenomenon. Therefore, the enforcement of bourgeois precepts of morality entailed close supervision of their children. Education by private tutors was one means, and the existence of a family and class enclave within a mixed neighbourhood was another. Stanley Mills's recollections of his upbringing in the area is pertinent both to an understanding of urban geography and to an appreciation of the intersection of social contact. Incidentally, it confirms the impression of a bourgeoisie that had an exceedingly earnest view of play.

Within this block (George, Queen, Main, and Ray) and fronting on the immediately adjacent streets and all within a stone's throw of one another, there lived during the years of 1870 to 1888 one hundred of the happiest boys ever raised in the Ambitious City, these hundred boys were much of the same age, and there was never any difficulty in getting a good lot together to play the good old games of cricket, Tom-Tom Pullaway, a paper chase or chalk the corner ... The early childhood games of those days taught the youth to climb fences and surmount obstacles ... Thus did those happy boys prepare themselves truly and yet unconsciously for their life's serious work. Their morals were also well looked after by their parents, for every mother's son of them had good church-going Christian parents who knew and realized their own responsibilities.[20]

Living in proximity to their tenants allowed the Mills to screen their neighbourhood. Almost certainly they exercised social control through the selection of tenants and by overt or tacit threats of eviction. In 1869 George appealed to another landlord in the area to remove from the neighbourhood those whom the family had determined were a bad influence.[21] The proximity of the family members was also a matter of business convenience. When brothers George and Samuel were not feuding, George could drop into his brother's home and conduct the latter's business affairs. Nelson did not collect rents in person, still his agent undoubtedly found it convenient to complete the rounds and walk just a few blocks to report to his employer. By the 1880s the pervasiveness of the family on city streets extended to a few street names: George, Nelson, Harriet, and Stanley.

Churches had a place in estate management in its broadest sense. It would be unfair to claim that the elements of faith and piety were not a part of the Mills family's association with Anglican church construction. Nevertheless, the fixations on property accumulation and enhanced property values came as a major consideration in sponsorship of Christ Church and later of All Saints. Architectural historians see in the Gothic church a structure designed to draw eyes upward, toward heaven and away from material concerns. To James Mills, whose vision was keenly pragmatic, the enhancement of Christ Church in the early 1840s came as a measure of civic adornment that would advertise Hamilton. To another great Hamilton businessman and promoter, the Presbyterian Isaac Buchanan, James explained in 1852 a practical reason for purchasing pews: "If you or your firm will give us a donation I can assure you it will be thankfully received, perhaps as a matter of speculation you will take a pew, I have myself taken three and consider them good property."[22] James had not exaggerated the investment potential, because when his son John Mills died the trustees of the estate collected a rent of $30 per annum for one half pew. In other words, a pew had the same income potential as a modest cottage.[23]

In 1872 the Honourable Samuel Mills (by then a Senator) had All Saints Anglican Church built. The church crowned the family estate and stood across the street from the old James Mills house. Aside from the clear message of power and influence that the spatial relationships proclaimed, they also encapsulated subtle practices for enhancing property values and strengthening community deference toward the Mills family. Particularly in the case of All Saints, Samuel's action embodied the intentions of someone who understood the social uses of aristocratic patronage. The spiritual comforts of religion cannot be measured; neither can the prestige and potential for social control that derived from exerting an extraordinary influence over a parish church. Christ Church was not near the estate. It had a prestigious membership that would have prevented it from being controlled by a single family. What Samuel attempted to do in 1872, therefore, was to forge an English-style connection between landed wealth and the parish by erecting a new structure. Samuel never donated the church to the archdiocese. He retained title and so did his heirs until 1888, when they negotiated a transfer to the Bishop of Niagara.[24]

Within All Saints the hierarchical ordering of society that figured in the residential contrasts of the estate was reinforced. During the 1870s and 1880s only families who owned or rented a pew were entitled to attend morning service. Additionally, All Saints offered a showcase for the family's rites of passage: Samuel was the first person to be buried from the church, and a half-dozen women in the third generation were married from All Saints between 1894 and 1911. Samuel's heirs had transferred ownership in 1888, but they imposed conditions that allowed them to retain their paramount influence on parish affairs. They held one pew rent free and had the right to nominate a rector for appointment by the Bishop in the event of a vacancy. The grip on the trappings of prestige within the Church of England, along with the family's adoption of conservative politics in the 1860s, pointed toward an effort to create a tradition of British-style aristocratic values.[25]

As a great landowning and landlord family, the Mills were politically active, changing parties to suit circumstances. From the 1830s to the 1860s, James's sons were, to varying degrees, spokesmen for reform politics. James and many of his neighbours had been emigrants from the United States, and Samuel's wife was American. The family was, then, familiar with the concepts of both elected local governments and a government executive responsive to democratic forces. Moreover, the struggle for responsible government pitted a development-conscious bourgeoisie against an official party of place-holders. The initial reform inclination of the Mills also tied the family into a network of other prominent families. Finally, reform had a popular appeal that defined the family as something more than a combination of landowners. The family had even suffered for its commitment, since James's eldest son, Michael Marcellius, fled the colony and went into exile in Illinois after the Rebellion of 1837.[26]

The reformers had a difficult time in the late 1830s, but their ascendency in the 1840s paid handsome dividends for a well-connected reform family. Samuel's partnership with the Dartmouth College-educated democrat George S. Tiffany was only one of several strong bonds in the reform fraternity around Burlington Bay. Two of Samuel's brothers were connected to the family of John Willson, an early reform spokesman in the Legislative Assembly of Upper Canada. Nelson Mills had married Cynthia Gage, Willson's grand-daughter. George Mills had articled at the law office of Willson's son, Hugh, in 1846. Their sister, Sarah Mills, had married William Smith, the editor of the *Hamilton Free Press*, a reform newspaper. Samuel benefitted from the reform party's gains most dramatically in January 1849 when he received his life appointment to the Legislative Council of Canada. However, both he and George moved toward the centre, as did many other successful urban-bourgeois leaders, including George-Etienne Cartier, who had assisted the Papineau forces in 1837. Significantly, Samuel named his fourth son after the most moderate and bourgeois of reformers – Francis Hincks. Samuel took part in the nation building of the 1860s by supporting Confederation. He saw the scheme as a measure that might restore the prosperity of Canada West.[27]

George Mills, as an alderman (1857–8; 1870–4; 1877), initially maintained a bourgeois liberal stance on a range of civic issues. If Samuel's colonial recognition reflected some lustre on the Mills, it was George who really kept alive the family's reputation locally. He advocated the selection of Hamilton's mayor by ballot rather than by alderman. During a famous Hamilton strike in 1872 when iron moulders demanded a nine-hour day, George supported their right to rent the Crystal Palace for a labour rally; he opposed the manufacturers on Council who argued against the workers' petition.[28] Stable relations with working-class tenants possibly recommended a cautious handling of a modest appeal from city artisans. The 1876 murder of his brother Nelson by a disgruntled tenant – an event to be examined later – and a labour-based social-reform movement in the 1880s made George a far more reactionary analyst of current events. His response to a world that seemed more menacing was to begin the family's cultivation of the loyalist epic and its passion for genealogy. The Knights of Labor movement in the late 1880s appears to have shaken George's once calm faith in his class's security. Labour, merely one of many social forces in 1872, had grown into a grave threat. Thus in April 1887 he wrote: "I see before me, and not in the remote future, great social and political disturbances. The working classes and the unemployed, assisted by adventurers, are everywhere organizing and clamoring for what they call their rights ... in a word it is capital against labour."[29]

George worked to establish patriotic and historical groups to "check [the] progress of discontent and elemental sedition of those who feel little interest in the country which the pluck and patriotism of the early pioneers preserved against great odds."[30] Among other things patriotic, the Mills family would sponsor the statue to the Loyalists that has stood in front of the Wentworth County Courthouses since 1933. The statue, which appeared on a Loyalist commemorative stamp issued by Canada in 1934, was the last measure of consequence undertaken by the Mills to reaffirm and justify their place in the social order. Let us consider what else they had done to encourage the city's residents to see them as natural leaders who worked for the interests of many others.

In addition to the initial identification of George and Samuel with reform politics, they courted respect through their attention to neighbourhood and city affairs. Samuel served as a path master in 1838 and 1841; he was a member of the town Police Board in 1839 and sat on City Council in 1847. George, of course, had a more intense identification with local affairs, including civic betterment ventures that also enhanced family real-estate values. It was the genius of the bourgeoisie in urban centres to work for the enhancement of neighbourhood and city. In so doing, they gained materially while their actions were touted in the newspapers as promoting the community interest. Too close an identification with a specific area brought the risk of turning civic politics into a sectional conflict that usually attracted unwanted newspaper exposure. This happened as early as 1845 when the east end of the city – "Conservative, Radical, Moderate, Yankee, Free-born and Slave" – amalgamated to oppose the domination of town affairs by west-end interests – Isaac Buchanan, George S. Tiffany, and the Mills. In a later episode, in 1877, George pressed for a civic franchise for a Dundas street railway to enter into his ward and family territory. On this occasion, he was resisted bitterly "by many members of the council, chiefly those who represented eastern wards."[31]

From the 1830s to the 1890s the Mills had undertaken assorted community activities that accented paternalism and community identification. Their actions and occasionally their words presented arguments that aimed at justifying economic prominence. They believed that they could demonstrate that their affluence and ambitions had aided the broader community. Presumably the family hoped that the public would esteem the Mills as individuals "to whom much is given, much is expected." The gift relationships between family and community may have been important in legitimizing the prevailing ownership relations in the eyes of non-owners. The Mills also felt obliged to produce a rational explanation for their privileged position. They eventually settled upon the myths of loyalist virtue and pioneering vigour. They adopted the cause of a strong imperial association, which from the 1880s to the 1920s had a local appeal comparable to that of the reform movement of the 1830 and 1850s. United Kingdom immigrants swelled the city's working-class population and Queen Victoria's last decade on the throne brought out a sentimental enthusiasm for the monarchy.

Marriage was considered an opportunity for material improvement that demanded appraisal and formal agreement. It also represented an alliance of status, and the bourgeoisie intermarried in ever-widening circles. The second generation of the Mills family married into local, colonial, and, in Samuel's case, American families. A sample of the wider connections of the third and fourth generations can be found in the Parish Registers for All Saints: in 1902 Irene Hesse Mills married a paper manufacturer from East Orange, New York; Katherine Cory Mills married a barrister from New Westminster, British Columbia. The most prestigious union – a supreme prize for the grand bourgeoisie – was marriage into the British aristocracy. Minerva Mills, a daughter of Senator Samuel, married Baron Robert Arthur Dillon in 1862, helping to stamp the Mills as something more than a family of local influence. The British class system was not universally popular in Canada, but old-country values held the loyalty of some, colonials and immigrants alike. Many British subjects, and Hamilton was very British, knew their place in the system and regarded it as natural. Marriage into the aristocracy told them where the Mills fitted into the social scheme. However, one traumatic episode revealed the fragility of deference.[32]

Deference for an elite was strong when the elite appeared to act generously and justly. Samuel and especially George had established themselves as public figures, and while their endeavours probably fell short of curtailing criticism of their wealth, their brother Nelson totally failed to build the image of benefactor. Nelson, according to the *Hamilton Spectator's* notice of his death, "was not without his faults."[33] He also had an angry and deranged tenant, Michael McConnell. Among the tacit practices of landlords, the granting of a period of grace in winter to enable delinquent tenants to retain shelter was commonplace. The generosity was practical – a vacant property deteriorated during the winter. Bad blood between Nelson and McConnell endangered the prospect of even such limited paternalism.

The winter of 1876 combined seasonal hardship with an economic slump. McConnell, a butcher, owed Nelson two months' back rent, a sum of $14. McConnell used a dispute about repairs as a pretext to defer payment. Nelson refused to see McConnell and sent around his rent collector to deal with the recalcitrant tenant. When Mc-Connell stopped by Nelson's house later that day to complain, Nelson shut the door on him. The following day, 5 January 1876, Nelson had a constable serve a distraining order for McConnell's household goods. An agitated Mrs McConnell rushed to her husband's stall in the city market. He closed up, sharpened a knife, and with the blade tucked up his coat sleeve went to pay a call at Nelson's house. He observed Nelson walking along the street and approached him, but Nelson pushed McConnell away. McConnell pursued and overtook Nelson, and stabbed him repeatedly. Nelson died of his wounds several days later.[34] On 14 March McConnell walked to the gallows declaring that Mr Mills should not have been so harsh to him: "I had been his tenant for four years, and when I was only two months behind with my rent he went and set the bailiffs on me to drag me to pieces. If he acted with less harshness the fatal occurrence would have not happened. I hope it will be a warning to all the people of Hamilton in the future to act differently from the way Nelson Mills did and to exercise a little moderation in all their transactions."[35] Newspapers condemned the attack and the *Canadian Illustrated News* produced a solemn tableau. Nelson was not the villain of the piece, but reporting exposed the sometimes coercive nature of landlord-tenant relations.[36]

THE ABSENCE OF ABSENTEE LANDLORDS

The study of concentration in the ownership of rental shelter in the era before the apartment building reveals an interesting mix of great landlords and petty owners. The close examination of one of the foremost families of landlords shows how the power implied in a landlord-tenant relationship could work out in an attempt to shape a corner of the city. The Mills were certainly the antithesis of absentee landlords. In their extreme identification with the locality of their operation, they may not have been very typical, although additional fragments of information about the rental market appearing in class-ified advertisements give an impression of considerable proximity between landlords and their tenants. The classifieds from the 1870s to the 1910s occasionally gave instructions to prospective tenants: inquire next door; apply to Mrs Leggo three doors south; key next door; inquire at brickhouse on corner.

From the 1866 sample, when we started to record the information on landlords' residences, up to the 1926 sample, when we last entered this variable, 80 to 90 per cent of tenants lived in accommodations owned by fellow Hamiltonians. The rental-shelter business was extremely localized. In fact, it would seem that some of the out-of-town landlords had once resided in Hamilton. This is inferred from the fact that during business slumps the number of rental dwellings held by out-of-town people increased noticeably. Seeking greener pastures, the owner of a house might have found it impossible to sell the dwelling in a depressed market. Landlords, then, knew the city and the overwhelming majority of rental dwellings were readily overseen by their owners.

Among other things this parochial focus made civic politics, with its local improvement dimensions, a matter of considerable business interest. Given the many small landlords and their status as Hamiltonians, the rental-shelter business becomes one of the contributing factors to the city's inwardness. Rental shelter was a locally based, massive, and somewhat diffused business that worked – along with many other factors – to support a city-centred layer of news and politics. Hamilton was not only a city of houses and home ownership, it also was a city with little evidence of absentee landlordship. While government housing programs and large corporate holdings of apartments by interests outside the city have taken away some of the localism in rental shelter since the 1950s, the concurrent rise in home ownership and a growth of rental flats may have kept alive important supports for interest in local issues. If self-interest in property ownership, either by local landlords or by a vast public of home owners supports local development, then research on the health of North American cities should include an analysis of the real-estate market. During the twentieth century, Toronto in many ways extended its metropolitan shadow over Hamilton, taking control of the smaller city's financial institutions. Home ownership, property development, local apartment-building management, and abundant flats in converted houses kept alive the city's validity as a political entity of interest to its residents, whether these were small property owners or major investors.

Localism – the civic consciousness of both landlords and home owners – certainly does not imply a basis for a consensus on the issues of local development nor does it imply social harmony. The city's intimacy, as evidenced by its local landlords, says nothing about the tensions and conflicts between landlords and their tenants. We saw the very worst that this connection could produce in the murder of Nelson Mills. Landlords and tenants almost always shared a civic identity and at times, for example in the Mills estate, they lived cheek by jowl. From 1866 to 1926 a quarter to a third of tenant households lived in the same block as their landlord. In his three turn-of-the-century areas in Toronto, Richard Dennis found that 30 to 40 per cent of an area's landlords lived in that study area.[37] Even the rise of the apartment building after 1926 and the decline of the rented house do not necessarily upset the picture of personal relations between landlords and tenants because a considerable share of the rental market was still composed of units in unconsolidated holdings. In 1956 and 1966 the samples disclosed numerous flats in houses and the presence of landlords in many of these converted dwellings. The potential for face-to-face friction, therefore, remained considerable. Sometimes, however, landlords made themselves scarce. A 1965 report on complaints from tenants in low-rent housing cited examples of elusive owners. "The owner is someone who is slippery and evasive as a ghost. I have signed a lease, through the caretaker [who is] not allowed to give out the phone number of the owners."[38] We can only speculate about the precise character and frequency of landlord and tenant meetings and about the differences in occupational stratification. Certainly more research on this important area of social relations would be welcome. Labour history has provided many significant models and case studies for relations at the workplace. Accounts of the city should do the same for rental shelter, given its obvious importance in social affairs.

Let us hazard an overview of landlordship for over one hundred years and follow up with detailed research where we have been able to conduct it. In Hamilton's earliest decades, from the 1820s or 1830s at least until the 1850s, propertied wealth was relatively concentrated. Of course, there were many small landlords, but ambitious men such as James Mills and Alan MacNab, a future Prime Minister, had proportions of the rental market that no successors, including modern corporations, could match. When their holdings, social prominence, and political power are added to those of the city's early merchants and manufacturers, a picture of concentrated class might is clear enough. Indeed, we have mooted elsewhere a two-class model for Hamilton in the 1830s based on a variety of sources. Michael Katz, using largely quantifiable sources has gone further and proposed a two-class model enduring even into the 1870s. The landed members of the dominant class, exemplified by the Mills, managed their relations with the tenantry of the working class in a largely personal fashion. George Mills handled the collections and evictions for his wealthy brother, the Hon. Samuel Mills. Professionalism and a more impersonal framing of rental management edged forward under Nelson Mills, though he was visible enough to be marked by his killer.

However, for those property owners wishing to disengage from personal connections between themselves and tenants, landed cap-

italism at mid-century had already introduced the property agent
into Hamilton. From 1850 to 1920 landlords were individuals; even
the pools of concentrated rental shelter were owned by individuals
and not as yet by corporations. The picture of landlordship is comparable to that of land development and mortgage lending – this
was the era of individualism. However, landlords could have recourse to agents or to lawyers. From the 1920s to the 1960s, the
corporation grew in importance as the apartment building entered
the shelter scene. However, no new phase in capitalism and shelter
swept away old practices. Therefore, even as the apartment towers
rose in the 1960s, small investors carved up homes into several rental
flats and kept alive the significant strand of rental shelter where
tenants brushed against their landlords. From the 1960s to the present, collectivism and government intervention on several fronts began to transform the character of the rental-shelter market. Urban
renewal in Hamilton's poorer "north end" brought a significant increase in government-sponsored social housing. Co-operative housing was introduced in the "north end" in the mid-1980s. Beginning
in the 1970s, tenants formed tenant associations. The Government
of Ontario passed rent-control legislation in 1975 and revised it in
1979, establishing a Residential Tenancy Commission with twenty
offices across the province. For many North American urban tenants, proximity to the landlord no longer exists. It has proven difficult to identify with a numbered company. In rapid succession,
rental housing has gone through phases of corporate consolidation
and state intervention, but both separately and now sometimes in
combination corporations and states continue to modify their positions in the rental market.

The above outline of long-term trends, like all areas of shelter
history that we comment on in this book, confirms the slowness of
change. The stages or scaffolding for urban history are visible in the
history of rental shelter as proposed above, but there is an intricacy
that upsets the purity of the stages – the sudden appearances of
interventionist practices by the federal and provincial governments.
But as Fernand Braudel has maintained, political and social history
move at different speeds. Moreover, the model of landlord and
tenant ties – personal shading into impersonal – really leaves unanswered any questions about the quality of the ties. Just exactly
what occurred between landlords and tenants? Conceivably, many
tenants got along famously with very considerate landlords who
wanted no friction. Given the spatial intimacy that persisted in many
tenancy situations such as the flat in a converted house, it is possible
that tenants greeted landlords daily over the fence, in the hallway,

on the way to the same streetcar stop, or, in more recent times, when pulling out of the driveway. These potential situations cannot be documented. It is instead the conflict situations that can be found in our one extensive set of landlord's records.

THE BUSINESS OF PROPERTY
MANAGEMENT, 1858–1919

The North American real-estate business has evolved slowly over a fairly long period of time. Historian Pearl Davies cites the Cruikshank Company, founded in New York in 1794, as the oldest existing real-estate firm in the United States, and suggests that New York was also the first city to obtain a real-estate board when the New York Real Estate Exchange was founded in 1847. By the 1880s many cities had similar organizations, which were often used by their members to enhance local community growth and prestige. But realtors were more than civic boosters even at that date. They quickly acquired an impressive list of duties and responsibilities including "appraisal, negotiating and making leases, collecting rents, making sales of real estate, and negotiating mortgage loans."[39] By the early years of the twentieth century some firms had branched out into the development of property as well. The business, then, underwent a long, slow, and gradual process of professionalisation.

Even the most fragmentary accounts of the real-estate agent's practices place the business in an important social position mediating between a *bourgeoisie* with funds to invest and tenants residing in the investment dwellings. Their handling of mortgage capital, investment in shelter development, the collection of rents and management of rental properties, and the issuing of eviction notices allows insight not only into business cycles and structural changes in capitalism, but also into important social relationships. Early real-estate agents figured as main actors in the investment and social history of urban North America; they often were the flesh and blood behind the distilled statements of land-use theory.

The fundamental obstacle to comprehending the real-estate agent's significance to the history of capital markets and to the social history of housing is the scarcity of archival records. The office records of one Hamilton firm for the period from late 1858 to early 1919 provide the necessary base for suggesting the general practices of the North American real-estate business during important phases in its history. Admittedly, in older and larger metropolitan centres, the timing of changes in business practices and the level of the agent's specialization may have differed. The real-estate and general-agency

Table 8.6
Moore and Davis's Outgoing Correspondence for Two Time Periods
(1858-1865; 1912-1914), Broken Down by Category

Category of Correspondence	1858-1865		1912-1914	
	N	%	N	%
Collections/Financial	542	37.4	647	48.0
Sale of Property/Land Values	127	8.8	78	5.8
Property Management	281	19.4	69	5.1
Rental Matters	209	14.4	220	16.3
Eviction Notices/Rent Increase Notices	48	3.3	70	5.1
Mortgages/Titles	106	7.3	150	11.1
City Property for Sale	16	1.1	36	2.6
Farm Property for Sale	26	1.8	2	0.2
State of Land Market	10	0.7	1	0.1
Bonds and Stocks Available	34	2.3	12	0.9
Sale of Miscellaneous Items	28	1.9	7	0.5
Character References	15	1.0	2	0.2
Estate Matters	12	0.8	50	3.7
Family Matters	4	0.3	5	0.4
Total	1,148	100	1,349	100

Source: Hamilton Public Library, Special Collection, Moore and Davis Records

business of William Pitt Moore and John Gage Davis opened in August 1858 in offices at King and James Streets, Hamilton's central intersection. In 1866 the firm relocated one block to the north in the new Lister Block, where it remained for more than a century. Records of the company's activities include letterbooks of correspondence sent from the firm, rent ledgers for properties managed by the firm, a mortgage ledger, general ledger books, estate books, diaries or day books, journals, and miscellaneous deeds and other documents. Davis left the partnership at an unspecified date and Moore retired in 1891. Moore's family have continued to manage the business to the present.[40]

Initially Moore and Davis had to look beyond the land market to secure enough accounts to keep them solvent. The letters sent from the firm to its clients over the first seven years of operation illustrate this point (see table 8.6). Collection notices, reports on the progress of collection activities, and notices of deposits made to clients' bank accounts constituted the largest single category of correspondence during those years. Other non-property-related matters included the location and sale of bonds and stocks, the sale of miscellaneous items such as leather (the Moore family had been associated with a tannery

since the 1830s), the provision of and request for character references, and estate and personal affairs. In all, about 44 per cent of
the letters written by the partners between 1858 and 1865 had nothing to do with real property. Later, much of the correspondence
dealt with the insurance business that today remains as the sole
activity of the firm.

Property was the major if not the only concern of the firm in the
early years. By 1863 the partners were publishing a monthly "land
circular" and the firm provided a growing array of property-related
services as time passed.[41] These included selling property, compos-
ing advertising copy, posting for-sale and for-rent signs on property,
letting buildings and houses and collecting rents, arranging mortgage money, preparing property valuations, recommending and
then supervising repairs to property in their care, and conducting
other property management duties such as the payment of taxes
and water rates. Their attention to detail in these matters was remarkable. In 1862, for example, the partners informed Joseph Sudborough of Toronto that "the roof of the house at the corner where
White lives is rotted so that we shall be obliged to have a new roof
and eve troughs [sic], and also the boards of cellar in Grayson's mill
will require to be new. It will cost about $27 or $30 to put them in
order. We have had a carpenter to look at them. Please let us know
by return Post as they should be done before the Equinoxial [sic]
Storms set in."[42] Such attention and service was especially popular
with non-resident property owners and the executors of estates, even
if Moore and Davis's advice was not always accepted at face value.

The firm examined rented dwellings when tenants vacated. If
there had been damage, Moore and Davis went after the former
residents. Having located Mrs H. Jagoe in May 1891, Moore demanded that she either replace the glass in the front door and return
a night latch or he would have it done "and look to you for payment."
Besides exercising vigilance with respect to tenants' conduct, the
property agent kept a watch on civic and provincial affairs insofar
as they impinged on real estate. The partners were among those
who petitioned the provincial legislature in 1873 to incorporate a
Hamilton Street Railway, and Davis served on the first board of
directors. They pushed the city to reduce taxes levied against their
clients and insisted that the municipal police force protect property
owners from the annoyance and expense of vandalism. The windows
of vacant houses, for example, had a terrible fascination that compelled "small boys" to throw stones. Moore and Davis were politically
active and knew enough about city neighbourhoods to be of some
use to the Hamilton Liberal party apparatus. For example, they

increased the number of eligible voters in a family by spreading out ownership on the assessment rolls to enable sons to meet the property qualifications.[43]

As Liberals they supported the moral reform measures of the Ontario government, such as a moderate liquor licence act, but they definitely drew the line at legislation touching upon property rights and inheritance. They had no compunction about pressing the Liberal MPP, who owed them political favours, especially on matters of property and tenant legislation. After the 1886 provincial election William P. Moore wrote to Nicholas Awrey MPP for South Wentworth: "I have noticed in the reports of proceedings in the house that there is a move made to do away with the Law of Distress for Rent. I know that the papers have been harping on the hardship of a Landlord having the power to Distrain for rent on poor people, particularly the Labour union has been at it for some time. My opinion is that they know very little about the matters."[44] Moore went on to argue that if landlords could not seize a tenant's chattels it would be necessary to demand rent in advance. "A great many poor people would go to the poorhouse as they would not have money to pay the rent in advance."[45] The clarity and directness of these epistles establish the firm's attention to general rental interests and not just to details about the management of parcels of property.

Overall, clients seem to have been satisfied with the services and judgment provided by the firm. Moore and Davis came to represent a number of important corporate clients in Hamilton real-estate matters, including the Colonial Securities Company, the Canada Permanent Building and Loan Company, and the Guardian Trust Company, all of Toronto; the Trust and Loan Company of Kingston; the Canada Life Assurance Company and the Gore Bank, both of Hamilton, and the Toronto, Hamilton, and Buffalo Railway. Fees were levied for all services rendered to corporate and individual clients. These were specified in detail in the monthly land circulars. The commission charged for property management usually varied according to some unstated appraisal of the degree of difficulty associated with collecting rents and attending to repairs. During the late nineteenth and early twentieth centuries, the firm charged a fee of 5 to 7 per cent.

The essential nature of Moore and Davis's business remained constant for half a century. Closer insight into the components of that business can be developed from a comparison of the contents of letters at the beginning and near the end of the extant letter books. All 1,448 letters leaving the firm from 1858 to 1865 were analysed; all 1,349 letters from 1912 to 1914 were assessed according to the same categories as the earlier correspondence. The technique is flawed in that it places no weight on the relative significance of each type of letter. For example, a letter requesting payment of an insurance premium entailed less preparation than a letter reporting on repairs to a rental dwelling owned by an absentee landlord. Moreover, the firm used the telephone in much of its intra-city business by 1900, though conversation seems to have been followed by confirmation in writing. Collections, including reports on mortgage instalments and insurance premiums, constituted about one-third to one-half of the outgoing letters. A comparison of the sample periods suggests an increase in collection activity. However, the most appreciable shift in business involved the decline in property management. Moore and Davis had been quite busy in the early period supervising the repairs of houses and shops owned by absentee landlords. Quite likely the early volume of letters on this subject stemmed from the drastic collapse of the city's economy and the exodus of perhaps 20 per cent of the populace from 1857 to 1864. Property owners who departed often found it impossible to dispose of their Hamilton assets in a failing property market. Moore and Davis provided a service for these former residents, but over the years more and more of the properties they managed were owned locally.

The two sample periods yielded a common pattern of destinations. This is especially evident when the drop in correspondence to Wentworth County locations is seen as a likely result of the city's annexation of county territory beginning in 1891 (see table 8.7). The city-bound letters constituted a special cluster because they were sent to tenants, mortgagors, and vendors of property. Letters sent outside the immediate area included reports to absentee landlords, statements for mortgagees, and communications to former residents seeking to liquidate Hamilton assets. The network of communications establishes the considerable economic and social interdependency among North American cities. For example, Toronto was the paramount destination for Moore and Davis letters sent outside the immediate vicinity because the neighbouring metropolis was the head office centre for the insurance company represented by the firm and for several trust companies with Hamilton property. Moreover, numerous Hamiltonians had moved to the larger centre, leaving behind unsold lots and dwellings. The steady 10 per cent flow of letters to the United States – frequently to places in Michigan or New York – confirms the well-known social history phenomenon of transiency as well as Hamilton's significant railways links with the border states. High turnover rates in the urban population, there-

Table 8.7
Outgoing Correspondence Broken Down by Destination

Destination of Correspondence	1858-1865		1912-1914	
	N	%	N	%
Hamilton	297	20.5	429	31.9
Wentworth County	196	13.5	18	1.3
Remainder of Ontario	751	52.9	726	53.8
Remainder of Canada	68	4.7	30	2.2
United States	136	9.4	143	10.6
Other	0	0.0	3	0.2
Total	1,448	100	1,349	100

Source: Hamilton Public Library, Special Collections, Moore and Davis Records

fore, contributed significantly to the need for agencies like that of Moore and Davis (see table 8.7).

Eviction notices served by the firm capture a sense of the friction in landlord–tenant relations. Working with city directories and assessment rolls, an effort was made to determine what happened to recipients of Moore and Davis eviction notices and to the property from which they may have been evicted. Of 567 residential and commercial eviction notices from 1898 to 1918, all but 72 (12.9 per cent) could be traced to yield data about the tenants and the outcome of the notice. In more than two-fifths of the cases, the original tenant stayed in the dwellings. There were only 24 instances (4.2 per cent of known outcomes) when a tenant seems to have been removed to allow the sale of the property. That represented the fourth most frequent conclusion to the range of possible actions initiated by a notice (see table 8.8). Furthermore, rental property was converted to building lots only 6 times (1 per cent). After the upsurge of concern about public health in the early twentieth century, it was occasionally civic authorities, not landlords, who forced the issue of reinvestment and intensified the housing crisis that afflicted the urban poor.

Many of the notices and related letters of advice to property owners suggest that landlords may have attempted to increase rents during periods of urban growth. Such threats of eviction were exercised vigorously during upswings in the business cycle and must be viewed as a sign of investor strategy. In certain cases, the notice contained a rent increase or threat of eviction. "If you are willing to pay the increased rental kindly call at our office and sign the lease. If you object to the payment of said increases in rent we wish you

Table 8.8
Outcomes of Eviction Notices Issued by Moore and Davis, 1898-1918

Outcomes as Traced in City Directories and Assessment Rolls	N	% all cases	% locatable cases
Original Tenant Stays[1]	244	43.0	49.3
New Tenant	134	23.6	27.1
House is Vacant	27	4.8	5.5
House is Sold to New Occupant	24	4.2	4.8
New Owner and New Tenant	17	3.0	3.4
Owner Moved in	14	2.5	2.8
Property Becomes Lots	6	1.0	1.2
New Owner and Vacant	4	0.7	0.8
Sold to Recipient of Notice	3	0.6	0.6
Other	9	1.6	1.8
Could not Locate	72	12.7	–
Total	567	100	100

Source: Hamilton Public Library, Special Collections, Moore and Davis Records
[1]This suggests the tenants complied with demands for higher rents.

to consider this as a notice to vacate on the said 6 day of June 1913."
The letters to tenants were not sufficiently detailed in all years to
state conclusively that rent increases were the overwhelming consideration. However in 1916 and 1917 when the letters included more
than the typical content, 32 out of 56 notices mentioned a rent
increase. Serial information of another sort strengthens the hypothesis. The peaks in eviction notice activity coincided with firm upward
moves in the local business cycle. In contrast, during the depressed
1890s, the brief recession of 1907–08, and the crash of 1914–15,
Moore and Davis rarely issued notices.

Of the threatened tenants who could be identified in the assessment rolls, 244 (49.3 per cent) remained tenants at the same location
after the date of the threatened eviction. That was by far the most
common outcome of the firm's action. It seems, therefore, that many
tenants complied with new rental charges. New tenants appeared in
only 134 cases (27.1 per cent), and vacant units figured in 31 cases
(6.3 per cent). Generally, recipients of notices were residential tenants. Property use could be determined definitely in 499 instances
(88 per cent), and of these known properties, 452 (90.6 per cent)
were residential.

The largest social groupings of tenants who had received a notice
were: labourers (81), widows (32), machinists (23), and teamsters

(21). Age could be determined in only 332 cases (59 per cent). The measures of central tendency indicate that the typical tenant affected by a notice was an unskilled or semi-skilled labourer of 40 years of age residing in a dwelling with a low assessed value of $600 – $700. By contrast, the assessed value of dwellings rented by clerks during the period stood at roughly $1500 – $2000.

Beyond raising rents in times of rapid population growth due to immigration, Moore and Davis could follow instructions about being selective with regard to the origins of tenants when the city was full of working-class families recently arrived and seeking shelter. The firm reported in August 1913 to Mrs Ellen Keyes of Springfield, Massachusetts: "Your favor of the 9th inst. to hand. The present tenants are English people and we have not at any time rented to foreigners. We always do our best with reference to the class of tenants."[46]

Property managers could not always select tenants in the manner suggested by Moore and Davis. The undulations in the business cycle and climate occasionally established conditions whereby landlords had to accept any tenants and reduce or even forgo rents. Until the 1920s multi-unit rental dwellings were rare in Hamilton. Their appearance in the form of rows of three-storey brick apartment buildings along major thoroughfares during the 1920s marked a new stage in the evolution of the shelter industry. Prior to that time cheaply constructed frame houses, rough-cast dwellings, and a few rows of terrace housing made up the bulk of rental accommodation. Such shelter deteriorated rapidly if left vacant; the vigilant maintenance of many scattered vacant houses was expensive and could not always prevent serious damage during the winter.

As winter approached in the depression year of 1891, Moore and Davis worried about the situation of many houses that they managed. They wrote to John Storie of New York that his house was still vacant. "We have offered to fix it up and rent it at a low price but it seems impossible to get it rented. There are a great number of vacant houses in the City."[47] During the subsequent depression years, especially in winter, Moore and Davis recommended lowering rents, permitting delays in payment, and making repairs to houses. The prospects of burst pipes, broken windows, and vandalism were real. The firm suggested assorted concessions as a means of keeping houses occupied and hence protected. When the firm first could do so, it had tenants sign a year's lease in April as a means of holding occupants over the following winter. In contrast, during boom eras, the preferred system was rental by the month, with increases levied in March and again in August or September. The following wisdom exemplified the firm's objective in the winter of hard times: "It is much easier to keep a tenant than to get a new one especially at this season of the year (21 November)."[48] Tenants recognized their leverage – they complained about the state of the dwellings, they moved out in search of better situations, and they withheld rent knowing that the firm preferred not to evict. In February 1893 Moore and Davis sent, in exasperation, a note to a tenant, but they did not threaten to evict. "There is now three months rent past due, on the No. 31 West Avenue. We request that you will kindly let us have the amount *in full*. Yours truly, Moore and Davis."[49]

When a depression struck the city in 1914, tenants once again appear to have recognized the existence of surplus housing and the landowner's vulnerability. Of course, they knew all too well about their falling and jeopardized incomes; they felt the need to roll back rents through threatened and actual moves to other dwellings. In a report to the Guardian Trust Company of Toronto, Moore and Davis noted that "the prospect of renting or selling are [sic] not very bright at present." One consequence was that "the tenants in Nos. 9 and 15 Windsor Street want their rent reduced to $20.00 per month each. We have spoken to Mr. Alexander about it and we thought to try to get $22.00 but if not to reduce to $20.00 for the present rather than have them move and the houses remain vacant."[50]

Not all absentee landlords appear to have listened to advice. M. J. Cashmen of Port Huron, Michigan, had to be given a second account of conditions: "we understand that the rent was to be increased after the winter ... You will find it very difficult to get a desirable tenant especially in the winter at an increased rent." The correlation of the eviction notice pattern with the business cycle has been mentioned, but there was also a seasonal trend. The firm was reluctant to precipitate a vacancy in winter. By inference this denotes a real if limited possibility for tenant leverage.[51]

The advent of apartment buildings involved considerations about land costs and required the participation of entrepreneurs and financial institutions willing to break with the well-established piecemeal gradualism of the old form of the residential rental business. From the landlord's perspective an additional feature was the eradication of unorganized tenants' power. When apartment structures sprouted across the city in the 1920s and again in the 1960s and 1970s, the capital demands necessitated a new level of sophistication in landlord activity. The nature of landlord-tenant relations changed. On the one hand, corporate owner-operators replaced the network of landlords acting through the medium of the property manager. On the other, the growth of flats put landlord and tenant under the same roof. In neither situation was the property manager who handled many accounts a much-needed intermediary. Moore and Davis continued to manage a few rental homes into the 1970s. Ultimately, the Moore family left the old business and concentrated on property insurance. To more fully appreciate the context for the demise of the property manager, it is necessary to look at a set of transformations in the mix of shelter units that have constituted the rental market since the 1920s.

TENANTS AND THE RENTAL MARKET, 1921–1966

In many passages in our analysis of shelter, continuity has been stressed. Admittedly, we treat it as an important insight into a variety of shelter-history topics. After all, the city carries forward its mass. Its stock of buildings, streets, and services persist, or change but slowly. Cultural predilections also have a way of entrenching themselves; the ideal of the single-family detached house and the habits of home construction have endured from the early or mid-nineteenth century. Yet there is room, in a subject as complex as housing, for change.

The history of shelter has never ceased to embody movement, although the phases in its transition often – though not always – have been unobtrusive. Almost intuitively we know that the era of the estate-owning Mills family, whose members lived on a hill above but amongst tenants, has gone. We can imagine that even the business of property management as practised by Moore and Davis would be a marginal enterprise at best in the late twentieth century. Behind the vanishing ways of shelter management embodied by the Mills family or the Moores, there moved a related, invisible, and fundamental reordering of the rental market. Like so many occurrences in shelter history this shift proceeded over many decades and could be demonstrated only by the data collected from more than a century of assessment rolls. The composition of shelter forms in the rental market has altered appreciably from the conditions prevailing up to World War I. Quite likely the specific dates in the sequence of change would vary from city to city, but the shift in the mix may well have occurred in many North American urban centres. That, at least, poses a working hypothesis, and the likely explanations for the observed changes in Hamilton's rental market reinforce the idea that these changes had the potential to be very widespread, particularly so for those cities that had an early attachment to the single-family detached dwelling.

Table 8.9
The Decline of Houses as an Element in Hamilton's Rental-Shelter Stock,
1921-1966

Year	Percentage of All Households Living in Rented Houses	Percentage of All Tenant Households Living in Rented Houses
1921	44.5	91.8
1926	38.8	70.0
1931	41.2	68.7
1936	45.3	70.3
1941	46.1	68.4
1956	16.1	31.3
1966	11.9	28.0

Source: Hamilton Assessment Rolls

The rental market began its diversification in the 1920s. Until that decade the house dominated the rental market in Hamilton. The term "house" refers to an imprecisely defined article. In our coding procedures, street numbers and the existence of statements on lot frontage were the components of the assessment rolls that permitted a property to be labelled a house. Extant large-scale maps sometimes provided clues about the shape and orientation of building units. These, along with photographs and surviving structures, make it clear that the houses in the samples included some semi-detached houses, a few row houses (three or more attached dwellings of low value), and a few terrace units (attached dwellings of a more elegant quality built for bourgeois tenants). In other words, not all of the so-called rental houses were detached dwellings. However, they were single-family dwellings and were not in a multi-storey arrangement. Houses, as determined by the above criteria, accounted for 9 out of 10 rental units as late as 1921 (see table 8.9). There was a very modest smattering of apartments in apartment buildings, apartments above commercial properties, and apartments – flats – in converted houses (see table 8.10). The 1926 sample revealed the first significant traces of a great alteration. Houses had slipped to three-quarters of the rental market. Some of this decline may be attributed to the re-markably inflated construction costs of the early 1920s. It is possible that soaring costs did not deter prospective home buyers, such was the culturally rooted demand for proprietorship aided now by the sophisticated hyping of suburban land and houses. However, soaring material and labour charges may have stimulated investors to in-vestigate alternative investments in the rental market. For example,

Table 8.10
Comparison of the Shares of the Rental Market Held by Houses, Apartment
Buildings, Apartments in Commercial Blocks, and Flats in Converted Houses,
1921-1966

	Rental Type			
Year	Houses Including Duplexes and Triplexes	Apartment Buildings	Apartments in Commercial Blocks	Flats in Converted Houses
1921	91.8	2.0	2.9	3.3
1926	70.0	4.7	11.3	14.0
1931	68.7	3.9	14.1	13.3
1936	70.5	7.9	11.7	9.9
1941	68.4	8.3	11.4	11.9
1956	31.3	14.3	24.4	30.0
1966	28.0	14.0	30.4	27.6

Source: Hamilton Assessment Rolls

the 1920s witnessed a boom in low-rise apartment building, the construction of two-storey brick commercial blocks with apartments
above shops along major arteries in the new suburbs, and a notable
conversion movement that increased the number of flats. Thus the
immediate victim of the cost escalations in the residential-construction industry during the early 1920s was the house built for rental
income. Not always a robust investment prospect and fraught with
management problems, the rental house also could not withstand
the pressure from a marketplace that had responded to higher construction costs with higher prices on resales of existing houses.

While the mix of structure types in the rental market during the
depression of the 1930s remained similar to what it had become in
the great shift of the 1920s, there were nuances. Many owners of
small businesses were forced to move into rental lodgings above
shops, possibly economizing in order to save their businesses. The
war also forced a variation from the basic mix of the 1920s. War
industries in Hamilton drew in thousands of new labourers and
precipitated a housing crisis. The crisis peaked in 1944, but the
sample reveals its traces as early as 1941. The 1936 sample contained
301 flats and a ratio of 1 owner to 3 tenant households; the 1941
sample contained 361 flats and a ratio of 1 owner to 5 tenant households. Evidently, the emergency conversion process produced higher
densities through a temporary crowding into existing structures not
originally built for the concentration of tenant households. Under-

standably, wartime housing and planning reports for Hamilton spelled out gloomy conditions.

Rented houses faded slightly in the late 1920s, stabilized at about 70 per cent of the rental market during the 1930s and into the war years, and collapsed as the paramount element in the rental market by the time of the 1956 and 1966 samples (see table 8.10). To a degree, the trend away from rental houses may be perceived as a counterpart to escalating home ownership. More than ever before homes had become liquid assets to be traded for capital gains or for retrieval of savings. Fewer houses remained in an owner's hands for rental income. The number of investment opportunities available to people with modest savings and little interest in property management had mounted throughout the century. Although the growing complexity of the capital market and especially its absorption of smaller investors and retirement funds requires a thorough historical treatment, it seems safe to claim that rental income had been displaced by a packet of paper investments promoted as less worrisome and supposedly offering stability through diversity. Quite likely too, mortgage innovations had made the house a more completely liquid asset than it had been before the 1950s. Other factors were at work, because not only were rental houses now a slender rod in the bundle of the city's total housing stock, shrinking to between 12 per cent (1966) and 15 per cent (1956) (see table 8.9), but they also made up a vastly diminished share of the rental market. Whereas houses had accounted for about three-quarters of rental units from the 1926 to the 1941 samples, they made up less than a third of rented units in the 1956 and 1966 samples (see table 8.10). One force contributing to this decline was the mobilization of large investments for concentrated properties – apartment buildings or mixed apartment and commercial blocks – rather than for scattered houses, which lacked economies of scale for maintenance. Another consideration was the growth of flats. Many immigrant and labouring families converted their houses into two-flat units.

In the 1956 sample, 1,393 households occupied flats either as tenants or as owner occupants – 22.8 per cent of all households in the sample. This significant showing for the humble flat may have had its root in unusual historical circumstances. A booming post-war industrial economy, the immediate post-war shortages of building materials, and the damage done to housing stock and building trades by the depression and war had combined to create a severe shortage of shelter in the early 1950s. The converted flats that had been one immediate response to the crisis continued to show up as an important component of Hamilton's rental housing stock in the

Table 8.11
The Rise of Flats as an Element in Hamilton's Rental-Shelter Stock, 1921-1966

Year	Percentage of All Households Living in Rented Flats	Percentage of All Households Living in Flats, Rented and Owned	Percentage of All Tenant Households Living in Rented Flats
1921	1.6	2.3	3.3
1926	7.8	10.2	14.0
1931	8.0	10.3	13.3
1936	6.4	8.5	9.9
1941	8.0	9.5	11.9
1956	15.5	22.8	30.0
1966	11.7	18.6	27.6

Source: Hamilton Assessment Rolls

1956 sample. Apartment building in the 1960s represented a late and dramatic reaction to the underbuilt rental market of the 1950s. The spate of apartment building and the popularity of a wrecking-ball approach to urban renewal, which cleared away some converted houses, lowered the number of households in flats within the sample blocks to 1,130 in 1966. Flats now housed – counting both tenants and owner-occupants – 18.6 per cent of all the 1966 sample households. Therefore, considerable scope remained in the city for face-to-face associations among landlords and tenants despite the post-war showing of apartment buildings. Investors had started to abandon the detached house, but a new form of revenue property had evolved to take its place and the makers of the flats included new social groups – the city's recent immigrants (see table 8.11).

A word should be said about social groups and rental accommodations. An examination of table 8.12 suggests that tenants in each of the four main social groups chose types of accommodation on the basis of socio-economic evaluation of several shelter forms. In other words, the mix of rental forms with tenant occupations reveals nuances of status and the ways in which status became worked into the urban material culture. The professionals and proprietors who at some point in their lives rented shelter have certainly participated in the long-term decline of the house as the commanding feature of rental accommodation. For well-heeled or aspiring households, tenancy was perceived as a very brief interlude. When they did rent, they occupied a greater proportion of rented houses than any other social group in the 1956 and 1966 samples. Concurrently, however, they started to flock to apartments. Expressive of their relative af-

Table 8.12
Changes in the Types of Rental Accommodations according to Occupational
Groups, 1921-1966

	1921	1926	1931	1936	1941	1956	1966
PROFESSIONALS AND PROPRIETORS							
House	90.5	70.4	62.6	63.0	67.0	49.8	47.8
Apartment	4.3	7.0	6.1	15.1	7.0	16.9	21.1
Commercial	0.9	14.8	17.1	20.0	18.5	9.7	9.4
Flat	4.3	7.8	14.2	7.5	7.5	23.6	21.7
WHITE-COLLAR							
House	89.9	73.4	71.6	74.7	66.8	32.3	36.7
Apartment	3.9	8.1	7.8	14.5	14.4	26.5	22.5
Commercial	1.6	3.6	4.9	1.1	4.4	8.3	10.2
Flat	4.6	14.9	15.7	9.7	14.4	32.9	30.6
SKILLED AND SEMI-SKILLED							
House	94.8	76.8	81.2	77.1	80.7	37.9	39.8
Apartment	1.5	3.3	3.1	6.5	1.1	15.6	15.1
Commercial	0.5	5.0	1.8	5.9	6.1	7.6	11.7
Flat	3.2	15.9	13.9	10.5	12.1	38.9	33.4
LABOUR							
House	96.8	84.4	77.3	81.5	37.5	35.6	29.7
Apartment	0.9	2.1	0.7	1.7	3.7	5.5	10.1
Commercial	0.0	0.4	4.9	3.8	3.3	4.7	8.9
Flat	2.3	12.7	17.1	13.0	55.5	54.2	51.3

Source: Hamilton Assessment Rolls

fluence and social aspirations, they became strongly committed to apartment living. In the 1970s this was translated into luxury condominium buildings that combined ownership and a good address. The entire subject of the cultural meaning of the apartment building is deferred to the next chapter. We note here, however, that apartment buildings have appealed to a young professional and white-collar clientele. Committed to keeping quality roofs over their heads and good addresses in the directories, the professionals and proprietors had relatively less recourse to apartments in commercial blocks or flats in converted houses. The important exception to this avoidance of less desirable refuges occurred during the depression when a great many shopkeepers occupied apartments above their stores.

Until the 1956 and 1966 cross sections, the sample white-collar heads of households succeeded in staying out of apartments in commercial blocks. Likewise they avoided flats. The sample data for these years establish indirectly the ambiguous position of white-collar employees. The majority of those who rented resided in houses and in the respectable apartment buildings. Indeed, an usually high percentage of white-collar employees lived in apartments. So, for most tenant white-collar employees, the residential trappings of bourgeois identification were supported. Nevertheless, a good many – roughly a third – rented flats. That proportion nearly approximates the level held by skilled and semi-skilled labour. What really separates these blue-collar groups from the white-collar employees in terms of their rental patterns was the apartment building. The blue-collar workers were always less likely to dwell there, preferring houses and flats. Rent may have been a consideration, because apartment units were no bargain insofar as cost per unit of area was concerned. In an apartment building, one paid for status, appliances perhaps, and a degree of segregation from the racket of the industrial city. The explicit locational considerations that apartment investors took into account were those that made the apartment building inconvenient for blue-collar workers. Apartments were to be placed in or near elite neighbourhoods and definitely away from railyards and heavy industries.

Common labourers took to rental flats more markedly than any other social group, particularly during and after World War II. Given the decline of the house for let, the deliberate remoteness of apartments from industry, and the relative exclusivity of apartments, the flat became the principal form of rental shelter for the city's poorer workers. This obvious association between occupation and character of housing leads into a final analysis of the landlord and tenant data. Returning to our speculation about how landlords and tenants have gotten along, we now consider who has let out dwelling units to whom, in an attempt to determine whether landlords and tenants knew one another simply through an economic association with an exploitative quality or whether more complex social relations were involved.

In one sense, occupational links between landlords and tenants had a modest presence – skilled and semi-skilled labourers who rented units let them out to a host of tenants, showing no great predilection for fellow blue-collar employees (see table 8.13). Neither had white-collar tenants especially shied away from blue-collar landlords. Other occupational groups likewise had mixed tenants. Regrettably, we cannot explore the possibility that landlords and tenants might have become associated through their workplace. Bulletin boards and the grapevine at the shop, factory, warehouse, or office may have performed a significant information service, supplementing the listings at real-estate agents and in the classified advertisements in newspapers. The existence of such information about the

Table 8.13
Percentage of Tenants within an Occupational Group Who Rented from
Blue-Collar Landlords, 1876-1966

Year	All Tenants	Professionals and Proprietors	White-Collar	Skilled Labour	Labour
1876	39.2	37.5	40.5	41.2	38.0
1886	32.9	31.1	38.8	34.8	29.7
1896	24.3	23.3	18.6	30.5	18.3
1906	33.5	21.6	30.2	24.3	37.8
1916	29.4	20.2	30.9	31.5	28.9
1926	23.6	15.5	20.0	26.3	26.6
1936	22.4	15.0	19.3	23.6	26.8
1956	35.0	22.0	30.1	39.0	57.9
1966	36.7	22.7	31.8	39.3	53.6

Source: Hamilton Assessment Rolls

rental market is inferred from the relative lack of notices about rental dwellings carried by Hamilton newspapers until the mid-twentieth century and from the fact that it would have enabled landlords to avoid agents' commissions. There is also additional evidence.

Denominational and ethnic networks surely functioned, helping to explain tendencies among landlords to rent houses to householders of the same religious or ethnic background. Landlords who elected to cite their religious affiliation – more than half did – had a notable inclination to let to co-religionists (see table 8.14). The possibility that this pattern had discriminatory overtones in no way precludes the slightly different likelihood that these landlords sought out tenants from members of their congregations. Congregations served to spread rental information to possible tenants and to give the landlords some assurance of reputable and acceptable tenants. Similarly, patterns of ethnic association appear in the Hamilton data. The data on rental dwellings that were both owned by and rented by Anglo-Celtics are not remarkable, because the group was by far the largest in the city at every cross section. The more interesting data demonstrate the over-representation of selected ethnic groups as tenants of dwellings owned by members of the same ethnic group. The tendency to rent a dwelling to someone of a like background was most pronounced when that shelter unit was a house or a flat, suggesting that petty landlordship often involved personal or at least cultural ties when the landlord came from a minority group (see table 8.15).

Table 8.14
The Networks within the Rental Market: Percentage of Houses Owned
by Landlords Declaring a Particular Religious Denomination and
Rented to Co-religionists, 1926-1966

| | Selected Denomination | | | |
Year	Anglican	Roman Catholic	United Church	Jewish
1926	59.3	53.5	–	26.1
1931	72.6	43.2	67.1	26.7
1936	71.8	44.7	75.0	42.9
1941	78.3	52.7	70.0	45.8
1956	62.7	58.8	78.2	77.7
1966	87.3	57.4	92.7	66.7

Source: Hamilton Assessment Rolls

Let us summarize the history of rental housing. First, changes in
the composition of the free-entreprise rental market established
status-associated choices for better-off tenant households only.
White-collar employees, for example, discovered and could often
take advantage of the convenience, novelty, and cachet associated
with residing in an apartment building. Second, the shifts in the
blend of shelter types in the free-enterprise rental market did very
little for the city's poorer households. They were merely moved out
of shoddy houses and into flats in a revamping of rental conditions
that took nearly fifty years. The great shift in the rental market
conveyed psychological and material benefits to a few occupational
groups, and may also have created new and less potentially vexing
investment outlets. It may have established – in the form of flats –
a means of accommodating a set of immigrant aspirations. In contrast to these positive or potentially positive results of major structural changes in the rental market, the situation for common
labourers remained essentially static. Third, one of the forces behind
the revamping of the rental market was the evolution of capital
mobilization. It hastened or even effected the demise of rental
homes. Modern capitalism, as it functioned in the housing sector,
made the house an article to be traded, not let. It also developed
techniques, discussed in chapter 9, that channelled resources into
new rental facilities primarily for more affluent households. In sum,
modern capitalism expanded the material or architectural options
for the prospering white- and blue-collar citizens while squeezing
the options of labourers and the poor. Conversion of large and older

Table 8.15
The Networks within the Rental Market: Percentage of Houses and Flats Owned
by Landlords of Particular Ethnic Groups and Rented to Tenants of the
Same Group, 1926-1966

Year	Anglo-Celtic		Italian		German		Eastern European	
	Percent Rented Within Group	Index[1] Number	Percent Rented Within Group	Index Number	Percent Rented Within Group	Index Number	Percent Rented Within Group	Index Number
HOUSES								
1926	92.0	102.9	33.3	1513.5	27.5	1100.0	33.3	1958.8
1931	92.4	105.0	14.3	595.8	–	–	–	–
1936	91.5	105.0	31.6	1316.6	18.2	674.1	25.0	892.9
1941	90.4	n.a.	43.9	n.a.	20.0	–	25.0	–
1956	83.5	115.5	47.1	841.1	21.2	493.0	12.9	174.3
1966	84.8	129.5	26.3	295.5	27.0	465.5	23.5	239.8
FLATS								
1926	87.8	98.2	–	–	–	–	–	–
1931	92.4	104.4	–	–	–	–	–	–
1936	88.9	102.1	–	–	–	–	–	–
1941	87.7	n.a.	–	–	–	–	–	–
1956	83.6	115.6	29.9	533.9	21.2	493.0	12.9	174.3
1966	80.6	123.1	50.5	567.4	29.2	503.4	19.2	239.8
APARTMENTS								
1956	86.9	120.2	–	–	8.6	200.0	–	–
1966	37.4	57.1	–	–	5.1	87.9	2.3	23.5

Source: Hamilton Assessment Rolls

[1]The index number was calculated by dividing the ethnic group's percentage of units let to
members of the same group by the ethnic group's percentage in the sample population. The
result was multiplied by 100.

single-family detached dwellings into flats provided a safety valve
when the detached house generally became a traded commodity
rather than a rented facility. However, the flats presented a vulnerable shelter arrangement. Zoning regulations and the age of the
converted dwellings were strictures on the growth of this rental form
and on its durability. In the 1960s the older homes susceptible to
conversion were often liable to demolition, and the land was subsequently used for apartments or non-residential purposes.

While low-rent flats ran into barriers, social housing began to
expand. Without wishing to minimize the sensitivity of housing reformers, the courage of officials who charted the programs initiated
by CMHC, and the elements of social democracy in Canadian political
life, the federal government's foray into social housing after 1964

seems to have been a timely move into the vacuum created by modern capitalism. Did government action come almost naturally as a reflex response to a structural change, to a drying up of free-market, low-cost shelter or, at least, to a free-market provision of low-cost shelter that was vulnerable to the politics of urban renewal? The relatively smaller commitment to social housing in the United States raises one of the more important barriers to a complete match-up between housing experiences in Canada and the United States. It appears quite unlikely that modern capitalism provided for American labourers and the poor more generously than it did for their counterparts in Hamilton. Thus Canada's social-housing activities were not just a natural response to a structural change in the rental market – assuming that Hamilton's case was unexceptional – but a significant political act conceivable because Canada was less slavishly bound to allowing the free marketplace alone to determine how its citizens lived.

But housing policy, considered in the chapter on home finance, needs no recapitulation here. We turn now to the apartment building. Somewhat like the house, though arising later and with less consensual support, the apartment building has been a North American artifact with a forceful visual presence. Beyond its visibility, its introduction and refinement – like the design of the modern suburb and modifications to the building materials industries – issued from modern capitalist innovations.

The Landed Banking and Loan Company was Hamilton's oldest mortgage firm and had originally built the Renaissance-style building in the
foreground. In 1905 it moved from the original site to a neo-classical
temple of lending that can be seen three buildings back.

The city's financial firms, servicing the mortgage market, are depicted here in a stage of historic institutional growth. The building of the Equitable Trust tower in 1974 required the demolition of the original home of the Landed Banking and Loan Company.

An assortment of poorer quality houses surviving in the area developed by the Mills family.

THE MOUNTAIN VIEW APARTMENTS

THE LARGEST AND FINEST IN ONTARIO

VERY MODERATE RENTS   READY OCTOBER 1

FURNISHED OR UNFURNISHED

Built on the most beautiful site in Hamilton, these magnificent apartments offer the very latest conveniences, including electric refrigeration, continuous hot water, heat, "Kernerator" for daily disposition of all garbage, elevator service, of course. A wonderful view of the city can be obtained from the roof garden, which is for free use of all tenants. A fully equipped laundry, with electric washing machines and electric driers, is another convenience offered. Another unusual service is garage accommodation right at the building. Rents, one room and bath, from $25.00; three rooms, from $50.00; four rooms, from $55.00; five rooms, from $75.00; six rooms, from $100.00. Be sure and see these early to secure choicest location, as over half these suites are now spoken for. Special terms for long leases. Apply any time day or night.

THE MOUNTAIN VIEW APARTMENTS

The Mountain View Apartments were somewhat more luxurious than contemporary buildings, but many of the conveniences were featured at other respectable apartment addresses in the 1920s and these amenities attracted "the newly wed and the nearly dead."

Rental housing from three eras: row houses in the foreground date from the 1850s, lowrises in the middle from the 1920s and 1950s, and the highrises were constructed in the 1960s.

Hamilton Mountain was a safety valve for working families in the post-1945 decades. The highrises and *cul de sacs* are hallmarks of urban development in the 1960s. Mohawk Road at Cardinal Drive in the theme subdivision known as Birdland. (Photograph by Tom Bochsler, neg. 25538G-9)

Corktown, ca 1881. The frame cottages are visible and there is a sense of the area's poor drainage.

The solid house on the corner was an expanded version of the basic two-and-half-storey brick dwellings that proliferated across Hamilton from the 1890s to the 1910s.

9 The North American Apartment Building as a Matter of Business and an Expression of Culture: A Survey and Case Study, 1900s–1980s

In the 1870s and 1880s apartments became established as a note-worthy shelter form in Boston, New York, and Chicago. A stylish mode of shelter then associated primarily with Paris, the apartment dwelling provided lavish quarters for the affluent in large sophisticated cities. Well-appointed French flats in New York occasionally contained as many as ten rooms per unit, and the buildings' ornate exteriors announced riches.[1] A trend toward multi-level accommodation for other than the rich in plush French flats or the poor in tenements caught on later. By the first decade of the twentieth century, apartment living had likely established a foothold in the shelter market of every North American city (see table 9.1). During a boom in apartment construction in the 1920s, the roots of the trend were placed around 1900. Some cities in that pre-1920 period displayed greater apartment-building construction than others. New York, Chicago, Detroit, Cleveland, Minneapolis, and Milwaukee were considered active centres; Philadelphia and Denver were not.[2] In a rare historical account of apartment architecture and its social significance, John Hancock identified three boom periods of apartment construction: 1890–1917, 1921–1931, 1960 to the present.[3]

By 1920 the apartment building had evolved into several forms and complemented the social segregation of North American cities. Hancock established a credible classification for the architectural and social types of apartment buildings for the first two booms: palatial apartments, luxury apartments, owner-occupied apartments for the rich and affluent, and efficiency apartments for the middle class. A

Table 9.1
Approximate Date of Construction of the First Apartment Buildings
in Selected Cities[1]

City	Year
Boston	1855
New York	1869
Chicago	1878
Los Angeles	1890
Detroit	1892
Cleveland	1896
Denver	1901
Minneapolis	1901
Atlanta	1903
Buffalo	1907
Philadelphia	1908
HAMILTON	1914

Sources: Various issues of *Buildings and Building Management; Hamilton Times,* 10
March 1914
[1]These dates are rough guides based on contemporary accounts and not upon
detailed architectural research.

fifth type, subsidized apartments for low-income and poor tenants, arose in more recent times, although there were philanthropic prototypes early in this century. Despite the complexity of form and the somewhat different agenda of its arrival among North American cities, the apartment building burgeoned as an especially middle-class shelter form in the early twentieth century. It could offer decent rental accommodation for the urban middle class, although Hancock has recorded critical evaluations. Sharp retrospective appraisals by architectural critics Carl Condit and Lewis Mumford decried the middle-class efficiency apartment for alleged cramped banality at high costs to its tenants. That may not have been the complete picture seen by occupants. The single-family detached dwelling has been the unshaken pinnacle of housing status, but the efficiency apartment supplied practical and psychological advantages for aspiring home owners or for those who had retired from that burden. The property industry ingeniously marketed apartment buildings as places of quality and independence, and it tackled the matter of quality in more far-reaching ways than critics appreciated.

The middle-class apartment also presented a set of management challenges to the ever-adaptive property industry. Signs of a coming-of-age for a new shelter form with special marketing and management tasks abounded in the property industry's professional circles between 1900 and 1920. At first, local apartment managers and owners met and shared concerns; soon national and international bonds of professional association developed. In 1902 Chicago owners and managers formed an association to deal with employees, lawmakers, city tax assessors, and the fire department. Before World War I there was a short-lived publication entitled *The Apartment House Magazine*. A more enduring herald of the trend toward apartment living and of a substantial field for entrepreneurial endeavour was the publication of *Building Manager and Owner* in Chicago beginning in 1906. Starting in 1913, it included an apartment house section. In 1908, also in Chicago, the National Association of Building Owners and Managers, including Canadian delegates, held its first annual convention. From 75 delegates, the association grew to over 1,600 by 1928, near the end of an exceptional surge of construction. Sinclair Lewis knew what he was doing when he had realtor George F. Babbitt travel in 1920 from Zenith to Monarch to address the State Association of Real Estate Boards about real estate being "like a reg'lar profession." The journal *Building Manager and Owner*, its several descendants, and the National Association devoted more attention to office buildings than to apartment management. However, until the 1930s about a third of both the articles published and the papers presented at annual meetings did concern apartment buildings.[4] The same distribution held true for the *Journal of Real Estate Management* founded in 1935. A new organization, the National Association of Apartment Owners, was created in 1938 to deal more exclusively with apartments, and especially to exchange information on rents and vacancies.[5]

Initially, apartment living provoked controversy. On the one hand, the apartment building had to overcome associations with the tenement building. On the other, sponsors had to convince potential tenants that apartment living was much more affordable than residence in an exclusive French flat. Of course, the single-family detached home, at the zenith of its affordability and popularity between 1880 and 1920, was often contrasted in glowing terms with the apartment. Boosters in many centres touted their particular community as "a city of homes." Possibly the Bellamyite social-reform movement of the 1880s and 1890s helped to foster acceptance of the apartment building, for according to the utopian prescription of Edward Bellamy, the exceedingly popular American writer of the 1880s, apartment living promoted harmony and foreshadowed the coming age of communal living.[6]

Behind these issues, which indicated a new trend in urban culture and engaged intellectual historians, there moved the vigorous city-shaping forces of investment, architecture, and social tastes. Apartment living surged ahead, developed and refined by people who had no concern for literary diversions about the alleged moral and social consequences of apartments. The apartment building was a business proposition – and soon a very major one. The New York based *Carpentry and Building* first remarked on the phenomenon in its city in 1880, noting the great interest that apartments stirred among capitalists and architects. This interest continued to provoke intermittent action, especially in the 1920s and the 1960s.[7] A significant feature of urban design, the middle-class apartment exemplified the interaction of capitalism, social values, and subservient aesthetics that characterized so much of the material culture of urban North America. Let us consider briefly each of these components, beginning with capitalism.

During the 1920s, in an atmosphere of speculative mania that swept first the Florida land market and then the stock exchanges, apartment construction lured individuals and institutions into a rash episode of easy money and remarkable building activities.[8] By 1930 investors had put $9 billion into apartment construction in United States cities. The apartment-rental business, in one estimation, had become America's fifth largest industry. Through the construction of new units it touched most corners of the North American economy. During the twenties the wave of apartment over-building, soon to be criticized, had produced about as much new accommodation as had the construction of new single-family dwellings during the same period. In 257 American cities in 1926 the construction of apartment units exceeded the construction of single-family dwellings for the first time. Depression, war, and wary investors, however, brought a retreat from abundant apartment construction until the 1960s. In 1962 apartments accounted for one-third of housing starts, the greatest proportion since the 1920s.[9] The highest point for the new boom was reached in the United States in 1969 when 42 per cent of new housing unit starts were in apartment buildings. In sum, apartment construction has swung upwardly more sharply than home construction during rises in the business cycle. Similarly, collapses in apartment construction have been extreme. Urban employment opportunities and available pools of investment capital may have precipitated the volatility of apartment construction.

Social values and aesthetic concerns also have a part in the making of efficiency apartments for the middle class. From about 1910 forward, specialists in the design, investment finance, and management of apartment buildings wrote extensively about the handling of this novel urban shelter. In the 1920s and 1930s their output became especially voluminous, presaging the technical but not always more insightful studies by academics who "discovered" the apartment building as a scholarly topic in the 1960s. The high-rise construction trend of the 1950s and 1960s, along with the maturation of the social sciences, stimulated sophisticated academic projects that tested and retested theories about apartment-building location, disclosing both patterns and complexities. The conclusions of the many academic studies would have been comprehensible to the property industry specialists who had informed the industry for nearly a half century. Much of the academic research simply rediscovered the common sense that had guided the builders and investors who erected apartment buildings with an eye to maximizing profits by securing and holding dependable tenants.[10]

A few of the social science studies went further than testing land-use theories – they elucidated a relationship between apartments and the family life cycle.[11] The high-rise phenomenon of the 1950s and 1960s, like the apartment buildings of earlier times, also provoked critical tracts that alleged excessive profits by developers and deplored the concept of high-density living through the architectural device of towers.[12]

Working exclusively with statistical data, much of it compiled from census reports, social scientists have examined multi-family housing rather than apartment buildings. The distinction is important. Robert Schafer noted that housing is "a very heterogeneous commodity."[13] To make use of national data for his major investigation, *The Suburbanization of Multifamily Housing*, Schafer treated housing "as divided into two homogeneous goods – multifamily and non-multifamily units."[14] He defined multi-family structures as those containing at least five units. Canadian census officials used a definition of at least six units. Since these purely statistical definitions could embrace tenements, converted homes, and units above commercial blocks, they preclude investigating social traits in relation to discrete architectural forms. That is unfortunate because the middle-class apartment building, in contrast to a catch-all statistical definition, has had a fairly precise identity in each of its several boom eras. It has had an architectural lineage, and was a recognized form that imparted more respectability and prestige than did multi-family dwellings in general. This lustre was apparent from its origins, although across a century of adaptation to North America the aura of apartment living has become somewhat diminished.

It is possible to see apartments not only as artifacts amenable to locational theory, but also as products of real actions much talked about by real-property specialists. Building-management journals, realtors' textbooks, architectural pattern books, and the structures

Table 9.2
Number, Scale, and Vacancy Percentage for Apartment Building Sample Drawn
from Hamilton Assessment Rolls, 1924-1969

Year	Number of Buildings in Sample	Total Units in Sample	Average Number of Units	Percentage Units Vacant
1924	13	107	8.2	7.0
1929	26	269	10.3	7.0
1934	26	330	12.7	16.0
1939	25	319	12.8	5.0
1944	31	429	13.8	0.7
1949	31	437	14.1	2.3
1954	38	526	13.8	1.8
1959	38	516	13.6	5.0
1964	47	760	16.2	6.0
1969	48	981	20.4	6.0

Source: Hamilton Assessment Rolls

themselves provide evidence for a rich account of building up and living in a component of urban North America. The same sources yield information about social values, but to probe more deeply into these matters – especially when census data are problematic – an additional research strategy was employed. An initial random sample of thirteen apartment buildings (1 out of 10) was drawn from the Hamilton city directory for 1924, a source that reliably distinguished middle-class apartment buildings from multi-family dwellings in general. A slight bias in social traits was produced because the sample also picked up several buildings that were close to luxurious. The initial thirteen structures remained in all subsequent five-year cross sections. When the directory indicated an increase in apartment structures a (1 out of 10) sample of the newcomers was added to the original group (see table 9.2). Once an apartment building joined the sample, its address was located in the assessment rolls and the following variables were coded for that building: number of units, number of vacancies at the time of assessment, name and social characteristics of the owner (occupation, religion, and ethnicity). Further variables related to the households: name of household head, age, sex, conjugal condition, occupation, religion, ethnicity, and size of household. By blending conventional sources applying to many North American cities with the local data, the interplay of capitalism, social values, and aesthetics can be spelled out more precisely.

INVESTMENT TRENDS TO 1930

Business calculations, guided by rules of thumb, determined whether or not to invest in apartments, the scale or quality of structure in which to invest, and where to locate the building. During the first decade of the twentieth century, as the construction costs of the single-family dwelling escalated, the demand for apartment accommodations encouraged strong investment optimism. A housing crisis of shortages and rising rents in the pre-World War I years inspired confidence in a net rate of return on investment equity that might reach 50–100 per cent above the current mortgage rates. Some reports did place profits at that enticing level. An 1899 review of the balance sheet for a 556-unit apartment complex meant to house Boston artisans boasted a return of 6.5 per cent after deducting operating expenses and payments into a sinking fund for repairs. Current interest rates were around 4 per cent. Not every report or projection was rosy. Cautious real estate operators, perhaps recalling periodic collapses of real property markets in the nineteenth century, warned of eventual overbuilding and depressed markets, but the optimism continued. In 1906 *Building Manager and Owner* discounted doomsayers' rumours: "the demand for apartments has more than kept pace with the supply and real estate men look upon the apartment building as profitable and stable investments [sic] which will continue to bring good returns."[15]

Another of the rare balance sheets to be published before 1920 indicated the investment attraction of the middle-class apartment building. Based on a 1912 survey of three types of Chicago buildings, the report noted the following rates of return: high-class old style, 4–5.5 per cent; less expensive buildings, 6–6.8 per cent; tenements, 4–5.2 per cent.[16] Investors were cautioned against working-class housing. As the post-1910 inflationary cycle gathered weight, the anticipated net rate of return on middle-class apartments escalated. Self-proclaimed "real estate speculator" Fred C. Kelly alleged in 1916 that "men who make a practice of putting their savings into apartment houses demanded a net rate of return of 10 per cent. I myself like more than that; I aim to get 12 per cent or just double the profit I would make if I merely let money out at interest."[17] His was an unusually sanguine view, even when taking into account a heady initial period of enthusiasm for a fresh investment outlet.

More reserved opinions appeared as some cities became overbuilt in the early 1920s. In 1921 the *National Builder* reported that a gross return of 10 per cent, a once commonly accepted standard for determining a profitable apartment building, was now too low. Overhead, it had calculated, now ran to 7.7 per cent of investment. Consequently, a gross return of 14 per cent was needed for a net return of 6 per cent or the going interest rate.[18] Trade publications nevertheless continued to hype the apartment building more frequently than they issued warnings against investment, at least until the mid-1920s.

Even the supposedly prudent banking community found the apartment building attractive. As an outlet for building loans, financiers preferred it to tenements because it retained its appraised value. Some thought it a better risk than the office building because the latter tended toward lavish excess. The usually cautious lending institutions relaxed their common stipulation that buyers of apartment buildings hold 50 to 60 per cent equity in the structure. Additionally, appraisals became inflated.[19] What could be better than a good revenue generating, possibly appreciating, and comparatively inexpensive structure that received generous terms from lending institutions? Furthermore, stories of quick gains and of men who controlled apartment building development in their cities helped to snowball the investment trend. The apartment investment market of the 1920s can be described as a crowd that, though it lacked central co-ordination, consisted of intercommunicating members who kindled each other's confidence – the emotions of each member were fanned by the expectations, perceptions, and behaviour of others. Therefore, especially during the 1920s, the investment that streamed into apartment construction was guided less by scrutiny of income records – for these were still infrequent and localized as well as historically shallow – than by expected income based on a scattering of generally optimistic appraisals. Investment expectations have often chased unusual stars, and in the 1920s the apartment building was one of these. Some of Babbitt's joyful shop talk at the state convention of real estate boards dealt with capital raising for apartment construction:

"How'd this fellow Rountree make out with this big apartment-hotel he was going to put up? Whadde do? Get out bonds to finance it?"
"Well, I'll tell you," said Babbitt. "Now if I'd been handling it – ".

We never learn what Babbitt would do or Rountree had done, but the sale of apartment bonds provided training for two extremely influential Canadians. E. P. Taylor, who went on to consolidate several Canadian industries as well as British brewing concerns, and later to develop Don Mills, one of Canada's post-World War II planned communities, worked the Ottawa market for a Toronto

Table 9.3
Number of Apartment Units in Hamilton Sample Broken Down by Scale
of Apartment Building, 1924-1969

Scale of Apartment Building	Units by Year									
	1924	1929	1934	1939	1944	1949	1954	1959	1964	1969
Fewer than 7 units	12	59	48	52	59	61	61	61	61	60
7 to 20 units	95	134	179	160	194	214	204	204	275	286
21 to 75 units	0	0	24	24	92	93	180	184	341	312
76 +	0	76	79	83	84	84	81	77	83	323
(Mean)	(8.2)	(10.4)	(12.7)	(12.8)	(13.8)	(14.1)	(13.8)	(13.8)	(16.2)	(20.4)
Total	107	269	330	319	429	437	526	516	760	981

Source: Hamilton Assessment Rolls

bond house. W. C. Clark, Canada's Deputy Minister of Finance from 1932 to 1952, rose to vice-president in a New York financial firm that specialized in arranging funding for office towers and apartment buildings. From the persistent salesman to the brilliant economist, the apartment construction boom presented opportunities to learn about the mobilization of capital.[20]

Enthusiasm ultimately created an investment crisis through overconstruction in the late 1920s, but throughout the early and seemingly irrational investment period the calculation of investment returns also built on certain rational dimensions of market analysis and land-use economics. The builders and realtors who erected the structures for eventual sale to investors recognized that buildings with smaller equities appealed to a broader base of potential investors than did large structures. Apparently the contractors were simply "speculative" builders who formerly had put up rows of houses. Such builders required a quick turnover. As a result, buildings of 8 to 12 units were pushed onto the market in an era when most prospective buyers were private investors rather than corporate entities.[21] Larger structures and more complex financing characterized the elevator apartments of Manhattan and the apartment hotels of Los Angeles by 1920. As in all aspects of the property industry, a few cities generated specialised housing forms or led the rest in devising financial instruments or design conventions.

In the subject city of Hamilton, the sample apartment buildings constructed in the early 1920s contained no fewer than 7 units and no more than 20 (see table 9.3). The mean in 1924 was 10 units. However, in the latter half of the decade the boom in apartment

Table 9.4
Ownership Characteristics of Apartment Dwellings in Hamilton by Year
(1924, 1944, 1964, 1987)

Description of Owner as Cited in Assessment Rolls	Percentage of Total Owners Mean Number of Apartment Units			
	1924	1944	1964	1987[3]
Individuals of various occupations	72.0% (9.7)	34.0% (10.5)	28.0% (10.5)	15.5% (26.4)
Builders and contractors	15.9% (8.0)	17.2% (16.1)	1.1% (9.0)	n.a.
Managers possibly working for true owner[1]	0.0% (0.0)	2.1% (9.0)	7.6% (16.7)	n.a.
Companies and trustees[2]	12.1% (13.0)	34.5% (50.0)	63.3% (48.4)	66.7% (92.4)
Numbered companies	0.0%	0.0%	0.0%	6.4% (89.5)
Governmental co-operative	0.0%	0.0%	0.0%	11.4% (142.0)

Sources: Hamilton Assessment Rolls; Hamilton-Wentworth Planning and Development Department, High Density Residential Development Study, Background Report (November 1987)
[1]Occupation of manager or president cited
[2]Managed by agent, company name, trust account
[3]The sample for this year consisted of 173 apartment buildings found in the central area of the city.

construction was accompanied by an increase in scale. The Waverly, erected in 1928, had about 80 units; the sample mean had risen to around a dozen units, where it remained until the 1960s. The modest size of the earliest structures did attract individual investors, for 72 per cent of all units in the 1924 sample were owned by individuals. That altered with the appearance of the Waverly because it had been a corporate investment. From 1924 forward, the percentage of units held by individual investors declined (see table 9.4).

The initial era of major private investment bequeathed an architectural legacy. Articles and pattern books – for example R. W. Sexton's American Apartment Houses of Today (1926) – broadcast the maxims of design so that great similarity prevailed among the apartment buildings across the continent. One critic of the practice saw the designing of apartment houses as "a game of 'follow the leader'" where "plans were copied blindly, or stereotyped plans were taken from the files and adapted to new plots without question."[22] Scattered throughout the North American city, along or near arteries once served by streetcars, there remain the sturdy three-storey structures of the 1920s boom: some with indented courtyards, many with decorative balconies and impressive neo-classical pillars, and all with more space in the individual units than those built in the 1950s or 1960s.

Their locations can be correlated roughly with models based on land-use economics, but it is more instructive to consider the locational rules propounded by the industry itself. Site was extremely important to the investor. If economy of investment was necessary, property experts recommended a reduced outlay on the building rather than on the lot.[23] Drawing on their experience with single-family dwellings, they stressed that location governed value and rent, and meant something beyond convenience to work and shopping. Neighbourhood quality, important in the detached-dwelling market, was now recognized in the apartment business. "Even the most stupid developer knows the value of restricting clients to maintain property values."[24] Rents were set, in part, by determining the average charged in the neighbourhood. The tone of the neighbourhood was a factor, as well as views, sunlight, parks, shopping, proximity to employment, and distance to the centre of the city. Ideally, investors hoped to secure locations within a half hour of the central business district. An elevated rail or subway station was perfectly situated at 3 or 4 blocks distance, but it was never to be adjacent to the building itself. Streetcar lines could be closer. Corner lots added distinction and permitted more sunlight to enter.[25] Not all newly developed areas of a city had the requisite cachet for the middle-class tenants so highly prized by investors. Consequently, even during the early 1920s, shrewd specialists advised others to follow their lead and seek out larger old homes in formerly fashionable areas. Furthermore, the heirs of an estate with a large dwelling, it was noted, made particularly good prospects for land deals; they would sell readily because cash was a more easily divided asset than real estate. The old home could be torn down and its superior site built upon. This happened even in young cities.[26]

A final cardinal rule stipulated that the wise builder or investor should discover plans for adjacent vacant lots. Social homogeneity was an expressed condition of successful apartment development just as it had become with fashionable suburban housing developments like Hamilton's Westdale area (see chapter 2). For the most part, the sample group of Hamilton apartment buildings conformed to conventional wisdom. The two major east-west arteries, King and Main Streets, attracted much of the apartment construction of the 1920s. Some buildings were clustered along the edge of a major east-end green space, Gage Park; a few were sprinkled into quality neighbourhoods; many sat on corner lots; still others sat side by side, protecting their flanks from unwanted non-conforming structures. Their siting was not dictated by zoning or a comprehensive city plan, for the city had neither. The industry itself affected a pattern that would not disturb the social segregation increasingly apparent in the distribution of single-family detached dwellings. A luxury apartment structure picked up in the sample was on the edge of one of the city's most fashionable areas. Several tatty buildings, also a part of the variation in the sample, were near a major factory. Rules of thumb, not formal planning, guided many independent decision makers into reinforcing the spatial dimensions of class. The end result, as Sam Bass Warner Jr has observed, was a form of "regulation without laws."[27]

Although the wave of construction in the 1920s seemed an uncoordinated collective dash of builders, investors, and lending institutions across the always tricky ice of real property, the apartment business had a few rough guidelines that held it to a cautious course on some matters. Location, of course, was one. However, before the added onset of the crises that led into the depression, most conventional locational wisdom focused on the decision making required for a single building or cluster of buildings or for an investor or group of investors. In the 1920s the overall organization of the industry for a particular city or region was ignored. The prevalent propositions about location were indicative of partial caution, evidence of industry planning but not industry-wide planning. Accounting practices too, though again treating only one structure, displayed solid conventional wisdom. By the era of vigorous investment in the 1920s owners were advised to set aside a depreciation fund. For accounting purposes the life of a frame building was estimated at twenty-five to thirty years, that of a brick building at thirty years. Furniture was generally assumed to have a life of eight to twelve years and appliances fifteen years.[28] In some instances, the mortgagee stipulated that a depreciation sinking fund be established as a condition of the financing of a building. Such cautious financial arrangements, which might have required the setting aside of as much as 3 per cent per annum of the building's valuation, created problems in the depression of the 1930s[29] when apartment owners had to meet both their mortgage payments and their sinking-fund obligations while income plummeted.

THE VACANCY QUESTION

The unpredictable element in any formula for success in apartment investment was the vacancy rate. Unlike site selection and depreciation calculations, vacancy could not be managed by the well-advised, prudent, solo entrepreneur or syndicate. Vacancy levels hinged on circumstances beyond the individual's control, such as matching up population growth with the building and advertising efforts of many entrepreneurs largely acting without coordinated restraint. John Kenneth Galbraith in his study *The Great Crash* suggested that "the high production of the twenties did not ... outrun the wants of the people." That might have been true for "automobiles, clothing, travel, recreation or even food."[30] However, in one of the leading elements for the economic boom – apartment construction – the Galbraith proposition was in error. Post-depression analyses set an 8 to 10 per cent vacancy guideline as a measure of profitability.[31] Builders in a city with a higher vacancy rate received the cold shoulder from lending institutions as a result of belated recognition that a market could be overbuilt. In Los Angeles, the market had suffered this state throughout the 1920s, because of a special rental circumstance – tourism and wintering northerners prompted the construction of numerous apartment hotels with seasonal fluctuations in occupancy.[32] For the bulk of urban America, the vacancy rate among apartment dwellings hovered between 15 and 20 per cent during the early 1930s (see table 9.5). Among the Hamilton sample apartments, vacancies rose from 7 per cent in 1924 and 1929 to 16 per cent in 1934.

Apartment management experts cited two explanations for the low occupancy levels. Both apartment units and single-family dwellings had been overbuilt, and contractors, lacking buyers, had turned to offering rental bargains. In addition, the depression simply knocked a number of families out of the conventional urban-housing market altogether. In Hamilton, as elsewhere, families doubled up. A number of occupants simply left the city, and between 1931 and 1936 Hamilton's population dropped by 1.7 per cent. As in most North American cities, vacancies in Hamilton had a seasonal rhythm, because owners and managers followed the standard practice of fixing lease expiry dates on May 1 or October 1. During the late 1920s the industry attempted to reduce the number of traditionally dated leases in an effort to alleviate pressure on the redecorating trades. However, the inconvenience and cost of seasonal vacancy and moving were nothing compared to the challenge of vacancy in the depression.[33] On the one hand, social activists among urban specialists saw the misery of the poor and the overcrowding produced by doubling up. In Canada, they lobbied for state housing and were thwarted by none other than W. C. Clark, the realty expert turned Ottawa mandarin. On the other hand, the property industry that influenced Clark could properly point to empty dwellings and warn

Table 9.5
Apartment Building Vacancy Rates (% of units vacant) in Selected Years and in Selected Cities and Regions, 1924-1959[1]

City or Region	1924	1929	1934	1949	1954	1959
HAMILTON	7.0	7.0	16.0	2.3	1.8	5.0
New York		5.0	7.0	14.0		
Chicago	10.0		14.8			
Los Angeles	29.0	19.0	25.0			
Pittsburgh			7.7	25.0		
St Louis			17.7			
Omaha	10.0	9.0				
New England			21.6			
United States	5.0-20.0			2.6		7.2

Sources: *Buildings and Building Management; The Proceedings of the 27 Annual Convention of the National Association of Building Owners and Managers (1934)*; Hamilton Assessment Rolls
[1]The selection criteria were purely functional. Many vacancy surveys were not used because they did not fit the years chosen for the Hamilton sample or they were compiled by management associations from data supplied by members. Such data reflected lower rates than those found in buildings not under professional management.

of the havoc that building more accommodation might precipitate. Here lay the paradox of poverty in the midst of plenty.

The apartment industry pursued two contradictory courses to counteract the extraordinary number of vacancies arising between 1929 and 1935. One line stressed vigorous competition; the other accented cooperation. In either case, the acknowledged problem of overbuilding stimulated an interest in the training of managers. Owners and mortgagees desperately sought out professional aid. Educational institutions responded to the emerging field of building management. Whereas only a handful of colleges or universities in the United States had any type of real-estate course in the 1920s, by 1932 one or more courses were offered in 62 institutions and by 1936 a total of 187 courses were presented in 73 colleges or universities. Apartment-management texts now proliferated: Alvin Lovingood, *Apartment House Management* (1929); Violet Rose Widder, *Apartment Hotel Management* (1930); Grace Perego, *Apartment House Ownership and Management* (1934); James R. McGonagle, *Apartment House Rental, Investment and Management* (1937).[34]

Perego's guide, the most detailed, supplied plentiful cunning hints on how to manage the vacancy crisis. Recognizing from her twenty-one years of experience in real estate that psychology had its place in the shelter industry, she recommended that the first step was to disguise the vacancies. Bogus names – Smith or Jones – were to be fitted into name plates and put on letter boxes. "The less your tenants know your vacancies the better."[35] Eliminating vacancies demanded exertion, pressure, and collecting information. Advertising was an obvious tactic. Additionally, building managers were urged to use the services of local real-estate offices whose commissions amounted to only about a week's rent. But the agency had to be contacted every few days, not merely to remind sales staff but also to discover the rents charged by competitors. "Make a friend out of a salesperson in each office," Perego encouraged, "and get him interested in you and your needs."[36] If, as a result of such information, it became apparent that the nearest competitors had cut rents, Perego counselled pressure to get them to stand fast. "With your analysis of income and expenses you can show them where they are hurting themselves and may be of mutual benefit."[37] It was not so easy. The advice must have seemed like telling a drowning man not to thrash about for fear of upsetting nearby life rafts. All manuals accented the need for high-pressure sales techniques to salvage the building. "Renting apartments, then, is not a waiting game. It demands push and action. The loss from each day's vacancy is so complete that you cannot afford to wait for the right prospect to come along."[38] In very serious circumstances, a few writers recommended rent reductions or other concessions,[39] but the dilemma of how best to seek survival – competition or regulation – was bitterly disputed within real-estate associations.

Faced with rent cutting to lure or hold tenants, city-wide or regional real-estate boards devised tactics to hold the line. Vilification set the tone of tough-minded campaigns to pressure a fragmented industry into a more unified and cautious front. *Apartment Management in New England* denounced Mr Chiseler, "that notorious character, that chesty man with the big hat and undersized brain, who was willing to fill his building at any cost." Perhaps there is no smoother advocate of cooperation than a struggling agent of free enterprise threatened by competition and consumer bargains. The journal, therefore, continued by lamenting that Mr Chiseler would "demoralize the industry at the expense of his brother owner or operator."[40] One, two, or three months of rent concessions and even the payment of moving expenses – these were dastardly inducements. Rent cuts were no small matter. In a 1936 survey allegedly based on the principal cities of the United States, rents were reported to have dropped to 59 per cent of those prevailing in 1926.[41] Price cutting was understandably "weak sister renting," "an evil."[42]

As soon as the overbuilt conditions became evident, local associations moved to control the marketplace. In early 1929 the Baltimore

Building Managers and Owners Association started to meet with the mortgage brokers and lending institutions in an effort to limit new building.[43] To try to enforce a united front, the Philadelphia Real Estate Board organized a January 1932 meeting with the owners, managers, and mortgagees of the city's apartment buildings. Twenty-six financial institutions were represented. As a result, a ten-man panel fixed rules for uniform policies in the rental and operation of apartments. Furthermore, they were to be consulted about the contemplation of any new apartment structures.[44] The large property management agency became another means of upholding a standard. In New York City, Wood, Dolson Company controlled the leasing and operation of 150 large and small apartment buildings with diversified ownership. The central office made periodic checks to study rental situations in each building.[45] The National Recovery Administration (NRA), that tangle of New Deal wage and price codes, fitted neatly in to these assorted flights from competition already underway in the shelter industry. California apartment managers and owners jumped on the NRA bandwagon and published *The Federated News* to promote cooperation.[46] It is impossible to determine the effectiveness of any of the numerous "distribution coalitions"[47] and, indeed, the NRA was found to be unconstitutional. However, by 1935 overbuilding and vacancies had ceased to afflict the industry.

The data on apartment construction in North America demonstrate indirectly that investors and financial institutions had clamped down the brakes with a fierce pressure that virtually brought apartment construction to a dead stop. While in 1932 annual construction of new urban single-family dwellings fell in the United States by a catastrophic 90.5 per cent from its high of 1925, apartment-building construction dropped 97.1 per cent from its 1925 peak (see table 9.6). In Canada the pertinent data were collected differently and presented only the estimated value of construction. The value of new urban single-family dwellings dropped by 75 per cent from its high in 1928 to its depths in 1932; during the same time the value of apartment construction collapsed by 97 per cent (see table 9.7). In Hamilton apartment building construction peaked in 1927 when, according to the city directory, 45 were built. Not one was raised in 1933.

In a swift *volte-face* during the mid-1930s, the building management experts turned from the challenge of rent slashing to the blissful prospect of rent increases. Managers tried to make escalations painless and subtle, recognizing that "painless rent-raising is like painless dentistry."[48] The manipulative calculations proposed by industry specialists were indefensible in their dishonesty and confirm

Table 9.6
Number of Urban Residential Buildings in the United States for Which Permits
Were Issued, 1921-1932

Year	Single-Family Houses	Apartment Buildings	Apartment Buildings with Stores
1921	132,121	4,924	571
1922	181,128	9,592	1,079
1923	211,235	12,925	1,271
1924	214,213	13,091	1,426
1925	234,899	15,109	1,771
1926	200,373	14,993	1,502
1927	164,268	13,663	1,783
1928	143,889	12,063	1,528
1929	104,718	6,662	565
1930	61,656	3,019	209
1931	55,010	2,177	99
1932	22,328	432	33

Sources: United States, Department of Commerce, Bureau of Foreign and Domestic Commerce, *Statistical Abstract of the United States* (Washington, D.C.: United States Government Printing Office, 1921-1933.) The data came from the number of buildings covered by building permits in all cities of 50,000 or more and some cities of 25,000 or more. In 1921 there were 262 cities and in 1932 there were 311. Therefore, the data are not completely suitable for comparison over time, but the imprecision is slight and actually minimizes the slide from boom to collapse.

the worst suspicions of any tenant who has received notice of an increase. No firm should initiate a story about rent hikes, more than one commentator warned. Real-estate boards would do the job by sponsoring stories blaming impersonal forces for driving rentals upward. Safe topics for apartment managers to discuss with newspapers included accounts of housing scarcity. It was recommended that notices of rent increases call attention to new touches added to the building – carpets in the hall, improved heating, or other improvements.[49] Indirect statements were preferred to blunt notices of fact: "The rate for the apartment you occupy has been established at $50 for the coming year" rather than "Your rent will be raised to $50."[50] If a tenant confronted a manager, the latter should remark that "it is competition that regulates prices and not any one building."[51] An alert and able manager should listen to tenants. That had been common wisdom. In other eras, the agent or manager had to guess when a tenant might skip out, avoiding a rent payment. "The agent," wrote one such specialist in 1913, "has seen the washings in the laundry and he knows who is wearing cheap and ragged underwear under the latest tailored outer garments."[52] When it became obvious

Table 9.7
Value of Construction Contracts by Housing Type in Canada, 1923-1932

Year	Single Family Residences	Apartment[1]
1923	$ 88,826,600	$ 8,818,600
1924	81,427,400	9,797,400
1925	83,766,300	12,723,600
1926	88,583,100	20,979,300
1927	98,957,800	25,981,800
1928	102,445,800	36,720,500
1929	106,374,100	22,527,200
1930	77,961,200	15,330,300
1931	65,482,100	16,202,200
1932	27,356,600	1,536,000

Source: Canada, Dominion Bureau of Statistics, General Statistics Branch, *The Canada Year Book* (Ottawa: King's Printer, 1929-1933)

[1] These figures may include flats in partly commercial structures. The source is not clear on this matter.

that tenants' employment conditions had improved, when they saw a slight ray of hope and felt that better times were in store, then the time was ripe for the manager to recommend a rent increase to the owner. If tenants accepted an increase without too much grumbling, it indicated that a further increase could be contemplated.[53]

ARCHITECTURAL CONVENTIONS

Like rules for location and the growth of management skills, structural design involved manipulative aims. By the 1920s architects had already worked out schemes for achieving decorative enhancement that would hold middle-class tenants. By continent-wide convention, apartment building exteriors strove for ageless simplicity. Fads in architecture told the age of a heavily stylized structure, but with upkeep, a basic brick structure trimmed with tasteful terra-cotta adornment and wrought-iron hardware never disclosed its age. Dignified by elegant entrances and windows dressed with fancy grills, these structures had practical qualities to recommend them. The lack of any historical style was thought an advantage because "one tenant might have a very decided preference for some one style, another for another; they cannot both be satisfied. Therefore, if kept nondescript, or neutral in character, it will not subject itself to criticism; if it is in good taste, it will be entirely unobjectionable, and there will be a greater chance of pleasing both."[54]

Still, the marketing psychology demanded some frills and attractions. Apartments were virtually tailored to satisfy youth's longing for independence, expressed in shelter. Evidence of this association between the urban youth culture of the 1920s and apartments is offered later, but a tie-in between age and style was commented upon in one of Charles E. White's monthly columns on apartment buildings for the *National Builder* in 1920. "Newly married couples like things a little more ornate than older people, as a rule, so you must cater to this taste by making your building especially attractive as to details."[55] The recognition of this penchant for attractive finishing helps to explain the admittedly economical use of terra cotta, wrought iron, and stained glass on building exteriors, as well as the wainscotting, picture railings, scroll woodwork, and chandeliers on interiors. These were commonplace with luxury apartment buildings, but some combination of "classy" finishing was also applied to the middle-class efficiency structures.

The middle-class apartment façades of the 1920s and 1930s connoted timeless respectability. Exteriors and adornments harmonized with the entire campaign to impress potential tenants with the elegance of apartment living. The profoundly entrenched appeal of single-family homes and the lingering association of apartment buildings with tenements required early apartment promoters to polish an image of distinguished accommodation. Like many writers on the subject of housing, apartment management expert James R. McGonagle comprehended the psychological features of shelter. "Man is a peculiar animal," he wrote in 1937, "he revels in the admiration, yes, even in the envy of his friends; and as this tendency increases, so too does the appeal of socially desirable districts increase, even though physical requirements can be satisfied elsewhere at a cheaper rate."[56] In a list of things desired by apartment tenants, one authority proposed "a neighbourhood that will provide me an address that I will not be ashamed to give – one that will indicate my true social status in the community."[57]

Along with address or site and the statements of architecture, the naming of apartments assigned distinction in a minor way. It helped expand acceptance and contributed to envy through something akin to a fashionable address. In the city of Zenith, George F. Babbitt installed the party-loving widow Tanis Judique in "the new Cavendish Apartments" in 1920. In Hamilton, culturally very much a British city, many names fixed on royalty, notable places and palaces, and the history of the old country. In the samples of apartments selected for 1924, 1929, and 1934, royalty was represented by Windsor, Queens, and Royal Alexandra; famous buildings – palaces –

were memorialized by Holyrood and Westminster; places by Leinster and Ulster; history by Lloyd George. While half of the names originated in the class-conscious setting of the United Kingdom, a few others – the Astoria and Alhambra, for example – denoted prestige based on more cosmopolitan tastes. Several buildings took their names from Arcadian features specific to the city. The Gage faced the 116-acre expanse of Gage Park. The Mountain View sat at the foot of the Niagara Escarpment. Names changed with the times. The elegance sought in the 1920s and 1930s was out of place during the war years. Builders responded to the remarkable housing crisis of the wartime boom and the immediate post-war years with tidy functionalism. Gone were the courtyards, ornate entrances, terra cotta and wrought iron fixtures. The names of the apartments, at least in the Hamilton sample, declared their emergency origins: Victory, Whitehall, Normandy, and De Gaulle.

During the 1950s and emphatically in the 1960s, apartment construction came back to levels like those experienced in the 1920s. In recognition of post-war family formation and later of new households formed by baby boomers, investors returned to apartments. Government-subsidized urban renewal projects contributed to high demolition losses among aging central-city accommodations. That tightened the market, reducing vacancies. The new buildings came dressed in new styles. Those erected in the 1950s accented efficiency, not elegance. "Efficiency is the new keynote," wrote one author in the *Journal of Property Management* in 1953. "Every part of the modern apartment is designed to cut costs to the builder and reduce the effort tenants need to expend in order to live in the apartment."[58] Units were smaller than in earlier structures and had sparse trimmings. Specifically, bedrooms were about half as large as those of the 1920s; living rooms were three-quarters the size of the older ones; dining rooms were reduced to alcoves within the living room; and ceilings were lowered. The stress on efficiency led to a realistic appraisal of lobbies with the result that they became bare passageways. Materials shortages contributed to this shrinkage. New electric gadgets replaced the former stress on space and fine finishing. Automobile parking assumed importance in site planning.[59]

Bursts of high-rise construction in the 1960s introduced a distinct phase within the revived industry. Aside from being much higher than the typical walk-ups of the 1920s and 1950s, the towers included community facilities such as patios, swimming pools, tennis courts, party rooms, and sun decks on the roof.[60] In the Hamilton samples of 1954, 1959, and 1964, the construction of larger units was striking. The total number of apartments supplied to the sample by buildings with between 21 and 75 units in 1969 was triple that of 1949. Units in buildings containing more than 75 apartments almost quadrupled during the same interval. There had been little change among the smaller sizes (see table 9.3). Corporate ownership accompanied the increased scale. In 1964 companies and trustees owned 63 per cent of the sample units. Managers, likely working for corporate owners, were cited in the assessment rolls for another 7.6 per cent of units (see table 9.4).

A SOCIAL PROFILE OF TENANTS

The amenities of the apartment towers appealed to the youth market in housing, a market fostered by the baby boom. Half of the male household heads in the 1964 and 1969 samples were between the ages of 15 and 34. In other eras, this age range had accounted for only a third of the male-headed households. As a university city, Hamilton experienced a growth in student and faculty numbers that accompanied the Canada-wide social revolution of increased access to post-secondary school education. In one new tower that housed 235 households in the 1969 sample, the following resided as household heads: 56 students, 11 engineers, 8 professors, and 24 teachers. Apartment living increasingly was seen as providing a care-free youth-oriented lifestyle at precisely the time when the baby-boom generation was maturing. Testifying to the success of this marketing strategy, the percentage of apartment units rented by bachelors in the 1969 Hamilton sample had nearly doubled since 1959, rising from 8.8 per cent to 16.7 per cent. Names for the structures reflected the image and reality of a society of youthful "cliff dwellers." Work on Camelot Towers had begun by 1964. Its name united the vistas and dominance of elevation – a tower – with the popular musical about a mythical kingdom. Camelot too had a strong association with the relatively young, stylish, and exuberant administration of President John F. Kennedy. The relaxed living projected by the Capri and Mountain Breeze certainly contrasted with the propriety sought in the staid labels and bland architectural forms of the 1920s. Social and sexual mores of the two eras had subtle manifestations in the forms of that type of residence which at all times appealed most to those wishing to minimize the cares of housekeeping and property maintenance.

Young married couples and singles were not alone in seeking apartment accommodations, casting off a tradition of residing with in-laws or renting a small house. More of the elderly now could live apart from their children, leaving behind the extended-family house-

Table 9.8
Selected Life-Cycle Characteristics of Male-Headed Households in the
Accumulated Sample of Apartment Buildings in Hamilton, 1924-69

Characteristics	Number	Percentage of Male-Headed Households
Head of household is under		
31 and there are 2 in household	388	13.4%
Head is 31 to 40 and there are		
2 in household	302	10.4
Under 31 with 3	222	7.7
31 to 40 with 3	181	6.3
41 to 50 with 2	220	7.7
41 to 50 with 3	139	4.8
41 to 50 with 4	62	2.1
51 to 60 with 2	206	7.1
61 and over with 2	224	7.7
Male-headed households	1944	60.2%

Source: Hamilton Assessment Rolls

hold that once had served as a means of accommodating widows and the aged. The increased number of pension plans, survivor benefits, the commonplace of life insurance, and the presence of women in the workforce augmented the ability of widows to live apart from their offspring. In the Hamilton study, of all female-headed apartment households, women over the age of 55 accounted for 32.5 per cent in 1954 and 50.5 percent in 1969. George F. Babbitt's mistress, the fast widow Mrs. Daniel Judique, was unusual in having moved into the Cavendish Apartments in 1920 when only in her early forties.

By combining the age of the head of the household with the other variables such as gender, marital status, and number in the household, it can be seen that apartment dwellers had a special profile in terms of a life-cycle (Tables 9.8 and 9.9). Of course, a collective portrait for all years obscures several of the subtle trends of demographic change. Among male heads, the dominant age cohort has been a youthful one of 25 to 34. Only in two cross-section years, 1934 and 1949, did the 35 to 44 cohort have greater representation (see table 9.10). In both of those years, problems with finance or building material supplies – war and a post-war boom – forestalled new home construction and locked many young families into apartment accommodations for longer periods than usual. If male heads of households were youthful, female heads were not, their mean age hovering around fifty.

Table 9.9
Selected Life-Cycle Characteristics of Female-Headed Households in the
Accumulated Sample of Apartment Buildings in Hamilton, 1924-1969

Characteristics	Number	Percentage of Female-Headed Households
Single 15 to 30 and living alone	48	5.6%
Single 31 to 40 alone	48	5.6
41 to 50 alone	73	8.5
51 to 60 alone	60	7.0
Widows 61 and over	133	15.5
Total households	342	42.2%

Source: Hamilton Assessment Rolls

Table 9.10
Distribution of Male Household Heads in Hamilton Apartment Sample
by Age Group, 1924-1969. (n = 2,890)

Group	1924	1929	1934	1939	1944	1949	1954	1959	1964	1969
15-24	3.3	6.5	5.4	2.6	1.8	3.2	3.8	6.9	11.2	15.4
25-34	45.9	43.0	31.2	29.6	34.2	23.4	30.0	26.0	30.0	32.5
35-44	26.2	26.3	33.2	29.1	24.0	28.8	21.0	17.3	20.0	16.4
45-54	16.4	14.0	21.8	21.9	22.2	21.9	16.9	15.9	15.6	12.7
55-64	4.9	7.5	3.5	10.2	10.2	13.7	18.4	17.7	12.6	11.7
65 +	3.3	2.7	5.0	6.6	7.6	9.0	9.9	16.2	10.6	11.4

Source: Hamilton Assessment Rolls

The city's household size has been declining since the 1920s, from a mean of 3.9 in 1921 to 2.7 in 1971. In contrast, apartment dwellers' households retained consistency in size from 1924 when the mean was 2.4 to 1969 when it was 2.2 (see table 9.11). The structural and psychological limits characteristic of apartment living engendered, as expected, a reasonably stable household size. Furthermore, a bias against children operated within the apartment industry itself. In a 1909 confessional report entitled "Being a Flat Landlord Isn't All Peaches," *Building Management* presented a classic industry statement on children: "Children? Well, that's a problem. Got any kids in your flat. Then you know something about it, or you should. Personally, I'm fond of children. I can't afford to have children in my flats."[61] Edwin S. Jewell, an Omaha apartment specialist repeated this point just as bluntly in 1935. He did not want "young children to mar the woodwork, soil the wallpaper, dig up the yard and litter up the halls."[62]

Table 9.11
Mean Household Size for Sample of Hamilton Apartment Dwellers
(1924-1969) and for Entire City (1921-1971)

Mean Household Size of Apartment Dwellers		Mean Household Size for Entire City	
		3.9	1921
2.4	1924		
2.4	1929	4.4	1931
2.5	1934		
2.2	1939	4.2	1941
2.3	1944		
2.2	1949	3.8	1951
2.1	1954		
2.0	1959	3.7	1961
2.2	1964		
2.4	1969	2.7	1971

Sources: Hamilton Assessment Rolls; John C. Weaver, *Hamilton: An Illustrated History* (Toronto: James Lorimer 1982), table 4, 197

Demographic trends describe important aspects of apartment-based social history, yet the industry's early campaigns to nurture respectability and quality, to elevate the prestige of apartment living to a cut above the commonplace, imply an occupational or status dimension to the prospective clientele. "Emphasize that you discriminate," advised Alvin Lovingood in 1929, "and you will impress the applicant with the desirability of the house."[63] Social homogeneity and respectable tenants (no prostitutes or gangsters, warned a Chicago writer) bolstered the social airs requisite for the maintenance of good rental income. The broad, bold strokes of generalization for the entire sweep of the sample data, from 1924 to 1969, portray a population of essentially middle-class households. Hamilton was a heavy-industry city throughout the period of study and the occupational traits of apartment household heads contrast with those of the workforce at large (see table 9.12). Whereas leading occupations among all city residents embraced sets of skilled and semi-skilled metal workers, among apartment residents the leading "occupations" recorded were, in descending order: widows, salesmen, clerks, spinsters, and gentlemen (see table 9.13).

General characteristics submerge the individual building and its distinctive community of tenants. Not surprisingly, buildings with a high assessed value per unit housed a disproportionate number of higher-status occupations. Furthermore, the few sample buildings erected during and immediately after World War II, a time of

Table 9.12
The Occupational Profile of Apartment Dwellers in Hamilton Compared with the Population at Large during the Initial Period of Apartment Building Construction

	Group 1¹ 1926	Group 2 1926	Group 3 1929	Group 1 1931	Group 2 1931	Group 3 1934	Group 1 1936	Group 2 1936	Group 3 1939
Professionals and rentier	3.3	3.9	**7.1**	3.2	**5.0**	**5.5**	4.4	4.4	**6.3**
Agents and merchants	7.5	7.9	8.2	6.4	6.3	6.1	5.7	4.4	5.3
Service and semi-professional	0.8	0.0	3.0	0.6	0.0	2.4	0.7	0.6	3.4
Business employees	12.5	**19.7²**	**27.5**	13.6	**26.3**	**27.0**	14.7	**26.7**	**32.6**
Government employees	2.9	3.9	4.8	3.2	3.8	3.0	3.1	3.3	2.5
Masters and manufacturers	0.7	1.3	2.2	0.8	1.3	2.4	0.7	1.7	2.2
Skilled labour	36.0	26.3	13.4	37.0	25.0	10.0	34.0	23.9	12.9
Transportation employees	5.8	0.0	0.1	5.5	2.5	0.1	4.9	3.3	1.9
Other workers	4.8	3.9	0.3	5.6	6.3	5.2	5.4	7.8	4.7
Labour	13.2	6.6	0.0	13.5	2.5	0.4	14.5	3.3	3.1
Female domestics	0.0	0.0	0.0	0.0	0.0	0.0	0.1	0.0	0.0
Female other	0.2	0.0	0.1	0.3	2.5	3.6	0.5	3.9	5.0
Agricultural proprietors	0.5	0.0	0.4	0.7	0.0	0.3	0.4	0.0	0.0
Other	1.0	6.9	2.2	0.9	0.0	1.5	1.0	1.1	0.6
Retired, widows, widowers	10.9	**23.7**	**21.9**	8.7	**18.8**	**28.8**	10.3	**15.6**	**19.4**
Total	100.0	100.0	100.0	100.0	100.0	100.0	100.0	100.0	100.0

¹Group 1 consists of all household heads for the entire city-wide sample drawn as part of the broader research project on the history of housing: N = 2,953 (1926); 3,383 (1931); 3,533 (1936).

Group 2 consists of apartment-dwelling household heads extracted from the city-wide sample. Apartments in this sample were defined as residential units in exclusively residential buildings having 3 or more units. The architectural quality and general tone is likely to have been lower than the apartment-building sample selected from the city directories: N = 76 (1926); 80 (1931); 180 (1936).

Group 3 consists of household heads living in the separate sample of apartment buildings drawn from the city directories: N = 269 (1929); 330 (1934); 319 (1939).

²Bold figures highlight over-representation of apartment dwellers.

Table 9.13
The Most Frequently Cited Labels Appearing under the Heading Occupation in the Assessment Rolls: Households in the Apartment Sample, 1924-1969 Inclusive

Label	Absolute Number	Percentage	Cumulative Percentage
Widow	350	8.0	8.0
Salesman	248	5.6	13.6
Clerk	238	5.4	19.0
Spinster	195	4.4	23.5
Gentleman	133	3.0	30.4
Steelworker	105	2.4	32.8

Source: Hamilton Assessment Rolls

materials shortages, perpetually housed much smaller households than other sample structures. Throughout the period covered by the study, the more compact units sheltered households with a mean of around 1.5. Nevertheless, the range of variation among sample buildings was never so great as to shatter the impression of small households drawn from the middle class.

Studies of immigrants and housing in other North American cities have demonstrated the apparent aversion to rental accommodations displayed by certain groups of Europeans.[64] We have observed that home ownership was an important goal for many Italian and Polish newcomers. There is also evidence that they actively avoided apartment living. Household heads identified as of Italian ethnic origin accounted for between 1 per cent and 2 per cent of the apartment sample in the 1950s and 1960s whereas their numbers in the city as a whole reached 4 per cent to 6 per cent. Largely from small towns and poorer rural areas, the post-war Italian immigrants were unfamiliar with and ill-disposed toward accommodations that could not house an extended family. Moreover, Italians tended to fill the labour positions under-represented among residents of middle-class apartments. Whether apartment managers, striving to generate respectability, welcomed or discouraged immigrants from the poorer regions of Italy is another question. Whatever the cause, there appears to have been a social boundary between such immigrants and established residents, and apartment shelter marked part of the limits.

Not all immigrant groups avoided or were steered away from apartment living. Germans, especially, figured as an over-represented ethnic group among apartment dwellers. For most cross sections from 1929 to 1969, they made up from 4 per cent to 5 per cent of the apartment sample, when their numbers in the city ranged from 1 per cent to 2 per cent. Germany not only had a proportionately large urban population – far more so than Italy – but it also had a tradition of apartment living that, like urban France, was certainly stronger than in England – another highly urbanized country – where row housing had been favoured.[65] It is reasonable to assume that culture predisposed German newcomers to accept apartment living.

THE APARTMENT BUILDING AS MIRROR TO URBAN CULTURE

An analytic history of the North American apartment building frames the following themes. First, as a business, the apartment industry grappled with economic crises. Competition, overproduction, and rent cutting reached levels dangerous to the many operators. Especially during the 1930s, assorted forms of rationalization and regulation came into use. The structure of the industry altered, moving away from an uncoordinated independence exercised by sophisticated investors who, by the 1920s, recognized axioms guiding site selection and architecture. They were forced to realize that, no matter how sound their individual judgments on matters of site and design, they were interdependent. A variety of institutional arrangements developed in the 1930s to check excessive construction or to warn away rash builders and investors. Also, as a further expression of the structural move away from individualism, corporate involvement began.

Second, in the search for legitimacy and in campaigns to secure and hold tenants, the apartment promoters adopted a view of rental income that showed an appreciation for psychological factors. They undoubtedly knew that they were not merely in the business of providing convenient accommodation. The industry had accumulated lore about linking image and rents. Site location and architecture certainly involved utilitarian considerations, but by the time of the great wave of construction in the 1920s material elements of location and form were also the consequences of psychological calculations. These psychological components always presented themselves overtly as very important determinants in an apartment's place, shape, and advertised attributes. However, the precise psychological notes being sounded changed over time and demarcate a definite periodization. Sedate respectability was the initial mood, sought during the heyday of the 1920s; economy and efficiency rose to prominence in the 1940s and 1950s; and youth and recreation directed design in the 1960s. Clearly, tenants consumed more than space, but to what degree had tastes been directly manipulated by the property industry, indirectly manipulated by a commercialized mass culture, or developed in consequence of demographic patterns and economic circumstances? Whatever the mix of factors, the importance of the intangibles of "quality" remains central to the choice of site and architecture.

Third, although marketing practices shifted over the decades, the middle class remained prominent in the plans of owners and managers. The true apartment building – not yesterday's tenement, not today's public housing project – has had a fairly consistent association with the middle class. Social segregation by housing stands as one expression of urban inequality. Yet such separation viewed from a high perch on the social ladder conveys distinction. Only the poor and socially conscious might have serious cause to dispute the notion that what keeps North Americans together is their ability to be separate. Again it can be stressed that residents of efficiency apartments secured more than confined and costly accommodation. Millions ultimately reached the goal of home ownership in fairly homogeneous neighbourhoods. Apartments provided what they may have felt were appropriate interim situations. The shelter industry was not just exploiting and manipulating, rather it was engaged in a two-way exchange where, concerning site and design at least, it paid heed to mass tastes, especially those of the substantial middle class. The NorthAmerican shelter industry has succeeded brilliantly in giving many families – possibly even most – what they have wanted. This makes the situation of "forever doomed outsiders" a well of frustration. Social-housing remedies have never fully succeeded in erasing a social dilemma in housing. The aspiring masses and the property industry, intrinsic to the sinews of culture and economy in urban North America, cannot part with a segregated city. Social housing in a segregated city never empties the well of frustration.

Fourth, while holding to a class course, the apartment building has also had a partial association with the greater freedom of North American youth in the twentieth century. In her account of American youth in the 1920s, Paula Fass has chronicled the loosening of ties that bound young people to the parental household. Urban families by the 1920s usually did not have a family work regime and elaborate household routines involving parents and children. There was altogether less interaction and a large measure of personal independence granted to urban children, who were given time and occasions to engage in extensive extra-family activities, particularly with peers. For youth so fully conditioned to independence by the 1920s, the apartment building offered an eventual extension of the relatively unfettered ways of modern urban family life. While the apartment building has housed a good many older folk, the youth market strongly affected the psychological appraisals made by the builders, owners, and managers when considering whether to build and how to build. It was in part for the young married couples, in the still morally restrained 1920s, that respectability had been burnished; it was largely for youth in the more free-wheeling 1960s that facilities for the breezy good life were assembled.

The apartment building stands as an artifact of a specialized business, responding to land-use economics, demographic projections, and hard-earned lessons accumulated during the phases of the business cycle. But that is not all – shelter is more than a functional form. In the relative richness of urban North America, in a context of an exceedingly style-conscious affluence, the single-family detached dwelling and home ownership have assumed tremendous cultural presence. It is not surprising, then, that the apartment building also expresses numerous North American cultural phenomena. It is possible to claim that vital traits of urban form are poorly explained by neo-classical economics and purely utilitarian calculations or neo-Marxist discussions of the urban question. These tools, often expressed in the forms of hypothetical models or hermeneutics, do not present more than superficial glances at the panorama of land and housing as they have been and continue to be. That is not to denigrate their value as intellectual exercises and as spurs to empirical inquiry, but to call for manifold historical reviews of the real interaction of capitalism, design skills, and social values. Such work, pioneered by Sam Bass Warner Jr and H. J. Dyos, still offers the best route to an understanding of how and why we have the shelter forms of today.

Reflections on very recent developments conclude this study of a particular housing form and tie it into the subjects and themes of the other chapters. Apartment buildings have been related to developments in housing more generally. They were a counterpart of the decline of the rental house, a point touched on in chapter 8. In terms of management and ownership, apartment buildings figure in the evolution of the shelter industry from individual to corporate enterprise, a process overtaking land assembly, a good share of current house construction, and mortgage lending. Moreover, the apartment building has also been affected by the era of state intervention. A 1987 city-planning study of high-rise apartment buildings in the central area of Hamilton reported ownership. The study defined apartment buildings as structures that have two or more residential units; it was not limited to high-rise buildings. Of the 46 structures that had more than 76 units, an individual held title to only one. The Mutual Life Assurance Company owned two. Most belonged to individually named or numbered companies; there was no concentration of ownership (see table 9.4). Among corporate bodies owning apartment buildings in 1987, but not in the last sample year of 1966, the Ontario Housing Corporation was a mammoth landlord. This government agency operated the largest apartment tower in central Hamilton (397 units) and four other large buildings in the same area for a total of 1,297 units. Government-operated housing in large concentrations, rising and falling in favour first in the United States and then in Canada, was assembled in Hamilton essentially between 1965 and 1975. Low-rise and townhouse developments have subsequently been preferred for public-housing units. However, the interventionist state continued to function powerfully in the apartment market. Ontario, as noted earlier, brought in rent controls in 1975, following a common paradigm for state and private sector entanglement.

Transportation history offers the usual examples of North American aid to and regulation of private enterprises, but the genius for such activities in the housing sector cannot be underestimated, as the last fifty years of mortgage history demonstrate. Coaxing, inducing, and restraining the private sector have fostered research and administrative specialties within state bureaucracies. In the early 1980s, in both the United States and Canada, the genius for indirect operations had a strenuous workout in the mortgage field. Rental housing promises to be just as challenging. The large apartment building, as a housing form, has begun to receive slightly more favourable attention after a decade of general criticism. The issues of revitalizing the stock of apartment buildings and stimulating construction in the rental market have been added to the questions of architectural form. Coincident with rent controls, building trailed off. Insurance companies, for example, started to devote greater attention to commercial property. A number of apartment buildings have been converted to condominiums, and vacancy rates have shrunk to almost invisible levels. The standard vacancy rate in Hamilton, prior to rent controls, had been around 5 to 7 per cent. After controls, it plummeted to below 1 per cent.[66] Middle-class demand for apartment buildings in highly desirable locations could grow as young families discover that they, unlike their parents, must defer home ownership. It is understandable, therefore, that the Ontario Ministry of Housing became very active in the late 1980s, encouraging municipalities to compile information on rental accommodation within their boundaries and to assist them with the preparation of local-housing policy statements. To help to alleviate the crisis of low vacancy rates, which its rent controls may have fostered, the province sought to preserve older rental stock. It initiated a low-rise rehabilitation program to upgrade 17,000 apartments from 1987 to 1990 and began a pilot high-rise rehabilitation scheme in metropolitan Hamilton, Ottawa, and Toronto.[67]

Government intervention has appended an imposing layer of immediate activities onto housing history. Political in origin, they come and go quickly. Programs are terminated and replaced. Many have sunset dates. Therefore, state activity has not only affected current shelter provision, it has also subtly influenced our culture's thinking about time, and perhaps led to a focus on the immediate. Prior to the era of government intervention, capitalist evolution pulled housing development, as it pulled much else in the city. Since capitalism worked against the grain of tradition, many themes in housing from roughly the 1850s to the 1920s expressed the slow change that has been a major theme in this study. However, governments have forced the pace of changes in housing, influencing design, mortgage lending, and the provision of rental dwellings. Whereas, in writing about housing before the 1930s, there was a tempo of gradual change chronicled in trade journals and routinely generated sources, today the speed of many changes requires a close following of government press releases. The volume of contemporary government activity has troubled classical economists and annoyed the property industry operators – who nevertheless are happy to find many useful government supports. We cannot share their critical appraisals, because the history of private-sector apartment operations, replete with chapters on manipulative tactics and expectations of gross gain, forfeits sympathy. Government activity, in Ontario at least, has been conducted openly, has worked to protect tenants, and has tried to extend fairness to landlords. There are also many arguments for greater state intervention. Our concerns about the state and housing are not so ambitious. We wonder about the character and quality of discussion and analysis of housing questions in an atmosphere bursting with political events. What will constitute the reference material for journalists, planners, civil servants, and politicians who interest themselves in shelter? The vigorous activities of governments during the last twenty or more years have created intergovernmental agreements, short-term and continuing policies, new ministries, and housing authorities. Against the immediacy and volume of a flood of necessary political information, it is difficult to advance the case for historical inquiry and study. However, without history's long view there can be little intelligent discussion about housing needs, the cultural or psychological burdens that have shaped these needs, the functioning of capitalism in shelter matters, and the quality of housing for different groups over time.

10 Housing Quality and Urban Environments, 1830s–1980s

A qualified thesis of democratization in shelter has crept, not haphazardly, into the preceding chapters. It is at the crux of the account of the will to possess; it is a conclusion to be drawn from the vigorous and sometimes creative architectural concern with the common house, amply depicted in late-nineteenth-century building journals. That part of the history of the apartment building chronicling how it offered some youths, and seniors, a specialized form of shelter intimates a change that freed a good many people at appropriate stages in the life cycle from residential dependency on parents or adult children, from the probably greater costs of house rental, or from house maintenance. Housing history, overlapping with a bounty of social questions, also has its grim chapters. Moreover, not all criticism should be reserved for the past.

In the mid-1980s, social activists saw in the consequences of youth unemployment, the de-institutionalization of mental health patients, and the decline of boarding houses the phenomenon of homelessness, culminating in 1987 with the United Nations' International Year of Shelter for the Homeless. Other critical observations have focused on the burden borne mainly by young families that have bought houses. Furthermore, whether affordable or not, the purchase of housing through long-term financing implies a debtor status somewhat at odds with the idea that increasing home ownership represents a democratization of opportunity. Therefore, the thesis of progress must be qualified, and treated warily. However, no attempt to tally up the costs and benefits of the shelter situation – to praise or criticize North American housing – can claim credibility by looking at parts of the picture, whether those parts be hostels or upper middle-class enclaves. The quality of housing must be assayed for all regions of a city, across many decades.

Home-ownership levels may go a long way toward advancing the theme of democratization, but they would be credible elements in such an argument only if the character of the city's shelter improved. Home ownership, as *Shaky Palaces*, a recent Boston-based study, maintained, may be a poor investment for blue-collar families. However, the Boston exercise insisted too emphatically on but one feature of housing demand – housing as an investment. Buying could have achieved an efficacious balance of goals: retaining equity, meeting culturally important psychological needs, and providing – vis-a-vis rental shelter – the better shelter option in terms of household needs for certain age groups. *Shaky Palaces* failed to evaluate the quality of owner-occupied housing versus that of rental shelter.

For many basic reasons, then, a fundamental question must be raised. How well have the city dwellers in Hamilton – or anywhere – been sheltered? To move intelligently into this topic requires confronting an assortment of formidable issues, several of which should be noted immediately to stress the limitations in existing literature and the philosophical and methodological challenges to be encountered in evaluating housing data historically. The methodological aspects of the following account may be as instructive as the conclusions. By working with a particular data set and asking a range of questions, we hope to encourage comparable and ultimately better studies.

Canadian inquiries into shelter lagged behind those in the United States where, in larger cities, housing reformers had been active in the 1870s and 1880s. In both countries, journalists and medical health officers in the late nineteenth-century described and publicized some of the continent's worst housing conditions. The major Canadian example of the exposé genre, Herbert Ames' classic, *The City Below the Hill* (1897), calibrated overcrowding and unsanitary conditions for a working-class area of Montreal.[1] The St Henri district, the city of Montreal itself, and the 1890s, are specific historical contexts that make Ames's study suspect as a representative example of Canadian working-class housing in an industrial city. As Montreal geographer David Hanna has discovered, labourers' housing in Montreal started to run on its own peculiar course of high-density multiple-family units in the 1850s, probably following examples of railway company housing, which in turn had been derived from a high-density form used in Newcastle-upon-Tyne.[2]

By World War I, municipal reformers and social workers throughout North America had probed into overcrowding in slums and
immigrant neighbourhoods, graphing and mapping pathological dimensions of the city. Reformers left a body of sensational literature.
A 1912 Hamilton investigation, for example, divulged examples of
overcrowded boarding houses occupied by sojourning industrial
workers from southern Europe. Out of 221 houses examined by a
special team from the health department, 65 were judged greatly
overcrowded and, continued the report with an air of reproach about
"foreigners," in one house of 7 rooms there were 45 male boarders.[3]
Drawing back from unglamorous details, city boosters sketched the
city from scenic outlooks. The writer of a 1918 description pointed
out the qualities of Hamilton from the mountain. The factories announced the Ambitious City's economic potential. The public buildings, acres of parks, well-paved streets, and beauty of bay and beach
indicated amenities. "But probably the greatest impression left with
the visitor is of the sweeping panorama of roofs of homes – the
homes of the people who make Hamilton the splendid city she is."[4]
Neither the account of cramped squalor nor the one of domestic
amplitude were false, but they were incomplete when taken alone
and without standards or context. Yet, even after a statistical analysis
of the whole city, the two realities will have confirmed status: poorer
labourers persistently occupied dreadful and life-shortening environments while the majority had very good shelter by any standard.

By the 1930s social workers had established standards for shelter
and had refined survey methods in order to report more dispas-
sionately, more scientifically, on housing conditions. Nevertheless,
traces of subjectivity lingered. Examinations of baleful dwellings in
dreadful times, coming as social science supports to prescriptive
efforts, command respect for what their authors were attempting to
achieve in detailed reporting on circumstances in limited areas and
for what they were hoping to realize, namely raising living conditions
among the poor. By the 1950s, at least in Hamilton, social service
agencies had begun to commission housing surveys that drew samples from a city's entire housing stock and by the 1960s there was a
rich North American literature on how to evaluate housing conditions.[5] At every step, the Canadian research followed the American
closely. Some agencies recommended that investigators examine the
dwellings; others relied upon census data. Questionnaires were considered problematic, although a 1955 Hamilton study had an 89 per
cent response rate and yielded significant information pertaining to
the will to possess as well as about housing conditions.[6]

Until the post-depression flood of investigations, housing reports were not normally meant to chronicle how working families, let alone most urban dwellers, lived, for they were deliberately localized. They assembled details from door-to-door surveys undertaken in a very few blocks chosen for their visibly neglected state. Study results did not offer glimpses into broadly-based urban settings, but they included details of a kind unavailable in our research. They captured particulars, but they lacked breadth and they presented no sense of how occupants of overcrowded districts coped. Reform inquiries did not report on what people did with their shelter to personalize it and make it comfortable in spite of its limitations and their limited personal resources. Regrettably, we shall always know too little about how families endeavoured to fashion their shelter into homes. So it must be recognized that the following struggle to reconstruct the quality of shelter itself and of its vicinity, from sources never meant to yield precisely such information, will fall short in some ways. But it will offer a basis for inter-city, intra-city, and temporal comparisons.

It is necessary to set out rough criteria for decent housing. Subsequently and by no means less controversially, the reconstruction of Hamilton's housing conditions will be inferred from assessment values. Then the yardstick and the interpolations of Hamilton's housing conditions can be brought together. One imperfect measure of shelter adequacy is that there should be more than one room per household member with bathrooms excluded. More refined standards have recommended a separate kitchen and sufficient bedrooms to provide for the appropriate separation of the sexes. Unfortunately, these more specific requirements are difficult to report on and rarely appear in housing statistics. The one room per person standard has had considerable application. Obviously it has drawbacks, failing, for example, to take into account size of rooms, lighting, ventilation, and heating. The Dominion Bureau of Statistics accepted the criterion in an important 1941 monograph on Canadian housing. The author cited the 1934 Real Property Inventory of the United States and a study of European housing conditions published by the International Labour Office as authorities for a rooms-per-person standard. Actually, the International Labour Office made the American standard of one room per person appear astonishingly liberal by considering two persons per room as adequate.[7] In the 1950s, when European immigration flowed abundantly into Hamilton, only half the countries of Europe had attained a mean better than one room per person. Italy's mean of 0.8 may add context to the Italian-Canadian embrace of home ownership. There are further possible implications. The overcrowding in the old world might have permitted newcomers to accept larger than usual household size in Canada. Larger households composed of boarders and adult family members might have helped with capital accumulation and home buying.

At a certain point, water and sewer connections came to define minimum standards of sanitation. The challenge is to fix a date. The English sanitary movement originated in the 1840s, but it would be unrealistic to expect a European or North American city before 1900 to have set a target of complete sanitary service connections. Large-scale concrete sewer tiling was perfected only around 1900. It greatly reduced sewer costs since formerly, excepting cheap wooden sewers that caused many later difficulties, they had been made of brick. Only in 1901 had Montreal made enclosed sewers obligatory.[8] Therefore, an extensive sewer network was something that a North American city could have and should have built or have begun to build by 1910. Other inquiries designed to gauge housing quality come to mind. The mean values of houses lived in by several occupational groups may be contrasted and changes across the decades detected. The existence of a 1936 housing survey, stating criteria for the determination of whether a dwelling was deemed satisfactory, has permitted a linking of our sample blocks with notions about satisfactory shelter. The estimation of housing quality is discussed in more detail later in this chapter.

CONTINUITIES AND LONG-TERM TRENDS TOWARD THE MEAN

The only path to reconstructing the nature of housing throughout the city at different times lies through the thickets of assessment rolls. Assessed values have been calculated in different ways during the history of Hamilton. From the town's founding until 1849, a flat rate of assessment applied to particular types of structures. There were eight categories. From 1850 to 1865 assessed value was supposed to have been established as an annual rack rent, fixed at 6 per cent of a property's value. Beginning in 1866, assessed value was to be estimated at the actual value of a property, as that cash value would have been appraised in payment of a just debt from a solvent debtor.[9] This awkward formula did not translate into market value. From 1868 to the 1930s assessed values ran roughly 20–50 per cent below market value, but after World War II market values soared beyond assessed values. Assessors who worked to revise the assessment rolls annually each summer used rules of thumb that took into account the number of rooms, type of construction, age and condition of the structure, and the size and location of the lot. During World War II, when municipal expenditures were curtailed, the local government still obtained sufficient tax revenues without reassessments. Consequently assessed values were permitted to lag behind current market values. Until 1964 the Ontario Ministry of Municipal Affairs officially retained 1940 as a base year for valuation purposes.[10] In Hamilton, during the early 1960s, assessed values on residential property were approximately a third of sale values. The 1940 base remains in place in many municipalities to the present, but some localities switched to market value assessment in the 1980s.

There are two straightforward and analytically useful methods to relate assessed values to deliberations about the character or quality of housing. Assessments can be indexed to the mean thereby providing a basis for discussing the spreads in indexed values. For example, to discover that in 1876 the indexed mean assessment for all dwellings inhabited by professionals and proprietors was 216.8 while that for common labourers was a mere 53.8 is to make a blunt observation about inequality, though not necessarily about the absolute quality of shelter for common labourers.

In order to focus on quality, a number of houses in each sample year from 1876 forward were found in classified advertisements in the *Hamilton Spectator*. If the advertisements had both a simple description and an address, they could further the inquiry because they could be linked to their entries in the rolls and thus to assessed values. Knowing the 1896 assessed values for a four-room frame cottage, a six-room brick house, a nine-room stone house, and so forth, permitted analysis of shelter occupied by different occupational groups. In the tables presenting the findings of this linkage exercise, the variety of housing forms has been compressed to four or five for the sake of brevity.

Imperfections flaw the linkage and compression efforts. Advertisements doubtless overstated size and quality, making estimates of housing stock slightly better than they really were. Sunrooms, summer kitchens, attic bedrooms, an original room partitioned to make two, or other considerations could have increased the advertised number of rooms, making such houses inflated but not greatly different from less modified versions of their prototypes. A further complication concerns apartments. Until the 1950s there were too few newspaper descriptions to allow apartment types to be linked with assessed values. However, it is safe to assume from abundant surviving samples that the majority consisted of four to five rooms.

Table 10.1
Comparison of Assessed Values of Owner-Occupied and Rental Housing,
1876-1966

Year	All Dwellings	Mean Assessed Values Expressed and Index Number for						
		Owned Residences Houses/Flats		Rented Residences Houses/Flats		Apartments	Above Shops	All rental
1876	100.0	113.6				89.6		
1886	100.0	126.6				84.1		
1896	100.0	128.1				85.9		
1906	100.0	128.8				82.2		
1916	100.0	125.5				81.2		
1926	100.0	128.8	81.5	84.8	103.0	59.3	89.2	87.1
1936	100.0	114.4	81.4	96.3	98.7	58.2	79.7	90.5
–	–	–	–	–	–	–	–	–
1956	100.0	119.9	61.1	88.0	61.1	78.1	78.7	80.0
1966	100.0	116.6	78.3	96.5	61.0	70.3	66.9	76.1
All	100.0	120.7		87.6				

Source: Hamilton Assessment Rolls

The survival of many examples of housing stock from the mid-nineteenth century and subsequent building periods have made it possible to check statistically derived observations with field work.

Before linking mean values and house examples to illustrate the character of housing for occupational groups, it is instructive to scan indexed mean values to explore the shelter inequalities in the city. Many questions come to mind. Have the indexed differences in shelter among various social groups altered in significant ways over nearly a century? Have there been glaring contrasts between the indexed values of rental shelter and owner-occupied dwellings? This last question should be addressed immediately, because of its association with the previous chapters on home ownership and tenancy. The will to possess will be more fully understood when the spread between the indexed values is considered, for the spread implies differences in scale and amenities. In other words, it is likely that more than psychological urges for independence, investment considerations, promotions of the property industry, and rent charges have encouraged families to secure their own houses (see table 10.1).

From 1876 to 1966, the indexed values for rented houses expressed a quite consistent ratio in relation to owner-occupied residences. The 25 to 35 per cent spread between the two reaffirms the advantage of looking at the city over the long term, for what the continuity testifies to is the endurance of a city's housing stock and the persistence of investors' strategy. To clarify these two elements, consider the housing market in 1906. The city had begun to attract new factories and to benefit from the expansion of existing plants. The prairie settlement and railway construction booms stimulated demand for labour in the industrial heartland. Housing became scarce. Over the next several years, the farm implements giant, International Harvester, constructed some company housing. The 2½-storey dwellings that it let out to skilled labourers were exemplary new housing for the period, but its new houses and others that found their way into the rental market could not alter the basic character of the enormous accumulated stock of dwellings for rent. The large property holders – the Mills come to mind – had constructed an assortment of small rental cottages beginning in the 1830s and had followed up with similarly modest cottages and row houses in the growth years of the early 1850s.

Meanwhile, another wellspring of rental units had added to the accumulation. During the depressed times of the 1860s and 1890s, a number of owner-occupants pulled up stakes and sought employment elsewhere. Their homes could not be sold readily in depressed times. Investors might pick up some, but they selected the cheapest homes, which had some prospect of a healthy return despite very deflated rents. Often the former occupants became absentee landlords. Both deliberate investment and construction as well as misfortune provided the city with a stock of rental houses that was too large to be altered suddenly.

On the other side of the tenant-owner divide, the stock of owner-occupied homes remained relatively large and distinctive. Investment caution and the accumulated housing stock established an equilibrium clearly expressed in the 1906 index values of 128.8 for owner-occupied houses and 82.2 for tenant-occupied houses. Rental houses were typically assessed at values that were 70 per cent of those for owner-occupied houses. Housing economist James R. Follain, using American data from the 1970s, found that renter-occupied and owner-occupied housing expressed distinct features, suggesting that the demographic characteristics of rental households tend to be apart from those of home owners. The mean number of rooms for rental units was about 70 per cent of that for owner-occupied dwellings (see table 10.2). The continuity in the Hamilton data and its coincidence with an appraisal of recent conditions in the United States call attention once again to persistent North American features in housing history.

Table 10.2
Comparison of Typical Renter-Occupied and Owner-Occupied Dwellings in the
United States, 1978

	Renter-Occupied	Owner-Occupied
Number of Bathrooms	1.1	1.58
Number of Rooms	4.1	6.16
Number of Bedrooms	1.8	3.1
Age of Dwelling	17.7	17.5
Length of Tenure	4.91	12.08

Source: James R. Follain Jr, "Cross-Sectional Indexes of the Price of Housing," draft cited in
Masaka Darrough, "The Treatment of Housing in a Cost-of-Living Index: Rental Equivalence
and User Cost," *Price Level Measurement: Proceedings from a Conference Sponsored by Statistics Canada,*
edited by W.E. Diewert and C. Montmarquette (Ottawa: Minister of Supply and Services, 1983),
616.

Departures from the equilibrium herald important disruptions.
In the 1936 and 1966 samples, the spread narrowed to under 20 per
cent. Ironically, the great depression temporarily generated a stock
of superior rental housing. The mean index value for rental units
shot up to 96.3. Obviously, this is not the picture of housing in the
depression that ordinarily comes to mind. Across Canada, during
the depression, urban housing studies depicted the deterioration of
inner-city low-rent districts. Slums became worse, and repairs were
deferred on all manner of dwellings – owned and rented. These
things happened. Their impact on Hamilton will be measured
shortly.

Nevertheless, rental bargains poured onto the market, though
their origins in heartbreaking loss return the account of shelter in
the depression to the annals of capitalism's most universally accepted
failure to date. The housing industry had overbuilt during the late
1920s and could not readily dispose of its inventory in the 1930s.
Concurrently, a few owners in the usually secure proprietor and
professional, white-collar, and skilled and semi-skilled labour groups
lost their homes, thereby swelling the rental market with better
houses than those usually available. If some households found themselves renting rather comfortable homes, they likely failed to see any
silver lining. Conceivably they had had to postpone home buying
which, for better or for worse, was loaded with great psychological
value.

The narrowing of the spread of indexed values between rented
and owned houses in 1966 occurred for distinctive reasons also, only
this time the phenomenon might have heralded a new structural

feature of the rental market. Urban renewal projects reduced the stock of older housing, a mix of owned and rented dwellings. The massive development of new suburbs and modest suburban homes held down the indexed value of owner-occupied houses. While the number of rental dwellings had continued to grow, the bulldozed units of cheap rental houses were not offset by the accumulation of other houses, but by apartments, public housing, and flats. The increasingly scarce rental house now lost its historic association with small cottages and houses, as these were subjected to demolition or restoration. Moore and Davis, the real estate and property management firm introduced in previous chapters, abandoned the management of house rental accounts in the mid-1970s. In sum, the rise in the indexed value of rented houses was a counterpart to a trend away from renting houses as an important source of investment income.

The historical spread between the assessed value of an owner-occupied house and a rented dwelling was maintained even in 1966, but now by the introduction of other rental forms. Ownership has always offered a direct route to better housing, insofar as a 25 or 35 per cent gap states something about size and structural quality. Against the backdrops of the culturally powered will to possess and the entrenched modesty of rental shelter, the rise of home ownership or home buying for all occupational groups bolsters a theme of social improvement. Undoubtedly, young families since the 1950s have had to carry larger debt burdens to become home buyers, but it is arguable that, in terms of a culturally supported notion of self-esteem and a set of real benefits bestowed in consequence of limitations in the rental market, the debts and service costs have secured things strongly desired if dearly paid for. And, if household heads have been carrying higher debt loads, the mortgage market, through lower down-payment requirements, had enabled earlier home ownership until the 1970s when rising prices took their toll. Thereafter, ownership has been exposed to stress.

In view of the preceding claims about the goals of tenants, it is fortunate that a 1955 Hamilton housing survey reported on an unusual finding. Researchers who had interviewed a sample of 1,000 households discovered that tenant families with better incomes had been more inclined than had lower income families to describe their current shelter as unsatisfactory. The reason for "this curious thing" appeared to be that most of the tenants in this more prosperous group wanted to own houses rather than to remain as tenants. In fact, three quarters of those who complained stated that their wish to become home owners was the reason for their dissatisfaction.[11]

Table 10.3
Breakdown by Occupational Group of Mean Assessed Values for All Housing
Forms Expressed as Index Numbers, 1876-1966

Year	Dwellings of All Groups	Professionals and Proprietors	White-Collar	Skilled and Semi-Skilled	Labour
1876	100.0	216.8	136.9	85.4	53.8
1886	100.0	186.4	127.9	75.9	55.1
1896	100.0	172.9	139.9	83.2	55.4
1906	100.0	181.5	144.6	86.4	53.7
1916	100.0	156.8	122.2	88.0	61.4
1926	100.0	150.0	126.9	87.0	63.4
1936	100.0	149.3	122.1	86.6	68.7
1956	100.0	120.5	113.5	95.3	77.1
1966	100.0	110.2	114.8	98.4	80.3
All	100.0	160.5	127.6	87.4	63.2

Source: Hamilton Assessment Rolls

The diffusion of ownership and the related psychological and tangible gains provide firm props for a claim of social progress within the capitalist city. The indexed assessment values, the linking of actual assessment values with existing housing examples, and an analysis of amenities fortify the hypothesis of improvement – the notion of a democratization of opportunity. When considered over time in relation to the major occupational groups used in the study, indexed assessment values call attention to a minimization of extremes, a long-term tendency toward the mean. From 1876 to 1966 the housing of white-collar groups drew closer to the mean and so did that of all blue-collar groups (see table 10.3). Owner-occupied dwellings and rented shelter of all types are included in this appraisal. Some of the trend toward the mean is to be explained by the rise of home ownership. It helps to account for the gains of the blue-collar groups, though not for the slide of mean values for the white-collar groups. What had happened to the latter? For the proprietors and professionals, the answer may well be found in the socio-economic and cultural distinctions between merchants in 1876 and those in 1966.

The city of 1876 had its grand retailers and wholesalers, its merchant princes. This was not the case in 1966. The ostentatious mansions of the merchant princes had had a social role. For the men, they served as places for cultivating prestige and respect, for hosting political, militia, and business dinners. Of course, men and women had separate domains: for men, the library and billiards room accommodated those male gatherings known as "smokers"; for women, the drawing room served. The elaborate social apparatus of the elite house depended on servants. The North American servant shortage of the early twentieth century forced residential compression.[12] Concurrently, the appearance of city clubs and golf and country clubs facilitated the corresponding shrinkage of the elite dwellings, for these places now provided the social space for the transaction of business. The slight fading of the index values for the white-collar employees is perplexing because their level of home ownership had been rising. It seems to have been a consequence of a change in taste and supply in the rental market. Possibly, until the early twentieth century, many members of this group sought social respectability by renting large homes beyond their purchasing power. As explained in chapter 9, in the 1920s many members of this occupational group could establish an acceptable address at one of the early apartment buildings designed for white-collar tenants.

HOUSING TYPES, HOUSING DENSITY, AND URBAN AMENITIES

The detour through examinations of indexed values has clarified further the attraction of home ownership and intimated an evolution towards the mean values of housing, for all occupational groups. In order to appraise the character of housing assessed values must be considered in a different light. Attempting to link houses in the assessment rolls with those clearly described in newspaper advertisements proved difficult. A scarcity of real-estate advertisements prior to the 1870s truncated the study. Even when they began to appear in Hamilton in the late mid-nineteenth century, they often failed to provide a precise address, inviting prospective clients to drop by the realtor's office or requesting them to pick up a key from a neighbour. However, when an exact link could be made, the assessment rolls themselves sometimes confirmed the connection, indicating that the property in question stood vacant. Selecting first from 1876 and then every ten years thereafter, we located ten houses of varying scale. Their assessed values in the year of their selection and, frequently, for two subsequent cross-sections were recorded. There were twenty to thirty examples in each cross section, from which four or five housing types were identified and a range of assessed values for each was adduced (see table 10.4). Compressed further into three clusters of housing types, the data furnished evidence of an expansion in the size of new housing during the 1920s, as reported at the end of chapter 5.

Table 10.4
The Approximate General Description of Houses according to Occupational Groups and the Common Occupations of Residents, 1876-1966

Year (Mean Value of all Houses)	General Description of Houses	Estimated Range of Assessed Values	Occupational Group[1] of Occupant (Mean Value for Group)	Common Occupations or Other Designations for Occupants (Mean Value for Occupation)
1876 ($950)	7 + Rooms Brick	Over $1,500	1 ($2,500)	Gentlemen ($2,350) Merchants (3,300)
	6 Rooms Brick	801 – 1,500	2 (1,500)	Clerks (1,000) Foremen (1,300)
	6 Rooms Frame	501 – 800	3 (900)	Machinists (650) Carpenters (750)
	Under 6 Frame	400 – 500	4 (500)	Labourers (500)
1886 (1,000)	7 + Rooms Brick	Over 1,500	1 (2,300) 2 (1,700)	Gentlemen (2,250) Clerks (2,300) Merchants (3,700) Foremen (1,100)
	6 Rooms Brick	1,101 – 1,500		Machinists (700) Carpenters (750) Moulders (850)
	6 Rooms Frame	601 – 1,100	3 (900)	Labourers (550)
	Under 6 Frame	400 – 600	4 (600)	
1896 (1,100)	7 + Rooms Brick	Over 1,500	1 (2,500) 2 (1,800)	Gentlemen (1,900)
	6 Rooms Brick	1,201 – 1,500		Clerks (1,500) Travellers (1,500)
	6 Rooms Frame	901 – 1,200	3 (950)	Machinists (1,100)
	Under 6 Frame	500 – 900	4 (700)	Labourers (700) Carpenters (750) Moulders (800)
1906 (850)	7 + Rooms Brick	Over 1,400	1 (1,900) 2 (1,500)	Gentlemen (1,000) Moulders (1,200)
	6 Rooms Brick	901 – 1,400		Machinists (650) Bricklayers (660) Carpenters (750)
	6 Rooms Frame	601 – 900	3 (900)	Labourers (500)
	Under 6 Frame	400 – 600	4 (500)	
1916 (1,200)	7 + Rooms Brick	Over 2,000	1 (2,200)	Merchants (2,100)
	6 Rooms Brick	1,401 – 2,000	2 (1,900)	Clerks (1,500) Salesmen (1,650) Gentlemen (1,800)
	6 Rooms Frame	801 – 1,400	3 (1,250)	Moulders (900) Machinists (950) Carpenters (1,000)
	Under 6 Frame	650 – 800	4 (750)	Labourers (750)

Year	House Type	Value Range	Class (Value)	Occupations
1926 (2,700)	7 + Rooms Brick	Over 3,000	1 (4,700) 2 (3,800)	Gentlemen (4,900)
	6 Rooms Brick	2,001 – 3,000	3 (2,550)	Moulders (2,100) Machinists (2,200) Carpenters (2,500)
	6 Rooms Frame	1,501 – 2,000	4 (1,850)	Labourers (1,850)
	Under 6 Brick	1,701 – 1,900	4 (1,850)	Labourers (1,850)
	Under 6 Frame	950 – 1,500		
1936 (2,500)	7 + Rooms Brick	Over 3,000	1 (4,400) 2 (3,350)	Gentlemen (4,150) Merchants (4,700)
	6-7 Rooms Brick	2,001 – 3,000	3 (2,400)	Machinists (2,000) Moulders (2,300) Clerks (2,250)
	6-7 Rooms Frame	1,400 – 2,000	4 (1,850)	Labourers (1,850) Steelworkers (2,000)
	Under 6 Brick	1,600 – 1,900	4 (1,850)	Labourers (1,850)
	Under 6 Frame	900 – 1,400		
1956 (3,000)	8 + Rooms	Over 6,500	1 (4,100) 2 (4,100) 3 (3,500)	Merchants (5,300) Travellers (5,300)
	7 Rooms	4,801 – 6,500	4 (2,850)	Machinists (3,400) Foremen (3,800) Salesmen (4,000)
	6 Rooms	3,301 – 4,800		Labourers (2,950) Bricklayers (3,000)
	5 Rooms	1,801 – 3,300		Carpenters (3,100)
	Under 5	900 – 1,800		
1966 (4,000)	8 + Rooms	Over 6,500	1 (9,200) 2 (4,500) 3 (3,900)	Merchants (5,100)
	7 Rooms	4,801 – 6,500	4 (3,350)	Clerks (3,800) Machinists (4,000) Steelworkers (4,100)
	6 Rooms	3,301 – 4,800		(6 Rooms Continued) Foremen (4,200)
	5 Rooms	1,801 – 3,300		Salesmen (4,700)
	Under 5	900 – 1,800		

Source: Hamilton Assessment Rolls

1 = Professional and Proprietor Group

2 = White-Collar

3 = Skilled and Semi-Skilled

4 = Labour

Table 10.4 summarizes who lived in the different categories of houses. Until 1926 the pattern declares the continuity of a particular hierarchy. Common labourers' households dwelt in the most miserable houses, but they seemed to have shared somewhat in the better housing stock arising from the 1920s construction boom. Paradoxically, this improvement in the stock of houses meant generally better accommodations for Hamilton households during the great depression than during a boom year such as 1906. The explanation overlaps with the earlier discussion about home ownership. A young migrant industrial labour force (1906) crammed itself into an unprepared city in the midst of a remarkable boom. A stable or even declining population (1936) found rental bargains amidst an oversupply of comparatively new larger houses, although the unemployed – the welfare recipients – had no possibility of access to the bargains. In the decades after World War II, there were improvements in the stock of houses and its distribution among occupational groups. These improvements were analogous to those of the 1920s.

By 1966 a further noteworthy levelling of house distinctions among occupational groups had occurred. In general, white-collar employees, steelworkers, machinists, carpenters, and even the always better-housed merchants occupied somewhat similar houses. This restates the diminishing spread in the indexed assessed values for houses. As a consequence of several factors, the spreads in the scale and quality of houses occupied by professionals and proprietors on the one hand and by semi-skilled and skilled labourers on the other had narrowed. The vanishing rental house had something to do with the trend. Some young blue-collar household heads in the 1956 and 1966 samples resided in flats and apartments and thus were not included in the appraisal of houses and occupants. Presumably, in earlier periods, they would have rented small houses and pulled down the mean. Household size too had been declining, alleviating the need for larger houses. In any event, the data describe a significant clustering of occupations in the six- or seven-room houses that constituted the bulk of the city's housing stock. We soon will estimate what this trend and the rise of the apartment building and flats have meant in terms of housing density calculated by rooms per person.

Despite limitations, the expression of housing density by either the number of rooms per person or its reciprocal, the number of persons per room, remains the common standard for international comparisons conducted by the European Economic Community and by the United Nations. Seldom routinely collected even as recently as the 1930s, information on rooms per person before 1900 rarely exists for any nation or city. Assessment rolls can provide an indirect means of reconstructing the number of rooms per person in Hamilton before 1900 as well as for later intercensal points. For most of our sample years, the assessment rolls reported on household size, serving as an annual civic census. Therefore, an estimate of the number of people residing in a dwelling is readily available. Undoubtedly, some boarding-house operators, fearful of public health inspectors, minimized household size. More generally, underenumeration is a greater likelihood than overenumeration. Although the data are not perfect and require careful handling, there is no compelling reason to ignore them. They can be sensibly and rewardingly interpreted.

Using the linkages between advertisements and assessments to estimate room counts proved a tricky exercise – ranges of assessed values were equated with ranges of numbers of rooms. Thus, for 1876, the assessment range up to $500 translated into dwellings five rooms and under; $501 to $1,500 into six to seven rooms; over $1,500 into more than seven rooms. Dividing the number of people residing in the dwellings of each assessment range by estimates of the total number of rooms in each range produced the quotients for the number of rooms per person.

The method has a weakness that hampers the formulation of a neat and definitive quotient summarizing the housing density in any given cross section. Unfortunately, there can be considerable play in the estimate of room numbers. Unable to make truly precise equations between rooms and assessments, there had to be recourse to ranges. Two of these have a wide latitude. How many dwellings assessed at under $500 contained three rooms rather than five? How many dwellings with seven or more rooms had seven and how many contained a dozen? Plausible worst and best case estimates on room numbers were made for the cheapest dwellings and all of the houses assessed at more than $1,500 were considered to hold just seven rooms (Table 10.5). The worst case estimates, skewed deliberately to yield high densities, suffer from an implausibility. For example, nearly half of the houses in the lowest range of assessed values were actually assessed at over $400. Few if any of these would have had merely three rooms.

Table 10.5 summarizes estimates on the number of rooms per person in Hamilton's houses at five cross-sections, each about a generation apart. Given that apartments formed a minor part of the city's stock of dwellings until the 1960s, these estimates deliver a reasonable picture of shelter conditions, embracing all important dwelling types: detached houses, duplexes, and row houses. A sixth cross-section (1974) incorporates apartment units into the picture.

Table 10.5
The Estimated Housing Density in Hamilton for Selected Cross Sections, 1876-1974

Description of House	1876 Estimated Rooms per Person (Percentage of Houses)	1906	1936	1966	Description of Dwelling¹	1974 Density: Rooms per Person (Percentage)
					All Dwelling Types	
5 ROOMS AND UNDER:						
If all were 3 rooms	0.72	0.68	0.73	0.77	3 ROOMS	2.54
If all were 4 rooms	0.95	0.91	0.97	1.02	Percentage of Housing Stock	(8.4)
If all were 5 rooms	1.19	1.13	1.22	1.28	4 ROOMS	1.87
Percentage of Houses	(47.2)	(48.0)	(18.6)	(35.3)	Percentage	(16.7)
6 ROOMS:	1.16	1.26	1.42	1.64	5 ROOMS	1.64
Percentage of Houses	(43.0)	(36.5)	(51.8)	(36.4)	Percentage	(22.4)
7 OR MORE ROOMS:					6 ROOMS	1.76
If all were 7 rooms	1.30	1.36	1.72	1.86	Percentage	(24.6)
Percentage of Houses	(9.8)	(15.5)	(29.6)	(28.3)	7 ROOMS	1.85
					Percentage	(15.4)
					8 AND MORE	2.03
					Percentage	(10.4)

Sources: Hamilton Assessment Rolls; Central Mortgage and Housing Corporation, *1974 Survey of Housing Units: Cross-Tabulation of Dwelling Units and Households, Survey Area, no. 6, Hamilton*, table 3.20

¹The 1974 survey by Central Mortgage and Housing Corporation included all dwelling types by household size and number of rooms. The assessment estimates only dealt with houses.

The table exposes a contrast between high density during the late nineteenth and early twentieth centuries and lower density during the second quarter of the twentieth century. In the earlier era, about half of the city's households seem to have occupied dwellings of five or fewer rooms. The mean density within this class of dwellings likely attained the standard of one room per person; if so, it was barely that, especially if there were any serious underenumeration of the population. From a modern perspective, these plentiful small houses would appear uncomfortably limited. Photographs, drawings, and surviving examples confirm a suspicion that the rooms were surely small. Row houses and urban cottages of four to five rooms were likely to have had no more than 500 to 700 square feet of living space. The families who lived in these dwellings may well have evaluated them positively, because by the yardstick of contemporary old country housing, Hamiltonians could have occupied quite decent quarters. The claim will be supported shortly. Furthermore, at least another half of the city's households had spacious accommodations. The ambiguities in interpreting the general character of housing density diminish by the mid 1920s. Smaller households and the substantial construction boom of the mid and late 1920s helped to provide roughly 80 per cent of Hamilton households with better than one room per person. An estimate of a room and a half for this considerable majority of households would not be too generous.

Significant and unusual when compared with the figures for cities in the United States, nearly a third of Hamilton's housing was classified as seven or more rooms. Even if the assessment linkage process erred and shunted a number of houses up from the six to the seven room group, the features of the blocks involved do favour the overall claim of larger houses. Most of the new and evidently larger homes appeared in the white-collar and industrial managerial areas of south-east Hamilton and the planned suburb of Westdale. In use today, that housing stock still seems relatively large, probably averaging between 1,200 to 1,800 square feet of living space. The prosperous industrial city supported an increasing number of owners, managers, and professionals associated with the iron, steel, and textile industries. That increase and the heady optimism of the 1920s stimulated a middle-class and upper-middle-class housing boom.

Thus far, we have discussed mean densities, which may be misleading, since mean values cancel out extremes. A city's housing stock may provide better than one room per person on average, but a substantial proportion of households still could be residing in dwellings with under one room per person. Fragments of international data from the 1930s and more consistent worldwide reporting from the 1950s forward suggest another way of looking at housing density. A few reports have indicated the percentage of dwellings that sheltered more than one person per room. Reworking the same raw data and assumptions about equating assessment values with numbers of rooms that were used in table 10.5, some interesting results are shown in column 5 of table 10.6. Depending on one's standards of evaluation, Hamiltonians in the late nineteenth century were either tightly packed in their dwellings or fairly fortunate in that more than half exceeded the high standard of one person per room density and a great many more than that surpassed a much lower standard of two persons per room.

During the twentieth century, the housing of Hamilton and of urban Canada was less densely occupied than shelter in most European countries. Only the prosperous or planned societies of Belgium, the Netherlands, Switzerland, and Denmark approximated Canadian conditions. However, a 1974 CMHC study revealed that the least densely occupied housing units in Hamilton, as measured by rooms per person, were apartments (see table 10.5). This finding is related to the life-cycle feature of apartment tenancy spelled out in chapter 9. Rowhouses and rented detached houses were the most densely occupied, which is explained by the circumstances of many of the city's common labourers. After the urban renewal projects of the 1960s and a subsequent period of apartment-tower construction, those poor who resided outside the city's social housing dwelt in surviving remnants of the nineteenth-century worker's cramped housing. While the bourgeoisie and skilled labourers had abandoned the rental-house market, the poor could not.

The declining size of households and the occasional periods of construction of larger homes have provided recent generations of Hamiltonians with generally greater privacy and space than their late nineteenth-century counterparts had. But urban housing of that era was not uniformly poor. Perhaps Herbert Ames's account of Montreal's miserable "city below the hill" has held the attention of Canadian social historians for too long. It was not the quintessential working-class housing experience for urban Canada. Hamilton stands as a band in the spectrum of national housing, but close to the Canada-wide patterns (see table 10.6). Even so, when contrasted with the many larger houses constructed during the late 1920s, much of the 1960s, and again in the late 1980s, nineteenth-century Hamilton housing seems cramped. But this sets the past against a very high standard. When measured against two international yardsticks, Hamilton's housing stock in the late nineteenth century looks better than passable.

Table 10.6

The Number of Rooms per Person in Hamilton Compared with Similar Measures of Density in Other Canadian and European Jurisdictions, 1876-1966

Estimated Range of Mean Number of Rooms per Person in Hamilton Houses by Year	Mean Number of Rooms per Person in Major Canadian Cities in 1931	Mean Number of Rooms per Person in Selected Countries around 1950	Mean Number of Rooms per Person in Selected Countries 1960	Estimated Range of Percentage of Dwellings in Hamilton with more than One Person/Room by Year	Percentage of Dwellings in Selected Countries with more than Two Persons per Room	
					Around 1930	Around 1960
1876 1.00-1.25	Halifax 1.23	Australia –	Australia 1.42 u	1876 35.1-45.8	Austria –	Austria 30.9 b
		Austria 1.10	Austria 1.00 b		Belgium 2.0	Belgium 1.1 b
1886	Montreal 1.18	Belgium 1.35	Belgium 1.66 u	1886	Denmark 10.0 b	Denmark 0.8 u
		Denmark 1.40	Denmark 1.42 u		Finland –	Finland 25.5 u
1896	Toronto 1.41	Finland 0.65	Finland 0.83 u	1896	France 8.9 u	France 15.7 u
		France 1.00	France 1.00 u		Greece –	Greece 36.7 Athens
1906 1.05-1.25	Hamilton 1.41	Greece 0.45	Greece 0.71 u	1906 14.8-25.0	Hungary –	Hungary –
		Hungary 0.65	Hungary 0.67 b		Ireland –	Ireland 9.8 u
1916	Ottawa 1.48	Ireland 0.95	Ireland 1.11 u	1916	Italy –	Italy 21.6 b
		Italy 0.80	Italy 0.90 b		Netherlands 3.0 b	Netherlands 4.3 b
1926	Winnipeg 1.19	Netherlands 1.20	Netherlands 1.25 u	1926	Norway 8.7 b	Norway 4.0 u
		Norway 0.96	Norway 1.25 u		Poland –	Poland 18.8 u
1936 1.20-1.35	Calgary 1.25	Poland –	Poland 1.00 u	1936 11.5-15.4	Portugal –	Portugal 23.7 u
		Portugal –	Portugal 1.00 u		Spain –	Spain 14.9 u
1956	Edmonton 1.22	Spain 0.90	Spain 1.11 u	1955 11.2	Sweden 10.0 u	Sweden –
		Sweden 1.00	Sweden 1.25 u		Switzerland –	Switzerland 1.16
1966 1.30-1.45	Vancouver 1.30	Switzerland 1.30	Switzerland 1.42 u	1966	UK –	UK –
		UK 0.78	UK 1.42 u		USSR –	USSR –
		USSR 0.65	USSR 0.77 b	1974 3.5	Yugoslavia –	Yugoslavia 22.0 u
		Yugoslavia 0.43	Yugoslavia 0.59 u			
		W. Germany 0.82	–		CANADA:	
		CANADA 1.27	CANADA 1.42 b		Montreal 3.7	
					Toronto 1.5	
					Winnipeg 3.6	

u = urban

b = both rural and urban

First, even with a pessimistic formula for the conversion of assessed values into numbers of rooms, the density of Hamilton's housing in the late nineteenth century was lower than that prevailing in many European countries even by the 1960s (see table 10.6). The destruction caused by two world wars cannot account for all of the high density countries of Europe, though war combined with poverty can. North American abundance did extend sufficiently across the society of one industrial city to provide better housing than that found in many nations, nations understandably prominent as twentieth-century centres of emigration. Second, a more significant benchmark for an appraisal of Hamilton housing, and probably that of many other North American cities, stems from the state of shelter in those parts of Europe that provided the vast majority of nineteenth-century immigrants to North America. Geographers and historians have sufficiently scrutinized the housing of assorted locales in the United Kingdom and Ireland between the early 1800s and the early 1900s to permit descriptions of how the poor and the not so poor lived.

In the English countryside, the better estates had provided tenants with cottages of about three rooms, offering 350 to 400 square feet of living area. Such estates were rare. Cottage speculators, who built most of the dwellings for rural labour, were less generous. Furthermore, by the 1880s the investors in cottages found the returns too small to warrant new construction. Old, unimproved, small cottages characterized the shelter of rural tenants. In Scotland tenants on estates often had to build their own dwellings. Highland tenants built the meanest of these. Their field stone "black houses," which had vanished by the 1880s, had one floor and about 450 square feet, including in that total an area for cattle. While there were great differences in rural housing, influenced by landlords and degrees of rural poverty, the rural cottages of England and Scotland can be characterized as having had two to three rooms and seldom more than 400 square feet of living area.[13]

Rural industrial housing was similar, except in the villages created by employers intent on providing model housing. Saltaire, founded in 1854 by Titus Salt, was the most famous. Here overseers had dwellings of from five to eight rooms; other employees had access to units with four to six rooms. In contrast, a great deal of English urban housing for labourers consisted of "back-to-backs." In Leeds such a housing unit would have had a cellar, a kitchen, and an attic chamber, for a living area of about 400 square feet. It was roughly the same in the industrial towns of West Yorkshire. Houses in Liverpool were usually ten to twelve feet square, containing a cellar, a ground floor room, and two chambers above. Shelter erected by philanthropic housing associations may have been marginally better, although their regulations excluded or deterred many potential tenants. With an average of just over five persons per inhabited dwelling for England and Scotland during most of the nineteenth century, an estimate of two people per room among the poor is not excessive. The high mean levels that persisted in British housing into the mid-twentieth century document an enduring hardship (see table 10.6). The mean density continued to tower above North American readings, but sanitation and construction improved. A sequence of legislation essentially beginning in 1919 with the Housing and Town Planning Act would replace squalid housing with better, if unattractive, council housing.[14]

The diverse late-Victorian reform sentiments that had inspired housing investigations and philanthropic activities in England had also touched Ireland, where dreadful rural conditions and the pressure of the Home Rule Movement brought remedial attention in 1891 through the creation of a Congested Districts Board. The board attracted attention, leading to contemporary descriptions of Irish housing conditions. The board's reports contained information on the old mud cabins and on the board's own standard designs, which had considerable rural application. Often in the western counties, the former had been two-room dwellings of approximately 200 square feet in floor area. The modest improved cottage had to have a kitchen and bedrooms sufficient in number for due separation of the sexes. Most had two bedrooms, producing a total floor area of 450 to 480 square feet. In the towns and cities, migration from the countryside exacerbated a housing shortage and neither the British government nor, after 1921, the Irish Free State seriously confronted the problem. An extreme example was Dublin at the end of the nineteenth-century, reputedly the worst-housed city in the United Kingdom, where an estimated one-third of the population lived in one-room tenements. Philanthropic and urban-district corporation housing provided the city with one-room and two-room tenements and a few rare three-room brick cottages.[15]

A reproach to British landlordism, the housing of commoners in the United Kingdom was an aspect of the old country gladly jettisoned by immigrants who, with few exceptions, found superior shelter in Canada and the United States. The contrasts in the quality of old-world and new-world accommodations, in something that was both a necessity and a meaningful symbol, could have contributed to an entrenchment of the will to possess in North American culture. The drabness and spartan compression of philanthropic housing likely strengthened the brief for the free market, even though the achievements in North American housing may have had more to do with a sheer volume of resources and construction innovations than with a system of delivery.

What had been comparatively decent housing stock in nineteenth-century Hamilton improved during the 1920s. The more spacious housing and its diffusion to labouring families followed other improvements very closely. Good sanitation has long been acknowledged as a basic requisite of decent urban shelter. Water and sewer lines, advancing both health and convenience, reached almost all of Hamilton's blocks during the 1920s (see chapter 2). Hamilton had constructed a sophisticated municipal waterworks in the 1850s. At first, the system served the major streets only. In the 1876 sample, water lines served barely half of the block frontages, but by 1896 the proportion had climbed to seven out of ten frontages. It is impossible to say how many houses made use of the utility at their doorsteps, but at least the city engineering and works department had been expanding the opportunity. Significantly, the locational disadvantages of common labourers resulted in their being somewhat less expeditiously served by water mains. Moreover, there was the problem of securing the compliance of landlords to connect dwellings with the sewers.

Sewer extensions, demanding greater civic investment, proceeded far more slowly than water mains. Complaints about drainage in the late 1840s had initiated the city's sewer network, but for the remainder of the century the municipal authorities sought cost-cutting measures. Concentrating its sewer construction work on the urban core, the city corporation left several parts of Hamilton ill-served. Approximately one house in ten had a sewer along its frontage in 1876. The sewer system initially had evolved to carry off surface water, but within two decades the scale and goals of the sewer network had grown. Better than half the sample houses had sewers along a frontage by 1896. Massive construction between 1911 and 1915 achieved the 1916 figure of better than nine out of ten houses. Of course, for as long as voluntary connections were required – a condition that seems to have ended around 1910 – it is impossible to say how many households took up the sewer option. Property agents Moore and Davis informed a number of their clients in 1910 and 1911 that the city's health department demanded that outdoor privies be removed, filled up, and replaced with indoor water closets. The costs were to be passed on to the tenants through higher rents. [16]

The progress of the city's sewer extensions dragged in areas typically occupied by common labourers. In a study of disease, mortality, and public health in Hamilton from 1900 to 1914, Rosemary Gagan found that in 1910 mortality tended to be lower in the two better quality wards (1 and 2) and higher in two blue-collar wards (5 and 6). The lack of age-specific data is a consideration likely to water down a conclusion, but the Medical Health Officer, James Roberts, also claimed that the highest mortality rates occurred where "overcrowding was very much in evidence, and the careless tendencies of the population were intensified by the lack of sewerage."[17] By 1912 Hamilton's mortality rate had started to drop to levels below that for the rest of urban Ontario. Gradually, but slightly unequally, the essentials for sanitary housing spread, and by the 1920s houses without water or with a privy pit were exceptional.

It is difficult to evaluate Hamilton's performance comparatively until the 1930s, when city-wide data on sanitary facilities began to be collected by many western countries. A 1939 housing study published by the League of Nations' Economic Intelligence Service assembled fragments of data from European nations and the United States. This information helps to situate the Hamilton record in the 1930s. According to the 1941 Canadian census, which reported extensively on housing conditions (conditions that had not improved greatly since the onset of the depression freeze), sanitary basics reached virtually all Hamilton dwellings (see table 10.7): running water was available in 99.8 per cent of the housing units; of homeowner households, 94.5 per cent had exclusive use of a flush toilet; 90.9 per cent of tenant households had the same amenity; outside privies served only 0.6 per cent of Hamilton households; and for the exclusive use of a bath or shower, the percentages were 91.6 and 83.3 for homeowners and tenants respectively.

Scandinavian countries, which suffered from some of the worst housing in Europe, according to the League of Nation's report, were far behind Hamilton – or for that matter all of Canada – in household sanitation facilities. In Helsinki 26.3 per cent of the houses lacked inside plumbing and 41.3 per cent lacked a water closet. In 1935 three-quarters of the dwellings in Denmark had no private bath and a third had no water closet. Fifty-two per cent of Oslo dwellings in 1934 had no water closets.[18] The tormented faces in paintings by Edvard Munch had a basis in experience. He had accompanied his father – a doctor – on visits to patients in a poor section of Christiania (Oslo). In view of the miserable quality of Scandinavian shelter, the region's later reputation for planning and social housing is understandable. Meanwhile, a 1934 real-property survey conducted in the United States considered sanitary facilities in 203 urban areas covering 40 per cent of the country's urban families. It found that 15 per

Table 10.7
Housing Conditions in Canadian Cities with More than Ten Thousand Dwellings,
1941

| Cities | Percentage of Overcrowded Dwellings[1] | Percentage of Dwellings Lacking or with Only Shared Use of: | |
		Flush Toilet	Bathtub or Shower
Halifax	26.1	17.6	27.3
Saint John	17.9	8.0	37.1
Montreal	24.4	3.6	16.1
Quebec	31.1	3.2	34.7
Verdun	19.9	0.4	3.1
Toronto	12.4	16.8	18.6
Hamilton	*10.7*	*7.4*	*13.0*
Ottawa	17.4	9.3	15.1
Windsor	17.0	6.8	11.3
London	8.9	10.5	18.0
Winnipeg	19.0	14.5	24.4
Edmonton	22.2	29.0	35.7
Calgary	18.5	20.0	24.4
Regina	24.0	29.3	35.9
Saskatoon	19.5	36.3	42.2
Vancouver	13.2	9.7	13.3
Victoria	11.1	9.5	13.0

Source: Canada, Advisory Committee on Reconstruction, *Housing and Community Planning, Final
Report of the Subcommittee* (Ottawa: King's Printer, 1944), tables 19 and 23
[1] More than one person per room

cent of these families had no private indoor flush toilet.[19] World
War II and the destruction of housing in central and eastern Europe
rendered European sanitation data for many years problematic;
nevertheless, the majority of European countries would not draw
close to Canadian and American urban standards until the 1970s
and 1980s.

Urban transportation presents another feature of housing and
conveniences, although it cannot be compared precisely with that of
other cities in assorted eras. Like streetcar lines across North America, the privately run Hamilton Street Railway (HSR) was criticized
for its crowded cars, high fares, and accidents. During a 1906 strike
the owners received little public sympathy. In 1912 and 1913 the
city pushed a reluctant company to consider extending its lines into

new suburbs. The company pleaded that the areas would not generate sufficient revenue.[20] In sum, the HSR behaved and was treated like most North American traction franchise operations of its day. Yet what sort of route service did it supply? How convenient were the lines? Who was most inconvenienced? From 1876 to 1896, as a horsedrawn service, the HSR lines were most convenient for the professionals and proprietors and the white-collar employees. Members of these groups could walk, on average, less than 200 yards from their homes and arrive at a line; skilled and semi-skilled labourers had to press on for an additional 50 to 100 yards; and common labourers generally had to go further still. The shades of distinction represent the company's recognition that short central lines with stretches through better quality neighbourhoods made money. Nevertheless, Hamilton as a small city afforded greater convenience to workers than did the greater metropolitan centres. Painter J. Holmes, who had worked in Brooklyn before coming to Hamilton, observed in 1888 that he could live in a house close to work in Hamilton, but in Brooklyn he lived at a distance in a tenement. "A man of this country naturally cannot make a home there, as regards comfort and everything of that kind."[21] Later, on the eve of World War I, the HSR introduced rush hour specials going to the factory district and charging lower fares. Overall and in contrast to that in large cities, Hamilton's urban transportation was fairly convenient.

The housing and services nexus in Hamilton at the turn of the century showed general though unequal and gradual progress, comparing favourably with United States cities and proving superior to European cities. Conditions improved significantly across the half century from the 1870s to the 1920s, though less swiftly for common labourers than for other occupational groups. Doubtless, many slum landlords postponed the installation of sanitary services long after the city had made these available along virtually all streets. The urban environment, moreover, encompasses more than the hard services of water, sewer, and transit. Poor drainage, noise, traffic, and hazards or parks, peace, and safety – the combination of natural and man-made features – have fashioned many shades of quality among urban environs. Perhaps in the allocation of these by market conditions, a proposition of persistent and debilitating inequality will find support. A foretaste of inequalities in the urban environment has already figured in the analysis of services. An account of the segregation of occupational groups and descriptions of associated neighbourhood qualities provides further evidence of the partiality of city building by capitalist operations.

OCCUPATIONAL CONCENTRATIONS AND
HOUSING ENVIRONMENTS

What have been the characteristics of occupational group segregation in the city? To chart the intensity of segregation, the alterations
to segregation over time, and the environmental meaning of segregation, the spatial situations of occupational groups were evaluated
at the following intervals: 1856, 1876, 1906, 1936, and 1966. Additionally, the young town was scrutinized for the year 1839, using
crude early assessment rolls, newspapers, and maps. Contrary to the
now hoary findings of American urban historians writing in the
1960s and 1970s, conclusions based on Hamilton undermine the
model of development that proposed decay from an idealized and
integrated pre-industrial city to an unfortunately more segregated
metropolis of recent times. The older interpretation, espoused for
example in Sam Bass Warner Jr's *The Private City*, proposed a considerable intermingling of social groups in the pedestrian city of the
eighteenth century and a spreading of segregation in the twentieth
century, encouraged by the automobile, which facilitated a middle-
class flight to the outer suburbs.

Hamilton does not conform precisely to this pattern, perhaps because Canadian cities have not yet had serious encounters with race
relations and their free market analogue. For while Canadian cities
have been ethnically and linguistically heterogeneous, until quite
recently they have remained overwhelmingly white. But this is not
the whole explanation for a difference in findings. The classic model
erred most seriously in its claims of an early environment of mixed
social and economic space. In applying Warner's eras to Hamilton,
that of the pedestrian city seems the most emphatically segregated.
One methodological problem with Warner's study of Philadelphia
was his recourse to ward level data, which failed to pick up nuances
in segregation. For example, housing segregation based on elevation
and drainage could vary within the boundaries of a ward and create
socially distinctive pockets, while a ward-level probe jumbled all
groups together.

The related propositions that social segregation in the city really
required the introduction of mass transit and that the pedestrian
city was fairly well integrated have been dispelled by American historians who have revisited the pre-industrial city. Canadian geographer Richard Harris has likewise reviewed the question of what
constitutes significant segregation and has wisely recommended sensitivity to "historical and geographic circumstances ... the unique
configurations of residential space in the local setting."[22] In an analysis of mid-nineteenth century Milwaukee, Kathleen Conzen dem-

onstrated patterns of segregation in a pedestrian city because of her use of areal units that were more refined than those formerly employed. She specifically detected the segregation of the poor Irish in low-lying portions of the city.[23] Interestingly, the identical spatial arrangement existed in Hamilton. An 1843 map designated a Corktown and set it in an area that early maps had shown as cut by ravines that carried both the run off from the Niagara Escarpment and drainage from the prominent ridge that ran into the escarpment just west of Corktown. Corktown spread over a conspicuous depression.

Testing for assorted forms of social segregation in the early town presents a methodological challenge. In the 1830s occupations were not cited on Hamilton's assessment rolls and the rolls normally failed to attach locational tags to households and properties. However, the 1839 roles unaccountably listed entries by street segments, but without any street numbers or indications of intersecting streets. Fortunately, many streets in the early years of the town extended only a few blocks before terminating or assuming a different name. Two maps drafted around 1830 and a map of 1851–2 that indicates the names of many property and shop owners proved useful in linking the 1839 assessment information to locations. Newspapers and local history accounts helped too. Thus all but 196 (16 per cent) of the 1,231 assessment entries could be slotted into eleven tracts created for the purpose of a spatial inquiry.

Taking the four most common housing types out of nine on the assessment rolls (eight plus the shanties deduced from Hamilton Police village minutes and the rolls), it is apparent that the town had concentrated areas of wealth and a quite pronounced region of poverty and of probable ethnic segregation. The town's most common housing type, the single-storey frame dwelling with no extra fireplace, was almost evenly spread throughout the eleven tracts. So too was the two-storey house with no extra fireplace. Clustering occurred with the two housing types that were polar opposites (see table 10.8). Dwellings of more than one storey with extra fireplace(s) were over-represented in the original townsite, the area around town founder George Hamilton's Bellevue estate, the Main Street West axis, and Peter Hunter Hamilton's survey. All of these town tracts were in the south half of the city and contained elevated blocks with superior drainage and views of the bay. A disproportionate number of better homes had been built on lots with elevations of 340 to 420 feet above sea level. Much of the town remained undeveloped.

Elevation in Corktown ranged from 300 to 320 feet above sea level and Corktown housing was by far the worst in town. Here shanties proliferated. Strictly speaking, shanties were not recognized on the

Table 10.8
Hamilton's Land Use (by Selected Assessment Categories) According to
Town Tracts and Aggregated Tracts, 1839

Tracts	Cultivated Lands as % of Entries for Tracts	Town Lots as % of Entries for Tracts	Two Storey Houses with Extra Fire- places as % of Entries for Tract	Shanties as % of Entries for Tract	Shops as % of Entries for Tract
CENTRAL TOWN					
Original Town	0.8	32.8	7.6	1.5	11.5
Market Area	4.1	32.4	3.2	1.8	3.2
King Axis	1.0	35.0	3.0	2.0	4.0
ADJACENT TO CENTRAL TOWN AND MARKET					
York Axis	0.0	60.0	6.0	0.0	0.0
Hughson Tract	1.5	31.2	7.5	1.5	0.5
EXTREME EASTERN FRINGE					
Ferguson Tract	50.0	0.0	0.0	0.0	0.0
LOW-LYING FRINGE					
North End	28.6	22.9	0.0	4.3	0.0
Corktown	6.7	30.5	1.0	20.0	0.0
HIGH SOUTHERN FRINGE					
Main West Axis	4.3	45.7	7.1	0.0	0.0
Bellevue	11.1	48.9	6.7	0.0	2.2
Peter Hunter Hamilton Tract	42.4	24.2	15.2	0.0	3.0
TOWN	5.4	35.1	4.8	3.4	2.7

Source: Hamilton Assessment Roll 1839

assessment forms, but they existed and could be discovered on the
rolls. The Hamilton Police Village Minutes suggested that if a shelter
had a mere dirt floor and lacked a shingle roof it would receive
exemption from assessment as a frame house. Occupants of the
qualifying shanties were assessed only for the lots they occupied.
When a household was assessed for a lot and the household could
not be located elsewhere on the rolls, then the chances were that it
inhabited an especially mean dwelling, labelled here as a shanty.
Usually such a household was enumerated as Roman Catholic and
Corktown accounted for a quarter of Hamilton's Roman Catholic
households but a mere tenth of the town's entire households. In
other words, early Hamilton had an enclave of poverty congruent
with an Irish Catholic ghetto and a miserable form of shelter comparable to that commonly found in Ireland. Regrettably, the absence
of an occupation variable restrains any attempt to link the social
segregation of 1839 firmly with that seen in subsequent cross sections. It is a plausible conjecture that the occupants of Corktown's

shanties were economically kindred to the common labourers who, in later periods, figured as most persistently concentrated in unsatisfactory environments.

Text and tables provide a monochrome of the town's housing in 1839. In 1846, Hamilton became a city and, not ten years after its elevation, artist Edwin Whitefield sketched it, leaving an undated record of showy streetscapes. By the 1860s, photography was thoroughly established and its otherwise splendid documentation projected an austere, colourless city. Whitefield's sketchbook restores a brighter side to even commonplace dwellings. The frame houses with single chimneys seen everywhere in 1839 were still to be found in the same areas around the 1850s when Whitefield walked about the city. He sketched these plain dwellings, depicting their few windows, simple pitched roofs, and sometimes the two front entrances that indicated a semi-detached. They had no verandas, rarely a front step, and sat close to the streets, which likely added grit to the household scene. Still, the houses were colourful. There was an abundance of white, yellow, and red, and here and there landlords or owner-occupants had added to the range of the civic palette with buff, brown, blue, dark grey, and green.

While the colourful quality of common housing is a bit startling, the contrast in housing in Whitefield's drawings is also vivid. The differences in the assessed values for bourgeois houses and labourers' dwellings, which were so extreme in the nineteenth century, are explained in rough illustrations. Whitefield had to spend more time on the details of the estates built on elevated land and corner lots. Many had outbuildings. The homes themselves had multiple chimneys, dormers, and verandas. "The Castle," hardware merchant Richard Juson's aptly named new gothic house on the heights at the head of James Street was given its due. And how did Whitefield present Corktown? He drew it honestly – his characterization fits with that furnished by the assessment rolls of about twenty years earlier. On a ridge just above Corktown, Whitefield recorded substantial houses on Catherine Street, but as he descended that street into Corktown he saw small frame houses wedged together. At one point, he wrote "small detached 1 story houses on this side." On Peel Street, parallel to Catherine, it was the same: "Small houses on this side." Apart from the awareness that colour, and with it some pride and individuality, was a quality of the urban past, Whitefield's legacy confirms the existence of environmental extremes in housing at mid-century.[24]

From 1856 to 1966 it was possible to investigate the over-representation of occupational groups by wards and even by city blocks. A number of research strategies were explored to express the over-

representation or, to put it in other ways, the segregation or concentration of occupational groups. Indices of dissimilarity were calculated to capture the social distinctions among the city's wards and
also the city's blocks. Another technique consisted of a trial-and-
error method of finding how concentrated a given occupational
group was in relation to the total number of blocks in which it could
be found. If half of a particular group could be located in 10 per
cent of the sample blocks, segregation must have been great. To
calculate such expressions of concentration a basic criterion was set
and adjusted if required.

The basic criterion selected was a 1.5 factor; a block with an over-
represented group was one with at least 1.5 times the city-wide percentage of that group. It became evident from scanning the block
data that for several groups higher or lower criteria had to be applied. For skilled and semi-skilled labourers who always made up
the city's largest occupational group among household heads, the
1.5 factor excluded a great many blocks. Because this group was
large and relatively evenly distributed across the city, its segregation
factor had to be reduced marginally to 1.4. However, the factor for
common labour had to be raised to a remarkably high 2.5, because
a considerable number of common labourers lived in blocks where
their percentage of household heads was two and a half times that
for the city. This change was extremely important in locating the
victims of the city's worst environments. These criteria values also
served independently as indicators of segregation (see table 10.9).

The patterns of segregation and of their changes over time seem
plain enough at first. From the very segregated city of the 1850s,
the broad trend has been one of apparent amelioration, although
in the 1936 cross section a resurgence of separation cropped up.
The 1936 about-face likely expressed deliberate efforts at discrimination by land developers in the 1920s – such as the syndicate that
planned the massive development of Westdale (see chapter 2).
Professionals and proprietors as well as white-collar employees had
become much less spatially exclusive in the late nineteenth and early
twentieth centuries. Some of this transition toward a social mix unquestionably originated in a shift in the composition of these two
occupational groups. Small retailers and members of new professions
inflated the sample size of the professionals and proprietors group
and correspondingly deflated its mean income. Similarly, low-paid
clerical workers joined the white-collar group. Therefore, the falling
indicies of dissimilarity and the readings on occupational groups'
over-representation must be treated cautiously, though the trends
seem strong – especially in the index of dissimilarity by block.

Table 10.9
The Spatial Segregation of Hamilton's Occupation Groups, 1856-1966

Index of Dissimilarity by		Criterion Values for Selection of Blocks with Over-Representation	Percentage of Sample Blocks in Which Group is Over-Represented	Percentage of Occupational Group in Over-Represented Blocks	Index of Over-Representation
Ward	Block				
(a)	(b)	(c)	(d)	(e)	c (e/d)
PROFESSIONALS AND PROPRIETORS					
1856	.300	.485	Blocks with	13.0	69.2	10.6
1876	.238	.355	2 times %	12.8	32.0	5.0
1906	.145	.211	of group	9.3	16.7	3.6
1936	.136	.263	found in	14.2	38.8	5.4
1966	.200	.288	all blocks	10.0	22.4	4.4
WHITE-COLLAR					
1856	.261	.575		4.0	26.6	11.0
1876	.200	.415	1.66	6.4	16.4	4.2
1906	.140	.296	times	20.3	39.1	3.2
1936	.150	.334		27.5	66.2	3.9
1966	.179	.280		28.0	33.1	3.0
SKILLED AND SEMI-SKILLED					
1856	.290	.370		23.1	53.7	3.3
1876	.198	.332	1.4	25.0	40.0	2.2
1906	.182	.250	times	14.8	26.8	2.5
1936	.100	.268		20.1	31.9	2.2
1966	.089	.215		16.7	20.8	1.7
LABOUR					
1856	.210	.520		11.5	40.9	8.9
1876	.192	.327	2.5	10.6	23.6	5.6
1906	.179	.302	times	9.3	25.5	6.8
1936	.237	.340		12.5	37.6	7.5
1966	.284	.358		9.3	40.2	10.8

Source: Hamilton Assessment Rolls

Ambiguous findings issue from the 1936 and 1966 cross sections. The problematic – but likely – downward trends in segregation from 1856 to 1906 reversed themselves by 1936 for all occupational groups according to every measurement used. It appears that by 1966 another wave of integration had begun for white-collar and skilled and semi-skilled household heads, a plausible trend considering the post-war industrial prosperity. Professionals and proprietors seemingly held onto exclusivity. Certainly the rich never significantly mixed their residential space with that of the city's common folk. Common labourers also remained relatively separate. To lower levels of home ownership and the poorer character of their dwelling units, we can add that common labourers endured inferior environments. Their areas of over-representation are described presently. In concluding this discussion of temporal trends, it should be stressed that evidence of a long-term and broad trend of integration, which flies in the face of some conventional wisdom, is problematic. It is not exactly blazing a course toward a socially mixed city. The very considerable segregation of 1856 melted, causing greater social-spatial complexity, but the 1966 measures of segregation are in themselves significant.

Like the persistence of the will to possess, a desire to buy or to rent as fine an environment as possible has continued throughout Hamilton's history as a demand factor continuously working to sort out the city's population. The capacity to seize the better sites has been spread unequally. As David Harvey noted in *Social Justice and the City*, higher income groups could bid up prices for better locations because they would not have to sacrifice anything like the share of consumer surplus available to a less affluent group.[25] Spreads between the costs of necessities and income forced differentials in buying or renting dwellings with distinct environments. Access to credit and credit ceilings have likewise varied among households, reinforcing the consumer surplus factor. Throughout the city's history, land owners selling building lots have recognized that environment is part of what is being sold and have acted accordingly (see chapter 1). The free market then has made for a segregated city. And again continuity marks the city-building process.

A painstaking reconstruction of Hamilton, like that completed for 1839, was not necessary for deducing where concentrations of occupational groups could be found in subsequent eras. Like tissue slides, cross sections from 1876, 1906, 1936, and 1966 – approximately generational samples – were studied at two different powers of magnification. The schematic diagrams in figure 10.1 represent the first power, furnishing very general pictures of occupational distinctions among large urban regions. Alone these would be misleading, because the diagrams and data indicate more integration than close consideration will support. Yet, since the lore of broad ecological differences has long had currency in Hamilton, the foundations of that lore merit attention. The east end of the city north of King Street has borne the working-class label since at least the 1920s. Similarly, the south west and far west have been regarded as basically white-collar districts. These intuitive characterizations are confirmed in figure 10.1; they have existed for a century and a half.

Beyond the social map that people carry in their attitudes, the use of grossly defined regions – really quadrants – has a topographical

North-west quadrant	North-east	0.00	Quadrant's percentage of city population
		0.00	Percentage of professionals and proprietors in quadrant
		0.00	Percentage of white collar
		0.00	Percentage skilled and semi-skilled
South-west	South-east	0.00	Percentage labour

+ 0.00 over-represented
− 0.00 under-represented

1876
JAMES

39.3		30.3
+46.6	−24.0	
+42.7	−24.6	
−33.2	+35.8	
+45.3	−24.5	

————————————KING

15.1		15.6
−12.0	+17.4	
−14.6	+18.1	
+15.7	−15.3	
+16.4	−13.8	

1906
JAMES

31.2		36.7
−30.6	−34.0	
−25.6	−28.3	
−30.9	+41.8	
+39.1	−32.1	

————————————KING

17.7		14.4
−12.5	+22.9	
+19.9	+26.2	
−16.9	−10.4	
−17.4	−11.4	

1936
WELLINGTON

24.5		41.5
−24.1	−27.2	
−22.3	−30.9	
−22.3	+47.4	
+34.3	+48.7	

————————————KING

20.2		13.7
+28.3	+20.4	
+24.3	+22.5	
−18.8	−11.5	
−12.5	−4.5	

1966
QUEEN   FERGUSON

15.9	25.8		36.2
+24.6	−22.9		+36.9
+19.9	−20.3		−28.0
−12.9	+23.7		+38.1
−10.8	+43.1		+40.4

————————————MOUNTAIN

11.5	11.3	
− 8.1	− 7.6	
+16.2	+15.6	
+12.8	+12.5	
− 3.2	− 2.5	

JAMES

Figure 10.1
Concentration of occupational groups by Hamilton's quadrants for 1876, 1906, 1936, and 1966.

The italic number in each quadrant expresses the percentage of the city's population in that quadrant. The four numbers below give the percentage distribution of occupational groups (see key) in the quadrant. The signs in front of these numbers indicate whether a group is over- or under-represented in relation to the quadrant's share of the city's population.

basis in Hamilton's case. The city's pronounced physical features –
in recent times partly below the Niagara escarpment and partly on
its plateau – and its well-defined periods of development recommend
the initial use of a low power of magnification. The quadrants, however, though showing some specialized traits, contained mixed populations by virtue of their very size. Turning up the magnification,
individual blocks can be analysed. After all, relatively small natural
and man-made features have affected the character of neighbourhoods. Both the general appraisals of quadrants and the fine tuning
by blocks will make clear the disparity of living conditions.

By 1876 the economic lifeblood of Hamilton coursed to an industrial beat. The Great Western Railway (GWR) yard and shop,
developed in the 1850s, remained a major employer. Built along the
city's western waterfront, the yard had attracted a number of railway
and industrial workers to nearby streets. On Mulberry Street, quite
near the yards, the Hamilton Light and Gas Company had built its
coal gas plant in 1851. Around the city's central core, especially to
the north east, manufacturers of stoves, sewing machines, ready-
made clothing, and shoes were concentrated. Farther north, the
Sawyer-Massey farm-implements plant had located along the GWR
line near the waterfront, pioneering a shift of industries to the north-
east fringe of the city that would reorder the spatial patterns of
employment beginning in the 1890s and continuing into the 1950s.[26]

A band of industrial operations and noxious activities stretched
across the north half of the city. At a time when streetcars had barely
been introduced into Hamilton, the swath of employers attracted
closely associated workforces. Common labourers were over-represented around the GWR yards in the north west and skilled labourers
near the foundries in the north east. The extreme north east was
forever rendered unattractive for high-quality housing by city-building events of the 1850s and 1870s. The GWR line was one factor; a
county jail built on Barton Street in 1875 was another; and the
placement of trunk sewer outfalls in ravines that emptied into Burlington Bay yet another. In the 1870s a major open spillway that
wound circuitously on its course to the waterfront was crossed in
three places by the Hamilton and Northwestern Railway. The railway's culverts slowed the movement of sewage, "leaving an accumulation of filth, the stench from which is too horrible to describe
and which ... endangers the life and health of the inhabitants."[27] A
nearby soap factory added to the noxious brew. Complaints about
the insalubriousness of the area continued into the 1890s, when in
addition to the concentration of sewage there were problems caused
by the collapse of old wooden conduits.

Oddly, but quite explicably, the north west, which had an over-representation of common labourers, also held a disproportionate percentage of professionals and proprietors as well as white-collar employees. The great sand ridge upon which several pioneer estates had been built cut diagonally across the north-west quadrant and still attracted the rich. About a century later, a number of the finer old homes, restored as carriage trade shops and professional offices, had a rebirth as Hess village. Today a glance down the village's streets still reveals its view, overlooking the city's core. A tour shows the descending scale and quality of housing associated with lower elevations. As the streets slope toward the waterfront, the older houses become smaller; bay windows vanish; ceilings are lower; single-storey dwellings are more frequent. Some of the physical evidence on the lower elevations has been obliterated by apartments built in the 1960s and 1970s. The urban renewal itself testifies to a poorer housing stock unable to retain its market value.

South of King Street there were few large employers. Sharply changing topographical features made for a mixed occupational composition in the sample blocks. In the south east, where the ridge fell off, sloping into ravines, tiny frame cottages – some belonging to the Mills family – sheltered common labourers in poorly drained depressions. Over a century later, the frame and brick cottages remain, providing rare urban landscapes redolent of nineteenth-century and contemporary poverty. Embracing both a valley of humility – Corktown – and a hilltop of affluence – the estates on the heights of James and John Streets south – the south east in 1876 possessed a nearly representative blend of occupational groups.

Among household heads in the spatial arrangements of 1876, common labourers endured the worst conditions. Aside from the unpleasantness of poor drainage that plagued the sites where they were often forced to concentrate, another burden accompanied several of their block locations. The cottage area of the old Mills estate in the city's south west was one of Hamilton's more remote corners in relation to employment opportunities in the pedestrian city. Corktown, though somewhat better placed, was less advantageously located than the areas where skilled and semi-skilled workers concentrated. The latter tended to dwell in pockets: around the foundries and other manufacturers still situated near the city's centre; and in the north centre of the city near the railway yards. Because they had to walk to work, the extensive group of skilled workers, representing the bulk of the city's labour force, generally lived in the blocks surrounding the firms where they were employed. In the absence of cheap mass transit, there was a search for convenient shelter. Common labourers lost out in the workers' pursuit of strategic residential locations.

In the full swing of a vigorous expansion, the industrial city of 1906 displayed several social-spatial changes from former eras. The new spatial traits were essentially products of very new urban and industrial environments. From 1906 forward, the north-east end of Hamilton had the highly visible hallmarks of an industrial district. The movement of industrial employment from the core of the city to the waterfront and railway lines accelerated as the factories of heavy industry went up or expanded at a remarkable rate. New housing followed the shift in employment location. In the north-east quadrant there still were professionals and proprietors – largely small retailers – but white collar employees were notably under-represented in 1906 and remarkably so in 1936. Blue-collar household heads, slightly over-represented within the north east in 1906, had become much more so by 1936. The east end south of King bore the stamp of industry and of blue-collar shelter. Created in the era of corporate involvement, specifically a product of factory employment and the massive land-development ventures described in chapter 2, the east end was the physical and social epitome of industrialization.

A 1944–45 planning study specified several quality implications that followed from the mix of industries and housing. Railways and major arteries for automobiles and trucks serviced the huge industrial district sprawling at the edge of a waterfront that was becoming inaccessible to the public. In the early 1940s the city had 64 level crossings – 48 in the north-east quadrant of the city. Of Hamilton's ten most dangerous crossings, seven were in that same region. By contrast, the entire west half of the city had merely 8 level crossings and the planned middle-class suburb of Westdale had none. A traffic survey undertaken over a 17-hour period on 17 May 1941 pinpointed five heavily used street intersections. The number of vehicles per hour ranged from 600 to 1,100; the mean for the five intersections was 840 vehicles per hour or 14 per minute from 6:00 a.m. to 11:00 p.m. All five were found in the north east and the three along Cannon Street dotted the quadrant's central axis.[28] The planning report compiled information on the poor sanitary conditions afflicting the fairly small proportion of the city's population in overcrowded dwellings, but it said little about the effect of traffic on the quality of living. It did cite accident statistics and remarked on the inconvenience – "due to lack of adequate provision of grade separating crossings" – experienced by industrial labourers driving to work.[29] Since the 1960s planners and governments alike have recognizcd that traffic interferes with a good living environment, affecting physical and mental health.[30]

On the well-founded concerns of a later generation about the discomfort and health risks of air pollution, especially endemic to the north-east quadrant, the report had nothing at all to say. However, the damage was obvious. For some time Hamilton had been losing its shade trees. The *Hamilton Times* reported in March 1916 that street paving and gases from the industries were affecting the city's trees.[31] Organized labour made a stir about the atmosphere in 1944, and health and welfare workers have recognized problems. Labour's representatives, meeting with the planning officials and consultants, stressed the need to improve transportation to the mountain. One union spokesman envisioned the mountain as "the ideal place to live away from the smoke, dirt and traffic, and children, he said, could play in the streets there."[32] Another union representative demanded that the mountain be developed as a residential area for the labourer otherwise the worker would "be forced to continue to live down among the dirt and the factories."[33]

A few days after the meeting between organized labour representatives and the planning committee, an official from the city's health department stated that Hamilton's worst health conditions occurred where "industries and houses were bound together ... Noise was also bad."[34] The voiced claims to the mountain and the perception that it was a clean and healthful refuge for workers' families were harbingers of a reordering of residential space easily detected in the 1966 cross section. But there is much more to be said about housing quality in 1936.

Providing data collected in the middle of the great depression, the 1936 cross section provides a rare statistical window on the social distress of that period. As is known well enough to Canadians studying the 1930s, only the prairie provinces received a census enumeration in 1936. Furthermore, Canadian personal experiences recorded in the interviews published in Barry Broadfoot's *Ten Lost Years* or in the letters to Prime Minister Richard B. Bennett published as *The Wretched of Canada* overwhelmingly originated outside the urban centres of central Canada.[35] Therefore, although it diverts the spatial inquiry, a tangential appraisal of housing in 1936 is called for.

In earlier discussions, declines in home ownership, rises in apartment vacancies, and the possibility of rental bargains for the employed were topics raised when the depression years came under review. Young people evidently had to shelve hopes of home ownership. The data allow us to weigh another consideration – over-

Table 10.10
Household Size by Type of Tenure and Dwelling for Hamilton, 1926 and 1936

	1926	1936
Average Household Size of House Owners	3.9	3.8
Average Household Size of House Tenants	4.3	4.5
Average Household Size of Apartment Tenants	2.7	2.8
Average Household Size of Commercial Block Apartment Tenants	3.6	3.2
Average Household Size of Tenants in Flats	3.2	4.7

Source: Hamilton Assessment Rolls

crowding. The doubling-up of households mentioned when noting apartment-building vacancy rates implies a greater density somewhere in the shelter market. This certainly was the case. Comparing household size for several shelter forms in 1926 and 1936, a pattern of concentration comes into focus (see table 10.10). The mean size of tenant households generally increased; that of homeowners stayed about the same. A moribund home-buying market meant that tenant households grew at a proportionately greater rate than the increase of all households. The sample of tenant households dwelling in houses rose from 1,134 (1926) to 1,600 (1936), outpacing growth in the overall sample size. Within that expansion of household tenants, households of 3 to 4 people increased substantially from 463 to 679, while those of 1 to 2 people remained stable at 204 (1926) and 240 (1936). The mean household size for tenants in houses edged up very slightly (see table 10.10). Since houses still accounted for the overwhelming stock of rental shelter, these figures fail to build a case for widespread and severe deterioration of shelter conditions. They do confirm some adjustments in household composition due to the postponement of home buying by growing families.

If serious overcrowding occurred, it happened in flats, that is rental units in houses – units that shared an address number but were neither in commercial blocks nor in multiple-unit buildings. The number of such units in the samples remained nearly identical at 227 (1926) and 226 (1936) or 7.8 and 6.4 per cent of all dwellings. Within these something happened that signals a crisis: the number of single-member households was halved; two-member households declined; and the mean size of households rose by nearly 50 per cent. In 1936 households of more than 4 people made up 31.4 per cent of the total tenant households in flats; the same relatively large households had constituted 20.7 per cent in 1926. Here, then, were the probable instances of acute distress leading to overcrowding.

The one shelter type most likely to have been haphazardly created and cramped at the best of times, the one with the lowest assessed values, also proved to have undergone the greatest escalation in density.

How many people might have been affected? Assuming that our sampling practices capture the overall traits in housing, it appears that perhaps a third of the units in Hamilton's potentially poorest housing form might have been overcrowded. Put another way, about 2 per cent of the sample housing stock seems to have been of very dubious quality, probably densely occupied, and rented. In May 1942, at the climax of a wartime shortage of housing, the city's Council of Social Agencies surveyed overcrowding in the central core and concluded that "as a condition affecting the city as a whole it is a limited problem ... concentrated in the central region."[36] It endangered the welfare of an estimated 1,500 people, less than 1 per cent of the city's population. Of course, it was entirely possible that many more households, though not overcrowded, resided in poor quality housing. Common labourers, always at the short end of the stick, were over-represented as the household heads of the crowded rented flats, as they had been in the 1936 sample.

The prime site of poor housing in Hamilton, as identified by contemporaries, attracted an investigation in the spring of 1936. The Family Service Bureau, inspectors from Hamilton's Public Health Department, and some sociology students from McMaster University scrutinized the quality of shelter in an area four blocks wide and five blocks long, north east of the city's central business district. The 20 block area, making up just over 1.5 per cent of the city's total number of blocks, had 450 families or roughly 1.5 per cent of the city's families. Most likely the area in question fell within or along the margins of the old port area known locally as the "the north end." Our sample blocks for that part of Hamilton support the judgment that here indeed was a region with poor housing.

The single-family detached dwellings and the row houses in and around "the north end" were often assessed at under $1,400, and typically consisted of frame houses with fewer than six rooms. To be more precise, around 40 per cent of the houses in the sample blocks fell below the $1,400 standard. There also were a few lower valued apartments in each of the blocks. These too had assessed values below $1,400. Thus a rough standard can be applied to the rest of the city. Neighbourhoods where approximately 40 per cent of the houses were valued below $1,400 or where an unusually large number of apartment units of a low assessment were located can be counted as poor housing districts. In addition to the study area, there were certainly another 20 blocks to the north and north east that contained cheap and deteriorating housing. Both the study area and the entire "north end" were adjacent to the city's major trunk railway line.

With a concept of poor shelter in mind we scanned the data from the block samples in the well-known and historically established cheap housing districts. Corktown proved not to be dramatically down at heel. Its old shanties had been replaced either by rows of brick houses or by the tracks and sheds of the Toronto Hamilton and Buffalo Railway, all constructed in the late nineteenth century. Close to the urban core with its employment opportunities, Corktown had become a workers' neighbourhood with fair housing. The smaller area of workers' cottages at the lowest elevation in the old Mills estate paralleled "the north end." Both had concentrations of small cheap houses. There had never been any locational advantages to stimulate an upgrading in the hollow of the Mills estate, where the cottage area covered 12 blocks at most.

Near the CNR railyards (formerly GWR and GTR) in the city's north west, another area of poor housing had developed and it took in roughly 10 blocks. Just to the immediate east of James Street North another 10 could be added. Strung out through the extensive industrial suburbs in the north-eastern section of the city along the CNR mainline, Barton Street, and Cannon Street, there were many small and cheap dwellings. However, the poorer housing of the industrial arteries was generally newer than cheap living quarters elsewhere. Our sample data show especially poor conditions reflected in low assessed values for housing in the blocks abutting the CNR mainline that serviced the industrial district. The city in 1936 had about 50 such blocks; dwellings in neighbouring blocks were not much better. Thus, out of the pandemonium and grime of the industrial corridor, another 100 blocks could be designated as having unsatisfactory housing. A generous estimate would place the total number of blocks with housing almost as bad as that in the region selected for a social work study in 1936, and including that region, at roughly 172.

Our sample data actually produced 18 blocks where roughly 40 per cent of the houses had assessed values in the under $1,400 range or where there were an exceptional number of cheap apartments. On the basis of a one in ten sample, that is equivalent to 180 blocks. By either estimate, approximately 15 per cent of Hamilton's blocks probably contained a considerable portion of housing deserving the label inadequate either under the terms of the 1936 study or because of the special hazards and torment of living cheek by jowl with railways and industries.

Undesirable at the best of times, much of the housing surveyed in the 1936 report had sunk further into dilapidation during the depression. Owner-occupants and landlords deferred repairs. In the 20-block survey area, about a quarter of the 450 families lived in what investigators referred to as apartments, meaning non-detached rental units in tenements, above commercial establishments, and in converted houses. This was a much higher proportion than for the entire city where only about 15 per cent of households would have been in anything remotely resembling slum apartments. That is to say, almost 85 per cent of Hamilton families resided in single-family detached and semi-detached dwellings. The slum study dealt with an exceptional region of the city. Evidence of deplorable circumstances accumulated as researchers went their rounds.

What did these researchers have as standards? The author of an article based on the group's research admitted that "the housing standards here used are higher than society is willing to guarantee as a minimum human right."[37] Examples will clarify what he meant. For a toilet to qualify as satisfactory, it had to be private, properly located (not in a shed, basement, off a kitchen, or off a bedroom), adequately ventilated, and with plumbing in good condition. By these measures 81 per cent of apartments and 45 per cent of houses had unsatisfactory facilities. That works out to about 250 households. Other standards included central heating, a private porcelain bath tub, adequate light in all rooms, and play space for the family. The results of the survey appear in table 10.11. While all of the above standards are and were desirable, freedom from vermin is basic. If freedom from cockroaches, bed bugs, and rats is a rock-bottom standard, then 38 per cent of the apartments and 15 per cent of the houses failed. In effect, 100 families out of 450 must have dwelt in truly atrocious shelter.

Taking the estimated number of blocks with significantly poor housing and considering the circumstances in the very worst of these blocks, it is possible to estimate the number of Hamilton households who endured very poor shelter. On the basis of the above discussion, and overestimating the extent of poor conditions, let us assume that 300 units in the study region were unsanitary. The 300 families living in very inferior conditions would have amounted to nearly 70 per cent of the study area's households. Our own data suggested maybe 40 per cent, based on linking assessed values to descriptions of dwellings in the press. Alone, the vermin criterion would have placed the figure just under 25 per cent and the author of the 1936 study report concluded that about 25 per cent of the area's shelters qualified for the label "unsatisfactory." Consequently, one can choose from a range of values when estimating how many dwellings in a blighted

Table 10.11
Housing Deficiencies in Housing Survey Area, Hamilton, Spring 1936

	Percentage Apartments Affected	*Percentage Houses Affected*
SERIOUS DEFICIENCIES WITH		
Toilet Facilities*	81	45
Kitchen Sink	15	21
Bath Tub	59	52
Heating	10	26
Light	52	10
Play Space	98	48
Walls	38	15
Vermin*	38	15
LESS SERIOUS DEFICIENCIES WITH		
Frame Construction	7	25
Outside Dilapidation	7	40
Ceiling	14	24
Floors	21	41
Halls and Stairs	55	32
Cellar	–	34
DWELLINGS (APARTMENTS AND HOUSES) HAVING 7 OR MORE DEFECTS*	25%	

Source: Leo Haak, "A Housing Survey in Hamilton," *Social Welfare* (March 1937)
*Standards mentioned in the text

block were poor. These several values may be used to estimate the city's stock of poor housing. By the lowest calculation, 3.5 per cent of Hamilton households lived in miserable housing in 1936; by the highest, 9.6 per cent were so burdened.[38] Statistics can camouflage. These percentages seem small, but what they mean on a human scale is that 5,200 to 14,500 people were in inferior dwellings and, if shelter reflects food, clothing, and ultimately health, their general predicament was serious. Many of them may have been among the 4,000 to 5,000 households receiving direct relief in the mid-1930s.

The worker's dream of escape from the assaulted environment of the city's north-east quadrant – a dream articulated in an urban-planning meeting of November 1944 but surely entertained for some time before that – came true for many. The conditions of the 1930s and 1940s changed greatly for the better, partly because of the sheer good fortune of Hamilton's unique setting. The difficulty of hauling industrial materials up and down the mountain had preserved this buffered expanse from factory development. A post-war housing construction boom converted the rural lands of the mountain into a new dormitory for employees in the city's factories and offices. In 1949, not long after the suburban boom began on the mountain, its

character as a vast tract for labouring families and the initally poor
state of services were perceived as engendering a potential political
menace. Reactionaries feared that a new mountain ward would
strengthen the "left wing" on council.[39] Rolling back the urban frontier southward across farmlands on Hamilton mountain, developers
definitely transformed the city's shape and its housing stock during
the 1950s and 1960s. The political impact has been more indeter-
minate, for the social composition of the mountain became mixed.

In 1966 Hamilton differed remarkably from the city of 1936,
requiring an altered diagram of large regions for spatial analysis.
The new layout with accompanying data proclaims two firm social
observations: first, the appearance of the mountain as a fairly prosperous bedroom region for well-paid industrial labourers and white-
collar employees and, second, the concentrations of common labour
in the central city and north-east quadrant. Data on the mountain
declare the essentially middle-class composition of the household
heads encamped on its many tracts. Middle class, a term studiously
avoided for reasons spelled out in chapter 7, seems to fit here in a
crude way, capturing the very middling occupational character of
household heads. White-collar employees and skilled and semi-
skilled labourers were somewhat over-represented; professionals
and proprietors were not especially plentiful and common labourers
were scarce. The diagram splits the mountain into east and west
halves at Upper Wentworth Street, but residents of Hamilton have
accepted the more westerly Upper James Street as the dividing line.
From there eastward lay "the east mountain" where, serviced by
access roads cut into limestone to connect with the lower city, blocks
were packed with skilled labourers. Some, but not all, of these same
blocks had many white-collar household heads. Household heads in
the latter group occasionally concentrated in scenic blocks on the
mountain brow overlooking the city, the older but highly desirable
Westdale area, and streets around the green expanse of Gage Park.

The 1966 illustration of occupational groups has a few elements
in common with the 1936 one. As we have stressed frequently, urban
change occurs within the massive inert bulk of the previously built
environment. Therefore, a large number of the city's industrial
workforce remained in older housing stock below the mountain. For
those who dwelt on this expanse of grid surveys created before 1914,
the environmental discomforts and hazards that stretched back a
half century had grown. Hamilton's two steel companies had expanded greatly since 1944, when the planning report had recorded
disenchantment with the environment. A few declarations about
east-end residents' affection for their homes and neighbourhoods

and their loathing of the local environment surfaced in the 1970s, both in a federal survey of public opinion and in a remarkable set of planning documents.

The short-lived Ministry of State for Urban Affairs sponsored a 1979 survey of public concerns in 23 Canadian urban centres. Inflation and unemployment headed the lists of concern everywhere, but in Hamilton air pollution place third. Hamiltonians responded positively about their city's cultural facilities, felt less enthusiastic about its appearance, and were distinctly unhappy about pollution. The fouling of the city's air was, like all hazards, spread unevenly amongst the city's neighbourhoods – in other words, amongst its occupational groups. In 1967 the Hamilton planning department divided the city into 136 planning areas. Planners worked with neighbourhood leaders and through public meetings to determine what the residents of the city wanted for their locales. The householders closest to the heavy industry belt, in about 1,500 affected houses, complained of air pollution levels that were far in excess of government standards. An analysis of mortality variations in the city during the 1980s reported the annual death rate for males 55–64 as over 30 per 1,000 in this region. It was merely 0–9 in Westdale, the exclusive middle-class district. Cautious about the implications of their study and its meaning in terms of the urban environment, the researchers simply – but dramatically enough – concluded that spatial variation in income was related strongly to spatial variation in the probability of dying.[40] A resident of the McAnulty Neighbourhood, one of the so-called enclaves in the industrial region and within a high mortality area, wrote that the northerly winds of winter made pollution so strong "that one can't breath at times." "Dust and dirt" followed by "bad smells" led a list of residents' complaints. Nevertheless, most residents hoped to preserve their homes and presented planners with a dilemma, because the city gladly would have sacrificed the enclaves to industrial expansion.[41]

A little further away, in the Homewood neighbourhood, residents contended with air pollution only slightly less oppressive. However, they experienced the heavy traffic that rumbled along the east-west arteries that characterized Hamilton because of its unusual topography.[42] To return to the mountain, the post-war flight to that sanctuary, whose surveys were designed to inhibit through-traffic, benefitted only a minority of households in the 1960s and 1970s. But planners, homebuilders, and shopping-mall developers realized that the city's residential future would depend on the mountain. In the 1980s it has been the site of virtually all residential construction. A product of corporate and governmental interaction, the mountain exemplified a post-war consensus. Radical politics never developed forcefully and the state delivered aid for home ownership and supervised development. The 1979 survey on the public concerns of urban Canadians indicated a quite high degree of satisfaction with new suburbs, which in Hamilton meant the mountain.[43]

Another vivid feature of the 1966 diagram concerns common labourers. Repeatedly noted as a much segregated group, they remained strikingly concentrated. They found the housing they could afford – though that affordability is debatable – in the traditionally inferior environment of the old city where, since the middle-class suburban boom of the 1920s, the city's smaller dwellings were concentrated. The housing of Corktown, much of which had been acceptable 30 years earlier, had slipped a notch, marking the area for eventual urban renewal. It was in a classic state of transition with all that implied about neglectful landlords. The "north end" had remained a region of deteriorating housing; a 1963 urban renewal report compiled evidence on its poor state. On the city's retail streets, a number of tenements and cheap apartments above shops and offices also housed some of the city's poorer workers. A 1965 brief from the city's health department estimated that at least 500 families (less than 1 per cent of the city's households), including low-income families, welfare recipients, and old age pensioners, lived in poor housing. Some of this accommodation was located over stores along commercial streets.[44]

Barton and Cannon Streets, remaining heavily travelled routes to and from the factories, were flanked by residential blocks in decline. The traffic counts on the major arteries like Main and Cannon Streets were about 1,000 vehicles per hour over 24-hour periods in 1985–6.[45] Bereft of significant public green space, the central city bounded on the diagram by Queen Street, the mountain, and Ferguson Avenue had a great number of dwellings in transition. Many of the subdivisions surveyed during the era of corporate development had begun to lose population in the 1950s, presumably to the new suburbs on the mountain and elsewhere. But the decline was balanced by several positive developments. From the 1960s to the 1980s government subsidized housing, neighbourhood planning initiatives, the construction of new schools, expansion of central-city recreational and cultural facilities, and debates about how to work to achieve social justice in social housing programs, demonstrated political recognition of neighbourhood and environmental problems affecting the poor.

In the 1980s some members of the city council tried to achieve a distribution of subsidized housing in assorted areas on the mountain so that poorer families had proper access to schools, parks, and services. Their efforts had imperfect results and have been hotly contested. Environmental inequalities have endured, but the deprivations of 150 years have been neither politically ignored nor readily defended in recent years. Redressing them has been quite another matter, for better-off citizens remain loath to see their neighbourhoods designated for new rent-geared-to-income housing units. Moreover, zoning by-laws and the shortage of available land in the old city effectively pushed the city's social housing first onto slum clearance tracts and then into the outermost ring of new suburbs. Zoning, which has consolidated existing conditions, and the city's inherited housing stock have sustained the basic patterns of historical segregation.

The environmental felicities of the city's south-west end, including the extensive green spaces of parks, a golf course, and university grounds, fortified its status as a region with preferred home areas for professionals and proprietors, notably in Westdale and on the short streets that ran south off Aberdeen Avenue and into the mountain's base, which since World War I has been a city-owned greenbelt. Other favourite professional and proprietor group districts included the Gage Park area and scenic lookouts on the mountain brow. Professionals and proprietors shared these site preferences with a number of white-collar household heads. Both the natural setting of the city – its exceptional array of vistas – and the creation of green space by park reformers from the early 1900s to the 1930s established tracts of favoured land. A 1923 pamphlet from the Canadian Department of Health, *How to Build the Canadian House*, stated a common piece of advice that very few city dwellers could afford to heed: "Parks, Playgrounds and Open Spaces: Buy your lot as near them as possible."[46] The more affluent have done precisely that.[47]

CONCLUSIONS

Our conclusions about housing quality may be characterized as ambiguous, debatable, and thoroughly tenable. They rest on problematic data, assorted contemporary reports, and common sense. A final distillation recommends the thesis of a democratization. On the positive side, the indexed housing values for occupational groups have moved toward the mean, a convergence that implies that socially the city had become a more homogeneous place. Until the 1960s the property industry had supplied new housing as a fairly homogeneous necessity. In terms of the number of rooms, general conditions, and amenities, for most of the twentieth century skilled and semi-skilled workers obviously attained housing superior to that acquired by their late nineteenth-century counterparts. The massive development of the mountain in the 1950s delivered a residential environment that further enhanced the city's stock of shelter. Much of that shelter was accessible to households headed by skilled and semi-skilled labourers. Rising home ownership, therefore, had been gained without any sacrifice of quality in neighbourhood amenities or environment. In fact, home buying secured better shelter, especially for households with children, than tenancy. For the numerous European and United Kingdom immigrants settling in Hamilton in the 1950s, the contrasts in shelter quality available to them at their points of origin and in a prosperous North American industrial city must have been conspicuous.

Here the uncomplicated paeans to progress must trail off. Imperfections are readily listed. The housing story continues, of course, and its next chapters might have to deal with crises effacing or revising the twentieth-century record of gains. Stacked against the advances of the past are more than the pessimistic speculations about the future. Housing history provides ambiguous observations and definite problem areas. The marketplace allocation of housing and housing environments has generated persistent inequalities, most emphatically bearing on the lives of common labourers – the working poor – and increasingly in the 1980s on single mothers. As a factor in their lives, the city's housing dross has been stubbornly significant. If sanitation improved for them in the early twentieth century, then dense factory smoke began to poison the air in those parts of the city where they would concentrate. For other occupational groups, in fact for a considerable majority, housing conditions have been good. This judgment can stand even during the worst economic collapse in this century.

In the middle of the 1930s – admittedly not the peak of the crisis, for that occurred in the spring of 1933 – overcrowding afflicted a fraction of households. Households in poor shelter, as defined by fairly high standards, perhaps numbered less than one in ten. That is not an inconsiderable body of people – the dirty thirties brought plenty of discomfort and discomfiture to Hamiltonians. Hundreds and probably thousands left the city, and family formation was deferred by many. But the depression did not plunge the city's multitudes into squalor. Additionally, the crisis was superseded by an outpouring of new and accessible housing in better urban environments. Marginalized groups have always lost out. Common labourers have persistently missed sharing in the improvements to the degrees experienced by skilled and semi-skilled labourers. In the last several decades, moreover, increasing numbers of family breakups have thrust many more women into positions as formal household heads. Lacking equal pay for equal work, they have often been denied access to home ownership, not to mention decent rental accommodations. Assorted categories of the urban homeless also have missed the gains.

The distribution of some of Hamilton's prosperity to many of its citizens, measured by home buying and shelter improvements, may have had a part to play in the largely classless politics of the city over roughly the last forty years. Considering a proposition similar to this for all of Canada, geographers Richard Harris and Chris Hamnett felt that theoretically there were grounds for assuming different political behaviour for home owners and tenants. Summarizing a number of studies, they concluded that locally, in terms of neighbourhood property values, and nationally, in terms of social policy, owners are likely to be more conservative than tenants.[48] They proposed this as a possibility. If correct, consider the parallel and reinforcing influence of a set of improvements in housing and environmental quality. Consider too that for many decades the vast majority of households have expected and, at some point in their life cycles, have attained home ownership. The world of work may act as the principal influence on class formation, but lives are not neatly divided between labour and domestic conditions. Rising ownership and improvements in housing quality must have helped to bolster social satisfaction. It may be, as Harris and Hamnett proposed, that home ownership is "one of the reasons why socialism has never gained a hold on the hearts and minds of the American working class".[49] Whether this assertion applies equally to Canada depends on whether one regards the historical positions of the Cooperative Commonwealth Federation and the New Democratic Party as indicative of a socialist attribute in Canadian political life. Harris and Hamnett may be wrong for another reason. To repeat a conclusion arrived at by sociologist Lois A. Vitt, working-class house owners in the United States do not appear to feel that proprietorship has placed them in the middle class. Leaving aside such controversies, we come to an important point for speculation about the future of housing.

The historical record is difficult enough to fathom and should make historians chary of crystal-ball gazing. However, throughout many chapters, we have introduced notes of apprehension about high expectations and the growth of recent constraints on the delivery of home ownership, on the extension of improved housing, and on neighbourhood environments. If housing tenure and character have fed social satisfaction among the majority, what might happen if, for many plausible reasons, thresholds of access rise significantly and quality cannot be maintained? Are the Hamilton and Toronto housing crises of the 1980s – first high interest rates (1981) and then soaring house prices (1987) – the first shocks of many to hit the young in middle-income groups? The shocks are not localized. That Canadian and American housing experts have been asking the same questions about the meaning of housing and finance crises in the 1980s is a further measure of parallels in the basic ways of supplying shelter in North America. And what becomes of conservatism if a bit of its bedrock is shaken up by what seems like a significant abandonment of the middling occupational groups by land developers and builders who have demonstrated a recent preference for "the up-scale market"?

The effort to examine the city-building process by an historical account of housing, essentially in one city, ends at a critical point. Perhaps housing as a social concern has been a more acute problem in the United States than in Canada, where there has been a more vigorous recent history of government support for social housing. Nevertheless, urban shelter in the two countries shares several challenges. One of these is not especially new. In this chapter, we noted that despite generally high standards of housing quality, the families of common laboureres have not fared well. As David Cannadine supposed, Divine authority and empirical observation agree on one point: the poor are always with us.[50]

Recently, possibly for the last ten to fifteen years, home ownership has been less attainable at an early stage in the life cycle than it was in the 1950s and 1960s. Also, the rental market has serious problems. Very low vacancy rates suggest a shortage of rental units for those prospective tenants who can afford them. Across North America, the public is becoming aware of a variety of needy or disturbed individuals who have been labelled "homeless." The title of a 1988 public television documentary program stated the condition of the low-income households and of the homeless: "Locked Out of the American Dream." But in both countries there is a widening discussion of housing problems and an astonishing assortment of private sector, service club, public sector, and mixed measures that reach a few of those in most need.

A great deal is happening, but what it amounts to is not clear. Few appreciate the details and scope of the vast field of housing, and the general public probably has only the vaguest notion of housing issues and programs, let alone a sense of the city-building processes that affect their lives. Many individuals, however, most assuredly have an acute sense of how housing affects their budgets, comfort, and expectations. Periodically in the twentieth century, Canadian governments have sponsored housing-policy appraisals. The most recent, the Hellyer task force, published its report almost twenty years ago. Its second principle ambitiously and courageously proclaimed that "every Canadian should be entitled to clean, warm shelter as a matter of basic human right."[51] With an expanding stock of academic literature available, public interest piqued by interest-rate fluctuations and rising prices, and media attention given to the homeless and to the rental market, a new public inquiry would seem both timely and appropriate. And such an inquiry must attempt to supply the public with the information that will enable it to examine its own obsession – the will to possess – as well as the technical features of development, construction, and finance. More than likely, in an area rife with tradition, continuity, and economic interests embedded in capitalism, the short term will bring ingenious technical fine tuning. That is, more episodes to fill out the theme of evolutionary gradualism already detected in a history of housing covering the last one hundred and fifty years.

From one angle, this conclusion casts the future in a grimly deterministic pattern. It suggests a crisis as the North American city falters in its delivery of the cultural icon of home ownership. It expresses impatience about a city-building process that confronts bottlenecks with tinkering rather than with bold initiatives, and indignation about the marginalization of the poor. However, it can also be seen to celebrate ingenuity. As many parts of the world are more painfully aware than North America, the provision of shelter – let alone improvement in its quality – is an awesome task. This book has documented efforts by many people who have had to cope with obstinate and evolving challenges in very different areas of an essential industry. Some were more admirable than others, but our few glimpses of their activities underscore cunning leading to invention. Ultimately, the ambivalence in this conclusion and in numerous specific points in the narrative and analysis is about more than past, present, or future housing. It is a wavering appraisal of North America in the age of capitalism.

The Assessment Roll Sample

Assessment rolls have long been recognized as important research tools in the study of the city-building process.[1] A rich, ubiquitous, and routinely generated data source, the annual municipal taxation records provide, in historical geographer Peter Goheen's words, "a systematic and comprehensive inventory of the economic and social characteristics of the population and its environment ... The Rolls provide a fairly comprehensive profile of social and economic data by [household]. Included, also, are data on land use, land value, and condition of tenure."[2] In our research project, assessment records have been used to provide insight into questions relating to several major topics – the land-development process, tenure characteristics, neighbourhood development, and housing quality.

For most of North American urban history, taxes on property have been the prime source of municipal revenue. The assessment rolls, which are generated every year and identify with a good deal of precision *all* of the properties subject to taxation, are the documents used for this tax collection. While primarily concerned with the taxable characteristics of property – location, lot size, and land use, the rolls also incorporated many other variables, though these varied over time (see table A.1). Other variables, such as more precise meanings of land use, could often be inferred from the information listed. For example, vacant lots could be identified because they had neither an address nor an occupant, while vacant houses had a street address but no occupant. In Ontario the assessment records were generally organized by ward. A separate volume was prepared for each of these electoral units, and within each book, entries were arranged according to street and address. Each line in the rolls, therefore, represented the assessment information for a single property.

Table A.1
Assessment Roll Variables

Variables	Comments/Restrictions
OCCUPANTS	
Name	all years
Age	all years, sometimes incomplete
Occupation	all years
Marital Status	from 1881
Tenure/Franchise	all years
Juror	from 1881
Household Size	from 1881
Children 5-21	1881–1931
School Support	Public/Separate, from 1881
Religion	from 1926
British Subject or Alien	from 1906
Income	1852 to 1899
OWNERS	
Name	all years
Occupation	from 1866
Residence	from 1866
Marital Status	from 1881
School Support	from 1881
Religion	from 1926
British Subject or Alien	from 1926
THE PROPERTY	
Ward	all years
Street Address	all years
Lot Number	all years
Assessor's Comments	all years, re land use
Lot Frontage	from 1926
Lot Depth	from 1926
Land Use	inferred from other variables
Building Material	1856–76, sometimes incomplete
Value of Property and Buildings	all years, in dollars from 1858
Exemptions	from 1896
Business Assessment	1906–1921
Value of Land	from 1911
Value of Buildings	from 1911

Hamilton's assessment records are extant from 1847, and the rolls from that date until 1973 occupy some 78 metres of library shelving.[3] Not surprisingly, most researchers have drawn samples from the annual municipal taxation records, and we have followed this procedure for our research. While our sample is random, it differs from most others in its spatial orientation. The basic foundation for our assessment-roll sample was the ordinary, four-sided city block. A 10 per cent sample of all blocks within Hamilton's city limits, excluding the forty-one blocks that constituted the

Table A.2
Size of the Assessment-Roll Sample, 1852–1966

Year	Blocks	Developed Properties		Undeveloped Properties	Total Assessments
		Residential	Other		
1852	50	152	6	251	409
1856	50	300	6	370	676
1861	50	391	8	388	787
1866	50	371	0	376	747
1872	50	468	7	252	727
1876	50	723	2	216	941
1881	50	827	26	176	1029
1887	50	1099	2	129	1230
1891	60	1081	5	135	1221
1896	60	1315	25	67	1407
1899	60	1290	22	209	1521
1906	60	1378	30	296	1704
1911	85	1793	42	818	2653
1916	91	2380	107	614	3101
1921	91	2544	114	668	3326
1926	104	3192	148	621	3961
1931	120	3551	176	738	4455
1936	120	3768	187	703	4658
1941	120	3956	214	565	4736
1956	178	6535	358	698	7591
1966	187	6116	732	149	6997

city's mid-nineteenth-century central business district (CBD), was randomly selected for several times between 1852 and 1966.[4] Until 1941 the samples were spaced at about five-year intervals, with the few exceptions caused by missing or damaged records. To reflect changes in the post-World War II period, the two most recent samples were drawn for 1956 and 1966 (see table A.2).[5] Throughout the sampling procedure, a conscious effort was made to ensure that a stratified random sample was produced for each year chosen.

For each of the twenty-one cross sections, assessment-roll data were gathered, coded, and converted to machine-readable form for every identifiable property within the boundaries of the blocks included in the sample.[6] In all, 53,877 properties or cases were included in these data sets. The number of properties per block ranged from a low of one, in the case of the presence of large estates, to a high of about 125 in densely developed areas of the inner city. Most blocks, however, contained between 30 and 50 assessed properties, with the average for 1966 being 37.4. Once a block was selected for inclusion in the sample, it remained in all subsequent cross sections, so we have up to 22 sets of information for a given piece of land. As new areas were annexed by the City of Hamilton, the blocks within these districts were subjected to the same 10 per cent, stratified, random sampling procedure that had been used for blocks inside the original corporate limits of the city. Our sample, therefore, grew in direct relation to the substantial areal expansion of Hamilton.

Several variables were added to some of the data sets at the coding stage. In some cases, this involved the manipulation of information that was already on the assessment rolls. For tenant-occupied dwellings, for example, a variable was created to indicate the proximity of the landlord's residence to his or her rental properties.[7] Our new variable distinguished among four types of landlords – those who resided in the same building, those on the same city block, those elsewhere in Hamilton, and those outside Hamilton. From 1926 onwards an attempt was made to record the ethnicity of the assessed residents of our sample blocks, through an analysis of surnames and religious affiliations carried out with the assistance of the Hamilton Multicultural Society.

Measures pertaining to the infrastructure of the city and culled from other sources, especially the *Reports* of the City Engineer's Department and various local transit companies, were added to a number of the data files. For the period from 1864 until 1921, the presence or absence of water and sewer lines along the frontage of each property was recorded as a binary variable.[8] Distance, in tens of metres, to the nearest street railway line was entered for each lot between 1876 and 1921. By this latter date, Hamilton's utility and transit systems were virtually complete (see chapter 2). It was felt that the effort required to include such measures in our data files was no longer warranted.

One final file was also produced to summarize the character of the blocks included in the sample. For this file, maps, aerial photographs, and planning documents were used to determine ten variables: the average elevation of the block; its area; the year in which it became 50 per cent developed; the year in which it became 80 per cent developed; and six measures of the proximity of the block to selected attributes of the city (the CBC, the closest secondary shopping area, the nearest industrial district, the nearest rail line, the nearest park, and the nearest sports field).

APPENDIX B

Leading Hamilton Builders, 1978–1984

Name of Firm	Year	Production	Total
Geofcott Properties (1977) Limited	1978	52	
	1979	129	
	1980	87	
	1981	77	
	1983	59	
	1984	141	545
C. Valery Construction Limited	1978	69	
	1979	59	
	1980	70	
	1981	58	256
George Wimpey Canada Limited	1981	70	
	1982	54	
	1983	122	
	1984	79	325
Alterra Developments Limited	1979	86	
	1981	65	
	1984	64	215
Eastcrest Homes II	1978	100	
	1979	175	
	1983	67	406
Empire Developments (Hamilton) Limited	1979	75	
	1983	99	
	1984	60	234
Laurier Homes Limited	1978	117	
	1979	73	
	1983	107	297

Name of Firm	Year	Production	Total
Picture Homes Development Inc.	1978	117	
	1979	73	
	1980	55	245
Victoria Wood Development Corporation Inc.	1979	132	
	1980	88	
	1984	91	311
Eagre Holdings Limited	1978	84	
	1979	107	191
Ennisclaire on the Lake	1981	170	
	1984	168	338
Greenleaf Investments Inc.	1978	73	
	1979	74	147
Raleigh Chase Developments Limited	1979	60	
	1980	56	116
Samuel Sarick Limited	1981	60	
	1982	94	154
Skopit Homes	1980	102	
	1981	123	225
Starward Homes Limited	1983	55	
	1984	52	107
A.B. Cairns Limited	1983	78	
Anprop Investments Inc.	1984	65	
Bezic Pressaco Limited	1981	66	
Cedar Mill Homes Limited	1984	61	
Coventry Group Limited	1979	78	
CMHC	1979	132	
Costain Limited	1984	63	
Fieldgate Developments	1979	54	
GRW Construction Limited	1979	216	
Guernsey Acres Limited	1978	53	
Kelvingway Investments Limited	1980	100	
Longmark Homes	1979	61	
Meadows Homes Limited	1981	225	
Pine Falls Investments Limited	1984	62	
Poplar (Sheridan) Developments Inc.	1980	59	
Provincial Bank of Canada	1979	57	
Revenue Properties Development Limited	1980	56	
Rob Geoff Properties Limited	1983	84	
Sandbury Building Corporation	1979	168	
Scanbark Construction Limited	1981	54	
Sifton Properties Limited	1979	69	
Stardale Developments Inc.	1983	157	
Stargrove Developments Inc.	1981	235	

Name of Firm	Year	Production	Total
Taro Properties Limited	1979	69	
T. Valeri Construction Limited	1983	61	
Yorkwood Investments Limited	1979	56	
337078 Ontario Ltd. c/o Northville Heights	1978	76	
408417 Ontario Limited c/o B. Harris Towers	1978	182	
502665 Ontario Limited	1982	109	
542842 Ontario Limited c/o Lakeview Estates	1983	85	
150 Wilson Street Limited	1984	55	

Source: Ontario New Home Warranty Program, 1978–1984.

Notes

INTRODUCTION

1 G. Herbert Palin, "The City Builder," Ontario Archives, Dinnick Papers, MU 919, Toronto real estate file.

2 Roy Lubove, "The Urbanization Process: An Approach to Historical Research," *American Urban History: An Interpretive Reader with Commentaries*, ed. Alexander Callow Jr (New York: Oxford University Press, 1969), 650.

3 H.V. Nelles and Christopher Armstrong, *Monopoly's Moment: The Organization and Regulation of the Canadian Utilities, 1830–1930* (Philadelphia: Temple University Press, 1986). John R. Stilgoe, *Metropolitan Corridor: Railroads and the American Scene* (New Haven: Yale University Press, 1983).

4 Allen D. Marvel, "Land Use in 106 Large Cities," *Three Land Research Studies*, research report no. 12 (Washington D.C.: United States Government Printing Office, 1968).

5 Kenneth Jackson, *Crabgrass Frontier: The Suburbanization of the United States* (New York: Oxford University Press, 1981).

6 Theodore Hershberg, Alan Burstein, Eugene Ericksen, Stephanie Goldberg, William Yancey, "A Tale of Three Cities: Blacks, Immigrants, and Opportunity in Philadelphia, 1850–1880, 1930, 1970," *Philadelphia: Work, Space, Family, and Group Experience in the 19th Century*, ed. Theodore Hershberg (New York: Oxford University Press, 1981), 476, 483–4.

7 Robert Fishman, *Bourgeois Utopias: The Rise and Fall of Suburbia* (New York: Basic Books, 1987).

8 Jackson, *Crabgrass Frontier*, 296.

9 Ibid., 288.

10 Canada, Advisory Committee on Reconstruction, Report 4, *Housing and Community Planning: Final Report of the Subcommittee* (Ottawa: King's Printer, 1944), 90.

11 Tracy Kidder, *House* (Boston: Houghton Mifflin, 1985).

12 Philip Langdon, "The American House," *The Atlantic* (September 1984): 74.

13 Canada, Statistics Canada, *Mortgagor Households in Canada: Their Geographic and Household Characteristics, Affordability and Housing Problems* (January 1988), 48–51.

14 A.J.C. King, *The Adolescent Experience* (Brampton: Ontario Secondary School Teachers' Federation, 1985).

15 Fishman, *Bourgeois Utopia*, 194–5.

16 Michael Katz, *The People of Hamilton, Canada West: Family and Class in a Mid-Nineteenth Century City* (Cambridge, Mass.: Harvard University Press, 1975); Michael B. Katz, Michael J. Doucet, and Mark J. Stern, *The Social Organization of Early Industrial Capitalism* (Cambridge, Mass.: Harvard University Press, 1982).

17 Bryan D. Palmer, *A Culture in Conflict: Skilled Workers and Industrial Capitalism in Hamilton, Ontario, 1860–1914* (Montreal: McGill-Queen's University Press, 1979).

18 Sam Bass Warner Jr, "If All the World Were Philadelphia: A Scaffolding for Urban History, 1774–1930," *The American Historical Review* 74 (October 1968): 26–43.

19 Marc H. Choko, *The Characteristics of Housing Tenure in Montreal*, research paper no. 164 (Toronto: Centre for Urban and Community Studies, 1987).

20 Robert S. Lynd and Helen Merrell Lynd, *Middletown: A Study in Contemporary American Culture* (New York: Harcourt Brace, 1929); *Middletown in Transition: A Study in Cultural Conflicts* (New York: Harcourt Brace, 1937); Theodore Caplow and Howard M. Bahr et al., *Middletown Families: Fifty Years of Change and Continuity* (Minneapolis: University of Minnesota Press, 1982).

21 Michael Goldberg and John Mercer, *The Myth of the North American City: Continentalism Challenged* (Vancouver: University of British Columbia Press, 1986).

22 Michael Doucet and John C. Weaver, "Town Fathers and Urban Continuity: The Roots of Community Power and Physical Form in Hamilton, Upper Canada in the 1830s," *Urban History Review* 13 (October 1984): 76.

23 J. Perry Lewis, *Building Cycles and Britain's Growth* (London: St. Martin's Press, 1965).

24 Fishman, *Bourgeois Utopias*, 111.
25 Ibid., 115; Carl E. Schorske, *Fin-de-Siècle Vienna: Politics and Culture* (New York: Vintage Books, 1981), 24–61.

CHAPTER ONE

1 On the evolution of Ontario's Land Registry system see Michael J. Doucet, "Building the Victorian City: The Process of Land Development in Hamilton, Ontario, 1847–1881," ph d thesis, University of Toronto, 1977, 111–18. The principle of *caveat emptor* applies to land sales even today. A recent case in Mississauga, Ontario, saw a builder sell unregistered lots to unwitting buyers. See Gary Webb-Proctor, "Ban Sale of Unregistered Lots, Protestors Tell Province," *Globe and Mail*, Monday, 24 April 1989, A14.
2 We, like Paul-André Linteau in "Canadian Suburbanization in a North American Context: Does the Border Make a Difference?" *Journal of Urban History* 13 (1987): 255, take for granted the presence of speculators in the urban land-development process. On the nature of nineteenth-century land speculation see Michael J. Doucet, "Urban Land Development in Nineteenth-Century North America: Themes in the Literature," *Journal of Urban History* 8 (1982): 299–342. On land speculation in Upper Canada see J.K. Johnson, "The Businessman as Hero: The Case of William Warren Street," *Ontario History* 45 (1973): 125–32 and Leo A. Johnson, "Land Policy, Population Growth, and Social Structure in the Home District, 1793–1851," *Ontario History* 43 (1971): 41–60.
3 Throughout the mid-1980s, for example, an extensive advertising campaign in both the electronic and print media urged Toronto-area residents to "Come to Life in Erin Mills." These ads and others like them had very little to do with actual housing units (some television commercials barely showed houses at all), but a great deal to do with the lifestyles that the new subdivisions offered. Early in 1990, residential-development giant Bramalea, Limited adopted a new slogan for its newspaper advertisements – "Building a Better Quality of Life." For a brief discussion of this lifestyle approach to development see Denis Grayhurst, "Teron's Philosophy: Build Communities and Not Just Houses," *Toronto Star*, Saturday, 30 August 1986, E1.
4 Warren Potter, "Your New House Stands at the End of a Long Paper Trail," *Toronto Star*, Saturday, 21 November 1987, E1.
5 Michael Doucet and John Weaver, "The North American Shelter Business, 1860–1920: A Study of a Canadian Estate and Property Management Agency," *Business History Review* 58 (1984): 234–62 and "Material Culture and the North American House: The Era of the

Common Man, 1870–1920," *The Journal of American History* 72 (1985): 560–87. On multifamily dwellings see Robert G. Barrows, "Beyond the Tenement: Patterns of American Urban Housing, 1870–1930," *Journal of Urban History* 9 (1983): 395–420.

6 Few full-length treatments of the residential-development process exist for North American cities for the period before 1880. The major exception is Doucet, "Building the Victorian City." However, development issues do receive some attention in a number of books, including Michael H. Frisch, *Town Into City: Springfield, Massachusetts, and the Meaning of Community, 1840–1880* (Cambridge: Harvard University Press, 1972) and Kathleen Neils Conzen, *Immigrant Milwaukee, 1836–1860: Accommodation and Community in a Frontier City* (Cambridge: Harvard University Press, 1976).

7 Louis Gentilcore and Kate Donkin, *Land Surveys of Southern Ontario. Cartographica*, monograph no. 8. (Toronto: Department of Geography, York University, 1973). Most of the area of the United States lying west of the Ohio River was surveyed into a grid based on one-mile by one-mile squares as a result of the passage by Congress of a land ordinance in 1785. A similar survey system was used on the Canadian Prairies. See Norman J.W. Thrower, *Original Survey and Land Subdivision: A Comparative Study of the Form and Effect of Contrasting Cadastral Surveys* (Chicago: Rand McNally, 1966) and John Langton Tyman, *By Section Township and Range: Studies in Prairie Settlement* (Brandon: Assiniboine Historical Society, 1972). A similar grid system also prevailed in Australia. See R.J. Johnston, "An Outline of the Development of Melbourne's Street Pattern," *Australian Geographer* 10 (1968): 453–65.

8 On early plans for North American cities see John W. Reps, *The Making of Urban America. A History of City Planning in the United States* (Princeton: Princeton University Press, 1965). In spite of the subtitle, Reps does deal with New France. See also Gilbert A. Stelter, "The Classical Ideal: Cultural and Urban Form in Eighteenth-Century Britain and America," *Journal of Urban History* 10 (1984): 351–82 and Gerald Hodge, *Planning Canadian Communities: An Introduction to the Principles, Practice and Participants* (Toronto: Methuen, 1986), 33–49.

9 Michael Doucet and John C. Weaver, "Town Fathers and Urban Continuity: The Roots of Community Power and Physical Form in Hamilton, Upper Canada in the 1830s," *Urban History Review* 8 (1984): 75–90. See also Isobel K. Ganton, "Land Subdivision in Toronto, 1851–1883," in G.A. Stelter and A.F.J. Artibise, eds, *Shaping the Urban Landscape: Aspects of the Canadian City-Building Process* (Ottawa: Carleton University Press, 1982), 200–31.

10 Wentworth County Land Registry Records, Registered Memorials, A-16, Margaret Rousseau, Executrix, to James Mills and Peter Hess,

dated 12 June 1816 and registered 13 July 1816; and A-18, A Memorial of a Deed of Partition, dated 24 June 1816 and registered 13 July 1816. The price paid for the land is recounted in Stanley Mills, *Genealogical Record of the Mills Family* (Hamilton, 1910). On the early history of Hamilton, see John C. Weaver, *Hamilton: An Illustrated History* (Toronto: James Lorimer and Co., 1982).

11 Ontario Archives, Genealogical Society of the Church of Jesus Christ of the Latter Day Saints, microfilm reel 1411, number 48 James Mills to Abraham Smith, 31 January 1824.

12 *Western Mercury*, 26 June 1834; Hamilton Public Library (hereafter, HPL) clipping file on Mills family, *Hamilton Spectator* (hereafter the *Spectator*), 15 July 1926, quoting the *Free Press* of 1821. The quotation is taken from Willson's advertisement in the *Spectator*, Wednesday, 26 July 1848, 4.

13 HPL, Hamilton Collection, *Marcus Smith's Map of Hamilton, 1851–2. Hamilton Journal and Express*, 23 January 1846. According to the assessment roll for 1852, some 228 lots remained vacant in the Mills surveys.

14 Prices were obtained from the Ontario Archives, Genealogical Society of the Church of Jesus Christ of the Latter Day Saints, microfilm reel 1411, Land Registry Memorials of Wentworth County, number 32 James Mills to Obediah Mills, 23 December 1846; number 38 James Mills to John W. Mills, 4 February 1840; number 91 James Mills to Carney Sanders, 23 December 1846; number 104 Nelson Mills to Allen Nixon, 15 June 1848; number 117 James Mills to James Cummings, 23 December 1846; number 120 James Mills to Nelson Mills, 7 October 1843; number 168 Samuel Mills to R.H. Bengough, 6 July 1847; number 243 James Mills to Alexander Durand, 9 February 1850; number 269 James Mills to David Mott, 31 October 1842; number 304 James Mills to Peter McCowen, 12 February 1838; and number 309 James Mills to Nelson Mills, 24 June 1842.

15 A data file on the holdings of the Mills family was assembled using the assessment rolls for the City of Hamilton. All properties belonging to anyone with the surname Mills were entered into the file and later cross checked against the Mills family tree found in Mills, *Genealogical Record*. See also HPL, Ledger Book of Real Estate of the Late John Walter Mills, 1–8. On wages in this period see James G. Snell, "The Cost of Living in Canada in 1870," *Histoire sociale – Social History* 12 (1979): 190–1.

16 Figures on the number of vacant lots were calculated from the City of Hamilton assessment rolls, MSS, for 1852 and 1881. Even in 1901 Hamilton's population stood at only 52,634, an increase of just 15,973 over the 1881 figure, in spite of the annexation of parts of Barton Township.

17 The figure for ex-urban lots was determined from an examination of all registered plans for subdivisions in Barton Township that lay within one-half mile of Hamilton's city limits. On the slowness of actual development in this area see Marjorie Freeman Campbell, *A Mountain and a City: The Story of Hamilton* (Toronto: McClelland and Stewart, 1966), 206. For a general account of the implications of premature subdividing in the United States see Philip H. Cornick, *Premature Subdivision and Its Consequences* (New York: Columbia University Press, 1938). For Canada, see Hodge, *Planning Canadian Communities*.

18 Moore and Davis letterbooks, 3 (11 July 1868), 935 and 5 (12 November 1873), 667. *Spectator*, Saturday, 26 September 1874, 3. On the scope and contents of the Moore and Davis records see Doucet and Weaver, "The North American Shelter Business."

19 The literature on the relationships between property development and economic cycles is extensive. See, for example, K.A.H. Buckley, "Urban Building and Real Estate Fluctuations in Canada," *Canadian Journal of Economics and Political Science* 18 (1952): 41–62; James S. Duesenberry, *Business Cycles and Economic Growth* (New York: McGraw-Hill, 1958); H.J. Habakkuk, "Fluctuations in House-Building in Britain and the United States in the Nineteenth Century," *Journal of Economic History* 22 (1962): 198–230; Homer Hoyt, *One Hundred Years of Land Values in Chicago* (Chicago: University of Chicago Press, 1933); C.E.V. Leser, "Building Activity and Housing Demand," *Yorkshire Bulletin of Economic and Social Research* 3 (1951): 131–49; J. Parry Lewis, *Building Cycles and Britain's Growth* (New York: St Martin's Press, 1965); Lewis A. Maverick, "Cycles in Real Estate Activity," *Journal of Land and Public Utility Economics* 8 (1932): 191–9; Helen C. Monchow, "Population and Subdividing Activity in the Region of Chicago," *Journal of Land and Public Utility Economics* 9 (1933): 192–206; Brinley Thomas, *Migration and Economic Growth: A Study of Great Britain and the Atlantic Economy*, 2nd ed. (Cambridge: Cambridge University Press, 1973); and J.W.R. Whitehand, "Building Cycles and the Spatial Pattern of Urban Growth," *Transactions of the Institute of British Geographers* 56 (1972): 39–55 and *The Changing Face of Cities: A Study of Development Cycles and Urban Form*, Institute of British Geographers, special publication 21 (Oxford: Basil Blackwell, 1987).

20 Ganton, "The Subdivision Process," 200–31.

21 In the 1970s the local land registry office finally dealt with the confusion associated with the areas covered by unregistered plans by creating new and registered plans, known as Registrar's Compiled Plans, based on currently established property lines. See Registered Plans 1467 to 1497, Wentworth County Registry Office.

22 Thomas Adams, *The Design of Residential Areas: Basic Considerations, Principles, and Methods* (New York: Arno Press, 1974 [1934]), 47.

23 There were, of course, slight deviations from the strict grid plan that were caused by certain physiographic features of Hamilton's site, especially the jagged shoreline, and by the diagonal nature of York Street, the main road to Toronto. The first truly non-grid plan in Hamilton seems to have been the Mountainview subdivision, which was touted as the next Rosedale (Toronto's elite district) and laid out shortly after 1900 (see chapter 2).

24 Ganton, "The Subdivision Process," 221.

25 For example, in 1891 Messrs Stuart and Scott registered a plan for the subdivision of land in the north-east section of Hamilton. This plan contained 367 lots, each with a frontage of 25 feet (Plan 209, Hamilton Registry Office). This area subsequently developed into a working-class district, and, in time, it came to house many of the city's steel workers. On variations in subdivision design in Toronto see Ganton, "The Subdivision Process."

26 Forty-two per cent of the 794 Toronto plans examined by Ganton for the 1851–1883 time period were resubdivisions of existing surveys. Ganton, "The Subdivision Process," 210.

27 On the relationship between the Hamilton City Council and land developers see Doucet, "Building the Victorian City," chapter 5.

28 The list of nonresident subdividers included the most prominent politician in the area in the 1850s, Sir Allan Napier MacNab, laird of Dundurn Castle, which was situated just west of the city; Malcolm Cameron, another politician of some repute; and Toronto politician John H. Cameron. On MacNab see Marion Macrae, *MacNab of Dundurn* (Toronto: Clarke, Irwin & Co., 1971).

29 Ganton, "The Subdivision Process," 228.

30 Michael B. Katz, "The Entrepreneurial Class in a Canadian City," *Journal of Social History* 8 (1975): 24–5.

31 Several other surveys also met these criteria, but were not included in the sample because records were incomplete or because considerable development had already taken place prior to the registration of the plan of subdivision. The surveys by J.H. Cameron and by W.L. Billings and J. Lister, which each contained slightly under 50 lots, were included in the sample to provide better areal representation of the various parts of the city.

32 Only MacNab and Cameron were not Hamiltonians. In the case of the former this is true only in a technical sense, for Sir Allan lived on the very edge of the city that he represented in the Provincial Legislature.

33 Obviously our information is not complete, for it has been based solely on lists of names published in announcements pertaining to newly chartered companies that appeared in the *Canada Gazette*, the official government newspaper. As time passed these lists of names

became less common, especially after legislation allowed the granting of charters without the passage of a special Bill.

34 See, for example, Howard L. Shannon and H. Morton Bodfish, "Increments in Subdivided Land Values in Twenty Chicago Properties," *Journal of Land and Public Utility Economics* 5 (February 1929): 29–47; W.N. Loucks, "Increments in Land Values in Philadelphia," *Journal of Land and Public Utility Economics* 1 (October 1925): 469–77; and G.B.L. Arner, "Land Values in New York City," *Quarterly Journal of Economics* 36 (August 1922): 545–80.

35 Shawn McCarthy, *The Sunday Star* [Toronto], 3 May 1987, F7.

36 John Herzog, "Structural Change in the Housebuilding Industry," *Land Economics* 39 (May, 1963): 139.

37 James McKellar, "Building Technology and the Production Process," in John R. Miron, ed., *Housing Progress in Canada Since 1945* (forthcoming), chapter 8. We wish to thank members of the Centre for Urban and Community Studies, University of Toronto, for allowing us to examine the manuscript version of this volume.

38 George Taylor Denison Account Books, 1840–1868, folio 118, MSS, Denison Papers, Toronto Public Library. The specified costs included expenses that were associated with the sale of the property ($22.10), taxes ($30), mortgage interest at 8 per cent paid to the estate ($80.19), and other expenses ($10). This left a net profit of only $72.70.

39 Only thirty-five, or 2 per cent, of the lots under study were seized from the original subdividers for this reason.

40 Robert P. Swierenga, *Pioneers and Profits: Land Speculation on the Iowa Frontier* (Ames: Iowa State University Press, 1968), 205–9.

41 Moore and Davis letterbooks 6 (9 August 1878), 895.

42 Banks, for example, were much less stable organizations than they are today, and the total loss of one's savings during a financial panic was all too common. For a discussion of the plight of Canadian banks during the last third of the nineteenth century see Tom Naylor, *The History of Canadian Business 1867–1914*, vol. 1 (Toronto: James Lorimer and Co., 1975), 118–54. According to Naylor (p. 89), in the 1870s most Canadian banks paid 5 per cent interest on savings accounts. This all changed in 1876, when collusion brought about an almost universal reduction to 4 per cent.

43 Based on figures in the Crown Lands Papers, Surveyors' Ledger, 1847–1871, folios 35, 69 and 146, MSS, Ontario Archives. For example, the 1850 town plot of Sydenham, Ontario, was surveyed in three days at a cost of about twelve dollars, while in 1853 some town lots in Peterborough were surveyed in five days at a total cost of about twenty-six dollars.

44 A.G. McMillan, *New Manual of the Costs, Forms and Rules in the Common*

Law Courts of Upper Canada (Toronto: Rolle and Adam, 1865), 109–10. Since the plans were all drawn by Provincial Land Surveyors (as set out in the Revised Statutes of Ontario, 1877, c. 146, s. 70), approval was no real problem for the subdividers.

45 McMillan, *New Manual*, 116–17, comments upon the variable rates that could be charged for such purposes by the members of the legal profession at that time. The figures suggested here derive from those in the George Taylor Denison Account Books, 1840–1868, folio 66, MSS, Denison Papers, Toronto Public Library.

46 Ann Durkin Keating, *Building Chicago: Suburban Developers and the Creation of a Divided Metropolis* (Columbus: Ohio State University Press, 1988), 6.

47 Moore and Davis letterbooks, 2 (5 October 1865), 472. See also the memorials requesting sidewalks in the Reports of Committees, 1859–1865, MSS, 536 and 722 (Report of the Board of Works, 10 November 1862 and 24 October 1864). Even this improvement was relatively inexpensive. For example, the Board approved the second petition cited above and, in so doing, authorized a 530-foot sidewalk on East Avenue (in survey 7), the cost of which was not to exceed $14. This works out to just 2.6 cents per linear foot. In other words, the cost of putting a plank sidewalk across the front of a 50-foot building lot was something in the order of $1.30.

48 Campbell, *A Mountain and a City*, 106–7. On the much greater attention paid by subdividers to services by the early twentieth century see chapter 2 and John C. Weaver, "From Land Assembly to Social Maturity. The Suburban Life of Westdale (Hamilton), Ontario, 1911–1951," *Histoire sociale – Social History* 11 (1978): 411–40; Ross Patterson, "The Development of an Interwar Suburb: Kingsway Park, Etobicoke," *Urban History Review* 13 (1985): 225–35; and Karina Bordesa, "A Corporate Suburb for Toronto: Lawrence Park 1905–1930." MA thesis, York University, 1980.

49 Moore and Davis letterbooks, 4 (8 May 1869), 163. *Spectator*, Wednesday, 15 May 1850. For an example of one of Willson's advertisements see the *Spectator*, Wednesday, 26 July 1848, 4. A recent study of the influence of trees on property values suggested that arboreal plantings could have as much as a 3.5 to 4 per cent influence on sale prices. See L.M. Anderson and H.K. Cordell, "Influence of Trees on Residential Property Values in Athens, Georgia (USA): A Survey Based on Actual Sales Prices," *Landscape and Urban Planning* 15 (1988): 153–64.

50 George Taylor Denison Account Books, 1840–1868, folio 120, MSS, Denison Papers, Toronto Public Library. According to Peter Spurr, *Land and Urban Development: A Preliminary Study* (Toronto: James Lori-

mer, 1976), 57–60, in the mid-1970s Canadian urban land developers regularly paid at least $10,000, and often as much as $20,000 to provide services for each of their building lots.

51 Moore and Davis letterbooks, 1 (30 March 1862), 378.

52 Most major breakthroughs in urban infrastructure occurred after 1880. For a discussion of these developments see Norman R. Ball, ed., *Building Canada: A History of Public Works*, sponsored by the Canadian Public Works Association (Toronto: University of Toronto Press, 1988).

53 This estimate is based on figures in Ralph Cassady, Jr, *Auctions and Auctioneering* (Berkeley: University of California Press, 1967), 82–111 and Ben Ward-Price, *The Book of Auctions* (Toronto: Ward-Price Ltd, 1970), 45, as well as on conversations with Mr Ronald McLean, FRICS, of Wadington, McLean and Co. Ltd, a Toronto auctioneering firm established in 1850.

54 In 1871, for example, advertising rates in the daily edition of the *Spectator* (see the issue of 1 May 1871, 4) were 8 cents per line for the first insertion and 2 cents per line for subsequent insertions. At these rates, a 40-line ad that ran for 50 consecutive issues would have cost $42.40. Most subdividers probably advertised in both the *Spectator* and the *Times*, so a more realistic estimate of advertising costs would fall in the neighbourhood of $80. On the role of the press in Hamilton land sales see Michael J. Doucet, "The Role of the *Spectator* in Shaping Attitudes Toward Land in Hamilton, Ontario, 1847–1881," *Histoire sociale – Social History* 12 (1979): 431–43.

55 George Taylor Denison Account Books, 1840–1868, folio 60, MSS, Denison Papers, Toronto Public Library.

56 Wentworth County Registry Office, Registered Plan No. 125. Blyth's name appears on most of the surveys of the period, including one for James Mills (Registered Plan No. 1435) that was drawn up in 1833. He was born in 1813 in Essex, England, apprenticed with a Mr Hawkins of Toronto, arrived in Hamilton in the early 1830s, and married the daughter of one of our subdividers, Peter Hamilton, in 1851. His influence on Hamilton's morphology was remarkable. He died in February 1873 and his obituary in the *Spectator*, Tuesday, 25 February 1873, noted that "as respect his usefulness in the county, words cannot portray. For over thirty-five years he was engaged in establishing boundaries, settling differences about disputed lines, and in no instance, as we are informed by those most capable of judging, have his surveys been successfully disputed." Typical of most civil servants of his stature, Blyth receives no mention in recent historical treatments of Hamilton.

57 Moore and Davis letterbooks, 4 (10 November 1870), 700.

58 Elizabeth Kates Burns, "The Process of Suburban Residential Devel-

opment: The San Francisco Peninsula, 1860–1970," PhD thesis, University of California, 1975, 24–5.

59 Jerome D. Fellmann, "Pre-Building Growth Patterns of Chicago," *Annals of the Association of American Geographers* 47 (1957): 59.

60 See, for example, Edward Goodale, "The Origin of Some Street Names of Hamilton," *Wentworth Bygones* 2 (1960): 24–9 and Mabel Burkholder, *The Story of Hamilton* (Hamilton: Davis-Lisson, 1938), 60.

61 As in Toronto, north-south streets, which followed the original lot lines, were more continuous in Hamilton than those that took an east-west bearing. The problem of discontinuous east-west roads appears to have been more acute in Toronto, however. For a discussion see Jacob Spelt, *Toronto* (Toronto: Collier-MacMillan, 1973), 44 and Ganton, "The Subdivision Process."

62 *Spectator*, Wednesday, 26 July 1848, 4.

63 Hoyt, *One Hundred Years*, 421 and Burns, "The Process of Suburban Residential Development," 157.

64 For more detailed figures see Doucet, "Building the Victorian City," 260–3.

65 Hoyt, *One Hundred Years*, 67–74.

66 The five surveys were those of P.H. Hamilton, R.G. Beasley, V.H. Tisdale, Billings and Lister, and the western-most subdivision of A.N. MacNab.

67 A complete listing of the occupations in each category can be found in Doucet, "Building the Victorian City," appendix 6.1.

68 Husband's occupation was listed for 59.6 per cent of the women. On the role of property ownership by wives and other relatives in the nineteenth century, see Clyde and Sally Griffen, "Family and Business in a Small City: Poughkeepsie, New York, 1850–1880," *Journal of Urban History* 1 (May 1975): 329–31.

69 J.A. Bryce, "Patterns of Profit and Power: Business, Community and Industrialization in a Nineteenth Century Canadian City," unpublished term paper, Department of History, York University, 1977, 26–8. Bryce based his study on a careful analysis of the R.G. Dun credit registers for the City of Hamilton. On the rise of industrial investment opportunities in late-nineteenth-century Canada see Naylor, *The History of Canadian Business*, vol. 2 and Gustavus Myers, *A History of Canadian Wealth*, introduction by Stanley Ryerson (Toronto: James Lewis & Samuel, 1972).

70 Doucet, "Building the Victorian City," 269.

71 For an analysis of Hamilton's nonresident land owners see Doucet, "Building the Victorian City," appendix 6.2.

72 For example, between 1873 and 1878 James C. Macklin registered three surveys, containing 738 lots, for property in the west end of the city.

73 On the broader implications of industrialization on Hamilton see Michael B. Katz, Michael J. Doucet, and Mark J. Stern, *The Social Impact of Early Industrial Capitalism* (Cambridge: Harvard University Press, 1982).

74 Hoyt, *One Hundred Years*, 368–423. Details on the variations in vacant lot prices in Hamilton at this time can be found in Doucet, "Building the Victorian City," 284–8.

75 In November 1865 alone some 1,280 Hamilton building lots were offered at public tax-sale auctions. Another 261 were offered in November 1869. On Hamilton-area tax sales in this period see Doucet, "Building the Victorian City," 176–180.

76 Moore and Davis letterbooks, 1 (15 September 1859), 132.

77 *Spectator*, Wednesday, 12 June 1872, 3. Italics added.

78 Moore and Davis letterbooks, 4 (26 July 1869), 246.

79 Moore and Davis letterbooks, 4 (28 September 1869), 308.

80 *Spectator*, Thursday, 24 April 1873, 3 and Tuesday, 12 May 1873, 3.

81 Moore and Davis letterbooks, 2 (29 April 1865), 360.

82 Moore and Davis letterbooks, 4 (12 February 1870), 415.

83 Calculated from figures published in the *Census of Canada*, appropriate years and volumes.

84 Donald R. Adams, Jr, "Residential Construction Industry in the Early Nineteenth Century," *Journal of Economic History* 35 (1975): 796–7.

85 Doucet and Weaver, "Material Culture," 565. See also Keating, *Building Chicago*.

86 Calculated from figures published in the *Census of Canada*, appropriate years and volumes.

87 H.J. Dyos, *Victorian Suburb: A Study of the Growth of Camberwell* (Leicester: Leicester University Press, 1961); R.M. Pritchard, *Housing and the Spatial Structure of the City* (Cambridge: Cambridge University Press, 1976); Peter J. Aspinall, "The Internal Structure of the Housebuilding Industry in Nineteenth-Century Cities," in James H. Johnson and Colin G. Pooley, eds, *The Structure of Nineteenth Century Cities* (London: Croom Helm, 1982), 75–105; and Sam Bass Warner, Jr, *Streetcar Suburbs: The Process of Growth in Boston 1870–1900* (Cambridge: Harvard University Press, 1962).

88 C.W. Chalkin, *The Provincial Towns of Georgian England: A Study of the Building Process 1740–1820* (London: Edward Arnold, 1974).

89 E.W. Cooney, "The Origins of the Victorian Master Builders," *Economic History Review* 8 (1955): 167–76.

90 This trend was certainly not restricted to Hamilton. For a discussion of the increasing attention being paid to decorative features in US homes at this time see Dorothy S. Brady, "Relative Prices in the Nineteenth Century," *Journal of Economic History* 24 (June 1964): 161–2.

91 Gwendolyn Wright, *Moralism and the Model Home: Domestic Architecture and Cultural Conflict in Chicago 1893–1913* (Chicago: University of Chicago Press, 1980).

92 According to the *City of Hamilton Directory, 1871–2* (Hamilton: J. Sutherland, 1871), 347, the Kempster firm occupied "very extensive premises, employing... 16-horse steam power. Sashes, doors, window-blinds, &c., are manufactured. The firm also carries on building." By 1881 this company was no longer listed in the directories.

93 *Spectator*, Tuesday, 19 March 1872, 3.

94 Moore and Davis letterbooks, 3 (3 February 1868), 784.

95 Moore and Davis letterbooks, 4 (12 December 1868), 41.

96 Calculated from records in the Wentworth County Registry Office.

97 *Spectator*, Friday, 23 November 1873, 3.

98 *Spectator*, Saturday, 29 October 1881, 4.

99 Pigott went on to found one of the leading heavy contracting companies in Ontario, Pigott Construction Limited, which remains in business today.

100 *Spectator*, Saturday, 29 October 1881, 4.

101 Cooney, "The Origins," 167–8.

102 Martha J. Vill, "Speculative Building Enterprise in Late Nineteenth-Century Baltimore," paper presented at the Conference on Comparative Canadian-American Urban Development, University of Guelph, August 1982.

103 David Hanna, "Montreal – The City of Row Houses: Speculative Builders in the Housing Boom of the 1870s," paper presented at the Conference on Comparative Canadian-American Urban Development, University of Guelph, August 1982. For another view on the typical Montreal housing being built at this time see Rejean Legault, "Architecture et Forme Urbaine: L'exemple du triplex à Montréal de 1879 à 1914," *Urban History Review* 18 (1989): 1–10.

104 Aspinall, "The Internal Structure of the Housebuilding Industry," 90–102.

105 C.G. Powell, *An Economic History of the British Building Industry, 1815–1979* (London: Methuen, 1980), 31 and 73.

106 Aspinall, "The Internal Structure of the Housebuilding Industry," 76–7.

107 H.J. Dyos, "The Speculative Builders and Developers of Victorian London," *Victorian Studies* 11 (1968): 678.

108 Warner, *Streetcar Suburbs*, 119–27.

109 *Census of Canada*, appropriate years and volumes.

110 Brady, "Relative Prices," 161–2. *Spectator*, Tuesday, 22 October 1878, 1; Moore and Davis letterbooks, 2 (4 November 1864), 220. House prices in Hamilton fluctuated less markedly than did prices

paid for vacant lots. Between 1853 and 1880 the average price paid for a house in the sample subdivisions fell between $1,000 and $1,500 in 15 of the 28 years. The average ranged from a high of $1,633 in 1859 to a low of $667 in 1869.

111 A similar pattern prevailed during the early years of the twentieth century. See Weaver "From Land Assembly to Social Maturity."

112 Dyos, *Victorian Suburb*, 84.

113 Powell, *The British Building Industry*, 1.

114 For definitions of the occupations associated with each category see Doucet, "Building the Victorian City," appendix 6.1.

115 On the use-value and exchange-value of property see David Harvey, *Social Justice and the City* (London: Edward Arnold, 1973), 153–94; *Class-Monopoly Rent, Finance Capital and the Urban Revolution*, in Papers on Planning and Design no. 4 (Toronto: Department of Urban and Regional Planning, University of Toronto, 1974); and *The Limits to Capital* (Oxford: Basil Blackwell, 1982).

116 For a discussion of a crude method for predicting when a given lot would be developed see Doucet, "Building the Victorian City," 317–30.

117 This list included lawyer, later judge, George W. Burton (13 lots in 2 surveys), broker and subdivider Charles M. Counsell (16 lots in 2 surveys), New York-based railway director and merchant William B. Hunter (15 lots in 3 surveys), industrialist and subdivider William Onyon (18 lots in 1 survey), merchant James Osborne (12 lots in 1 survey), lawyer and subdivider James D. Pringle (12 lots in 2 surveys), banker and subdivider Thomas Stinson (25 lots in 1 survey), and the Hamilton Real Estate Association (22 lots in 3 surveys).

118 Warner, *Streetcar Suburbs*, 117 and 184.

119 Dyos, *Victorian Suburb*, 126.

120 Alan Gowans, *Building Canada: An Architectural History of Canadian Life* (Toronto: University of Toronto Press, 1966), 65–129. See also A.G. McKay, *Victorian Architecture in Hamilton* (Hamilton: Hamilton-Niagara Branch, Architectural Conservancy of Ontario, 1966); Elizabeth Smith Vickers, "The Victorian Buildings of Hamilton," *Wentworth Bygones* 7 (1967): 46–56; Marion Macrae and Anthony Adamson, *The Ancestral Roof: Domestic Architecture of Upper Canada* (Toronto: Clarke-Irwin, 1963); John Blumenson, *Ontario Architecture: A Guide to Styles and Building Terms 1784 to the Present* (Toronto: Fitzhenry & Whiteside, 1990); and Ralph Greenhill, Ken Macpherson, and Douglas Richardson, *Ontario Towns* (Toronto: Oberon, n.d.). For a useful discussion of changes in residential architecture in Hamilton see Murray W. Aikman, "A Study of House Types in Hamilton in the 19th Century and Their Relationship to the Historical Growth of the City," MA thesis, University of Waterloo, 1972.

121 According to the City of Hamilton assessment roll, 1881, MSS, there were 1,275 dwellings along the streets of the selected surveys by that date. In addition, the 14 subdivisions contained a total of 4 unfinished houses, 94 vacant houses, 19 offices, and 98 commercial or industrial properties.

122 *Spectator*, Monday, 18 March 1872, 2.

123 *Spectator*, Friday, 25 October 1878, 2.

124 *Spectator*, Saturday, 11 May 1872, 3 and Moore and Davis letterbooks, 3 (10 July 1867), 571.

125 W.F. Tye and N. Cauchon, *The Railway Situation in Hamilton, Ontario* (Montreal: n.p., 1913), n.p., HPL.

126 The occupational categories in this table derive from those found in Theodore Hershberg et al., "Occupation and Ethnicity in Five Nineteenth-Century Cities: A Collaborative Inquiry," *Historical Methods Newsletter* 7 (1974): 174–216. On Hamilton's working-class citizens see Bryan D. Palmer, *A Culture in Conflict: Skilled Workers and Industrial Capitalism in Hamilton, Ontario, 1860–1914* (Montreal: McGill-Queen's University Press, 1979).

127 On the relationships between home and workplace in the Victorian city see, for example, James E. Vance, Jr., "Housing the Worker: The Employment Linkage as a Force in Urban Structure," *Economic Geography* 42 (1966): 294–325 and "Housing the Worker: Determinative and Contingent Ties in Nineteenth Century Birmingham," *Economic Geography* 43 (1967): 95–127; A.M. Warnes, "Early Separation of Homes from Workplaces and the Urban Structure of Chorely," *Transactions of the Historic Society of Lancashire and Cheshire* 122 (1970): 105–35; Theodore Hershberg, Harold E. Cox, Dale B. Light, Jr, and Richard R. Greenfield, "The 'Journey-to Work': An Empirical Investigation of Work, Residence and Transportation: Philadelphia, 1850 and 1880," in Theodore Hershberg, ed., *Philadelphia: Work, Space, Family and Group Experience in the 19th Century* (New York: Oxford University Press, 1981), 128–73; Stephanie W. Greenberg, "Industrial Location and Ethnic Residential Patterns in an Industrializing City: Philadelphia, 1880," in Hershberg, *Philadelphia*, 204–32; Ian Davey and Michael Doucet, "The Social Geography of a Commercial City, ca. 1853," in Michael B. Katz, *The People of Hamilton, Canada West: Family and Class in a Mid-Nineteenth-Century City* (Cambridge: Harvard University Press, 1975), 319–42; Barbara Sanford, "The Political Economy of Land Development in Nineteenth Century Toronto," *Urban History Review* 16 (1987): 17–33; and Michael J. Doucet, "Working-Class Housing in a Small Nineteenth-Century Canadian City: Hamilton, Ontario, 1852–1881," in Gregory S. Kealey and Peter Warrian, eds, *Essays in Canadian Working Class History* (Toronto: McClelland and Stewart, 1976), 83–105.

128 On the meaning and use of nineteenth-century assessed values see chapter 10. Other useful discussions include Doucet, "Building the Victorian City," 118–21; Peter G. Goheen, *Victorian Toronto, 1850–1900: Pattern and Process of Growth*, research paper no. 127 (Chicago: Department of Geography, University of Chicago, 1970); Robin S. Holmes, "Ownership and Migration from a Study of Rate Books," *Area* 5 (1973): 242–51 and "Identifying Nineteenth-Century Properties," *Area* 6 (1974): 273–7; Gregory J. Levine, "Criticizing the Assessment: Views of the Property Evaluation Process in Montreal 1870–1920 and Their Implications for Historical Geography," *Canadian Geographer* 28 (1984): 276–84; A. Gordon Darroch, "Occupational Structure, Assessed Wealth and Homeowning During Toronto's Early Industrialization, 1861–1899," *Histoire sociale-Social History* 16 (1983): 381–410; and Paul André Linteau, *The Promoters' City: Building the Industrial Town of Maisonneuve, 1883–1918*, trans. Robert Chodos (Toronto: James Lorimer & Co., 1985), 38.

129 Goheen, *Victorian Toronto.*

130 Walter Van Nus, "The Fate of City Beautiful Thought in Canada, 1893–1930," in Gilbert A. Stelter and Alan F.J. Artibise, eds, *The Canadian City: Essays in Canadian History* (Toronto: McClelland and Stewart, 1977): 166.

131 Powell, *The British Building Industry*, 2.

132 Dyos, "The Speculative Builders."

133 On the revolution in municipal government attitudes that brought about the provision of urban services see Frisch, *Town Into City.*

134 Even as late as 1900, Hamilton had fewer than six public parks. See Kathryn L. Deiter, "The Development of Public Parks in Hamilton, Ontario, 1816–1941," unpublished term paper, Department of History, McMaster University, n.d.

135 David Ward, *Cities and Immigrants: A Geography of Change in Nineteenth-Century America* (New York: Oxford University Press, 1971), 14.

136 Elizabeth Blackmar, *Manhattan for Rent, 1785–1850* (Ithaca: Cornell University Press, 1989).

137 Cooney, "The Origins," 168.

CHAPTER TWO

1 There are several excellent full-length studies of the residential-development process during this period, including H.J. Dyos, *Victorian Suburb: A Study of the Growth of Camberwell* (Leicester: Leicester University Press, 1961); Sam Bass Warner, Jr, *Streetcar Suburbs: The Process of Growth in Boston, 1870–1900* (Cambridge: Harvard University

Press, 1962); Roger Simon, *The City Building Process: Housing and Services in New Milwaukee Neighborhoods, 1880–1910* (Philadelphia: American Philosophical Society, 1978); Paul-André Linteau, *The Promoters' City: Building the Industrial Town of Maisonneuve, 1883–1918*, trans. Robert Chodos (Toronto: James Lorimer & Co., 1985); Marc A. Weiss, *The Rise of the Community Builders: The American Real Estate Industry and Urban Land Planning* (New York: Columbia University Press, 1987); and Ann Durkin Keating, *Building Chicago: Suburban Developers and the Creation of a Divided Metropolis* (Columbus: Ohio State University Press, 1988).

2 For an interesting fictional account of the operations of a small builder in the 1980s see Tracy Kidder, *House* (Boston: Houghton Mifflin, 1985). A more scholarly treatment of the small builder can be found in Sherman J. Maisel, *Housebuilding in Transition* (Berkeley: University of California Press, 1953), 31–66. In Canada in 1983, 85 per cent of house-building firms had revenues of less than $500,000, indicating annual production levels of only two or three units. Fewer than 75 firms in the entire country built more than 100 houses in that year. Canada Mortgage and Housing Corporation (CMHC), *Housing in Canada 1945 to 1986: An Overview and Lessons Learned* (Ottawa: CMHC, 1987), 20–1.

3 "Moore-Davis Firm Marks Centenary," *Hamilton Spectator* (hereafter *Spectator*), 26 August 1958, Hamilton Public Library (HPL), real estate clipping file; *Mitchell and Company's County of Wentworth and Hamilton City Directory*, 1865–6 (Toronto: Mitchell & Co., 1864), 234; and *City of Hamilton Seventh Annual Directory 1880–1* (Hamilton: Irwin & Co., 1881), 327.

4 Pearl J. Davies, *Real Estate in American History* (Washington: Public Affairs Press, 1958), 36. In Hamilton prior to 1900 the *City Directory* listed those who dealt in land under the heading "Agents (Land and Estate)." After that date they could be found under "Real Estate."

5 "Real Estate Outlook," *Hamilton Herald* (hereafter *Herald*), Saturday, 9 August 1913, 8–10.

6 *Herald*, Saturday, 9 August 1913, 8.

7 Weiss, *The Community Builders*, passim.

8 John H. McDermid, *Historical Notes: A Condensed Historical Memory of the Ontario Association of Real Estate Boards, 1922–72* (Toronto: Ontario Association of Real Estate Boards (OAREB), 1972), 2–4. In 1925 F. Kent Hamilton became the first Hamiltonian to serve as president of the OAREB.

9 Weiss, *The Community Builders*, chapter 2. On the declining use of auctions see Helen C. Monchow, *Seventy Years of Real Estate Subdividing in the Region of Chicago* (Chicago: Northwestern University Press, 1939),

149–51. In the late 1980s charges of rate-fixing were levied against Canadian realtors. See Mary Gooderham, "Consumer Group Wants Provinces to Place Limits on Real-Estate Fees," *Globe and Mail*, Wednesday, 11 May 1988, A4.

10 On the impact of these changes see Gwendolyn Wright, *Moralism and the Model Home: Domestic Architecture and Cultural Conflict in Chicago 1893–1913* (Chicago: University of Chicago Press, 1980); Keating, *Building Chicago*; Dolores Hayden, *The Great Domestic Revolution: A History of Feminist Designs for American Houses, Neighborhoods, and Cities* (Cambridge: The MIT Press, 1981); and Daniel P. Handlin, *The American Home: Architecture and Society, 1815–1915* (Boston: Little Brown, 1979).

11 The six founders were Reginald Kennedy (merchant); Hugh Cossart Baker (banker, founder of the Canada Fire and Marine Insurance Company, co-founder of the Hamilton Street Railway, developer of Canada's first telephone exchange, and a son of the founder of the Canada Life Assurance Company); John W. Burton (coal merchant); Lyman Moore (esquire); John Eastwood (stationer); and Lewis Springer (esquire, of Barton township). Kennedy and Baker were listed as the principal shareholders (eighty shares each) at the time of incorporation. On the life of Hugh C. Baker see L.M. Shaw, "Two Hamilton Families," *Western Ontario Historical Notes* 16 (1960): 5–8.

12 Records of the Provincial Secretary, Charter Book of Companies, 4, folio 27, MSS, RG-8, Ontario Archives. *Prospectus of the Hamilton Real Estate Association* (Hamilton: Times Printing Company, 1875), 2.

13 Keating, *Building Chicago*, 67–8.

14 Weiss, *The Community Builders*, 38.

15 Ibid., 156.

16 Humphrey Carver, *Houses for Canadians: A Study of Housing Problems in the Toronto Area* (Toronto: University of Toronto Press, 1948), 62–9. For a more recent commentary on the scale of the Canadian house-building industry see CMHC, *Housing in Canada*, 20–2 and Frank A. Clayton, "The Single-Family Homebuilding Industry in Canada 1946–2001" in Tom Carter, ed., *CMHC and the Building Industry: Forty Years of Partnership*, occasional paper no. 18 (Winnipeg, Institute of Urban Studies, University of Winnipeg, 1989), 5–11.

17 Records of the Provincial Secretary, Charter Book of Companies, 17, folio 10, MSS, RG-8, Ontario Archives.

18 "Builder's Exchange: Organization Completed at a Meeting held Last Night," *Spectator*, Wednesday, 30 November 1898, 5.

19 Michael Doucet and John Weaver, "The North American Shelter Business, 1860–1920: A Study of a Canadian Real Estate and Property Management Agency," *Business History Review* 58 (1984): 247–53.

Weiss notes in *The Community Builders*, 38–9, that as late as 1929 operative builders (those who constructed buildings on their own account rather than on contract) were rare among large US building firms.

20 Terry Naylor, "Ravenscliffe: Housing in an Elite District," unpublished term paper, Department of History, McMaster University, 1978. Naylor suggests that each of the houses that he examined had different builders.

21 On the nature of Hamilton's building industry in the 1920s and 1930s see John C. Weaver, "From Land Assembly to Social Maturity. The Suburban Life of Westdale (Hamilton), Ontario, 1911–1951," *Histoire sociale – Social History* 11 (1978): 411–40 and John C. Weaver and J. Martin Lawlor, "The Making of a Suburb: Westdale, Hamilton, 1910–1950," paper presented to the fall meeting of the Ontario Historical Geographers, Toronto, 1976.

22 The other members of the Five Johns were Sir John M. Gibson, John Moodie, John S. Sutherland, and John Dickenson. According to Naylor, "Ravenscliffe," Moodie lived in the prestigious Ravenscliffe area and some of the houses in the district had been designed by Thomas Patterson's son, William. Information on John Patterson was taken from articles contained in the HPL, Five Johns Scrapbook: 47–47g and John M. Mills, *Cataract Traction: The Railways of Hamilton*, Canadian Traction Series, vol. 2 (Toronto: Upper Canada Railway Society and Ontario Electric Railway Historical Association, 1971). The Hamilton, Waterloo, and Guelph Railway was never built.

23 Their surveys, which were all registered prior to 1888, totalled only 165 building lots, with the largest containing 59 lots and the smallest just 3. Several of the surveys were resubdivisions of earlier parcels. For details see Registered Plans 131, 137, 139, 155, 161, 162, 171, 178, 236, 239, 249, and 261 in the Wentworth County Registry Office.

24 *Industries of Canada, Historical and Commercial Sketches of Hamilton* (Toronto: M. G. Bixley & Company, 1886), 57 and HPL, *Times* Scrapbooks, vol. 01, part 1, Monday, 27 January 1913. The estimate given of the house-building prowess of the firm was probably overstated.

25 Moore and Davis letterbooks, 9 (30 March 1888), 861.

26 "A Grand Scheme," *Spectator*, Tuesday, 12 July 1887, 3.

27 "Left Over $400,000," *Spectator*, Thursday, 18 February 1915, HPL, Five Johns Scrapbook, 47g.

28 *Advice to Tenants by Patterson Bros. The Builders of Homes* (Hamilton: A.A. Leishman, n.d.), 6, Moore and Davis Collection, McMaster University.

29 E.W. Cooney, "The Origins of the Victorian Master Builders," *Economic History Review* 8 (1955): 167–76.

30 Ibid.
31 On the rise of exclusive residential areas see Robert Fishman, *Bourgeois Utopias: The Rise and Fall of Suburbia* (New York: Basic Books, 1987).
32 *Advice to Tenants*, 6.
33 Carver, *Houses for Canadians*, 63.
34 *Spectator*, Wednesday, 6 August 1873, 3. The best lots on the property were priced at $150. A 10 per cent discount was offered for full cash payment. No down payment was needed if the purchaser arranged to build a house in the fall of 1873, in which case up to 36 monthly, interest-free payments were offered. Material on the Hamilton Real Estate Association was taken from the firm's *Prospectus*, 1–4.
35 John R. Stilgoe, *Borderland: Origins of the American Suburb, 1820–1939* (New Haven: Yale University Press, 1988).
36 On Harmon see Stanley L. McMichael, *Real Estate Subdivisions* (New York: Prentice-Hall, 1949), 10–13 and Monchow, *Seventy Years*, 142–8.
37 *Spectator*, Wednesday, 5 February 1896, 8.
38 Wright, *Moralism and the Model Home*, 83–4.
39 Ibid., 40–5.
40 *Advice to Tenants*, 5.
41 Ibid., 3 and 7.
42 Ibid., 7. For two views on the merits of home ownership see Michael B. Katz, Michael J. Doucet, and Mark J. Stern, *The Social Organization of Early Industrial Capitalism* (Cambridge: Harvard University Press, 1982), 131–57 and Matthew Edel, Elliot D. Sclar, and Daniel Luria, *Shaky Palaces: Homeownership and Social Mobility in Boston's Suburbanization* (New York: Columbia University Press, 1984). See also the discussion in chapter 4.
43 Richard Harris, *The Growth of Home Ownership in Toronto, 1899–1913*, research paper no. 163 (Toronto: Centre for Urban and Community Studies, University of Toronto, 1987). In contrast, owner-occupancy levels remained stable in Montreal during this period. See Marc H. Choko, *The Characteristics of Housing Tenure in Montreal*, research paper no. 164 (Toronto: Centre for Urban and Community Studies, University of Toronto, 1987). See also Marc Choko and Richard Harris, "The Local Culture of Property: A Comparative History of Housing Tenure in Montreal and Toronto," *Annals of the Association of American Geographers* 80 (1990): 73–95.
44 Weiss, *The Community Builders*, 32.
45 Calculated from figures in M.C. Urquart and K.A.H. Buckley, *Historical Statistics of Canada* (Toronto: Macmillan, 1965), 255–7.

46 Ibid.

47 John C. Weaver, *Hamilton: An Illustrated History* (Toronto: James Lorimer & Co., 1982), 199.

48 J. David Hulchanski, *The Evolution of Ontario's Early Urban Land Use Planning Regulations*, research paper no. 136 (Toronto: Centre for Urban and Community Studies, University of Toronto, 1982), 12–19. On the general controls over subdivision that were in place in North America by the 1930s see Thomas Adams, *Design of Residential Areas: Basic Considerations, Principles, and Methods* (New York: Arno Press, 1974 [1934]), 45–61.

49 On the nature of building and development cycles see Homer Hoyt, *One Hundred Years of Land Values in Chicago* (Chicago: University of Chicago Press, 1933) and J.W.R. Whitehand, *The Changing Face of Cities: A Study of Development Cycles and Urban Form.* Institute of British Geographers, special publication 21 (Oxford: Basil Blackwell, 1987).

50 See Marjorie Freeman Campbell, *A Mountain and a City: The Story of Hamilton* (Toronto: McClelland and Stewart, 1966), 206. Gage died in 1927, by which time he had laid out eleven surveys containing 3,187 building lots. According to his advertisements, Gage also maintained sales offices in Montreal, Rochester, and Buffalo and he developed at least one survey adjacent to Model City in Montreal. *Spectator*, Tuesday, 13 August 1913, 6 and Thursday, 25 July 1912, 12.

51 "Building Outlook," *Herald*, Saturday, 9 August 1913, 11.

52 "Development of the City," *Herald*, Saturday, 9 August 1913, 2.

53 As quoted in Campbell, *A Mountain and a City*, 206.

54 Plans 444 and 446, Wentworth County Registry Office. On the early settlement of the Mountain see HPL, Hamilton Mountain Scrapbook, vol. 1, 195, Bruce Murdoch, "First Mountain Expansion in 1910," *Spectator*, Saturday, 30 April 1955 and Mabel Burkholder, *Barton on the Mountain* (Hamilton: n.p., 1956). The Mountain area posed some serious problems of access that have taken decades to resolve. See, for example, HPL, Hamilton Mountain Scrapbook, vol. 1, 249–50, Gordon Hampson, "Getting Up the Mountain," *Spectator*, Tuesday, 17 November 1959 and Wednesday, 18 November 1959.

55 *Labour Gazette* 10 (June 1910): 1371.

56 Monchow, *Seventy Years*, 39–40.

57 Gerald Hodge, *Planning Canadian Communities: An Introduction to the Principles, Practices, and Participants* (Toronto: Methuen, 1986), 93–4. By the start of World War I, enough vacant land had been platted in Calgary, Edmonton, Saskatoon, Vancouver, and Ottawa to accommodate between 6.5 and 62.5 times their 1914 populations.

58 *Labour Gazette* 15 (January 1915): 780 and (February 1915): 901. For

an analysis of this depression see Dianne Vachal, "The 1913–15
Depression in Hamilton," unpublished term paper, Department of
History, McMaster University, 1976.

59 *Labour Gazette* 15 (August 1914): 168–169; (December 1914): 658;
and (January 1915): 779.

60 Weaver, *Hamilton*, 99 and 197.

61 On Chicago see Monchow, *Seventy Years*, 9 and on Los Angeles see
Weiss, *The Community Builders*, 112. On the boom in real estate in Canadian cities see John C. Weaver, *Shaping the Canadian City: Essays on
Urban Politics and Policy, 1890–1920*, monographs on Canadian Urban
Government – no. 1 (Toronto: The Institute of Public Administration
of Canada, 1977), 5–24. Studies of boosterism in Western Canada are
numerous and include Alan F.J. Artibise, *Winnipeg: A Social History of
Urban Growth, 1874–1914* (Montreal: McGill-Queen's University Press,
1975); Max Foran, *Calgary: An Illustrated History* (Toronto: James Lorimer & Co., 1978); and David Burley, "The Keepers of the Gate: The
Inequality of Property Ownership During the Winnipeg Real Estate
Boom, 1881–82," *Urban History Review* 17 (1988): 63–76.

62 See, for example, Jack De Lamere, "Opmistic [*sic*] Reflections of Operators of Note," *Spectator*, Saturday, 19 December 1914, 8. The *Census* of 1931 reported a population of 155,547 for the city.

63 Diana J. Middleton and David F. Walker, "Manufacturers and Industrial Development Policy in Hamilton, 1890–1910," *Urban History Review* 8 (1980): 20–46.

64 H.M. Marsh, *Hamilton, Canada: The City of 400 Varied Industries* (Hamilton: Commissioner of Industries [1913]), 26 and Department of Industries and Publicity, *Hamilton: The City Beautiful and Industrial Hub of
Canada* (Hamilton: The Reid Press, [1930]), 6–7.

65 *Spectator*, Tuesday, 13 August 1913, 5–6. See also Weaver, *Hamilton*,
chapter 3.

66 *Labour Gazette* 12 (February 1912): 746 and 13 (February 1913): 833.
For an excellent discussion of another Canadian city that was experi-
encing rapid industrialization at this time see Linteau, *The Promoters'
City*. Maisonneuve was known as the Pittsburgh of Canada, while
Hamilton was referred to as the Birmingham of Canada.

67 Walter Van Nus, "The Fate of City Beautiful Thought in Canada,
1893–1930," in Gilbert A. Stelter and Alan F.J. Artibise, eds, *The Canadian City: Essays in Urban History* (Toronto: McClelland and Stewart,
1977), 173.

68 "A Review of the Year's Real Estate Operations," *Herald*, Saturday,
21 December 1912.

69 On Llewellyn Park see John W. Reps, *The Making of Urban America. A
History of City Planning in the United States* (Princeton: Princeton Uni-

versity Press, 1965), 339–40 and on Riverside see Albert Fein, *Frederick Law Olmsted and the American Environmental Tradition* (New York: George Braziller, Inc., 1973), 32–5 and Fishman, *Bourgeois Utopias*, 126–33 and "American Suburbs/English Suburbs: A Transatlantic Comparison," *Journal of Urban History* 13 (1987): 237–51. Fein (p. 32) describes Riverside as "one of America's foremost examples of nineteenth-century community design and a clear, early roadmark in the development of the garden city movement throughout the western world."

70 G. Adam Mercer, *Toronto Old and New* (Toronto: The Mail Printing Company, 1891), 51.

71 Weaver, *Shaping the Canadian City*, 15–16. The plan for Leaside covered an area of about 1,025 acres, that for Mount Royal encompassed 1,500 acres, and that for Port Mann occupied 2,200 acres. In Leaside a concerted effort was made to separate land uses, with areas set aside for industrial, commercial, and residential activities. On the history of Leaside see Charles Clay, *The Leaside Story* (Leaside: Leaside Town Council, 1958). The activities of the Dovercourt Land, Building, and Savings Company are revealed in the Dinnick Papers, Ontario Archives MU 919.

72 Reps, *The Making of Urban America*, 348; C.N. Forward, "The Immortality of a Fashionable Residential District: The Uplands," in C. N. Forward, ed., *Residential and Neighbourhood Studies in Victoria*, Western Geographical Series, vol. 5 (Victoria: Department of Geography, University of Victoria, 1973), 3–10; and Weaver, *Shaping the Canadian City*, 15–17.

73 Fishman, *Bourgeois Utopias*, 121–33. Fishman argues that the roots of the North-American picturesque suburb can be traced back to English villa developments of the early nineteenth century. On the ethos of suburbanization see John Archer, "Ideology and Aspiration: Individualism, the Middle Class, and the Genesis of the Anglo-American Suburb," *Journal of Urban History* 14 (1988): 214–53; Mary Corbin Sies, "The City Transformed: Nature, Technology, and the Suburban Ideal, 1877–1917," *Journal of Urban History* 14 (1987): 81–111; and Richard A. Walker, "The Transformation of Urban Structure in the Nineteenth Century and the Beginnings of Suburbanization," in Kevin R. Cox, ed., *Urbanization and Conflict in Market Societies* (Chicago: Maaroufa Press, 1978), 165–212. On suburban development in the United States see Kenneth T. Jackson, *Crabgrass Frontier: The Suburbanization of the United States* (New York: Oxford University Press, 1985).

74 Reps, *The Making of Urban America*; Mel Scott, *American City Planning Since 1890* (Berkeley: University of California Press, 1969); Michael

Hugo-Brunt, *The History of City Planning: A Survey* (Montreal: Harvest House, 1972); Daniel Schaffer, ed., *Two Centuries of American Planning* (Baltimore: Johns Hopkins University Press, 1988); Jonathan Barnett, *The Elusive City: Five Centuries of Design, Ambition, and Miscalculation* (New York: Harper & Row, 1986); Anthony Sutcliffe, ed., *The Rise of Modern Urban Planning 1800–1914* (New York: St Martin's Press, 1980); and Anthony Sutcliffe, *Towards the Planned City: Germany, Britain, the United States, and France, 1780–1914* (New York: St. Martin's Press, 1981). On the history of planning in Canada see Ian Cooper and J. David Hulchanski, *Canadian Town Planning, 1900–1930: A Historical Bibliography*, 3 vols. (Toronto: Centre for Urban and Community Studies, University of Toronto, 1978) and Hulchanski, *The Evolution*.

75 The Hamilton Golf and Country Club, designed by H.S. Colt, opened for play in 1914. The club hosted the Canadian Open in 1930. Pat Ward-Thomas et al., *The World Atlas of Golf* (New York: Gallery Books, 1976), 266.

76 The Home and Investment Realty, Limited, *Mountainview Survey: Hamilton's New High-Class Residential District* (Hamilton: The Home and Investment Realty, Limited, [1913]). See Plan 575, Wentworth County Registry Office.

77 "Development of the City," *Herald*, Saturday, 9 August 1913, 2.

78 Herbert Lister, *Hamilton, Canada: Its History, Commerce, Industries, Resources*, issued under the auspices of the City Council in the Centennial Year 1913 (Hamilton: Spectator Printing Office, 1913), 247–8.

79 Plans 297, 393, 443, 482, 566, 568, 575, 576, and 595, Wentworth County Registry Office.

80 Cul-de-sacs would be showcased in 1929 during the planning and development of Radburn, New Jersey, which was touted as the first town designed for the motor age. Radburn covered 1,005 acres and was designed for 25,000 people. See "The Dead-End Street: It Has Charm, Economy, and Safety," *Building Developer* 3 (January 1929): 50–3.

81 Hamilton was not unusual in this regard. An examination of any set of large-scale maps reveals that massive areas of most Canadian cities have been laid out by the grid method.

82 Examples of such texts include John Nolen, ed., *City Planning: A Series of Papers Presenting the Essential Elements of a City Plan* (New York: D. Appleton and Company, 1916); Charles Mumford Robinson, *City Planning with Special Reference to the Planning of Streets and Lots* (New York: G. Putnam's Sons, 1916); and Adams, *Design of Residential Areas*. Robinson (pp. 163–4) recommended a width of 33 1/3 feet for "lot platting for humble homes" because it would provide adequate air and

light on either side of a detached house and it could not be readily halved.

83 Keating, *Building Chicago*, 65–8.

84 Weaver, *Hamilton*, 99.

85 *Herald*, Saturday, 4 June 1912.

86 *Labour Gazette* 13 (March 1913): 953.

87 Keating, *Building Chicago*, 5–6 and 71–8; Peter W. Moore, "Public Services and Residential Development in a Toronto Neighborhood, 1880–1915", *Journal of Urban History* 9 (1983): 445–71; Roger D. Simon, "Housing and Services in an Immigrant Neighborhood: Milwaukee's Ward 14," *Journal of Urban History* 2 (1976): 435–58 and *The City-Building Process: Housing and Services in New Milwaukee Neighborhoods, 1880–1910* (Philadelphia: The American Philosophical Society, 1978); and Christine Meisner Rosen, "Infrastructure Improvement in Nineteenth-Century Cities: A Conceptual Framework and Cases," *Journal of Urban History* 12 (1986): 211–56.

88 A.D. Theobald, *Financial Aspects of Subdivision Development*, research monograph no. 3 (Chicago: The Institute for Economic Research, 1930), 35 and Adams, *The Design of Residential Areas*, 178–80. See also Stanley K. Schultz and Clay McShane, "To Engineer the Metropolis: Sewers, Sanitation, and City Planning in Late Nineteenth-Century America," *Journal of American History* 65 (1978): 389–411.

89 Baldwin Room, Metropolitan Toronto Library, Home Smith & Company Papers, s212, "Home Smith's Legacy for Living," *Toronto Telegram*, Saturday, 16 May 1970, 21. Sies, "The City Transformed," 91–93, reports similarly high servicing costs for us suburbs like Roland Park, Baltimore ($100,000), Rochelle Park, New York ($76,000), and Riverside, Chicago ($2,000,000).

90 Dinnick Papers, Ontario Archives, MU 919, miscellaneous file. This amount also included the cost of grading the streets and allowed a 10 per cent cushion for contingencies. The concrete roadways laid in the subdivision were only fourteen feet in width. Assuming a 25-foot frontage, the improvement costs per linear foot amounted to $7.34.

91 A.G. Dalzell, "Housing in Relation to Land Development," in *Housing in Canada* (Toronto: The Social Service Council of Canada, 1927), 24–5. The total for Shaughnessy Heights included the following costs, all expressed per net acre: original clearing – $520, rough grading of lots – $300, sewer system – $792, water system – $322, rough grading of roads – $375, surfacing of roads – $392, sidewalks and curbs – $780, and boulevards and parkettes – $462.

92 *Labour Gazette* 10 (January 1910): 762.

93 Figures for 1905 were taken from *Hamilton: The Electrical and Natural Gas City of Canada* (Hamilton: Times Printing Company, [1906]), 6

and 8. Those for 1911 were taken from Marsh, *Hamilton, Canada*, 5.
Detailed yearly accounts of the progress of infrastructure construction
in Hamilton can be found in the *Annual Reports of the City Engineer*,
HPL, RG 16, series A. These reports begin in 1898. On the early development of Hamilton's water system see William James and Evelyn M.
James, *"A Sufficient Quantity of Pure and Wholesome Water": The Story of
Hamilton's Old Pump House* (London, Ontario: Phelps Publishing,
1978). A useful account of sewer development in Hamilton at this
time can be found in Ian A.R. Graham, "Sewage Line Distribution as
an Indicator of Urban Expansion: Hamilton, 1900–1920," unpublished term paper, Department of History, McMaster University,
1984. On the development of street railways in Hamilton see John M.
Mills, *Cataract Traction: The Railways of Hamilton*, Canadian Traction
Series, vol. 2 (Toronto: Upper Canada Railway Society and Ontario
Electric Railway Historical Association, 1971).

94 Local Improvement Books, HPL, RG 6, series J.

95 *City Council Minutes 1905*, Report of the City Engineer, 22 March
1905. Information was gathered by the City Engineer on the following North-American cities – London, Montreal, Ottawa, St Thomas,
Woodstock, Cleveland, Niagara Falls, NY, Rochester, Syracuse, San-
dusky, Toledo, and Geneva, NY.

96 Collector's Rolls – Local Improvements, 1894–1909, HPL, RG 8,
series H.

97 *Labour Gazette* 15 (June 1915): 1389.

98 This estimate is based on an analysis of the outcomes of the 749 petitions between the years 1914 and 1943 filed for streets that began
with the letters B, K, P, and W . By-laws were recorded beside 548 or
73.2 per cent of these petitions. See the Local Improvement Book,
1914–1943, HPL, RG 6, series J.

99 *Labour Gazette* 13 (October 1912): 324; 15 (July 1914): 28; and 16
(November 1915): 556. The latter citation indicated that one quarter
of a million dollars worth of road paving had been indefinitely postponed because it was felt that the property owners in the affected
areas would not be able to pay for the work.

100 Dinnick Papers, Ontario Archives, MU 919, Lawrence Park Estates
file. The total value of the subdivision was listed as $457,420.50.
Most lots were fifty feet wide, and the modal price in the survey was
$38 per front foot. The seven quality types were very good, good,
poor, high lots, fair lots, low lots, and bad unless filled.

101 *Spectator*, Friday, 27 May 1910, 10.

102 *Spectator*, Wednesday, 2 April 1913, 18.

103 Lister, *Hamilton*, 235–9. On other aspects of the cost of living in Canadian cities at this time see Edward J. Chambers, "A New Measure

of the Rental Cost of Housing in the Toronto Market, 1890–1914,"
Histoire sociale-Social History 17 (1984): 165–74; "New Evidence on
the Living Standards of Toronto Blue Collar Workers in the Pre-
1914 Era," *Histoire sociale-Social History* 18 (1985): 285–314; and "Ad-
dendum on the Living Standards of Toronto Blue Collar Workers in
the 1900–1914 Era," *Histoire sociale-Social History* 20 (1987): 357–62.

104 "Real Estate Values Reflect City's Growth in the Last Ten Years,"
Spectator, Tuesday, 13 August 1913, 6. The noted planner Thomas
Adams suggested an allowance of 20 per cent of the land area for
streets in his classic work *The Design of Residential Areas*, 200–36.

105 Edel, Sclar, and Luria, *Shaky Palaces*, 213. The Ottawa and Barton
Street area was near the race track of the Hamilton Jockey Club and
had been served by public transit since about 1893.

106 *Herald*, Saturday, 9 August 1913, 8–10. According to an advertisement in Lister, *Hamilton*, 305, Metherell had placed over 2,000 lots
on the market by 1913.

107 Lister, *Hamilton*, 311.

108 *Spectator*, Thursday, 19 May 1910, 9; Monday, 23 May 1910, 9; and
Saturday, 19 April 1913, 11. For a discussion of the use of inducements in Toronto at this time see John T. Saywell, *Housing
Canadians: Essays on the History of Residential Construction in Canada*,
discussion paper no. 24 (Ottawa: Economic Council of Canada,
1975), 109.

109 *Labour Gazette* 9 (June 1909): 1305. For a good discussion of the relationship between services and annexation in Hamilton see "Annexation Would Prove Great Boon," *Spectator*, Friday, 29 February 1924
in HPL, Hamilton Mountain Scrapbook, vol. 1, 29. On annexations in
the United States see Jackson, *Crabgrass Frontier*, 138–56; Keating,
Building Chicago, 98–119; David G. Bromley and Joel Smith, "The
Historical Significance of Annexation as a Social Process," *Land Economics* 49 (1973): 294–309; and Arnold Fleischman, "The Territorial
Expansion of Milwaukee: Historical Lessons for Contemporary Urban Policy and Research," *Journal of Urban History* 14 (1988): 147–
76. On annexation in Canada see Walter van Nus, "The Role of Suburban Government in the City-Building Process: The Case of Notre
Dame de Grâces, Quebec, 1876–1910," *Urban History Review* 13
(1984): 91–103; Moore, "Public Services," 455–65; and Michael J.
Doucet, "Politics, Space, and Trolleys: Mass Transit in Early Twentieth-Century Toronto," in G.A. Stelter and A.F.J. Artibise, eds,
Shaping the Urban Landscape: Aspects of the City-Building Process
(Ottawa: Carleton University Press, 1982), 356–81.

110 Keating, *Building Chicago*, 71–8.

111 *Spectator*, Friday, 3 June 1910, 9.

112 *Spectator*, Friday, 27 May 1910, 10.

113 *Spectator*, Friday, 27 May 1910, 9.

114 *Spectator*, Tuesday, 12 August 1913, 26.

115 The Home and Investment Realty, Limited, *Mountainview Survey*, 17.

116 Kathryn L. Deiter, "The Development of Public Parks in Hamilton, Ontario: 1816–1941," unpublished term paper, Department of History, McMaster University, n.d., 14–20.

117 The only extant issue, for May 1911 is in the Hamilton Public Library. In addition to advertisements for Metherell's surveys, it contained philosophical words of wisdom, poems, and advertisements for local merchants.

118 Dinnick Papers, Ontario Archives, MU 919, Dovercourt – Financial 1914–1915 file.

119 Charles S. Ascher, "Private Covenants in Urban Redevelopment," in Coleman Woodbury, ed., *Urban Redevelopment: Problems and Practices* (Chicago: University of Chicago Press, 1953), 228–30. According to Helen Monchow, *The Use of Deed Restrictions in Subdivision Development* (Chicago: The Institute for Research in Land Economics and Public Utilities, 1928), 2, deed restrictions were first used in the United States in 1749 when William Penn's sons used restrictive clauses to ensure the prompt construction of properly aligned brick or stone dwellings in Philadelphia. We wish to thank Dr Marc Weiss for providing us with a copy of the now rare Monchow monograph.

120 Patricia Burgess Stach, "Deed Restrictions and Subdivision Development in Columbus, Ohio, 1900–1970," *Journal of Urban History* 15 (1988): 45–6.

121 See, for example, registered deed no. G 703, Kerr, McLaren and Street to Samuel G. Patton, 30 September 1853, Wentworth County Registry Office.

122 Ascher, "Private Covenants," 231.

123 Monchow, *The Use of Deed Restrictions*. A useful summary of Monchow's monograph can be found in Stach, "Deed Restrictions," 45–8. See also J. C. Nichols, "A Developer's View of Deed Restrictions," *Journal of Land and Public Utility Economics* 5 (1929): 132–42 and Gwendolyn Wright, *Building the Dream: A Social History of Housing in America* (New York: Pantheon, 1981), 200–12.

124 Baldwin Room, Metropolitan Toronto Library, Home Smith & Company Papers, s212, "Humber Valley Surveys," pamphlet. For an analysis of Home Smith & Company see Ross Patterson, "The Development of an Interwar Suburb: Kingsway Park, Etobicoke," *Urban History Review* 13 (1985): 225–35. Around 1913 Wilfred Dinnick had employed a similarly comprehensive set of restrictions in the Lawrence Park Estates subdivision in North Toronto. See Dinnick Pa-

pers, Ontario Archives, MU 919, Lawrence Park Estates file, typescript.

125 According to Monchow, *The Use of Deed Restrictions*, 56–8, a duration of 30 to 35 years was typical.

126 Weiss, *The Community Builders*, 3–4.

127 John Delafons, *Land-Use Controls in the United States*, 2nd ed. (Cambridge: MIT Press, 1969), 85–6. On the rationale for residential zoning by-laws see Adams, *The Design of Residential Areas*, 62–75.

128 *Spectator*, Thursday, 20 June 1907, 8.

129 *Spectator*, Friday, 27 May 1910, 10 and Saturday, 14 May 1910, 10.

130 *Spectator*, Tuesday, 12 August 1913, 26.

131 *Spectator*, Thursday, 25 July 1912, 12.

132 Lister, *Hamilton*, 317.

133 The Home and Investment Realty, Limited, *Mountainview Survey*, 19–20.

134 *Herald*, Saturday, 11 November 1922.

135 Interestingly, Middleton and Walker in "Manufacturers," 24, found that the proportion of individuals on Hamilton's City Council with backgrounds in either real estate and finance or building and contracting remained constant at about 17 per cent between 1890 and 1909. Only merchants and manufacturers had a better record of representation on Council during this interval. The presence of "real estate interests" on municipal councils is an enduring element in Canadian urban history. For early examples see chapter 1, for more recent times see James Lorimer, *A Citizen's Guide to City Politics* (Toronto: James Lewis & Samuel, 1972), 95–128.

136 Calculated from information presented in "Real Estate Outlook," *Herald*, Saturday, 9 August 1913, 8–10.

137 "Building Outlook," *Herald*, Saturday, 9 August 1913, 11.

138 *Labour Gazette* 7 (September 1906): 249.

139 *Labour Gazette* 12 (February 1912): 746. In 1911 Bennet and Thwaites had registered a subdivision containing 45 lots (Plan number 473) in the Wentworth County Registry Office.

140 *Labour Gazette* 13 (March 1913): 953.

141 Carolyn Gray, *Historical Records of the City of Hamilton, 1847–1973*. (Hamilton: Department of History, McMaster University, 1986), 71.

142 "Building Work has Increased," *Spectator*, Thursday, 26 November 1903, 5. Fire limits were used to define areas of the city within which frame buildings were not allowed. According to Gray, *Historical Records*, 50, Hamilton's first fire limits were established in 1847.

143 *Labour Gazette* 15 (June 1915): 1388.

144 See the testimony of Hamilton construction workers in *Report of the Royal Commission on the Relations of Capital and Labor in Canada:*

Evidence – Ontario (Ottawa: Queen's Printer and Controller of Stationery, 1889), 866 and 887.

145 *Labour Gazette* 7 (February 1907): 249.

146 *Labour Gazette* 9 (January 1909): 676.

147 *Labour Gazette* 20 (February 1920): 93.

148 On the historical development of infrastructure in Canadian cities see Norman R. Ball, ed., *Building Canada: A History of Public Works*, sponsored by the Canadian Public Works Association (Toronto: University of Toronto Press, 1988).

149 The decline in the proportion of serviced lots in the suburbs between 1916 and 1921 was due to the inclusion of a large number of Westdale lots in the sample beginning in 1916. These remained unserviced until after 1921.

150 Moore, "Public Services," 459–64 and Simon, *The City-Building Process*. Moore also examined the impact of street railway lines on the timing of residential development. While we did collect data on the development and growth of Hamilton's street railway system, we did not analyse this information for this volume.

151 *Hamilton Times*, Tuesday, 12 April 1910 and Friday, 5 July 1912, HPL, *Hamilton Times* Scrapbooks, vol. HO.1. See also Rosemary R. Gagan, "Mortality Patterns and Public Health in Hamilton, Canada, 1900–14," *Urban History Review* 17 (1989): 161–75.

152 *Herald*, Thursday, 23 May 1912 and *Labour Gazette* 12 (May, 1912): 1044. On the Toronto Housing Company see Shirley Spragge, "A Confluence of Interests: Housing Reform in Toronto, 1900–1920," in A.F.J. Artibise and G.A. Stelter, eds, *The Usable Urban Past* (Toronto: Macmillan, 1979), 247–67.

153 "Sounds Death-Knell of the Slum Districts," *Spectator*, Wednesday, 23 July 1913, 1.

154 *City Council Minutes*, 12 May 1914 and *Revised By-Laws of the City of Hamilton, 1899–1910*, By-Law 41.

155 *Labour Gazette* 16 (October 1915): 392 and Hulchanski, *The Evolution*, 12–21. On early planning documents for Hamilton see Gray, *Historical Records*, 102–6.

156 *Herald*, Saturday, 13 November 1915. One of the purposes of Adams's visit was to obtain support for the 1912 Act. On Adams see John David Hulchanski, *Thomas Adams: A Biographical and Bibliographic Guide*, papers on planning and design no. 15. (Toronto: Department of Urban and Regional Planning, University of Toronto, 1978).

157 Edel, Sclar, and Luria, *Shaky Palaces*, 192.

158 The chi-square statistic for this cross-tabulation of area by occupation was 183.2 with 42 degrees of freedom, which was significant at

the 0.0000 level. In fact, statistically significant differences existed
for similar tables for every cross section between 1886 and 1921. For
another study of this type see G. Levine, R. Harris, and B. Osborne,
The Housing Question in Kingston, Ontario 1881–1901 (Kingston: Department of Geography, Queen's University, 1982).

159 Gagan, "Mortality Patterns," 169–73.
160 In 1921 Hamilton was home to just 4.1 per cent of Ontario's 2.8 million people.
161 *Hamilton Times*, Saturday, 24 October 1903, 1.
162 Olivier Zunz, *The Changing Face of Inequality: Industrial Development
and Immigrants in Detroit* (Chicago: University of Chicago Press,
1982), 152–6; Deryck W. Holdsworth, "Cottages and Castles for
Vancouver Home-Seekers," *BC Studies* nos. 69/70 (1986): 11–32; and
Simon, "Housing and Services." James McKellar, "Building Technology and the Production Process," in John R. Miron, ed., *Housing
Progress in Canada Since 1945* (forthcoming), 6.
163 *Labour Gazette* 5 (February 1905): 817–18 and 10 (May 1910): 1230.
164 Maureen Whelan, "The Housing Shortage in Hamilton, 1900–1915,"
unpublished term paper, Department of History, McMaster University, 9–11.
165 Saywell, *Housing Canadians*, 138–9.
166 Eugenie Ladner Birch and Deborah S. Gardner, "The Seven-Percent
Solution: A Review of Philanthropic Housing, 1870–1910," *Journal
of Urban History* 7 (1981): 403–38.
167 Newspaper clipping file, HPL, "One-Day House Brought City Fame,"
Spectator, Saturday, 24 November 1957.
168 Martin J. Daunton, "Cities of Homes and Cities of Tenements: British and American Comparisons, 1870–1914," *Journal of Urban History*
14 (1988): 283–319.
169 Saywell, *Housing Canadians*, 150–6.
170 Ibid., 157.
171 Ibid., 166.
172 *Guide Book and Street Directory of the Beautiful City of Hamilton* (Hamilton: Hill the Mover, [1925]). Our assessment roll samples uncovered
no apartment buildings until the 1921 cross section. Since blocks in
the CBD were not included in any of the samples, our figures probably understate the percentage of the population residing in apartment buildings. Toronto experienced a similar apartment-building
boom in the early twentieth century. See Richard Dennis, *Toronto's
First Apartment-House Boom: An Historical Geography, 1900–1920*, research paper no. 177 (Toronto: Centre for Urban and Community
Studies, University of Toronto, 1989). For a broader view see John
Weaver, "The North American Apartment Building as a Matter of

Business and an Expression of Culture," *Planning Perspectives* 2 (1987): 27–52.

173 Much of the material in the discussion of Westdale first appeared in Weaver, "The Suburban Life of Westdale."

174 The only rival for subdividing supremacy in Hamilton during this period was a Crown Corporation, Wartime Housing Limited, which developed six surveys containing 760 (largely 40 by 100 foot) lots between 1943 and 1945.

175 J.C. Nichols, *Real Estate Subdivisions: The Best Manner of Handling Them* (Washington: American Civic Association, 1912), 6, as quoted in Marc A. Weiss, "The Rise of the Community Builders: The American Real Estate Industry and Urban Land Planning," paper presented in the Urban History Workshop Series, Centre for Urban and Community Studies, University of Toronto, 28 April 1988, 1.

176 The quotation is from Miles L. Colean, *American Housing: Problems and Prospects* (New York: The Twentieth Century Fund, 1944), 97.

177 Weiss, *The Community Builders*, 45.

178 F. Kent Hamilton, *Beauty Spots in Westdale* (Hamilton: n.p., [1928]), 7.

179 "Action Requested to Promote Old Suburbs," undated clipping, Scrapbook on Real Estate, HPL.

180 See the "Copy of Agreement among McKittrick Properties, the City of Hamilton, and Ancaster Township presented to the Ontario Railway and Municipal Board, 26 January 1914," *Hamilton City Council Minutes*, 1914.

181 *Labour Gazette* 14 (March 1914): 1035.

182 *Spectator*, December, 1915 and January, 1916. The vote count was Booker – 4,738 (36%), Cooper – 3,945 (30%), and Morris – 4,479 (34%). *Spectator*, Tuesday, 2 January 1917, 9.

183 McDermid, *Historical Notes*, 2.

184 F. Kent Hamilton Papers, McMaster University. See William Lyon Somerville, Robert Anderson Pope, and Desmond McDonaugh, "Report of Survey and Recommendations, McKittrick Properties," 1 February 1919 and Somerville to Hamilton, n.d.

185 Scott, *American City Planning*, 96–9. See also Schaffer, *Two Centuries of American City Planning* and Sutcliffe, *Towards the Planned City* and *The Rise of Modern Urban Planning*.

186 *Spectator*, Saturday, 7 May 1921, 19.

187 City of Toronto, *Report of the Lieutenant Governor's Committee on Housing Conditions in Toronto* (Toronto, 1934).

188 Sommerville et al., "Report of Survey and Recommendations." For similar fears as an impetus to housing reform see John C. Weaver, "Reconstruction of the Richmond District in Halifax: A Canadian

Episode in Public Housing and Town Planning, 1918–1921," *Plan Canada* 16 (1976): 36–47.

189 Hamilton Real Estate Board, Kent Hamilton Scrapbooks, Scrapbook no. 2, 11. Clipping dated June, 1922.

190 *Spectator*, Saturday, 7 May 1921, 19.

191 Supreme Court of Ontario, in Bankruptcy, Depositions of F. Kent Hamilton, p. 11.

192 Southam Papers, J.P. Mills, Secretary, Westdale Properties to F.I. Ker, 19 November, 1928. We are indebted to Mr St Clair Balfour and his secretary, Mrs M. Shano, for providing us with copies of the Southam correspondence concerning Westdale.

193 Bureau of Municipal Affairs, "Report Re Housing for 1919," *Ontario Sessional Papers, 1920* (Toronto: King's Printer, 1920). For details of the federal and provincial programs see *Labour Gazette* 18 (August 1918): 858 and (December 1918): 1104.

194 Dominion Bureau of Statistics, *Housing in Canada*, Census Monograph no. 8 (Ottawa: King's Printer, 1942), 104.

195 Forward, "The Uplands," 22.

196 Supreme Court of Ontario, in Bankruptcy, Re: McKittrick Properties, Affidavit of Frank Ernest Roberts, 19 March 1928, 6.

197 Supreme Court of Ontario, in Bankruptcy, Re: McKittrick Properties, Petition of Charles Delamere Magee, 29 June 1926.

198 Supreme Court of Ontario, in Bankruptcy, Re: McKittrick Properties, Report of the Official Receiver, 28 June 1926.

199 Supreme Court of Ontario, in the Matter of the Bankruptcy of McKittrick Properties, Debtor, filed 6 July 1926.

200 Supreme Court of Ontario, Depositions of Frank E. Roberts, taken before John Bruce, Special Examiner, 25 October 1927.

201 HPL, West-End and Westdale Scrapbooks (1926–28), passim. On this park and the rest of Hamilton's park system during the era of corporate involvement see Deiter, "Public Parks in Hamilton." For a discussion of the influence of parks on property values see Thomas A. More, Thomas Stevens, and P. Geoffrey Allen, "Valuation of Urban Parks," *Landscape and Urban Planning* 15 (1988): 139–52.

202 Memorandum on Board of Directors' Meeting, Westdale Properties, 1 November 1928.

203 Charles M. Johnson, *McMaster University*, vol. 1, *The Toronto Years* (Toronto: University of Toronto Press, 1970), 204–36.

204 Baptist Archives, McMaster University, Removal to Hamilton/Hamilton Chamber of Commerce file, F. Kent Hamilton to F.P. Healey, Secretary, Chamber of Commerce, 3 October 1921; W.W. McMaster to Chancellor A. L. McCremmon, 21 February 1922. For the contributions made by syndicate members see New McMaster file, W.J.

Westaway to Chancellor H.P. Whidden, 13 June 1928. Altogether, those associated with Westdale Properties pledged $171,000.

205 Southam Papers, J.P. Mills, Secretary, Westdale Properties to F.I. Ker, *Spectator*, Monday, 19 November 1928.

206 Weiss, *The Community Builders*, passim.

207 *Spectator*, Friday, 12 May 1922, 29.

208 The material cited here is taken from Registered Instrument 327114, Wentworth County Registry Office. The racial clause was standard throughout Westdale. Monchow, *The Use of Deed Restrictions*, 47–50, found racial clauses in 40 of the 84 deeds that she examined in the late 1920s. Their use was escalating, for 38 of the 40 racial clauses were associated with areas that were developed after 1920.

209 Hamilton, *Beauty Spots*, 22.

210 Interview with contractor Thomas Casey, 24 September 1976.

211 J.M. Mattila, *An Economic Analysis of Construction* (Madison: University of Wisconsin Press, 1955), 14.

212 *Spectator*, Friday, 30 April 1926.

213 F. Kent Hamilton Scrapbooks, Scrapbook 2, *Spectator*, Monday, 17 November 1924.

214 Casey was interviewed in September of 1976 when he was a lucid 97 years of age.

215 Charles F. Johnson, one of the subdividers from Columbus, Ohio, studied by Patricia Stach was active in 26 separate real-estate or land-development companies between 1907 and 1923. Stach, "Deed Restrictions," 49.

216 Ellis had been in business since at least 1913. In 1922 he and some partners placed a small east-end survey, Ashford Place, (Registered Plan No. 653) on the market. The term realtor could only be used by members of a real-estate board.

217 Weaver, "The Suburban Life of Westdale," 426.

218 Ibid., 429–33.

219 In 1922 architect Chas. S. Sedgwick prepared plans for four houses – an English half-timber design, an economical cottage, an economical square house, and a brick and stucco bungalow – that proved to be very popular in Hamilton. These were widely known, as they were published in "Plans for the Home Builder," *Spectator*, Friday, 12 May 1922, 31. The definitive work on the residential architecture of this period is Alan Gowans, *The Comfortable House: North American Suburban Architecture, 1890–1930* (Cambridge: MIT Press, 1986). See also Clifford E. Clark, *The American Family Home, 1800–1960* (Chapel Hill: University of North Carolina Press, 1986) and Handlin, *The American Home*.

220 On the costs of housing in Canadian cities during the early years of
the Depression see Richard Harris, "Working-Class Home Ownership and Housing Affordability Across Canada in 1931," *Histoire
sociale – Social History* 19 (1986): 121–38.

CHAPTER THREE

1 Jill Wade, *Wartime Housing Limited, 1941–1947: An Overview and Evaluation of Canada's First National Housing Corporation*, Canadian Planning Issues, no. 13 (Vancouver: School of Community and Regional
Planning, University of British Columbia, 1984), 1. A briefer, but
more accessible version of this study is found in Jill Wade, "Wartime
Housing Limited, 1941–1947: Canadian Housing Policy at the Cross-
roads," *Urban History Review* 15 (1986): 40–59.

2 Dominion Bureau of Statistics, *Hamilton Housing Atlas* (Ottawa: King's
Printer, 1946), 5. These figures were based on the results of the 1941
census. For the as yet unannexed portions of Ancaster, Barton, and
Saltfleet townships, the figures for services were lower – 93.3 per cent
with electric lighting but only 42.1 per cent with running water.

3 For a useful chronology of housing-related events and organizations
see John R. Miron, "On Progress in Housing Canadians," in John R.
Miron, ed., *Housing Progress in Canada Since 1945* (forthcoming).
When published, this book will be the definitive work on housing in
Canada in the post-World War II period. On trends in the United
States see Marc A. Weiss, *The Rise of the Community Builders: The Ameri-
can Real Estate Industry and Urban Land Planning* (New York: Columbia
University Press, 1987).

4 For a discussion of the situation in Toronto see Peter W. Moore,
"Zoning and Planning: The Toronto Experience, 1904–1970," in
Alan F.J. Artibise and Gilbert A. Stelter, eds, *The Usable Urban Past:
Planning and Politics in the Modern Canadian City* (Toronto: Macmillan,
1979), 316–41.

5 After March of 1946, subdividing in Ontario was regulated by Section
25 of "An Act Respecting Planning and Development," chapter 71,
Statutes of the Province of Ontario, 1946 (Toronto: King's Printer, 1946).

6 Peter Keating, ed., *Into Unknown England, 1866–1913: Selections from
the Social Explorers* (Manchester: Manchester University Press, 1976);
Herbert Brown Ames, *The City Below the Hill*, introduction by P.F.W.
Rutherford (Toronto: University of Toronto Press, 1972 [1897]); and
Jacob A. Riis, *How the Other Half Lives: Studies Among the Tenements of
New York* (New York: Dover Publications, 1971 [1890]). A useful re-
cent work in this area is David Ward, *Poverty, Ethnicity, and the Ameri-*

can City, 1840–1925: Changing Conceptions of the Slum and Ghetto
(Cambridge: Cambridge University Press, 1989). On the condition
of the poor in two Canadian cities see Terry Copp, *The Anatomy of
Poverty: The Condition of the Working Class in Montreal, 1897–1929*
(Toronto: McClelland and Stewart, 1974) and Michael J. Piva,
The Condition of the Working Class in Toronto, 1900–1921 (Ottawa:
University of Ottawa Press, 1979).

7 J. David Hulchanski, "The 1935 Dominion Housing Act: Setting the
Stage for a Permanent Federal Presence in Canada's Housing Sector,"
Urban History Review 15 (1986): 35.

8 Calculated from figures in Royal Bank of Canada, *Royal Commission on
Canada's Economic Prospects: The Canadian Construction Industry* (Montreal: Royal Bank of Canada, 1956), 86.

9 See Margaret Hobbs and Ruth Roach Pierson, "'A kitchen that wastes
no steps...' Gender, Class, and the Home Improvement Plan, 1936–
40," *Histoire sociale-Social History* 21 (1988): 9–37.

10 Albert Rose, *Canadian Housing Policies (1935–1980)* (Toronto: Butterworths, 1980) and John C. Bacher, "Keeping to the Private Market:
The Evolution of Canadian Housing Policy," PhD thesis, McMaster
University, 1985. More accessible versions of Bacher's work can be
found in John C. Bacher, "Canadian Housing 'Policy' in Perspective,"
Urban History Review 15 (1986): 3–18 and "W.C. Clark and the Politics
of Canadian Housing Policy, 1935–52," *Urban History Review* 17
(1988): 5–15. Other useful studies of the evolution of Canadian housing policy include Hulchanski, "The 1935 Dominion Housing Act,"
19–39; John C. Bacher and J. David Hulchanski, "Keeping Warm
and Dry: The Policy Response to the Struggle for Shelter Among
Canada's Homeless, 1900–1960," *Urban History Review* 16 (1987): 147–
63; and Tom Carter, ed., *Perspectives on Canadian Housing Policy*, occasional paper no. 17 (Winnipeg: Institute of Urban Studies, University
of Winnipeg, 1989).

11 CMHC, *Housing in Canada 1945 to 1986: An Overview and Lessons
Learned* (Ottawa: CMHC, 1987).

12 On the current state of planning in Canada see J. Barry Cullingworth,
Urban and Regional Planning in Canada (New Brunswick, NJ: Transaction Books, 1987).

13 Wade, *Wartime Housing Limited*, 18. WHL became integrated with
CMHC in 1947. On Hamilton's experiences with WHL see John C.
Weaver, *Hamilton: An Illustrated History* (Toronto: James Lorimer &
Co.), 145–8 and G. Wright, "Hamilton Housing: City Council Strat-
egy, 1940–1955," unpublished term paper, Department of History,
McMaster University, n.d. As elsewhere, Hamilton's WHL-built homes
were intended to be temporary and many had been set on posts

rather than permanent foundations. In the mid-1950s some 450 of the dwellings were moved to permanent foundations in the Mohawk Garden Survey on the Mountain.

14 Wade, *Wartime Housing Limited*, 19.

15 CMHC, *Housing in Canada 1945 to 1986*, 12–14. A good sense of the range of these post-war government programs is provided in Rose, *Canadian Housing Policies*; Miron, "On Progress in Housing Canadians"; and Bacher, "Keeping to the Private Market."

16 See Rose, *Canadian Housing Policies*; Bacher, "Keeping to the Private Market"; and Larry S. Bourne, *The Geography of Housing* (London: Edward Arnold, 1981), 191–234. For a broader perspective see Graham Hallett, ed., *Land and Housing Policies in Europe and the USA: A Comparative Analysis* (London: Routledge, 1988).

17 Important publications on this topic include *Building Standard (excluding apartment buildings): Minimum Requirements for Planning, Construction, and Materials for Building Upon Which Loans are Made Under the National Housing Act, 1954* (Ottawa: CMHC, 1955); *Small House Designs: Two-Storey and One-and-one-half Storey Houses* and *Small House Designs: Bungalows and Split-Level Houses* (Ottawa: CMHC, 1954); *Small House Designs* (Ottawa: CMHC [1965]); *Modest House Designs, Metric Edition* (Ottawa: CMHC, 1978); *Site Planning Handbook* (Ottawa: CMHC, 1973); and *Canadian Wood-Frame House Construction* (Ottawa: CMHC, [1975]). Recent catalogues of CMHC publications and videos run to at least 30 pages. For an analysis of the progress made by CMHC see *Housing in Canada 1946–1970. A Supplement to the 25th Annual Report of Central Mortgage and Housing Corporation* (Ottawa: CMHC, 1970) and *Housing A Nation: 40 Years of Achievement* (Ottawa: CMHC, 1986). Since its inception, CMHC has issued an annual statistical review of Canada's housing achievements. At present, this publication is called *Canadian Housing Statistics*.

18 James McKellar, "Building Technology and the Production Process," in John R. Miron, ed., *Housing Progress in Canada Since 1945* (forthcoming), 10.

19 For discussions of Canada's post-1945 housing progress see Miron, "On Progress in Housing Canadians" and CMHC, *Housing in Canada 1945 to 1986*.

20 Studies of modern residential development in urban North America are numerous. See, for example, Bourne, *The Geography of Housing*; F. Stuart Chapin, Jr and Shirley F. Weiss, *Factors Influencing Land Development* (Chapel Hill: Institute for Research in Social Science, University of North Carolina, 1962); Marion Clawson, *Suburban Land Conversion in the United States: An Economic and Governmental Process* (Baltimore: Johns Hopkins University Press, 1971); Wallace F. Smith,

Housing: The Social and Economic Elements (Berkeley: University of California Press, 1971); Michael A. Goldberg, "Residential Developer Behavior: Some Empirical Findings," *Land Economics* 50 (1974): 85–9; Peter Spurr, *Land and Urban Development: A Preliminary Study* (Toronto: James Lorimer & Co., 1976); James Lorimer, *The Developers* (Toronto: James Lorimer & Co., 1978); Larry R.G. Martin, *Urban Dynamics on the Toronto Urban Fringe*, map folio no. 3 (Ottawa: Environment Canada Lands Directorate, 1975) and "Land Dealer and Land Developer Behaviour on the Rural-Urban Fringe of Toronto," in L. H. Russwurm, R.E. Preston, and L.R.G. Martin, eds, *Essays on Canadian Urban Process and Form*, Department of Geography publication series no. 10 (Waterloo: Department of Geography, University of Waterloo, 1977), 289–368; Douglas J. Dudycha, "A Simulation Model of Residential Development: Kitchener-Waterloo, 1960–1971," in R.E. Preston and L.H. Russwurm, eds, *Essays on Canadian Urban Process and Form II*, Department of Geography publication series no. 15 (Waterloo: Department of Geography, University of Waterloo, 1980), 431–51; and William L.C. Wheaton, "Private and Public Agents of Change in Urban Expansion," in Melvin M. Webber, ed., *Explorations Into Urban Structure* (Philadelphia: University of Pennsylvania Press, 1964): 154–96. A useful early bibliography on housing and development can be found in Joan Wirth D'Ambrosi, *The Housing Industry: A Bibliography*, exchange bibliography 148 (Monticello: Council of Planning Librarians, 1970). For an interesting picture of the residential development process in contemporary Britain see John R. Short, Stephen Fleming, and Stephen J.G. Witt, *Housebuilding, Planning, and Community Action: The Production and Negotiation of the Built Environment* (London: Routledge & Kegan Paul, 1986). On Australian development practices see John Hutton, *Building and Construction in Australia* (Melbourne: Institute of Applied Economic Research, University of Melbourne, 1970).

21 One Canadian resource book published in the mid-1970s identified about one hundred municipal, provincial, or federal government agencies that dealt with assisting and regulating the availability of and accessibility to shelter. Many of these agencies collect data and issue publications. A Statistics Canada publication series that dealt with housing numbered 17 by 1975. See Community Planning Association of Canada, *Canadian Housing Resource Catalogue* (Ottawa: Community Planning Association of Canada, 1975), 47–8. The quotation is from Melvin Charney, *The Adequacy and Production of Low-Income Housing* (Ottawa: CMHC Task Force on Low-Income Housing, 1971), 123.

22 These include Humphrey Carver, *Houses for Canadians: A Study of Housing Problems in the Toronto Area* (Toronto: University of Toronto

Press, 1948); CMHC, *Housing and Urban Growth in Canada: A Brief from Central Mortgage and Housing to The Royal Commission on Canada's Economic Prospects* (Ottawa: CMHC, 1956); Royal Bank of Canada, *The Canadian Construction Industry*; Charney, *The Adequacy and Production*; *Report of the [Ontario] Advisory Task Force on Housing Policy* ([Toronto: Queen's Printer for Ontario], 1973); *Report of the Royal Commission on Certain Sectors of the Building Industry* (Toronto: Queen's Printer for Ontario, 1974); A.D. Boyd and A.H. Wilson, *Technology Transfer in Construction*, Science Council of Canada, background study no. 32 (Ottawa: Information Canada, 1974); B.A. Keys and D.M. Caskie, *The Structure and Operation of the Construction Industry in Canada* (Ottawa: Economic Council of Canada and Information Canada, 1975); Ludwig Auer, *Construction Instability in Canada* (Ottawa: Economic Council of Canada and Information Canada, 1975); Joseph Chung, *Cyclical Instability in Residential Construction in Canada* (Ottawa: Economic Council of Canada and Information Canada, 1976); Larry S. Bourne and John R. Hitchcock, eds, *Urban Housing Markets: Recent Directions in Research and Policy* (Toronto: University of Toronto Press, 1978); *A Strategy for Ontario's Building Industry* ([Toronto]: Ministry of Housing, [1987]); Marion Steele, *The Demand for Housing in Canada*. Statistics Canada catalogue no. 99–763E (Ottawa: Minister of Supply and Services Canada, 1979); Miron, *Housing in Postwar Canada*; Marc H. Choko, Jean-Pierre Collin, and Annick Germain, "Le Logement et Les Enjeux de la Transformation de l'espace urbain: Montréal, 1940–1960," *Urban History Review*, 2 parts, 15 (1986): 127–36 and 15 (1987): 243–53; and J. David Hulchanski and Glen Drover, *Housing Subsidies in a Period of Restraint: The Canadian Experience, 1973–1984*, research and working paper 16 (Winnipeg: Institute of Urban Studies, University of Winnipeg, 1986).

23 According to McKellar, "Building Technology," 2, residential construction accounted for 6.6 per cent of GNP and 38 per cent of all construction in 1976 and 4.3 per cent of GNP and 30 per cent of all construction activity in 1983.

24 Not only did Canada become a more urban nation, but the urban population became increasingly concentrated in a relatively small group of metropolitan centres. Statistical reports were modified in the post-war years to reflect this metropolitanization. Beginning with the 1951 census, data were reported for both census metropolitan areas (CMAs) and for smaller spatial units, called census tracts, which were defined for each CMA. Hamilton was defined as a CMA for the 1951 census.

25 Nationally, 7.5 per cent of the housing stock was occupied by more than one family in 1941. According to Wade, *Wartime Housing Limited*,

6, figures were higher in most large urban centres: Toronto, 19.1 per cent; Halifax, 17.2 per cent; Winnipeg, 15.1 per cent; and Hamilton, 12.4 per cent. For a discussion of demographic trends see CMHC, *Housing in Canada 1945 to 1986*, 6–8 and Miron, *Housing in Postwar Canada*. According to figures published in CMHC, *Housing in Canada 1946–1970*, 37, the number of Canadian families living either doubled up or in dwellings in need of major repair fell from 976,000 to 238,000 between 1945 and 1970. According to research by Trevor Whiffen, "Postwar Suburban Expansion: An Analysis of Its Causes and Evolution in Hamilton and Ancaster," unpublished term paper, Department of History, McMaster University, 1979, average annual births in Hamilton increased from 2,876 to 5,168, marriages from 1,402 to 2,491, and immigrants from 242 to 2,956 between the 1930–39 and 1946–55 time periods. Hamilton's birth rate rose from 17.9 per 1,000 of population to 26.6 between 1935 and 1955.

26 The comparable figures for Hamilton were 28 per cent in need of major repair in 1941 and 5.4 per cent in 1981.

27 Initially, the switch was to oil; but more recently a preference has been shown for natural gas. The Federal Government, in response to supply problems created by members of OPEC, administered the Canadian Oil Substitution Program (COSP) between 1980 and 1985. In 1986, 53.8 per cent of Ontario dwellings were heated with piped gas, 20.4 per cent with electricity, and 17.2 per cent with oil – calculated from figures in Statistics Canada, *Population and Dwelling Characteristics: Dwellings and Households: Part 2*, catalogue 93–105 (Ottawa: Statistics Canada, 1989), table 3. Virtually all new homes built in Southern Ontario today are heated with natural gas if it is available.

28 Calculated from figures in Royal Commission on Canada's Economic Prospects, *The Canadian Construction Industry*, 86. More than 38 per cent of the loans approved in the first decade after World War II were granted in 1954 and 1955 alone. On North-American suburbanization at this time see S.D. Clark, *The Suburban Society* (Toronto: University of Toronto Press, 1966) and Peter O. Muller, *Contemporary Suburban America* (Englewood Cliffs: Prentice-Hall, 1981).

29 CMHC, *Housing and Urban Growth in Canada*, 16.

30 Ibid., 16–19.

31 McKellar, "Building Technology," 4.

32 Frank A. Clayton, "The Single-Family Homebuilding Industry in Canada 1946–2001," in Tom Carter, ed., *CHMC and the Building Industry: Forty Years of Partnership*, occasional paper no. 18 (Winnipeg: Institute of Urban Studies, University of Winnipeg, 1989), 7.

33 Charney, *The Adequacy and Production*, 124 and 128.

34 CMHC, *Survey of the Home-Building Industry* (Ottawa: CMHC, 1972), 17–22; as quoted in McKellar, "Building Technology," 6. See also Lorimer, *The Developers*, 83–128 and Spurr, *Land and Urban Development*, 183–361.

35 McKellar, "Building Technology," 7.

36 Clayton Research Associates, Ltd, *Ontario Building Industry Highlights 1988* (Toronto: Clayton Research Associates, Ltd, 1989), 12.

37 CMHC, *Housing in Canada 1945 to 1986*, 20.

38 Ibid., 20 and McKellar, "Building Technology," 5–6.

39 McKellar, "Building Technology," 1.

40 A. Cardoso and J.R. Short, "Forms of Housing Production: Initial Formulations," *Environment and Planning* A 15 (1983): 917–28.

41 Paul Wright, "Restrictions Driving Up Land Prices," *Spectator*, Saturday, 23 October 1976, 65.

42 The figures for 1945–1960 were taken from Weaver, *Hamilton*, 175; that for 1986 was calculated from Statistics Canada, *1986 Census of Canada, Census Tracts – Hamilton: Part I*, catalogue 95–113 (Ottawa: Minister of Supply and Services, 1987).

43 Weaver, *Hamilton*, 201, suggests that automobile ownership in Hamilton in 1968, the last year for city-specific data, stood at 96,706 motor vehicles. This represented almost a threefold increase since 1949.

44 The local press even encouraged readers to consider these distant locales. See, for example, Paul Wright, "Brantford Cuts Home Prices," *Spectator*, Saturday, 6 May 1978, 70.

45 Whiffen, "Postwar Suburban Expansion" and Lynn Livesey, "A Study of the Spatial and Physical Characteristics of Urban and Suburban Growth in Hamilton from 1946 to 1954," unpublished term paper, Department of History, McMaster University, n.d.

46 Livesey, "A Study of the Spatial and Physical Characteristics." Of the surveys platted between 1945 and 1954, only the Mount Royal Court (614 lots) and the Huntington Park (540 lots) surveys contained more than 350 lots.

47 Climate continued to thwart year-round construction activity, but by the late 1950s portable heaters and plastic enclosures were being introduced to Canadian building sites. CMHC, *Housing in Canada: A Brief From Central Mortgage and Housing Corporation to the United Nations* (Ottawa: CMHC, 1958), vi–3.

48 Registered Plans 1265 and M-95, Wentworth County Registry Office.

49 Bramalea Limited cut its teeth on a huge residential project called Bramalea, which lay to the north west of Toronto and was touted as Canada's first satellite city. Covering more than 5,000 acres, it was begun in 1958 and when completed contained industrial parks, a major

shopping centre, and more than 20,000 dwellings. For details see *Bramalea: Canada's First Satellite City* (Toronto: Bramalea Consolidated Developments Limited, 1958). *Annual Reports* issued by Bramalea Limited during the 1980s indicate that the company built between 1,000 and 1,600 single-family homes each year in Ontario, California, and Florida. Production in Ancaster amounted to between 14 and 50 units per year from 1982 to 1985.

50 Plans M-294, M-305, M-306, M-326, M-341, M-364, M-374, M-383, and M-419, Wentworth County Registry Office.

51 In 1956, 29 per cent of the population in Census Metropolitan Areas and 41 per cent of the population in Major Urban Areas lived in "largely unsewered municipalities." CMHC, *Housing in Canada*, ii-9.

52 CMHC, *Housing in Canada*, vi-5–6. Such houses contained an average of 1,000 to 1,100 square feet of space. CMHC required a minimum frontage of 40 feet and a minimum lot area of 4,000 square feet (7,500 if a septic tank was used) for NHA-financed houses at this time. On the history of the bungalow see Anthony D. King, *The Bungalow: The Production of a Global Culture* (London: Routledge & Kegan Paul, 1984). On current and future house design trends see Diane Sawchuk, "Decades of Design Shape Modern House," *Toronto Star*, Saturday, 26 March 1988, E1 and E4.

53 Weaver, *Hamilton*, 199.

54 For an analysis of the growth of housing stock in Canada after 1945 see Miron, *Housing in Postwar Canada*, 148–91.

55 Lorimer, *The Developers*, 102.

56 Of course, not every subdivider was a developer. In 1979 Marvin and David Wasserman claimed to have sold more than $750,000 worth of lots in their three fully serviced east Mountain-area subdivisions. Their full-page advertisement, however, was specifically directed to the attention of builders, who were urged to "Act Now to Secure your block of lots [and] join successful builders in this area. Builders Terms available." *Spectator*, Saturday, 12 May 1979, 67. A study in the mid-1970s found that most of Canada's large builders also were large land developers. The developers, however, could be broken down into three groups – those who did not build at all (Markborough Properties, Carma Developers, Abbey Glen), those who sold some lots to other builders and built on some themselves (Cadillac Fairview, Sifton Properties, Richard Costain, Victoria Wood), and those who built on all or almost all of the lots that they developed (Bramalea Corporation, Nu-West Developments, Campeau Corporation). See Ira Gluskin, *Cadillac Fairview Corporation Limited: A Corporate Background Report*, study no. 3 (Ottawa: Royal Commission on Corporate Concentration, 1976), 150–1.

57 *Spectator*, Saturday, 7 June 1952, 27.

58 John Sewell, "Don Mills: E.P. Taylor and Canada's First Corporate Suburb," *City Magazine* 2 (January 1977): 28–38. When completed in the early 1960s, Don Mills contained 8,121 dwelling units and housed 28,426 people. The practice of rationing building lots to builders continues to this day. In California, Bramalea Limited adopted the following strategy for the development of 2,500 acres of land in the Los Angeles area that it had purchased in 1982: "to service lots for sale to local builders, to construct homes for sale on our own account, and to maintain overall control of community development and design." *Bramalea Limited Annual Report* 1984, 22.

59 Even in the 1980s the use of multiple builders was considered the norm, so the presence of a single-builder was worth noting. In 1981 Mapledene Estates, a 15-lot development in Stoney Creek, was advertised as "a one-builder subdivision guaranteeing a planned, attractive community." *Spectator*, Saturday, 16 May 1981, 41. On a much grander scale, Glenway Builders is the sole contractor for Glenway Estates, an 800-lot project in Newmarket (north of Toronto). By April 1988 more than 400 homes had been built, but this had taken the firm four-and-a-half years. Pat Allossery, "Newmarket Builder Gives Homes a Cohesive Look," *Toronto Star*, Saturday, 23 April 1988, E8.

60 Urban Land Institute and National Association of Home Builders, *New Approaches to Residential Land Development: A Study of Concepts and Innovations*, technical bulletin no. 40 (Washington: Urban Land Institute and National Association of Home Builders, 1961), 11.

61 Sewell, "Don Mills," 37.

62 The 6,000 acres of land for Erin Mills had been assembled by E.P. Taylor in the mid-1950s. It was developed by Cadillac Fairview in the early 1970s. See Gluskin, *Cadillac Fairview Corporation Limited*, 135–158.

63 For details concerning current planning ideas for Heritage Green see John G. Williams Associates Limited, *Heritage Green Community: Secondary Plan. An Application to Amend the Official Plan of City of Stoney Creek* (Toronto: Ontario Land Corporation, 1985).

64 HOME was introduced in 1967 to help people cope with rising housing prices. Under this program, people purchased dwellings but leased the land on which the dwellings sat from an agency of the provincial government (initially the Ontario Housing Corporation and later the Ontario Land Corporation). After five years, they could buy the lot at its market value. The assumption was that after five years they would be earning more and could afford to pay more for housing. See Rose, *Canadian Housing Policies*, 118. On problems with the basic assumptions underpinning the HOME program see Bruce Campion-Smith,

"House Arrest: HOME's Hidden Price," *Toronto Star*, Friday, 17 March 1989, A22.

65 The OLC, which was merged with the Realty Group of the Ministry of Government Services in the mid-1980s, formerly serviced land and then called for tenders from builders on blocks of lots. The tendering procedure is still in effect, but now builders are sold draft-approved lots, which they must service themselves. The OLC, like most large developers, provides financing for its builders of up to 75 per cent of the land costs for a period of up to two years. Much of the information on Heritage Green was provided in interviews with Peter W. Scott, Coordinator, Land Development, Western Ontario Region, Ministry of Government Services, Realty Group, 26 July 1988 and 17 August 1990.

66 While it awaited approval of the expressway, the OLC devoted most of its attention to another Hamilton Mountain project, Highridge Estates. Ibid. Heritage Green still is lacking in shopping facilities and public transportation, and secondary plans must be approved for most of the area. The Red Hill Creek Expressway received final approval from the Ontario Cabinet in March 1987, but it was cancelled for fiscal and environmental reasons by Ontario's new NDP government in December 1990. See Kevin Marron, "Hamilton Freeway Plan Approved: $150 Million Cost," *Globe and Mail*, Saturday, 14 March 1987, A11 and "Province kills $187-million Expressway," *Toronto Star*, Tuesday, 18 December 1990, A1.

67 According to information collected by the Ontario New Home Warranty Program, only a handful of large builders who were not locally based were active in Hamilton between 1978 and 1984. These included Victoria Wood Development Corporation, Inc., G.R.W. Construction Limited, George Wimpey Canada Limited, Sandbury Building Corporation, The Coventry Group Limited, Lehndorff Construction Limited, Monarch Construction Limited, Costain Limited, Cadillac Fairview Corporation, A.B. Cairns Limited, and Revenue Properties Development Limited. Together, these eleven firms built 1,791 or 10 per cent of the 17,954 dwelling units constructed in Hamilton between 1978 and 1984. It was not always possible to determine the headquarters location of the firms listed in the file, so this should be viewed as a rough estimate of the activity that was not locally based. A safer figure would probably be closer to 15 per cent. We are grateful to Dr Barbara Wake Carroll of the Department of Political Science at McMaster University for providing us with access to the New Home Warranty Program data base.

68 Annual growth figures were far more spectacular in several other large Canadian cities, especially Calgary, Edmonton, and Toronto. In

the latter, increases of more than 50,000 per year have been common since 1951. The average annual increase in the Hamilton CMA for the most recent inter-censual period, 1981–1986, was just 2,987, compared to 59,335 in the Toronto CMA. Between 1976 and 1986, Hamilton experienced the fifth lowest rate of population growth, 6.1 per cent, among Canada's 23 Census Metropolitan Areas. Megabuilders, capable of erecting more than 500 homes in a year, have emerged in the most dynamic Canadian markets, and especially in the Toronto region. These include Canada Homes and Greenpark, both of which have built more than 2,000 homes in a single year in the Toronto-area. Neither has been active in Hamilton according to our research, though Greenpark did purchase some land in Stoney Creek in 1988. (On Greenpark see Warren Potter, "One-Stop Sales Centre Gets Crowds," *Toronto Star*, Saturday, 21 November 1987, E1-E2. In 1987 Greenpark was involved in twenty different sites in the Toronto area and a total of about 3,000 homes). Coscan (formerly Costain) Limited, part of the real-estate empire of Peter and Edward Bronfman, built and sold more than 11,000 houses in Toronto (Central Ontario), Ottawa, Alberta, Florida, Arizona, and Washington, DC between 1977 and 1987 and another 2,000 in 1988. (Coscan Development Corporation, *1987 Annual Report* (Toronto, 1988), 30–1 and Marian Stinson, "Coscan Looks West and South," *Globe and Mail*, Friday, 28 April 1989, B9). According to data from Ontario's New Home Warranty Program, Coscan did not enter the Hamilton market until 1984, when it built 63 homes. Figures for Ontario home builders for 1989 revealed ten firms, none based in Hamilton, with annual output of at least 430 homes. Only two builders (Greenpark Homes, 1,593 and Canada Homes, 1,477) managed to put up more than 700 detached and semi-detached houses in 1989. Warren Potter, "Greenpark Heads List of 1989 Home Builders," *Toronto Star*, Saturday, 10 February 1990, E17.

69 See, for example, Tracey Tyler, "Guelph and Kitchener Lure Metro Home Buyers," *Toronto Star*, Saturday, 28 May 1988, E1.

70 In fact, a failure to secure adequate services at a reasonable cost could scuttle a development. See HPL, Hamilton Housing Scrapbook, vol. 9, 33, Jerry Rogers, "Builder Cries Foul Over Cost of Sewers," *Spectator*, Thursday, 23 August 1984. On the importance of sewers in the Toronto area see Jim Byers, "$40 Million Sewer Clears Way for 250,000 New Residents," *Toronto Star*, Thursday, 25 May 1989, A7.

71 On the history of planning legislation in Ontario see Ontario Economic Council, *Subject to Approval: A Review of Municipal Planning Legislation in Ontario* (Toronto: Ontario Economic Council, 1973).

72 Hamilton City Clerk's Office, Agreements File, file 38–56.

73 Hamilton City Clerk's Office, Agreements File, file 45–56.

74 Hamilton City Clerk's Office, Agreements File, file 193–66.

75 Hamilton City Clerk's Office, Agreements File, file 191–66.

76 Lorimer, *The Developers*, 102–9.

77 Miron, *Housing in Postwar Canada*, 259–60.

78 Paul Wright, "Here's Why That New House Will Cost Upwards of
$30,000," *Spectator*, Saturday, 20 November 1976, 70. Ancaster was
home to some very prestigious developments by the late 1980s. For
example, according to brochures obtained by John Weaver in spring
1989, Senator Homes was building homes of from 1,850 to 4,000
square feet, priced from $199,900 to $369,900, in a 206-lot project
called Governor's Lane. Twelve models were offered for 50-foot lots
and another five for 45-foot lots. The Country Club collection of
homes was available for the largest lots in the development with five
models for 60-foot lots and two models for 70-foot lots.

79 Paul Wright, "Hidden Costs Raise Prices in Subdivision," *Spectator*,
Saturday, 2 August 1975. See also Paul Wright, "Storm Sewer Plan
Could Reduce Costs," *Spectator*, Saturday, 8 January 1977, 54; "Small
Lot Housing Gaining in Popularity," *Burlington Post*, Wednesday,
23 May 1979, RE-16; "A Little Less Luxury," *Spectator*, Monday,
15 March 1976, 7; Paul Wright, "Developers Seek to Cut House
Costs," *Spectator*, Saturday, 27 March 1976, 61; and "'Ridiculous'
Standards Help Boost Land Costs," *Burlington Gazette*, Tuesday,
2 December 1975, 17.

80 Wright, "Hidden Costs."

81 HPL, Hamilton Housing Scrapbooks, vol. 10, 24, Bob Melnbardis,
"Parkland Fund Hikes New Home Cost: Builders," *Spectator*, Thursday, 30 April 1987.

82 See Warren Potter, "GST and Levies Will Hurt Business, Building
Suppliers Say," *Toronto Star*, Saturday, 27 January 1990, E1-E2. The
combination of the GST and the new educational levy was predicted to
add almost $15,000 to a new house priced at $310,000. New-home
builders were not happy with the GST, which was not to apply to re-
sale homes. See Allan Thompson, "Home Builders Fire New Volley
Against Goods and Services Tax," *Toronto Star*, Thursday, 28 December 1989, C3; Keith Bolander, "Experts See Housing Disaster in
Goods and Services Tax," *Toronto Star*, Saturday, 14 October 1989,
E1-E2; and Licia Corbella, "How Taxes Affect the Price of Housing,"
Toronto Star, Saturday, 18 November 1989, E1.

83 Lorimer, *The Developers* and Spurr, *Land and Urban Development*. See
also Gluskin, *Cadillac Fairview*, 141–58.

84 To its credit, Ontario's Ministry of Municipal Affairs and Housing
studied the servicing question and recommended that a reduction in

lot size combined with a reduction in some servicing standards could lower the cost of new homes. See Ontario Ministry of Municipal Affairs and Housing, *Urban Development Standards* (Toronto: Ministry of Municipal Affairs and Housing, 1976). Between 1974 and 1978 the Ontario Land Speculation Tax was in effect, but it had little impact on speculative profits. See Lawrence B. Smith, "Canadian Housing Policy in the Seventies," *Land Economics* 57 (1981): 338–52 and "The Ontario Land Speculation Tax: An Analysis of an Unearned Increment Land Tax," *Land Economics* 52 (1976): 1–12. According to Gluskin, *Cadillac Fairview*, 149, the Ontario Land Speculation Tax had little impact on the activities of large developers. Calls for a tax on land speculation were renewed in the late 1980s. See, for example, Warren Potter, "Broadbent's Plans to Tax Speculators Make Good Sense," *Toronto Star*, Saturday, 11 June 1988, E1 and Michael Smith, "Speculation Tax Urged to Deflate House Prices," *Toronto Star*, Friday, 14 April 1989, A6.

85 See Ontario Ministry of Municipal Affairs and Housing, *Handbook for Energy Efficient Residential Subdivision Planning: Design Implications* (Toronto: Ministry of Municipal Affairs and Housing, 1984). The Planning Act has also been overhauled recently. See Ontario Ministry of Municipal Affairs, *The Planning Act, 1983. Subdivision/Condominium Approval Procedures: A Guide for Applicants* (Toronto: Ministry of Municipal Affairs, 1986).

86 Sherman J. Maisel, *Housebuilding in Transition* (Berkeley: University of California Press, 1953). On changes in the scale of the construction industry in Canada see Christian Marfels, *Concentration Levels and Trends in the Canadian Economy, 1965–1973: A Technical Report*, study no. 31 (Ottawa: Royal Commission on Corporate Concentration, 1976), 56–60.

87 Urban Land Institute, *The Community Builders Handbook* (Washington: Urban Land Institute, 1947); *Home Builders Manual for Land Development* (Washington: National Association of Home Builders, 1950); *New Approaches to Residential Land Development*; *Large-Scale Development: Benefits, Constraints, and State and Local Policy Incentives* (Washington: Urban Land Institute, 1977); and National Association of Home Builders, *Land Development Manual* (Washington: National Association of Home Builders, 1969). Each of these items was available in the Architecture Library at the University of Toronto soon after publication.

88 Ned Eichler, *The Merchant Builders* (Cambridge: The MIT Press, 1982), 112–13. On Levitt see Herbert J. Gans, *The Levittowners: Ways of Life and Politics in a New Suburban Community* (New York: Vintage Books, 1967).

89 CMHC, *Housing in Canada*, 21, noted in 1987 that "only a handful of firms builds as many as 1,000 to 2,000 units." While the number of firms that built at least 100 units stood at less than 75 in 1985, they accounted for about 30 per cent of all new single-detached houses.

90 HPL, Hamilton Mountain Scrapbook, vol. 1, 121, "Permits Issued for 61 Houses," *Spectator*, Wednesday, 1 June 1949.

91 D. Klashoff, "The Small Hamilton Contractor during the 1980s Real Estate Boom," unpublished term paper, Department of History, McMaster University, 1988, n.p. This firm's developments include Dewitt Heights, Glendale Estates, Strawberry Hill, Lakewood Landing, Glenbrae Estates, Blossom View, and Wellington Chase. According to brochures gathered by John Weaver in 1989, Sinclair was the largest builder in Wellington Chase with 77 lots, or 39.1 per cent of the properties. He was joined by six other builders (Rosemount Homes – 23, Sandona Homes – 27, Covington Homes – 16, Westpark Developments – 13, Scarlett Homes (Hamilton, Inc.) – 18, and Sam Chiarelli Homes – 23 lots). Sinclair's houses in Wellington Chase ranged in size from 2,000 to 2,325 square feet and in price from $188,900 to $205,900. Sales were being handled by HomeLife/State Realty Limited.

92 See "Hamilton Man Heads Builders' Group," *Spectator*, Tuesday, 8 April 1952 and "Hamilton Man Elected Head of Builders," *Hamilton News*, Wednesday, 9 April 1952. The quotation is taken from Grisenthwaite's obituary, "Developer Built Much of Hamilton," *Spectator*, Saturday, 9 February 1980. Many prominent builders have claimed to have started in the same way as Grisenthwaite. See, for example, Pat Brennan, "Builder Iggy Kaneff Puts a Lot Back Into the Land That Gave Him So Much," *Toronto Star*, Monday, 6 February 1989, B3. Kaneff started by building one house in 1951, borrowed against it to build another and eventually built more than 12,000 houses in the Toronto area.

93 Our records on Grisenthwaite, unfortunately, are not complete. Four of the surveys that we examined (762, 786, 792, and 1187), containing 630 lots, were registered by Grisenthwaite. A sampling of May and June newspaper advertisements taken at roughly five-year intervals suggests that he was involved in about one subdivision a year – 1954 (Brucedale and East 41st Street), 1956 (East 45th and Mohawk), 1961 (Sherwood Village and Westchester Park), and 1966 (Cherry Heights). If each survey averaged 100 lots and all of the homes were built by Grisenthwaite, then the firm could have been responsible for more than 2,000 houses on Hamilton's Mountain.

94 "Developer Built Much of Hamilton," *Spectator*, Saturday, 9 February 1980.

95 Ibid.

96 *Spectator*, Saturday, 5 June 1954, 41.

97 *Spectator*, Saturday, 20 May 1961, 45.

98 Eichler, *The Merchant Builders*, 94–7. On the use of other types of inducements during this era see Warren Potter, "Giveaways Tried to Help Hustle Homes: But None Matches Old Sales Stunts of Housing Circus: Free Car in Drive, Flamethrower Displays, Sit-In Bomb Shelter," *Toronto Star*, Saturday, 1 July 1989, E1 and E6.

99 *Spectator*, Friday, 18 May 1958, 59.

100 Ibid. and *Spectator*, Saturday, 20 May 1961, 44. Both Abbotsford and Sunshine remained in business in 1989.

101 See, for example, Klashoff, "The Small Hamilton Contractor."

102 Another example is Toronto's Hurlburt family. See Michael Salter, "Hot, Hot Houses," *Report on Business Magazine* 5 (August 1988): 43.

103 By 1951 the firm was listed in the annual *City Directory*.

104 *Spectator*, Saturday, 2 July 1988, H11 and H15.

105 Information on Campbell has been taken from Sherry Sleightholm, *Hamilton: A Community in Symphony* (Burlington: Windsor Publications in Co-Operation with the Hamilton and District Chamber of Commerce, 1986), 31; HPL, Hamilton Mountain Scrapbook, vol. 2: 92–4, Jerry Rogers, "Construction Hot as Building Fever Grips Mountain," *Spectator*, Tuesday, 20 May 1986; and HPL, Hamilton Housing Scrapbook, vol. 10, 79, Jennifer Gray-Grant, "Strong '88 Housing Market Predicted," *Spectator*, Saturday, 14 November 1987. Subdivision ownership information was extracted from data provided by the Subdivision and Condominium Administration of the Regional Municipality of Hamilton-Wentworth.

106 Interview with Peter W. Scott, Coordinator, Land Development, Western Ontario Region, Ministry of Government Services Realty Group, 26 July 1988.

107 According to Klashoff, "The Small Hamilton Contractor," n.p., Nello Carnicelli followed a similar path into homebuilding. After 24 years in bricklaying and assorted building partnerships, he founded Carriage Gate Homes in 1987. The company built 5 single-detached houses in 1987, and during the 1988 season expanded to put up 11 detached and 10 townhouse units. The fortunes of most small builders fluctuate from year to year. Another small Hamilton builder, Ross Bomarito, who has averaged 11 homes per year over two decades, told Klashoff that "during slumps in the housing market the company dissolves and [he] supplements his income through home renovation and bricklaying."

108 *Spectator*, Saturday, 2 July 1988, H5. Brochures obtained by John Weaver in spring 1989 revealed that DeSantis houses in the two-

phased, 187-lot Highridge Hills development on the Mountain ranged in size from 1,035 to 2,450 square feet and in price from $154,900 to $234,900. About 15 different designs were offered to prospective purchasers. Aldo DeSantis Realty served, as always, as the exclusive agent for the project.

109 Sleightholm, *Hamilton*, 32–33; HPL, Hamilton Housing Scrapbook, vol. 7, 104, Denis LeBlanc, "Realtor Laments Demise of Loan," *Spectator*, Friday, 31 December 1982; and HPL, Hamilton Housing Scrapbook, vol. 10, 87, Jennifer Grey-Grant, "Incredible Shrinking House: Smaller Homes 'Selling Like Hotcakes,'" *Spectator*, Saturday, 28 November 1987.

110 Between 1955 and 1972, NHA-financed dwellings accounted for 39 per cent of all housing starts in Hamilton. See CMHC, *Canadian Housing Statistics*, appropriate years and volumes.

111 Interview with Joachim Schwarz, 11 May 1988.

112 Calculated from figures in CMHC, *Canadian Housing Statistics*, appropriate years and volumes.

113 Jeremy R. Rudin, *The Changing Structure of the Land Development Industry in the Toronto Area*, major report no. 13 (Toronto: Centre for Urban and Community Studies, University of Toronto, 1978), 36. In 1961 such builders had accounted for only 34.3 per cent of Toronto's NHA-financed houses. On Winnipeg builders in the early 1980s, see Lynda H. Newman, *Structural Change in the Housing Industry*. report no. 1 (Winnipeg: Institute of Urban Studies, University of Winnipeg, 1984).

114 CMHC, *Canadian Housing Statistics* (Ottawa: CMHC, 1974), 87.

115 Lorimer, *The Developers* and Spurr, *Land and Urban Development*. Spurr's study was commissioned by CMHC, which refused to publish it. Corporate concentration applied to the building materials side of the residential construction business as well. For example, in July 1988 it was revealed that four firms controlled 92 per cent of ready-mix concrete sales in the Toronto region. The four charged identical prices for their products, a practice that some claimed resulted in an extra cost of about $1,000 for every new house that was built. See Jock Ferguson, "4 Concrete Firms' Set Prices Add $1,000 to House Costs," *Globe and Mail*, Thursday, 21 July 1988, A1 and A19.

116 Miron, *Housing in Postwar Canada*, 266–7.

117 HPL, Hamilton Mountain Scrapbook, vol. 1, 305, "A Plan to Save the Mountain from Monotony," *Spectator*, Monday, 24 April 1972.

118 HPL, Hamilton Housing Scrapbook, vol. 9, 25, Suzanne Bourret, "Look-alike Homes Come Under Attack," *Spectator*, Thursday, 31 May 1984.

119 Barbara Wake Carroll, "Market Concentration in a Geographically Segmented Market: House Building in Ontario, 1978–1984," forthcoming.

120 A total of 1,090 builders registered for Hamilton with the New Home Warranty Program between 1978 and 1984. Only 891 of them actually built any dwelling units – 451 of them built 1–5 units (accounting for 3.1 per cent of all production), 270 built 6–25 units (15.7 per cent), 76 built 26–50 units (14.3 per cent), 49 built 51–100 units (17.8 per cent), and 45 built more than 100 units between 1978 and 1984 (47.1 per cent). Those building at least six units over the period averaged 5.9 units per year, while those building more than one hundred units averaged 28.6 dwellings per year.

121 See, for example, Gwendolyn Wright, *Building the Dream: A Social History of Housing in America* (New York: Pantheon, 1981), 248 and Dolores Hayden, *Redesigning the American Dream: The Future of Housing, Work, and Family Life* (New York: W.W. Norton, 1984).

122 Rose, *Canadian Housing Policies*; Bourne, *The Geography of Housing*, 191–214; Bacher, "Keeping to the Private Market"; and Miron, *Housing in Postwar Canada*, 238–67.

123 Carver, *Houses for Canadians*, 69.

124 Lorimer, *The Developers*, 258–68. See also Susan Goldenberg, *Men of Property: The Canadian Developers Who Are Buying America* (Toronto: Personal Library, 1981).

125 On the history of planning in Hamilton see J.T.C. Waram, "City Planning in Hamilton," *Wentworth Bygones* 4 (1963): 36–43 and City of Hamilton Planning Department, *City of Hamilton: General Background Information* (Hamilton: City of Hamilton Planning Department [1972]).

126 Carver, *Houses for Canadians*, 62–3.

127 McKellar, "Building Technology," 10.

128 Concrete examples in the residential development industry are not easily located. It is known, however, that through rigorous scheduling procedures and the use of new technology in elevators, privately-owned development giant Olympia and York was able to save an estimated 1.3 million person-hours in the construction of the 72-storey First Canadian Place office building in downtown Toronto. Only a firm of this size would be able to harness this potential. Peter Foster, *The Master Builders: How the Reichmanns Reached for an Empire* (Toronto: Key Porter Books, 1986), 27–31.

129 Klashoff, "The Small Hamilton Contractor," n.p.

130 Spurr, *Land and Urban Development*, 186–8.

131 Statistics Canada, *1981 Census of Canada: Census Tracts – Hamilton*,

catalogue 95–952 (Ottawa: Minister of Supply and Services Canada, 1983), 1–3.

132 Perhaps the most thorough statement about red tape in property development can be found in A.H. Oosterhoff and W.B. Rayner, *Losing Ground: The Erosion of Property Rights in Ontario* (Toronto: The Ontario Real Estate Association Foundation, 1980).

133 Warren Potter, "Suburbs Might Look at Small Lots as Key to Affordability," *Toronto Star*, Saturday, 16 January 1988, E1.

134 AHOP used a combination of direct grants and repayable interest reduction loans to reduce required down payments to as little as 5 per cent of the purchase price. It was predicated on the twin assumptions that interest rates would remain stable and the economy would remain healthy, thus permitting incomes to rise. Neither assumption proved to hold true by the early 1980s, and hundreds of foreclosures resulted when residents were unable to carry renewed mortgages at rates of interest that topped 20 per cent. Rose, *Canadian Housing Policies*, 55–6 and Smith, "Canadian Housing Policies," 343–7.

135 HPL, Housing File, vol. 8, 80–1, Linda Jacobs, "Those Back-Breaking Rates: Is There a Way to Ease Homeowners' Pain?" *Spectator*, Saturday, 5 November 1983.

136 Albert Rose, *Regent Park: A Study in Slum Clearance* (Toronto: University of Toronto Press, 1958).

137 Weaver, *Hamilton*, 173.

138 Data supplied by the Ontario Ministry of Housing, Hamilton office.

139 Major critical analyses of Canadian housing policy include Michael Dennis and Susan Fish, *Programs in Search of a Policy: Low Income Housing in Canada* (Toronto: Hakkert, 1972) and Lawrence B. Smith, *Anatomy of a Crisis: Canadian Housing Policy in the Seventies* (Vancouver: The Fraser Institute, 1977).

140 In the summer of 1988 Hamilton City Council was exploring the possibility of a new by-law to limit to five the number of unrelated individuals who could live in a rented single-family dwelling. Aimed at university students, similar legislation already had been enacted by four Ontario municipalities – Sudbury, Guelph, Waterloo, and London. See Kevin Marron, "Hamilton Debating Controls on Single Family Home Rentals," *Globe and Mail*, Wednesday, 27 July 1988, A14.

141 On the problem of homelessness in Ontario see Ontario Ministry of Housing, *More Than Just a Roof: Action to End Homelessness in Ontario*, Final Report of the Minister's Advisory Committee on the International Year of Shelter for the Homeless (Toronto: Ministry of Housing, 1988).

142 For examples of this resistance in the Toronto area see Sue Pigg, "Lots of Subdivisions, Little Affordable Housing," *Toronto Star*, Monday, 23 March 1987, A13.

143 Ontario Ministry of Housing, *More Than Just a Roof*, 38.

144 HPL, Hamilton Housing File, vol. 9, 99, "Housing Bind is Serious, Report Says: Not Enough Affordable Survey Finds," *Spectator*, Wednesday, 9 April 1986.

145 HPL, Hamilton Housing Scrapbooks, vol. 11, 15, "Subsidies 'A Drop in the Bucket': 345 Housing Units Approved," *Spectator*, Tuesday, 2 February 1988. So far, the Hamilton units had been granted to Hamilton Baptist Non-Profit Homes, Inc. (111 units), Hamilton's Municipal Non-Profit Housing Corporation (92 units), Hamilton Portuguese Community Homes (5 units), Liuna Hamilton Association (67 units), and the Sons of Italy Foundation (10 units).

146 Adele Freedman, "Cul-de-Sick: Urban Sprawl and the Suburbs: The Trouble with the Town of Vaughan," *Toronto Life Homes* 22 (1988): H7–H12.

147 This phrase is taken from Malvina Reynolds's song, "Little Boxes," as quoted in Matthew Edel, Elliott D. Sclar, and Daniel Luria, *Shaky Palaces: Homeownership and Social Mobility in Boston's Suburbanization* (New York: Columbia University Press, 1984), 175–6. Written in 1962 and allegedly inspired by suburban San Francisco, the most famous rendition of the song remains that of Pete Seeger.

148 As quoted in Freedman, "Cul-de-Sick," H12.

149 Kathy Dermott, "Builders are Courting the First-Time Buyer," *Toronto Star*, Saturday, 11 November 1989, E1 and E4. Dermott cited this as a new trend in the Toronto area.

150 See Diane Sawchuk, "Workers Who Go the Distance: Today's Commuters Travel from St. Catharines and Ajax to Get to Their Jobs in Metro," *Toronto Star*, Saturday, 6 May 1989, G1 and Maureen Murray, "Where Will We Be Living in 20 Years?" *Toronto Star*, Saturday, 22 April 1989, E1 and "An *Affordable* Home? Get Out of Town!" *Toronto Star*, Saturday, 13 January 1990, E1–E2.

151 Warren Potter, "Immediate Action Needed to Help Out First-Time Buyer," *Toronto Star*, Saturday, 19 March 1988, H1.

152 Klashoff, "The Small Hamilton Contractor."

153 Michael A. Goldberg, *The Housing Problem. A Real Crisis?: A Primer on Housing Markets, Policies, and Problems* (Vancouver: University of British Columbia Press, 1983). On recurring affordability scenarios see Warren Potter, "Remember the Housing Crisis of 1967? People Despaired of Buying Homes as Lot Prices Climbed to $10,000 and Houses Soared to $25,000," *Toronto Star*, Saturday, 29 April 1989, E1.

154 Larry S. Bourne, "Choose Your Villain: Five Ways to Oversimplify the Price of Housing and Urban Land," *Urban Forum* 3 (1977): 16–24.

155 Gord Kolle, "Mortgage Plans May Have to Change," *Toronto Star*, Saturday, 24 March 1990, F1-F2.

156 David Israelson, "Skyrocketing Prices Destroy Couple's Dream of Buying First Home," *Toronto Star*, Tuesday, 24 May 1988, A1 and A21. See also David Israelson, "Houses with Frills at No-Frills Prices," *Toronto Star*, Saturday, 8 April 1989, D5. Two Montreal architects have designed a 1,000 square foot house that can be built and sold for about $50,000, exclusive of the land cost ($30,000 in construction costs, $10,700 in financing, and $11,400 in overhead and profit). Other articles on this theme include Keith Bolender, "The Struggle to Build Affordable Homes," *Toronto Star*, Saturday, 3 June 1989, E1 and "Europe May Teach Lesson in Housing," *Toronto Star*, Saturday, 24 June 1989, G1-G2; Victoria Stevens, "Will the Trend Turn Toward Smaller Homes?," *Toronto Star*, Saturday, 10 June 1989, G1; John Terauds, "Future Trends in Housing," *Toronto Star*, Saturday, 13 May 1989, F1; and Warren Potter, "Home Show Will Display 'European' Affordable Home," *Toronto Star*, Saturday, 16 September 1989, E1-E2. The *Report of the Federal Task Force on Housing and Urban Development* (Ottawa: Information Canada, 1969), 23, gave as its second principle the idea that "*every Canadian should be entitled to clean, warm shelter as a matter of basic human right*."

157 Royson James, "Frantic Buyers Grab 259 Homes in Just One Hour," *Toronto Star*, Saturday, 4 October 1986, A6. Perhaps the most amusing Toronto-area incident occurred in April 1988. About 100 frenzied would-be purchasers, many of whom had arrived in Jaguars, Porsches, and other expensive automobiles, fought in the mud of rural York Region over the right to purchase fifty estate-sized homes, priced between $500,000 and $900,000, in an as-yet unapproved subdivision. Sterling Taylor, "Frenzy Hits as Buyers Rush Luxury Rural Homes," *Toronto Star*, Saturday, 30 April 1988, A1.

158 See Michael Davie, "Home Ownership Dream Dimmer But Picture Still Bright in Our Area," *Spectator*, Tuesday, 9 June 1987, A10 and "This Area Leads Nation in House Price Increases," *Spectator*, Saturday, 16 July 1988, H1; and Rowena Moyes "Economic Woes Grip Market," *Ontario Home Builder Magazine* (June/July 1990): 18–23.

159 Warren Potter, "Housing Industry Needs More Skilled Building Tradesmen," *Toronto Star*, Saturday, 2 April 1988, E1. Potter notes that the building industry, with 350,000 workers in 80,000 firms, is

Ontario's largest employer. The ten-year retirement rate captures
about half of these workers.

160   Warren Potter, "Ontario Plans New Crackdown on Shoddy Homes,"
Toronto Star, Saturday, 20 February 1988, E1. On problems with
dwellings built by Starward Homes see Klashoff, "The Small Hamilton Contractor." Complaints in 1987 included cracking basements,
missing options, and defective porches.

161   Warren Potter, "Warranty Program Prints New Guide," *Toronto Star*,
20 January 1990, E8. Only Kitchener, at 16.3 per cent, had a higher
rating. The Hamilton area (which extends around the western end
of Lake Ontario from Oakville to the Niagara Region in recent
ONHWP reports) had 1,167 registered builders as of 1 August 1989.
Only 731 (62.6 per cent) of these had been active in residential construction between 1986 and 1988, with 169 (23.1 per cent of all active builders and 73.1 per cent of the 237 firms large enough and/or
long enough established to qualify for a rating) rated as either above
average or excellent builders. An excellent builder is one who has
built and completed possessions on at least 10 homes in the past
three years and against whom chargeable conciliations have been levied at a rate no higher than the equivalent of one in every 75 completed possessions. Calculated from figures in *1989 Home Buyer's
Guide to After Sales Service* ([Toronto]: Ontario New Home Warranty
Program, 1989).

162   In Toronto an attempt was made to start 41,000 units in 1987. A
more realistic capacity for the industry is 33–35,000 units. See Warren Potter, "Less Red Tape Means Greater Affordability, Builder
Says," *Toronto Star*, Saturday, 26 December 1987, F1. Potter used figures obtained from Robert Hume, President of the Toronto Home
Builders' Association. Hamilton housing starts in 1987 stood at
5,619 units, the highest total since 1975 and the sixth highest total
since 1952.

163   Clayton, "Single-Family Homebuilding Industry," 10. Some builders
are already seeking new markets within Canada's shifting demo-
graphic structure. See, for example, Donn Downey, "Builders Target
Elderly as Hot New Market," *Globe and Mail*, Friday, 2 October 1987,
B1 and Cindy Kleiman, "Builders Strive to Meet Housing Needs of
Seniors: Experts Predict That 36 Years from Now the Number of
People Aged 65 or Over Will Equal One-Quarter of Our Popula-
tion," *Toronto Star*, Saturday, 17 June 1989, F1.

164   Building Industry Strategy Board, *From Fragmentation to Consolidation*
([Toronto]: Building Industry Strategy Board, 1989), 1. Appendix F
of this volume lists some 230 active building-industry associations.

165 Hussein Rostrum, *Human Settlements in Canada: Trends and Policies 1981–1986*, Presentation to the United Nations Economic Commission for Europe (Ottawa: CMHC, 1987), 58–60.

166 Jennifer Gray-Grant, "A Piece of Ballinahinch Castle: Condo Features Some Classic Lines," *Spectator*, Saturday, 26 November 1988, H16–H17. Located at 316 James St South, Ballinahinch was built in the Italianate-Revival style in 1849–50 for dry-goods merchant Aeneas Sage Kennedy. It was renovated into six condominium units in 1980.

167 Such "monster home" developments were not always viewed positively. See, for example, "Etobicoke Needs Monster Home Law, Residents Charge," *Toronto Star*, Wednesday, 31 January 1990, A26.

168 Rostrum, *Human Settlements*, 57.

169 Frank Clayton and Robert Hobart, "The Supply of New Housing in Canada into the Twenty-First Century," in Michael A. Goldberg and George W. Gau, eds, *North American Housing Markets into the Twenty-First Century* (Cambridge: Ballinger Publishing, 1983), 127–38. On the problems associated with supplies of serviced building lots see Diane Sawchuk, "The Desperate Search for Serviced Land: Metro Lot Prices Have Risen from $1,200 to $6,000 per Lineal Foot," *Toronto Star*, Saturday, 15 April 1989, E1.

170 On the expected demographic changes in Canada see Philip W. Brown, "The Demographic Future: Impacts on the Demand for Housing in Canada, 1981–2001," in Goldberg and Gau, eds, *North American Housing Markets*, 5–31.

171 CMHC, *Housing in Canada*, 29.

172 *Service to the Public: Housing Programs in Search of a Balance* (Ottawa: Ministry of Supply and Services Canada, 1986) as quoted in Carter, *CMHC and the Building Industry*, 2.

173 In 1988 the OLC sold some of its land in the Toronto suburb of Scarborough for $4,000 a front foot, fully one-third above prevailing prices. This made the cost of a 35-foot building lot $140,000. See Warren Potter, "Government Action Helps to Perpetuate High Housing Costs," *Toronto Star*, Saturday, 6 August 1988, E1. Already some baby boomers are beginning to question the quality of their lifestyles compared to that enjoyed by their parents. Expressed in constant dollars, average individual income for Canadians increased by 72 per cent between 1957 and 1987. The price of a house in Toronto's Bloor and Bathurst area increased by 322 per cent, that of a house in Winnipeg's River Heights area by 85 per cent, over the same period. See Rona Maynard, "Were They Better Off Than We Are?" *Report on Business Magazine* 4 (May 1988): 38–47.

CHAPTER FOUR

1 Nancy G. Duncan, "Homeownership and Social Theory," *Housing and Identity: Cross-Cultural Perspectives*, ed. Nancy Duncan (London: Croom Helm, 1981), 98.

2 Constance Perin, *Everything in Its Place: Social Order and Land Use in America* (Princeton, N.J.: Princeton University Press, 1978), 83.

3 Richard F. Muth, "The Demand for Non-Farm Housing," in *The Demand for Durable Goods*, ed. Arnold Harberger (Chicago: University of Chicago Press, 1960), 66–74.

4 Ibid., 30.

5 Ibid., 58.

6 Richard F. Muth, "A Numerical Model of Urban Housing Markets with Durable Structures," in *Research in Urban Economics: A Research Annual*, ed. J. Vernon Henderson (Greenwich, Connecticut: JAI Press, 1982), 95; "The Allocation of Households to Dwellings," *Journal of Regional Science* 18 (August 1978): 159–78.

7 David Harvey, *Social Justice and the City* (Baltimore: Johns Hopkins University Press, 1975), 157–76. Useful collections of essays on the theme of the manipulated city include Dimitrios Roussopoulos, ed., *The City and Radical Social Change* (Montreal: Black Rose Books, 1982) and Stephen Gale and Eric G. Moore, eds, *The Manipulated City: Perspectives on Social Structure and Social Issues in Urban America* (Chicago: Maaroufa Press, 1975).

8 Michael Doucet, "The Role of the *Spectator* in Shaping Attitudes toward Land in Hamilton, Ontario, 1847–1881", *Histoire sociale/Social History* 12 (November 1979): 442.

9 Manuel Castells, *The City and the Grassroots: A Cross-Cultural Theory of Urban Social Movements* (Berkeley: University of California Press, 1983), xvi.

10 David Harvey, "Labor, Capital and Class Struggle around the Built Environment in Advanced Capitalist Societies," in Peter Saunders, *Social Theory and the Urban Question* (London: Hutchinson, 1981), 222–31. *Urbanization and Conflict in Market Societies*, ed. K. R. Cos (Chicago: Maaroufa Press, 1978), 17–21; On home ownership in Europe see Michael Harloe, *Private Rented Housing in the United States and Europe* (New York: St Martin's Press, 1984), 65–6.

11 Duncan, "Homeownership," 105.

12 Ibid., 104. Also see Harvey, "Labor,Capital and Class Struggle," 15.

13 Duncan, "Homeownership," 106. For the trend toward home ownership in Europe see Michael Harloe, *Private Rented Housing in the United States and Europe*, 65–6.

14 See John C. Bacher, "Keeping to the Private Market: The Evolution of Canadian Housing Policy: 1900–1949," unpublished PhD dissertation, 1985, 1–18.

15 Duncan, "Homeownership," 105.

16 Matthew Edel, Elliott D. Sklar, and Daniel Luria, *Shaky Palaces: Homeownership and Social Mobility in Boston's Suburbanization* (New York: Columbia University Press, 1984), xix-xx.

17 Ibid., 107.

18 Ibid., 223.

19 Ibid., 264.

20 Ibid., 265–90.

21 Miles L. Colean et al., *American Housing: Problems and Prospects* (New York: The Twentieth Century Fund, 1944), 19.

22 The information on Hamilton's street layouts came from McMaster University, Lloyd Reeds Map Library, Hamilton Subdivision maps and from Hamilton-Wentworth Planning and Development Department, *Hamilton Zoning By-Law: Neighbourhood Maps* (Hamilton, 1981).

23 Mel Scott, *American City Planning since 1890*(Berkeley and Los Angeles: University of California Press, 1969), 10–15; 89–91.

24 John C. Weaver, "From Land Assembly to Social Maturity: The Suburban Life of Westdale, Hamilton, 1911–1951," 414.

25 John Sewell, "Don Mills: E. P. Taylor and Canada's First Corporate Suburb," *City Magazine* 2 (1977): 28–38.

26 John Burnett, David Vincent, and David Mayall, *The Autobiography of the Working Class: An Annotated Critical Bibliography*, vol. 1, *1790–1900* (New York: New York University Press, 1984).

27 United States, *Seventh Census of the Unites States, 1850* (Washington: Robert Armstrong, 1853), 11–976; *Eighth Census of the United States, 1860* (Washington: Government Printing Office, 1864), 674–675.

28 John McAdam to his mother, quoted in *Autobiography of John McAdam (1806–1883) with Selected Letters*, ed. Janet Fyfe (Edinburgh: Printed for the Scottish History Society by Clark Constable, Ltd, 1980), 102.

29 George Martin to his father, 7 June 1837, quoted in *Invisible Immigrants: The Adaptation of English and Scottish Immigrants in Nineteenth-Century America*, ed. Charlotte Erickson (London: Weidenfeld and Nicolson, 1972), 286.

30 HPL, Hope Papers, Adam Hope to his father, 1 March 1837.

31 James Thomson to Alexander Thomson, editor, 25 September 1854, quoted in Richard A. Preston, *For Friends at Home: A Scottish Immigrant's Letters from Canada, California and the Cariboo, 1844–1864* (Montreal: McGill-Queen's University Press, 1974), 224.

32 Erickson, *Invisible Immigrants*, 253.

33 Thorstein Veblen, *Theory of the Leisure Class, An Economic Study of Institutions* (New York: The Macmillan Company, 1902), 110.

34 Ibid., 134.

35 Robert D. Lynd and Helen Merrell Lynd, *Middletown: A Study in American Culture* (New York: Harcourt, Brace and World, Inc., 1956, original 1929), 103.

36 W. Lloyd Warner, *Yankee City* (New Haven: Yale University Press, 1966), 66.

37 Art Gallaher, *Plainsville Fifteen Years Later* (New York: Columbia University Press, 1961), 93.

38 Samuel Taylor Coleridge, *On the Constitution of the Church and State according to the Idea of Each* (London: William Pickering, 1839), 31.

39 Mordecai Richler, *The Apprenticeship of Duddy Kravitz* (Don Mills: André Deutsch, 1959), 49, 100.

40 Perin, *Everything in Its Place*, 32.

41 Robert Lynd quoted in Ibid., 46.

42 Ibid., 53.

43 Ibid., 66.

44 Beverley W. Bond, *The Quit-Rent System in the American Colonies* (Gloucester, Massachusetts: Peter Smith, 1965, reprinted edition of 1919 Yale publication).

45 Richard B. Morris, *The American Revolution Reconsidered* (New York: Harper and Row, 1967), 81.

46 David Maldwyn Ellis, *Landlords and Farmers in the Hudson-Mohawk Region, 1790–1850* (New York: Octagon Books, 1967), 226–312.

47 Brian Young, *George-Etienne Cartier, Montreal Bourgeois* (Montreal: McGill-Queen's University Press, 1981), 97–100.

48 Francis W. P. Boler, ed., *Canada's Smallest Province* (The Prince Edward Island Centennial Commission, 1973), chapters 3,4,5.

49 Stephen O. White and Richard T. Vann, "The Invention of English Individualism: Alan Macfarlane and the Modernization of Pre-Modern England," *Social History* 8 (October, 1983): 345–63.

50 Evans, *The Emigrants' Directory*, quoted in Bruce S. Elliott, *Irish Migrants in the Canadas* (Kingston and Montreal: McGill-Queen's University Press, 1988), 69.

51 George Martin to his father, quoted in Ericson, ed., *Invisible Immigrants*, 286.

52 John Burnett, *A Social History of Housing, 1815–1985* (London: Methuen, second edition, 1986), 95.

53 Dorothy Thompson, *The Chartists* (London: Temple Smith, 1984), 304.

54 Elizabeth Blackmar, *Manhattan for Rent, 1785–1850* (Ithaca: Cornell University Press, 1989), 108.

55 Paul Wallace Gates, *Fifty Million Acres: Conflict over Kansas Land Policy, 1854–1890* (Ithaca, New York: Cornell University Press, 1954), 245.

56 John Thomas, *Alternative America: Henry George, Edward Bellamy, Henry Demarest Lloyd, and the Adversary Tradition* (Cambridge: Belknap Press, 1983), 5–34.

57 Henry George, *Social Problems* (New York: Doubleday and McClure Company, 1898, originally published in 1883), 239.

58 Henry George, *Our Land and Land Policy: Speeches, Lectures and Miscellaneous Writings* (New York: Doubleday and McClure Company, 1901, originally published in 1871), 101.

59 Henry George, *Progress and Poverty: An Inquiry into the Cause of Industrial Depression and of Increase of Want with Increase of Wealth: the Remedy*, vol. 1 (New York: Doubleday and McClure Company, 1898), 127.

60 George, *Progress and Poverty*, 73.

61 "Why Work is Scarce, Wages Low, and Labour Restless," a speech delivered at the Metropolitan Temple, 26 March 1878, quoted in Henry George Jr, *The Life of Henry George* (New York: Doubleday and McClure Company, 1900), 203.

62 For George's itinerary in Canada see George Jr, *The Life of Henry George*, 350–1.

63 George, *Progress and Poverty*, 403.

64 Ibid., 127.

65 Ibid., 293.

66 Jayne Synge, *Family and Community in Hamilton, 1900–1930* (unpublished manuscript), 30, 59.

67 United States, Department of Commerce, Bureau of Foreign and Domestic Commence, *Real Property Inventory of 1934: Summary and Sixty-Four Cities Combined* (Washington, DC, 1934), 20–31.

68 John Stilgoe, *Common Landscape of America, 1685 to 1845* (New Haven: Yale University Press, 1982), 342.

69 *Labor Union*, 20 January 1883.

70 *Palladium of Labor*, 12 October 1884.

71 *Palladium of Labor*, 4 August 1884.

72 *Palladium of Labor*, 4 September 1886.

73 HPL, *Hamilton Housing: Scrapbook of Clippings, 1905–47*, clipping from *Spectator*, 29 June 1912.

74 Lois A. Vitt, "The Social Psychology of Homeownership in US Society," unpublished paper presented at Les enjeux urbains de l'habitat, Paris, 3–6 July 1990.

75 John S. Adams, "The Meaning of Housing in America," *Annals of the Association of American Geographers 74 (1984): 515.*

76 Warren Potter, "There's No Quick Fix for Metro Renters Longing for Homes," *Toronto Star*, 13 February 1988; Social Planning Council of Metropolitan Toronto, "Housing Affordability in Metropolitan Toronto, 1987," *Social Infopac* 6 (1987).

77 "Home Ownership Dream is Dying, Gallup Finds," *Toronto Star*, 8 August 1988.

CHAPTER FIVE

1 Manuel Gottlieb, *Long Swings in Urban Development* (New York: National Bureau of Economic Research, 1976), 44–5. *Globe and Mail*, 3 February 1988, B4.

2 *The Small Home* 2 (November 1923): 3–4; William Haber, *Industrial Relations in the Building Industry* (Cambridge, Mass.: Harvard University Press, 1930), 52.

3 Gottlieb, *Long Swings in Urban Development*, 9–10.

4 Ned Eichler, *The Merchant Builders* (Cambridge, Mass.: The MIT Press, 1982); James Lorimer, *The Developers* (Toronto: James Lorimer & Co., 1978).

5 Eichler, *The Merchant Builders*, 112–13.

6 Stephen Kern, *The Culture of Time and Space, 1880–1918* (Cambridge, Mass.: Harvard University Press, 1983), 152.

7 US Department of Commerce, *Historical Statistics of the United States, Colonial Times to 1970*, part 2 (Washington, DC: United States Government Printing Office, 1975), series N62–65, N111–117, N156–169, N216–231; F.H. Leacy, M.C. Urquhart, and K.A.H. Buckley, *Historical Statistics of Canada* (Ottawa: Minister of Supply and Services Canada, 1983), SS220–224.

8 US Department of Commerce, *Historical Statistics of the United States, Colonial Times to 1970*, part 1, series D735–6. Michael J. Doucet and John C. Weaver, "The North American Shelter Business, 1860–1920," *Business History Review* 58 (Summer 1984): 245–7.

9 Olivier Zunz, *The Changing Face of Inequality: Urbanization, Industrial Development, and Immigrants in Detroit, 1880–1920* (Chicago: University of Chicago Press, 1982), 152–6; John Bodnar, Roger Simon, and Michael P. Weber, *Lives of Their Own: Italians, Blacks, and Poles in Pittsburgh 1900–1960* (Urbana: University of Illinois Press, 1982), 153–7.

10 *Canadian Architect and Builder* 1 (August 1888): 6.

11 Nancy Byrd Turner, "Home," *Keith's Magazine* 55 (February 1926): 109.

12 *Historical Statistics of the United States, Colonial Times to 1970*, N238–245; US Department of Commerce, Bureau of the Census, *Statistical Abstract*

of the United States (Washington: Government Printing Office, 1981), table no. 1377. Canadian figures were calculated by the authors from Statistics Canada, *Perspectives Canada III* (Ottawa: Minister of Supply and Services, 1980), tables 11.25 and 11.27.

13 On the difference in estimated costs of construction see *Bungalow Magazine* (Seattle) 5 (December 1916): 749–54.

14 Sam Bass Warner Jr, *Streetcar Suburbs: The Process of Growth in Boston, 1870–1900* (Cambridge, Mass.: Harvard University Press and The MIT Press, 1962), 127.

15 On early Canadian plans in American journals see *Carpentry and Building* 2 (February 1880): 26. On the bungalow in Canada see Charles A. Byers, "The Ideal Bungalow," *Maclean's Magazine* (June 1912): 59–60 and Deryck W. Holdsworth, "House and Home in Vancouver: Images of West Coast Urbanism, 1886–1929," in G.A. Stelter and A.F.J. Artibise, eds, *The Canadian City: Essays in Urban History* (Toronto: McClelland and Stewart, 1977), 186–91.

16 A list of Hodgson's publications was derived from the Library of Congress card index.

17 David R. Goldfield and Blaine A. Brownell, *Urban America: From Downtown to No Town* (Boston: Houghton Mifflin, 1979), 146–7; David B. Hanna "Creation of an Early Victorian Suburb," *Urban History Review* 9 (October 1980): 38–64.

18 Robert Barrows, "Beyond the Tenement: Patterns of American Urban Housing, 1870–1930," *Journal of Urban History* 9 (August 1983): 395–420.

19 John McAdam to his mother, 20 October 1833, *Autobiography of John McAdam* (Edinburgh: Printed for the Scottish Historical Society by Clark Constable, 1980), 103. Sigfried Giedion, *Space, Time and Architecture: The Growth of a New Tradition* (Cambridge, Mass.: Harvard University Press, 1978), 347. On the history of nails see Amos J. Loveday Jr, *The Rise and Decline of the American Cut Nail Industry: A Study of the Interrelationships of Technology, Business Organization, and Management Techniques* (Westport: Greenwood Press, 1983), 3–30.

20 Quoted in Giedion, *Space, Time and Architecture*, 349.

21 Roger Simon, "Expansion of an Industrial City, 1880–1910" (University of Wisconsin, PhD thesis, 1971), 90–105, 123–6.

22 "Building Contracts," *Building and Wood-Worker* 6 (1885): 57; Haber, *Industrial Relations in the Building Industry*, 57. E. W. Cooney, "The Origins of the Victorian Master Builders," *Economic History Review* 8 (1955): 167–76.

23 Haber, *Industrial Relations in the Building Industry*, 280.

24 *American Builder* 10 (November 1874): 250.

25 M.A. Pigott, letter to the editor, *Canadian Architect and Builder* 2 (March 1887): 3.

26 *Canadian Architect and Builder* 10 (April 1897): 74. *Building and Wood-Worker 30 (1894): 13–14; National Builder* 43 (October 1906): 246; *Small Home* 1 (July 1922): 22.

27 The autobiography of Charles Emory was made available by his grandson George Emory of the History Department at The University of Western Ontario. William Moore's operations were discussed in Doucet and Weaver, "The North-American Shelter Business, 1860–1920," *Business History Review*, 250. Thomas Casey's life was profiled in Weaver, "From Land Assembly to Social Maturity. The Suburban Life of Westdale (Hamilton) Ontario, 1911–1951," *Histoire sociale – Social History* 2 (November, 1978): 422–4.

28 On construction stunts see H.B. Van Sickle, "Another Impossible Feat," *National Builder* 58 (July 1916): 41–3; John C. Weaver, *Hamilton: An Illustrated History* (Toronto: James Lorimer & Co., 1982), 119. On wartime and immediate post-war housing construction methods see *National Builder* 60 (August 1918): 52–5; 63 (August 1920): 22–9, 30–6; 64 (July 1921): 22–4.

29 Examples of discussions on the manufacturing and labour savings of concrete are found in *Carpentry and Building* 14 (March 1892): 77; *Canadian Architect and Builder* 17 (February 1904): 25. Demands for plans for concrete dwellings were frequently cited in the *Canadian Architect and Builder*. Also see *National Builder* 42 (May 1906): 24.

30 Thomas Edison quoted in "Poured Concrete Bungalows," *Bungalow Magazine* (Seattle) 2 (July 1913): 37.

31 Ibid., 39. See also "The Edison Concrete House," *Journal of Modern Construction* 6 (August 1909): 43.

32 *National Builder* 58 (February 1916): 44–7.

33 *Official Report*, Second Annual Convention of the National Builders of the United States of America (February 7–9, 1888), 34.

34 Alfred Crossley, *Bricks and Brick Making* (Ottawa, Illinois: The Brick, Tile, and Pottery Gazette, 1889), 15.

35 Ibid., 31.

36 *Brick, Tile and Metal Review* 2 (August 1882): 4; *Carpentry and Building* 5 (February 1883): 34.

37 *Brick, Tile and Metal Review* 2 (April, 1882): 6.

38 *American Builder* 10 (July 1874): 90.

39 W.A. Eudaly, *Practical Thoughts for Brick-Makers* (Cincinnati, 1897): 5–69.

40 *Builder and Woodworker*, 30 (January 1894), 74.

41 Kern, *The Culture of Time and Space*, 116; "Building Construction in the United States as Viewed by an Englishman," *Carpentry and Building*

14 (February 1892): 45–6. On the Master Model Homes see *Building Developer* 3 (October 1928): 5; *Building Economy* 7 (May 1931): 15.

42 *Canadian Architect and Builder* 7 (August 1894): 108; *National Builder* 27 (January 1897): 320; 50 (November 1908): 19; 54 (January 1912); 55 (April 1913): 9.

43 A. D. Boyd and A. H. Wilson, *Technology Transfer in Construction: Science Council of Canada Background Study No. 32* (Ottawa: Information Canada, 1974), 145–7.

44 By a working man [James B. Turnbull], *Reminiscences of a Stonemason* (London: John Murray, 1908), 119.

45 "Work and Wage," *American Builder* 10 (November 1874): 238.

46 On Brunel see "Origins of the Circular Saw," *American Builder* 10 (December 1874): 265; *Builder and Woodworker* 6 (1885): 157, 159.

47 *American Builder* 15 (April 1879): 77.

48 *American Builder* 10 (June 1874): 125.

49 *American Builder* 23 (September 1887): 223.

50 Discussion following Wells I. Bennett, "The Threat of Technological Progress in Housing," *Michigan Business Papers* no. 5 (May 1939): 10.

51 *Carpentry and Building* 3 (November 1881): 204; 14 (May 1892): 128. The estimate on the cost of tools is based on *Canadian Architect and Builder* 10 (April 1897): 77.

52 *National Builder* 52 (March 1911): 72–3.

53 *Toronto Star*, 13 January 1990, E10.

54 Gilbert Herbert, *Pioneers of Prefabrication: The British Contribution in the Nineteenth Century* (Baltimore: Johns Hopkins University Press, 1978), 54.

55 Cited in Giedion, *Space, Time and Architecture*, 351.

56 D.N. Skillings and D.B. Flint, *Illustrated Catalogue of Portable Sectional Buildings* (Boston: Andrew Holland, n.d.), 56.

57 *American Builder* 19 (June 1883): 117.

58 *Brick, Tile and Metal Review* 3 (March 1883): 17.

59 G.E. Mills and D.W. Holdsworth, "The B.C. Mills Prefabrication System: The Emergence of Ready-made Buildings in Western Canada," in National Historic Parks and Sites Branch, Parks Canada, Indian and Northern Affairs, *Canadian Historic Sites: Occasional Papers in Archaeology and History*, no. 14 (Ottawa, 1975), 127–70; *National Builder* 55 (July 1913): 162.

60 HPL, Special Collections, *The Open Door to Better Homes at Lower Cost* (1925).

61 Comments made by the Common Brick Manufacturers' Association of America in its Cleveland-based journal, *Building Economy*, were typical. See editorials entitled "Button, Button, Who's Got It?" *Building Econ-*

omy 5 (December 1929): 1 and "Mass Home Production," *Building Economy* 6 (April 1930): 1.

62 Pat Brennan, "Factory-Built Homes Losing Their Old Image," *Toronto Star*, 2 July 1988. Canada's largest prefab company, Royal Homes of Wingham, Ontario, has built 2,500 houses since 1971, an average of about 110 per year. The firm offers about two dozen floor plans, but also can provide customized computer-drawn designs. Once on the site, the houses, which range in size up to 2,500 square feet, can be put together in five hours. For an account of "stress skin" panels, see Susan Doran, "Home on the Hinge," *Select Homes & Food*, July/August 1990, Ont. 8–11.

63 Advertisement in *Bungalow Magazine* (Seattle) 3 (July 1914): 11-A.

64 "The Decline of the Adze," *Building and Wood-Worker* 6 (1885): 129; *American Builder* 12 (July 1876): 167.

65 "The Model Foreman," *Carpentry and Building* 2 (September 1880): 173.

66 "Modern Mechanism," *American Builder* 10 (December 1874): 269–70.

67 *Carpentry and Building* 15 (January 1893): 11.

68 *National Builder* 44 (April 1907): 28.

69 Zunz, *The Changing Face of Inequality*, 140–76.

70 *American Builder* 15 (March 1879): 67.

71 *American Builder* 18 (May, 1882): 82.

72 Haber, *Industrial Relations in the Building Industry*, 270–6.

73 Carpenters and Joiners' National Union of the United States of America, *Proceedings of the First Annual Convention of New York* (New York, 1865).

74 Haber, *Industrial Relations in the Building Industry*, 271.

75 Ibid., 161, 280.

76 Ibid., 161.

77 Doucet and Weaver, "The North American Shelter Business, 1860–1920," 246.

78 Walter J. Mead, *Competition and Oligopsony in the Douglas Fir Lumber Industry* (Berkeley: University of California Press, 1966), 1, 38–9.

79 HPL, Special Collections, *Spectator*, 1903 Summer Carnival Edition, 104; W. D. Flatt, *The Trail of Love* (Toronto: Wm. Briggs, 1916), 127–83.

80 Mead, *Competition and Oligopsony*, 75–85; Joseph Zaremba, *Economics of the American Lumber Industry* (New York: Robert Speller and Son, 1963), 79.

81 *National Builder* 50 (February 1910): 343.

82 For the early lumber associations see Ralph W. Hidy, Frank Ernest Hill, and Allan Nevins, *Timber and Men: The Weyerhaeuser Story* (New

York: Macmillan, 1963); Charles W. Twining, *Downriver: Orrin H. Ingram and the Empire Lumber Company* (Madison: The State Historical Society of Wisconsin, 1975), 215, 218–19, 248. On hardwood see *Hardwood* 2 (25 June 1892).

83 *Brick* 5 (September 1896): 71; 5 (August 1896): 35; *Clay Record* 2 (12 January 1893): 14; 2 (28 March 1893): 9; 2 (28 April 1893): 13.

84 *Carpentry and Building* 18 (February 1896): 189; *Canadian Architect and Builder* 20 (April 1907): 66. The quote appears in ibid., 17 (April 1904): 59.

85 Clarence D. Long, *Building Cycles and the Theory of Investment* (Princeton: Princeton University Press, 1940), 195.

86 US, Department of Commerce and Labour, *The Lumber Industry*, part 1, *Standing Timber* (Washington: Government Printing Office, 1913), 111–12.

87 "Activity among Builders' Exchanges," *Canadian Architect and Builder* 21 (December 1907): 24. On Montreal see *Carpentry and Building* 21 (February 1899) and *Canadian Architect and Builder* 22 (January 1908): 20.

88 *Canadian Architect and Builder* 12 (March 1899): 60.

89 "Building-Up Hints for Builders' Exchanges," *Canadian Architect and Builder* 20 (March 1907): 43.

90 For builders' exchanges in the United States see *Official Report*, Third Annual Convention of the National Association of Builders of the United States of America (12–14 February 1889), 25–30; *Official Report*, Fifth Annual Convention... (9–14 February 1891), 31, 166; *Official Report*, Ninth Annual Convention... (15–18 October 1895), 25; *Carpentry and Building* 14 (February 1892): 36; *National Builder* 62 (September 1919): 55; 62 (October 1919): 14–15. On the builders' exchanges and unions, see Clarence E. Bonnett, *Employers' Associations in the United States: A Study of Typical Associations* (New York: Macmillan, 1922), 172; Mancur Olson, *The Rise and Decline of Nations: Economic Growth, Stagflation, and Social Rigidities* (New Haven: Yale University Press, 1982).

91 Fred T. Hodgson, *Practical Bungalows and Cottages for Town and Country* (Chicago: Frederick J. Drake and Co., 1906), 3. See also Una Nixon Hopkins, "The Explanation of the Bungalow Craze," *Keith's Magazine* 15 (April 1906): 193–7 and E.W. Stillwell, "What Is a Genuine Bungalow?" *Keith's Magazine* 35 (April 1916): 273–6. On the history of this housing form see Anthony D. King, *The Bungalow: The Production of a Global Culture* (London: Routledge and Kegan Paul, 1984).

92 Zebulon Baker, *The Cottage Builder's Manual* (Worcester, 1856), 16.

93 George E. Woodward, *Woodward's National Architect* (New York, 1869).

94 *Builder and Wood-Worker* 1 (1880): 222.

95 "Suggestions to Artisans," *American Builder* 8 (January 1873): 32.

96 *American Builder* 10 (March 1874): plates 9, 10, 12; 18 (June 1882): plate 41; 18 (August 1882): plate 57.

97 "A Small Frame House," *Carpentry and Building* 2 (February 1880): 26.

98 "Cheap Frame Houses," *Carpentry and Building* 6 (March 1884): 47.

99 *Carpentry and Building* 15 (April 1893): 99.

100 *Carpentry and Building* 20 (December 1898): 287.

101 *Canadian Architect and Builder* 1 (January 1888): 4.

102 *American Builder* 17 (March 1881): 43.

103 Walter J. Keith, *The Building of It* (n.p. 1898), advertisement at the end of the volume; Hodgson, *Practical Bungalows*.

104 Quoted in *National Builder* 28 (January 1898): 3, 7.

105 Gwendolyn Wright, *Building the Dream: A Social History of Housing in America* (New York: Pantheon, 1981), xvii.

106 Gwendolyn Wright, *Moralism and the Model Home: Domestic Architecture and Cultural Conflict in Chicago, 1873–1913* (Chicago: University of Chicago Press, 1980), 244.

107 *Bungalow Magazine* (Los Angeles) 1 (March 1909): advertisement.

108 *Bungalow Magazine* (Los Angeles) 1 (August 1909): 190.

109 "The Architect as Evangelist," *Bungalow Magazine* (Los Angeles) 1 (December 1909): 285.

110 *Bungalow Magazine* (Seattle) 7 (February 1918): 33.

111 *Prize Winning Small House Plans of the National Architectural Competition 1921* (New York, 1921).

112 See, for example, *The Small Home* 1 (1922): various issues.

113 "Why the Colonial is Popular," *The Small Home* 1 (December 1922): 9.

114 See, for example, Thomas S. Hines, *Richard Neutra and the Search for Modern Architecture* (New York: Oxford University Press, 1982); John Sergeant, *Frank Lloyd Wright's Usonian Houses* (New York: Whitney Library of Design, Watson-Guptill Publications, 1976).

115 *The Small Home* 1 (1922), various issues.

116 *The Small Home* 2 (April 1923): 12.

117 *The Small Home* 2 (March 1923): 8; 2 (April 1923): 12.

118 *The Small Home* 1 (March 1922): 26.

119 CMHC, *Small House Designs: Bungalows* (Ottawa, 1949); *Small House Designs: 1 1/2 Storey* (Ottawa, 1949); *Small House Designs: 2 Storey* (Ottawa, 1949); *House Designs* (Ottawa, 1971). Anthony Downs, *The Revolution in Real Estate Finance* (Washington: The Brookings Institute, 1985), 137.

120 Eichler, *The Merchant Builders*, 68.

121 Ibid., 69.

122 *National Builder* 60 (August 1918): 52–5.

123 Eichler, *The Merchant Builders*, 288–9.
124 Wright, *Building the Dream*; Dolores Hayden, *Redesigning the American Dream: The Future of Housing, Work and Family Life* (New York: W. W. Norton, 1984).
125 Zaremba, *Economics of the American Lumber Industry*, 99–100.
126 Gilbert Herbert, *The Dream of the Factory-Made House: Walter Gropius and Konrad Wachsmann* (Cambridge: The MIT Press, 1984).
127 Tracy Kidder, *House* (Boston: Houghton Mifflin, 1985), 135.
128 Jan Cohn, *The Palace or the Poorhouse: The American House as a Cultural Symbol* (East Lansing: Michigan State University Press, 1979).
129 Sam Bass Warner Jr, *The Urban Wilderness: A History of the American City* (New York: Harper and Row, 1972), 201.
130 David Englander, *Landlord and Tenant in Urban Britain, 1838–1918* (Oxford: Clarendon Press, 1983), 3–4.

CHAPTER SIX

1 Tracy Kidder, *House* (Boston: Houghton Mifflin Company, 1985), 38.
2 Robert H. Pease and Homer V. Cherington, *Mortgage Banking* (New York: McGraw-Hill, 1953), 15–23.
3 J. E. Smyth and D. A. Soberman, *The Law and Business Administration in Canada* (Scarborough: Prentice-Hall, 1976), 577; 35 Geo III c.V.
4 Concerning Robinson and the Court of Chancery question from 1822 to 1828, see the following: Great Britain, Colonial Office, Original Correspondence, 1700–1922, CO 42 series, vol. 385, John Beverley Robinson to Lieutenant-Governor Sir Peregrine Maitland, 12 May 1828, 234–44; Upper Canada, *Journal of the 4th Session, 9th Parliament*, Robinson to Maitland, 15 February 1828, 59.
5 J. D. Falconbridge, *The Law of Mortgages of Land* (Toronto: Canada Law Book Limited, Reprint of 1942 edition, 1970) is a useful guide, but only by reading the reports of the cited cases can the social, intellectual, and political activity surrounding the mortgage be appreciated. For the lengthy account of the revealing 1846 case, Simpson v. Smyth, see *The Upper Canadian Jurist*, vol. 2, 1846–8 (Toronto: Henry Rowsell, 1848). The position taken by Robinson on the use of a writ of *fieri facias* was confirmed by later scholarship. See, for example, R. W. Turner, *The Equity of Redemption: Its Nature, History and Connection with Equitable Estates Generally* (Cambridge: Cambridge University Press, 1931), 161.
6 On the introduction of the equity of redemption by statute see 4 Wm IV c.I and 4 Wm IV c.16. For the opinions on a Court of Chancery held by John Beverley Robinson and the colony's other leading jurists see Upper Canada, *Journal of the Legislative Council of*

Upper Canada, First Session of the Thirteenth Provincial Parliament (Toronto: Robert Stanton, 1837), appendix L. Rolph's support for the resolution of gaps in Upper Canadian law by statutes and his opposition to a Court of Chancery were reported in *Christian Guardian* 15 (February 1837).

7 City of Toronto Archives, George Weymouth Schreiber Papers, series C, box 1, file 1, Catherine Hill to Mr Barwick, 19 July 1872. The note was enclosed in a mortgage dated 24 February 1870. Further information was provided by Stephen Thorning of Elora.

8 Meyer vs Harrison (1850), *The Canadian Abridgement*, vol. 26, 472.

9 M. Morris, "The Land Mortgage Companies, Government Savings Banks and Private Bankers of Canada," *Journal of the Canadian Bankers Association* (April 1896): 238.

10 McLaren vs Fraser (1870); Faulds vs Harper (1886); Prentice vs Consolidated Bank (1886); Aldrich vs Canadian Permanent, etc. Company (1887), *The Canadian Abridgement*, vol. 26, 436–41, 455, 481. The Hamilton Provident and Loan Company case involved a Manitoba farm. Carruthers vs Hamilton Provident Soc. (1898), *The Canadian Abridgement*, vol. 26, 441.

11 42 Vic. c. 20, s.6; RSO, c. 279, s. 31.

12 The Housing Committee of the Twentieth Century Fund, *American Housing: Problems and Prospects* (New York: The Twentieth Century Fund, 1944), 243–6.

13 James E. Hatch, *The Canadian Mortgage Market* (Toronto: Queen's Printer for Ontario, 1975), 149–53; Canada, *Report of the Royal Commission on Banking and Finance* (Ottawa: Queen's Printer, 1964), 173–4; Canada, *Report of the Royal Commission on Banking and Currency in Canada* (Ottawa: King's Printer, 1933), 15.

14 E. J. Cleary, *The Building Society Movement* (London: Elek Books, 1965), 9.

15 Ibid., 11–13.

16 Ibid., 59–60.

17 Morris, "The Land Mortgage Companies, Government Savings Banks and Private Bankers of Canada," 229. *Debates of the Legislative Assembly of United Canada, 1841–1869*, general editor Elizabeth Gibbs (Montreal: Centre de Recherche en Historie Economique du Canada Français) vol. 5, part 1 (1846), 277, 302, 344, 387, 855.

18 Pease, *Mortgage Banking*, 266.

19 Morris, "The Land Mortgage Companies," 229.

20 Economist W. A. Mackintosh, a member of the Royal Commission on Banking and Finance, wondered why Canadian Building Societies had not developed into Building Societies like those operating in the United Kingdom. See Canada, *Royal Commission on Banking and Fi-*

nance Hearings Held at Ottawa, vol. 48, 5862. In fact, the Canadian ones did become economic ventures like those in the United Kingdom. The belief that British societies were significantly different from those developed in Canada was ill-founded. A similar misconception was raised by the federal task force on housing in 1969. See Canada, *Report of the Federal Task Force on Housing and Urban Development* (Ottawa: Queen's Printer, 1969), 27.

21 7 Vic. c. 63.

22 *Debates of the Legislative Assembly*, vol. 11, part 3 (1852–53), 867–79. The loan and trust company is discussed on 874–9.

23 51 Geo. III c. 9. On circumventing the law, see *Debates*, vol. 11, part 2, 761, 872, 875.

24 *Debates*, vol. 14, part 2, 763, 1273; 13 & 14 Vic. c. 138.

25 *Debates*, vol. 11, part 2, 757–62; vol. 11, part 3, 1890–99.

26 *Debates*, vol. 11, part 2, 757–60; 860–80.

27 Ibid., 759.

28 Ibid., 757.

29 Ibid., 761.

30 Ibid., 763.

31 Ibid., 874.

32 Ibid., 874.

33 Ibid., 876.

34 Ontario, Sessional Papers (no. 11), *Loan Corporation Statements: Being Financial Statements Made by Building Societies, Loan Companies, Loaning Land Companies, and Trust Companies for the Year Ending 31st December 1909* (Toronto: King's Printer, 1910), 350. Hereafter these statements will be cited as *Loan Corporation Statements*.

35 Cleary, *The Building Society Movement*, 31–42.

36 9 Vic. c. 90.

37 13 & 14 Vic. c. 79.

38 Ontario, Sessional Papers (no. 36), *Loan Corporation Statements for Year Ending 31st December 1897* (Toronto: Warwick Bro's and Rutter, 1898). For recent data see Ontario, Ministry of Financial Institutions, *Report of the Registrar, Business of 1985, Loan and Trust Corporations, 89th Report*, 16–17.

39 22 Vic. c. 45.

40 City of Toronto Archives, George Weymouth Schreiber Papers, series A, box 1, file 8 (1860s), Frances W. Kestermann (?) to Weymouth George Schreiber, 17 July 1860.

41 37 Vic. c. 50.

42 45 Vic. c. 24.

43 Canada, *Debates of the Dominion of Canada Fourth Session Third Parliament* (Ottawa: MacLean, Roger and Co., 1877), 1144.

44 39 Vic. c. 27; 39 Vic. c. 32; 40 Vic. c.22.

45 The published statements that building societies and other provincially chartered lending institutions made to the Ontario government included their legislative histories, name changes, and amalgamations.

46 *Loan Corporation Statements.*

47 *Palladium of Labor,* 19 January 1884.

48 *Palladium of Labor,* 1 March 1884.

49 Canada, *Report of the Royal Commission on the Relations of Labor and Capital in Canada: Evidence – Ontario* (Ottawa: Queen's Printer, 1889), 736–38.

50 *The Hamilton Provident and Loan Society: Building Society Statutes and Rules* (Hamilton: Times, 1880), 8–10.

51 Canada, *Report on the Affairs of Building Societies, Loan and Trust Companies in the Dominion of Canada, for the Year 1909 with Comparative Tables of the Chief Items for Years from 1867 to 1909 Inclusive* (Ottawa: King's Printer, 1910), viii-x. The role of British capital in Canadian loan company activities almost a century earlier was noted by the President and General Manager of the Huron and Erie Mortgage Company in 1962. *Royal Commission on Banking and Finance, Hearings Held at Ottawa,* vol. 48, Testimony of J. Allyn Taylor, 5862.

52 Morris, "The Land Mortgage Companies," 243.

53 *Loan Corporation Statements for Year Ending 31st December 1909,* 350.

54 Ontario, Ministry of Consumer and Commercial Relations, *Report of the Special Examination by James A. Morrison F.C.A. of Crown Trust Company, Greymac Trust, Seaway Trust Company, Greymac Mortgage Corporation, and Seaway Mortgage Corporation to the Honorable Robert G. Elgie M.D., Minister of Consumer and Commercial Relations, Province of Ontario* (Toronto: Touche Ross, 1983), 239–43. *Report of the Registrar, Business of 1985, Loan and Trust Corporations, 89th Report. Report on Business 1000* (Toronto: Report on Business Magazine, 1987), 53.

55 John C. Weaver and Peter De Lottinville, "The Conflagration and the City: Disaster and Progress in Nineteenth-Century British North America," *Histoire sociale – Social History,* 26 (November 1980): 441–8.

56 *Royal Commission on Banking and Finance, Hearings Held at Ottawa,* vol. 49, Testimony of David Mansur, 6163.

57 Ontario, Sessional Papers (no. 9), *Report of the Inspector of Insurance and Registration of Friendly Societies 1897* (Toronto: Warwick Bro's and Rutter, 1898), B31-B222. (Hereafter cited as Report of the Inspector of Insurance, 1897).

58 12 Vic. c. 168.

59 Canada, Sessional Papers (no. 9), *Statements Made by Insurance Companies in Compliance with the Act 31 Vict. Cap. 48, Sec. 15* (Ottawa: King's Printer, 1872), 9.

60 The information on John M. Gibson and the Landed Banking and Loan Company was furnished by Carolyn Gray from her forthcoming dissertation, *Enterprise and the "Electric City": John Morrison Gibson and his Interests, 1842–1929*.

61 Canada, Sessional Papers (no. 4), *Report of the Superintendent of Insurance of the Dominion of Canada for the Year Ended 31st December 1897* (Ottawa: King's Printer, 1898), 111–607.

62 *Report of the Inspector of Insurance 1897*.

63 Hatch, *Mortgage Banking*, 257; Canada, *Report of the Royal Commission on Banking and Finance* (Ottawa: Queen's Printer, 1964), 243; *Royal Commission on Banking and Finance, Hearings Held at Ottawa*, vol. 48, Testimony of J. E. Fortin, Associate Secretary, Dominion Mortgage and Investment Association, 5839–5843.

64 Hatch, *Mortgage Banking*, 257; Lawrence Berk Smith, *The Postwar Canadian Housing and Residential Mortgage Markets and the Role of Government* (Toronto: University of Toronto Press, 1974), 95–103.

65 John C. Bacher, *Keeping to the Private Market: The Evolution of Canadian Housing Policy* (McMaster University, PhD dissertation, 1985), 436–9.

66 W. E. McLaughlin, "Mortgage Lending by Canadian Banks: Its Background and Introduction," *The Canadian Banker* (spring 1955): 54–55.

67 Ibid., 57–61. Neil J. McKinnon, President, Canadian Imperial Bank of Commerce, "Submission to the Royal Commission on Banking and Finance," *Royal Commission on Banking and Finance*, vol. 57, A3–A16.

68 McLaughlin, "Mortgage Lending," 64.

69 *Royal Commission on Banking and Finance, Hearing Held at Ottawa*, vol. 61, Testimony of R. D. Mulholland, 7900.

70 "Submission," *Royal Commission on Banking and Finance*, vol. 57, A16; see Mulholland's testimony, *Royal Commission on Banking and Finance, Hearings at Ottawa*, vol. 61, 7901–7914.

71 Hatch, *The Canadian Mortgage Market*, 134.

72 Canada, Department of Finance, *Release*, 86–210; 67–72; 116; 135–44; Canada, Department of Finance, *Securing Economic Renewal: Budget Papers* (May 1985), 76.

73 Hatch, *The Canadian Mortgage Market*, 213–31.

74 Ibid., 67–9.

75 *Report of the Task Force on Housing and Urban Development*, 73.

76 *Report of the Royal Commission on Banking and Finance*, 243; Hatch, *The Canadian Mortgage Market*, 21. Particularly useful in tracing the features of Ontario mortgages by year and locality are Ontario, Ministry of Treasury, Economics, and Intergovernmental Affairs, *Realty Mortgage Loans Registered in Ontario*. These reports began in 1969 and continued until 1975. For the estimates on non-institutional mortgage

lending in the United States see Kent W. Cotton, "Financial Reform: A Review of the Past and Prospects for the Future," *Housing and the New Financial Markets*, ed. Richard L. Florida (New Brunswick, New Jersey: Rutgers, Center for Urban Policy Research, 1986), 108. Canada, Department of Finance, *Improved Security for Homeowners: Proposals for a Fairer and More Flexible Mortgage Market* (February 1984), 1.

77 Schreiber Papers, series A, box 2, file 3C (1870s), Moore and Davis to Schreiber, 23 October 1871. For the study of American mortgage agents see Allan G. Bogue, *Money at Interest: the Farm Mortgage on the Middle Border* (New York: Russell and Russell, 1968).

78 Schreiber Papers, biographic notes included with the finding aid.

79 Schreiber Papers, series A, box 1, file 7 (1850s),? de Lisle to Schreiber, 22 August 1856; Robert Woodhouse Mortimer, 27 August 1856; Thomas Schreiber to Schreiber, 20 March 1857; series B, box 1, file 3, notebooks.

80 Schreiber Paper, series A, box 1, file 8 (1860s), Frances W. Kestermann to Schreiber, 17 July 1860.

81 Schreiber Papers, series A, box 2, file 4 (1870s), Frances W. Kestermann to Schreiber, 28 August 1877.

82 Ibid.

83 Schreiber Papers, series G, box 1 (Loan and Investment Companies).

84 For an account of the Moore and Davis mortgage business, see Michael Doucet and John C. Weaver, "The North American Shelter Business, 1860–1920: A Study of a Canadian Real Estate and Property Management Agency," *Business History Review* 58 (summer 1985): 243–47.

85 Interview with Mr D'Arcy Lee, 6 May 1988.

86 *Royal Commission on Banking and Finance, Hearings Held at Ottawa*, vol. 49, Testimony of David Mansur, 6179.

87 Robert Fishman, *Bourgeois Utopias: The Rise and Fall of Surburbia* (New York: Basic Books, Inc., 1987), 165.

88 *Royal Commission on Banking and Finance, Hearings Held at Ottawa*, vol. 49, Testimony of A. D. Wilson, 6095–6096.

89 Falconbridge, *The Law of Mortgages of Land*, 837.

90 See the testimony of W. M. Somerville, National Construction Council, *Royal Commission on Banking and Currency Proceedings* (typescript, 1933), vol. 5, 2719.

91 Bacher, *The Evolution of Canadian Housing Policy*, 77–84.

92 Ibid., 84–8.

93 Canada, *Royal Commission on Banking and Currency, Proceedings*, Testimony of W. M. Somerville, vol. 5 (1933), 2719–20. HPL, The Housing Commission of the City of Hamilton, *Record of Advances Made on Loan*

as per Progress Certificate Issued (ledgerbook). On defaults see, HPL, *Hamilton Housing: Scrapbook of Clippings*, vol. 1 (1905–47), *Spectator*, 12 September 1939; 19 December 1946.

94 Bacher, *The Evolution of Canadian Housing Policy*, 93–4. For a revealing discussion of the impact of the depression on valuation see Maurice R. Massey Jr, "The Influence of FHA on Mortgage Lending," *Michigan Business Papers* (Proceedings of the First Mortgage Study Conference, Ann Arbor, 19 March 1939), no. 4 (April 1939), 2–9. The issue of location and appraisal was outlined by David Mansur who, as a former Sun Life mortgage officer and CMHC official, was well placed to undertstand lending practices, *Royal Commission on Banking and Finance, Hearings Held at Ottawa*, vol. 49, Testimony of David Mansur, 6129–6135.

95 Bacher, *The Evolution of Canadian Housing Policy*, 135–48.

96 Ibid., 148–53.

97 Ibid., 153–4.

98 Ibid., 209–21; 225–7.

99 Ibid., 229. The characteristics of regional discrimination were described by David Mansur in *Royal Commission on Banking and Finance, Hearings Held at Ottawa*, vol. 49, 6129.

100 Hatch, *The Canadian Mortgage Market*, 16–32.

101 Ibid., 54–5.

102 Ibid., 64; *Royal Commission on Banking and Finance, Hearings Held at Ottawa*, vol. 49, Testimony of Stewart Bates, 6032, 6034, 6075; Testimony of David Mansur, 6153–55. Canada, *Canada Mortgage and Housing Corporation, 1984*, 14–15. On the Federal Housing Administration's insurance operations see William H. Wilson, *Coming of Age: Urban America, 1915–1945* (Toronto: John Wiley and Sons, 1974), 188; Twentieth Century Fund, *American Housing*, 250–1.

103 Hatch, *The Canadian Mortgage Market*, 16–32.

104 On the advantages of a secondary market to institutional investors see ibid., 233. CMHC's early attempts to promote such a market are reported in *Royal Commission on Banking and Finance, Hearings at Ottawa*, vol. 49, Testimony of Stewart Bates, 6099–6115; and R. Adamson, 6125.

105 Daniel F. Williams "The Secondary Mortgage Market," *Housing and the New Financial Markets*, 37–8.

106 *Royal Commission on Banking and Finance, Hearings at Ottawa*, vol. 49, Testimony of Stewart Bates, 6061.

107 Canada, *Report of the Federal Task Force on Housing and Urban Development* (Ottawa: Queen's Printer, 1969), 27.

108 Hatch, *The Canadian Mortgage Market*, 226.

109 On the various innovations with mortgages see George Fallis, *Hous-*

ing Economics (Toronto: Butterworths, 1985), 113–14; George G. Kaufman and Eleanor Erdevig, "Improving Housing Finance in an Inflationary Environment: Alternative Residential Mortgage Instruments," *Housing and the New Financial Markets*, 359–71.

110 Shirley Spragge, "A Confluence of Interests: Housing Reform in Toronto, 1900–1920," *The Usable Urban Past: Planning and Politics in the Modern Canadian City*, ed. Alan F. J. Artibise and Gilbert A. Stelter (Toronto: Macmillan, 1979), 247–67.

111 Vera Chouinard, *State Formation and Housing Policies: Assisted Housing Programmes and Cooperative Housing in Postwar Canada* (McMaster University, PhD dissertation, 1986), 65–7.

112 Chouinard, *State Formation and Housing Policies*, 115–37. For protests over the location of public housing units in Hamilton see HPL, *Hamilton Housing: Scrapbook of Clippings*, vol. 4 (1968–1972), *Spectator*, 18 December 1972.

113 Spragge, "A Confluence of Interests," 255–9; Bacher, *The Evolution of Canadian Housing Policy*, 57–8.

114 John C. Weaver, "Reconstruction of the Richmond District in Halifax: A Canadian Episode in Public Housing and Town Planning, 1918–1921," *Plan Canada* 16 (March 1976): 36–47.

115 Bacher, *The Evolution of Canadian Housing Policy*, 250–63.

116 John C. Bacher, "Canadian Housing Policy in Perspective," *Urban History Review* 15 (June 1986): 5–9; Bacher, *The Evolution of Canadian Housing Policy*, 151–4; J. David Hulchanski, "The 1935 Dominion Housing Act: Setting the Stage for a Permanent Federal Presence in Canada's Housing Sector," *Urban History Review* 15 (June, 1986): 19–39.

117 Bacher, *The Evolution of Canadian Housing Policy*, 294.

118 Ibid., 302–10. HPL, Special Collections, Joseph Pigott clipping file.

119 Bacher, *The Evolution of Canadian Housing Policy*, 310–19. For an excellent general account of WHL, see Jill Wade, "Wartime Housing Limited, 1941–1947: Canadian Housing Policy at the Crossroad," *Urban History Review* 15 (June 1986): 41–59.

120 Bacher, *The Evolution of Canadian Housing Policy*, 368–375. Canada, Advisory Committee on Reconstruction, *Housing and Community Planning: Final Report of the Subcommittee* (Ottawa: King's Printer, 1944).

121 Bacher, *The Evolution of Canadian Housing Policy*, 377.

122 Ibid., 393–5.

123 Albert Rose, *Canadian Housing Policies, 1935–1980* (Toronto: Butterworths, 1980), 30.

124 HPL, clipping files, Mohawk Gardens, *Spectator*, 7 October 1955, 6 January 1956, 13 June 1972, 22 August 1975.

125 Chouinard, *State Formation and Housing Policies*, 57.

126 Ibid., 58.
127 *Report of the Federal Task Force on Housing and Urban Development*, 19;
     Chouinard, *State Formation and Housing Policies*, 137. The data on so-
     cial housing were provided by the Hamilton-Wentworth Housing
     Authority (August 1987) and the Southern Ontario regional office of
     the Ontario Ministry of Housing (February 1988).
128 Chouinard, *State Formation and Housing Policies*, 65.
129 See, for example, the following: Canada, CMHC, *NHA Loan Insurance
     Handbook* (a collection of notices retained in binder format), IV 5, V
     1; Canada, *Canada Mortgage and Housing Corporation* (a collection of
     notices retained in binder format), no. 47, no. 57.

CHAPTER SEVEN

1 Michael R. Haines and Allen C. Goodman, *Buying the American Dream:
   Housing Demand in the United States in the Late Nineteenth Century* (National Bureau of Economic Research, Working Paper Series on Historical Factors in Long Run Growth, 1989), 11.
2 Marion Steele, *The Demand for Housing in Canada* (Ottawa: Statistics
   Canada, 1979), 16.
3 Debra Frazer and Janet McClain, *Credit: A Mortgage for Life: A Report
   on Consumer Debt* (Ottawa: The Canadian Committee on Social Development, 1981), 100.
4 John Miron, *Housing in Postwar Canada: Demographic Change, Household
   Formation, and Housing Demand* (Kingston and Montreal: McGill-
   Queen's University Press, 1988), 271.
5 Tracy Kidder, *House* (Boston: Houghton Mifflin Company, 1985), 6.
   For Perin's general argument see *Everything in Its Place* (Princeton,
   N.J.: Princeton University Press, 1977), 32.
6 *Report of the Royal Commission on the Relations of Labor and Capital, Evidence – Ontario*, 747–8.
7 Hamilton-Wentworth, Planning and Development Department, *High
   Density of Residential Development Study, Background Report*, (November
   1987), 14.
8 Canada, *Report of the Federal Task Force on Housing and Urban Development* (Ottawa: Queen's Printer, 1969), 32.
9 *Royal Commission on Banking and Finance, Hearings Held at Ottawa*,
   vol. 49, Testimony of Stewart Bates, 6083.
10 Michael B. Katz, Michael J. Doucet, Mark Stern, *The Social Organization of Early Industrial Capitalism* (Cambridge, Mass.: Harvard University Press, 1982), 14–63; Bryan D. Palmer, "Emperor Katz's New
   Clothes; or with the Wizard in Oz," *Labour/Le Travail* 13 (spring
   1984): 192–4; *A Culture in Conflict: Skilled Workers and Industrial Capitalism in Hamilton, Ontario, 1860–1914* (McGill-Queen's University

Press, 1979), xii-xvi. Stuart Blumin, "The Hypothesis of Middle-Class Formation in Nineteenth-Century America: A Critique and Some Proposals," *America Historical Review* 90 (April 1985): 318–33.

11 Eric Olin Wright, *Class, Crisis, and the State* (London: NLB, 1978), 63; Anthony Giddens, *The Class Structure of the Advanced Societies* (London: Hutchinson, 1981), 182.

12 Giddens, *The Class Structure*, 193–5; Eric Olin Wright and Bill Martin, "The Transformation of the American Class Structure, 1960–1980," *American Journal of Sociology* 93 (July 1987): 15–25.

13 On the Toronto Polish community's home ownership see Ewa Morawska, Benedykt Heyden Korn, and Rudolf Kogler, *Poles in Toronto* (Toronto: Futura Graphics Inc., 1982), 44. The Canada-wide data for 1971 is discussed in Anthony H. Richmond and Warren E. Kalback, *Factors in the Adjustment of Immigrants and their Descendants* (Ottawa: Statistics Canada, 1980), 403–8. For examples of American studies that include explanations for the high ownership levels among certain ethnic groups' households, see Ewa Morawska, *For Bread with Butter: The Life-Worlds of East Central Europeans in Johnstown, Pennsylvania, 1890–1940* (Cambridge: Cambridge University Press, 1985), 144–8; John Bodnar, Roger Simon, and Michael P. Weber, *Lives of Their Own: Blacks, Italians, and Poles in Pittsburgh, 1900–1960* (Urbana: University of Illinois Press, 1982), 153–80; Olivier Zunz, *The Changing Face of Inequality: Urbanization, Industrial Development and Immigrants in Detroit, 1880–1920* (Chicago: University of Chicago Press, 1982), 152–61. Using the household survey data collected for 6,809 industrial workers in the United States for 1889–90, two economists have observed that relative to the native born, Irish, German, and Slavic immigrants had higher probabilities of ownership, controlling for other variables. See Haines and Goodman, *Buying the American Dream*, 24.

14 Bruno Ramirez, *The Italians in Canada* (Canadian Historical Association, 1989), 14.

15 Alessandro J. J. Roncari, "Economic and Cultural Contributions Made by the Italian Immigrant in the Hamilton and Surrounding Area," *Symposium '77 on the Economic, Social and Cultural Conditions of the Italian Canadian in the Hamilton-Wentworth Region*, ed. Mary Campanella (Hamilton: Italian Canadian Federation of Hamilton Inc., 1978), 15.

16 Calvin Goldscheider and Frances K. Goldscheider, "Moving Out and Marriage: What Do Young Adults Expect?" *American Sociological Review* 52 (April 1987): 283.

CHAPTER EIGHT

1 See, for example, David Gagan, *The Denison Family of Toronto, 1792–1925* (Toronto: University of Toronto Press, 1973), 3–4; Brian

Young, *George-Etienne Cartier: Montreal Bourgeois* (Montreal: McGill-Queen's University Press, 1982), 24–7.

2 Paul-André Linteau and Jean-Claude Robert, "Landownership and Society in Montreal: An Hypothesis," *The Canadian City: Essays in Urban History*, ed. Gilbert A. Stelter and Alan F. J. Artibise (Toronto: McClelland and Stewart, 1977), 17–36; Louise Dechêne, "La rente du faubourg Saint-Roch à Québec, 1750–1850," *Revue d'histoire de l'amérique française* 34 (mars 1981): 569–96.

3 Gregory J. Levine, Richard S. Harris, and Brian D. Osborne, *The Housing Question in Kingston, Ontario, 1881 to 1901* (Kingston: Department of Geography, Queen's University, 1982), 93.

4 Michael J. Doucet, *Building the Victorian City: The Process of Land Development in Hamilton, Ontario, 1847–1881* (University of Toronto: PhD thesis, 1977), 221.

5 Richard Dennis, *Landlords and Rented Housing in Toronto, 1885–1914* (paper prepared for the Housing Tenure Workshop, Centre for Urban and Community Studies, 27 February 1987), 35.

6 Brian Young, *George-Etienne Cartier*, 133.

7 David Gagan, *The Denison Family*, 4.

8 Charles Durand, *Reminiscences of Charles Durand of Toronto, Barrister* (Toronto: Hunter, Rose, 1897), 219.

9 The information on the number of Mills properties came from the data file on Mills family collected from assessment rolls of the City of Hamilton. All properties belonging to anyone with the surname Mills were entered into a data file and later cross checked against *Genealogical Record of the Mills Family* (Hamilton, 1910) and its listing of Christian names. For the rental data see HPL, Ledger Book of Real Estate of the Late John Walter Mills, 1–8; National Archives of Canada, NA, RG5, 321, vol. 1, Answers for the Information of Emigrants with Capital Intending to Settle on Land [1843]; R. H. Rae, Emigration Agent at Hamilton, to A. C. Buchanan, Statement showing the demand for the different classes of labor in the various sections of western country, visited ... in March 1865; James G. Snell, "The Cost of Living in Canada in 1870," *Histoire sociale – Social History* 12 (May 1979): 190–1.

10 *Genealogical Record*, 5.

11 HPL, George Sylvester Tiffany Papers, no. 419, Indenture of Bargain and Sale from Nathaniel Hughson and His Wife to G. S. Tiffany and Samuel Mills, 4 December 1838; no. 514, Samuel Mills to G. S. Tiffany, 23 March 1848: no. 1050, Deed, Samuel Mills to G. S. Tiffany, 10 December 1850; HPL, George Mills Letterbooks, letterbook 1, letters to Samuel Mills 1860.

12 HPL, Tiffany Papers, no. 419. The quote is from OA, RG22, series I, Probate Court Records, 1793–1850, Will of James Mills Jr dated 10 December 1842 and probated 21 August 1852.

13 HPL, Ledger Book of Real Estate of the Late John Walter Mills, passim.

14 HPL, George Mills Letterbooks, letterbook 1, George Mills to Peter Balfour, February 1860; letterbook 2, George Mills to "Dear Brother," 29 May 1869.

15 HPL, George Mills Letterbooks, letterbook 1, George Mills to the Sheriff of Halton County, 26 March 1860.

16 HPL, George Mills Letterbooks, letterbook 1, George to Samuel, 27 January 1860; George to Samuel, 13 July 1860.

17 *Genealogical Record*, passim.

18 Ibid., 42.

19 Ibid., 42–3.

20 HPL, clipping file on Mills family, article by Stanley Mills, unidentified clipping.

21 HPL, George Mills Letterbooks, letterbook 2, George Mills to Henry – , 10 November 1869.

22 NA, MG24, D16, Isaac Buchanan Papers, no. 43483, Samuel Mills to Isaac Buchanan, 3 November 1842.

23 HPL, Ledger Book of Real Estate of the late John Walter Mills, 8.

24 McMaster University, Mills Library, Church of England, Archdiocese of Niagara, All Saints' Church Parish Registers; Roy Wiles, *All Saints' Church: The First Hundred Years, 1872–1972* (Hamilton, 1972).

25 Mills Library, McMaster University, Special Collections, All Saints' Church Parish Registers, vols 1 and 2, passim.

26 For Michael Mills see Colin Read, *The Rising in Western Upper Canada, 1837–8; The Duncombe Revolt and After* (Toronto: University of Toronto Press, 1982), 162.

27 HPL, George Mills, "The Two Political Leaders" (unpublished address, 1886).

28 On George's political career see George Mills, *Life Memories* (undated typescript), 5–21; *Spectator*, 31 January 1872, 13 February 1872, 7 May 1872.

29 OA, George H. Mills, "Random Sketches, 1887–1889," 6 April 1887.

30 Ibid., 20 January 1889.

31 Mills, *Life Memories*, 15.

32 *Genealogical Record*, passim. Also see Mills Library, All Saints' Church, Parish Registers, vol. 2, notes 1897–1904.

33 *Spectator*, 10 January 1876.

34 Ibid., 14 March 1876.

35 Ibid., 14 March 1876.

36 *Canadian Illustrated News*, 9 January 1876.

37 Richard Dennis, *Landlords and Rented Housing*, 28.

38 McMaster University, Urban Documentation Centre, Social Planning Council of Hamilton and District, "Eastwood Centre," *Housing in Ham-*

ilton: A Preliminary Survey. Problem Areas: Briefs from Community Agencies and Other Background Material (November 1965), 10.

39 Pearl J. Davis, *Real Estate in American History* (Washington: Public Affairs Press, 1958), 39.

40 For a history of the firm see Michael Doucet and John Weaver, "The North American Shelter Business, 1860–1920: A Study of a Canadian Real Estate and Property Management Agency," *Business History Review* 58 (summer 1984): 234–62.

41 Ibid., 238.

42 HPL, Special Collections, Moore and Davis Records (hereafter M and D Records), letterbook 1, Moore and Davis to Joseph Sudborough, 13 September 1862.

43 Doucet and Weaver, "The North American Shelter Business," 239.

44 HPL, Special Collections, M and D Records, letterbook 9, W. P. Moore to N. A. Awrey, 24 March 1887.

45 Ibid., letterbook 11, 31 March 1891.

46 Ibid., letterbook 16, Moore and Davis to Ellen Keyes, 11 August 1913.

47 Ibid., letterbook 11, Moore and Davis to James Storie, 21 July 1891.

48 Ibid., letterbook 10, Moore and Davis to Mr. A. Irving, 21 November 1890.

49 Ibid., letterbook 11, Moore and Davis to John Bennett, 9 February 1893.

50 Ibid., letterbook 16, Moore and Davis to Guardian Trust Company, 19 August 1914; letterbook 11, Moore and Davis to Messrs Cohen and Sugarman, 10 September 1914.

51 Ibid., letterbook 16, Moore and Davis to M.J. Cashman, 28 February 1913.

CHAPTER NINE

1 "French Flats and Apartment Houses in New York," *Carpentry and Building* 2 (January 1880): 2. For a background discussion of French Flats see J. N. Tarn, "French Flats for the English in Nineteenth-century London", in Anthony Sutcliffe, ed., *Multi-Storey Living: The British Working-Class Experience* (London: Croom Helm, 1974).

2 R. W. Sexton, *Apartment Houses of Today* (New York: Architectural Book Publishing, 1926), ii; *Buildings and Building Management*, 20 (12 January 1920): 37; John Hancock, "The Apartment House in Urban America," *Buildings and Society: Essays on the Social Development of the Built Environment*, ed. Anthony D. King (London: Routledge and Kegan Paul, 1980), 158.

3 Hancock, "The Apartment House," 158–79.

4 *Building Management* 2 (13 August 1908): passim; ibid., 13 (February

1913): 41; *The Proceedings of the Twenty-Seventh Annual Convention of the National Association of Building Owners and Managers* (18–21 June 1934), xx. Babbitt's treatise is discussed in Sinclair Lewis, *Babbitt* (New York: New American Library, 1980), 130–2.

5 *Buildings and Building Management* 39 (November 1939): 35.

6 *Canadian Architect and Builder* 10 (May 1897): 86.

7 Tarn, "French Flats and Apartment Houses in New York," 2.

8 Lawrence A. McKinley, "Factors Causing Distressed Apartment Buildings," *Proceedings of the Twenty-Third Annual Convention of the National Association of Building Owners and Managers* (9–13 June 1930), 487–90.

9 *California Housing (Apartment Journal)* (September 1936): 7; Catherine E. Martini, "Vacancy Rates and the Outlook for the Rental Market," *Journal of Property Management* 28 (fall 1962): 7; Hancock, "The Apartment House," 159.

10 See the review of the early literature in John Mercer, "Some Aspects of the Spatial Pattern of Multiple Occupant Residential Structures in Hamilton," (MA thesis, McMaster University, 1967), 1122.

11 Robert Schafter, *The Suburbanization of Multifamily Housing* (Lexington: Lexington Books, 1974), 35–49.

12 Graham Barker, Jennifer Penney, and Wally Seccombe, *Highrise and Superprofits* (Kitchener: Dumont Press Graphix, 1973).

13 Ibid., 13.

14 Ibid., 13.

15 *Canadian Architect and Builder* 10 (May 1897): 86; Frederick J. Murdock, "About Apartment Houses," *The Building Manager and Owner* 2 (May 1906): 189–90.

16 Chester A. Patterson and William C. Lengel, *Scientific Building Operation* (Chicago: Patterson Publishing, 1913), 129–33.

17 Fred C. Kelly, "When I Owned an Apartment Building," *Buildings and Building Management* 16 (April 1916): 26–8.

18 *National Builder* 24 (January 1921): 18.

19 Walter Stabler, "Office Building and Apartment House Building Mortgage Loans," *Proceedings of the Fifteenth Annual Convention of Building Owners and Managers* (19–24 June 1922), 126–7; H. L. Stevens, "Reasons for the Success and Failure of Apartment Properties," *The Proceedings of the Twenty-Sixth Annual Convention of the National Association of Building Owners and Managers* (9–22 June 1933), 222.

20 Lewis, *Babbitt*, 135; Richard Rohmer, *E. P. Taylor: The Biography of Edward Plunket Taylor* (Halifax: Formac Publishing Company Limited, Goodread Biographies, 1983), 36; J. L. Granatstein, *The Ottawa Men: The Civil Service Mandarins, 1935–1957* (Toronto: Oxford University Press, 1982), 46–7.

21 "Evolution of the Modern Apartment," *Buildings and Building Management* 15 (July 1915): 45–6.

22 R. W. Sexton, *Apartment Houses of Today*, passim; E. K. Van Winkle, "Future Profits Depend on Better Planning," *Buildings and Building Management* 32 (21 March 1932): 43.

23 Grace Perego, *Apartment House Ownership and Management: Purchasing, Leasing and Managing* (Rochester: Rochester Alliance Press, 1937), 2–3.

24 W. J. Palmer, "Plan, Construction and Operation of Small Apartments," *The Proceedings of the Fifteenth Annual Convention of the National Association of Building Owners and Managers* (19–24 June 1922), 135.

25 Robert S. De Golyer, "Apartment House Construction from the Standpoint of the Investor," *Buildings and Building Management* 17 (October 1917): 42.

26 Palmer, "Plan, Construction and Operation of Small Apartments," 134–5.

27 Warner, *Streetcar Suburbs*, 117–52.

28 Alvin Lovingood, *Apartment House Management* (1929), 114; Perego, *Apartment House Ownership and Management*, 19.

29 Stevens, "Reasons for the Success and Failure of Apartment Properties," 223.

30 J. K. Galbraith, *The Great Crash, 1929* (Boston: Houghton Mifflin, 1961), 178.

31 James C. Downs, "Testing the Rent Market," *The Journal of Real Estate Management* 2 (February 1936): 32.

32 C. F. Schmidt, "Report of Rental Survey Committee," *The Proceedings of the Eighteenth Annual Convention of the National Association of Building Owners and Managers* (8–12 June 1925), 271; Violet Rose Widder, *Apartment Hotel Management* (Los Angeles: Wetzel Publishing, 1930), n.p.

33 H. R. Crow, "What Tenants Should Know," *Building Management* 6 (January 1908): 10; L. L. Newton, "Flattening Out the Rental Peaks," *Buildings and Building Management* 29 (11 March 1929); Edmond H. Sentenne, "Minimizing Moving Day Chaos," *Buildings and Building Management* 30 (7 April 1930): 90.

34 Daniel D. Gage, "Status of Collegiate Real Estate Training," *The Journal of Certified Property Managers* 11 (March 1946): 175–6.

35 Perego, *Apartment House Ownership and Management*, 57.

36 Ibid., 72.

37 Ibid., 99.

38 James. R. McGonagle, *Apartment House Rental: Investment and Management* (New York: Prentice-Hall, 1937), 24.

39 Lovingood, *Apartment House Rental*, 24.

40 *Apartment Management in New England* 1 (15 March 1935): 6.

41 Ibid., 11 (January 1937): 21.

42 O. E. Tronnes, "When Your Apartments Won't Rent – Study Your Renting Methods!" *Buildings and Building Management* 30 (14 July 1930): 69.

43 "Report on the Meeting of the Middle Atlantic Conference, Building Owners and Managers Association," *Buildings and Building Management* 29 (11 February 1929): 42–3.

44 *Buildings and Building Management* 32 (22 February 1932): 33.

45 Charles J. Quinlan, "Renting and Operating Practices in a Large Management Agency," *Buildings and Building Management* 30 (25 August 1930): 66–70.

46 Elena Hixson, "Apartment Management in Depression," *The Federated News* 1 (October 1933): 13.

47 For the concept of distributional coalitions see Mancur Olson, *The Rise and Decline of Nations: Economic Growth, Stagflation, and Social Rigidities* (New Haven: Yale University Press, 1982).

48 On rent increase generally during this period see "Why Not Raise Your Rents?" *Apartment House News* 6 (April 1936): 14. For the quote see Roy Hiddenburg, "Tenant Relations in a Rising Market," *Journal of Real Estate Management* 2 (December 1936): 301.

49 Ibid., 302.

50 Ibid., 303.

51 Ibid., 303.

52 E. S. Hanson, "'It's Not Our House Anyway' – Renters' Motto," *Building Management* 13 (March 1913): 43.

53 E. A. Printz, "How High Can Rents Go?" *Journal of Real Estate Management* 3 (May 1937): 25.

54 Sexton, *Apartment Houses of Today*, iii.

55 *National Builder* 63 (May 1920): 21.

56 McGonagle, *Apartment House Rental*, iii.

57 Anthony James, "The Case Today for Older Apartments," *Journal of Real Estate Management* 19 (September 1953): 16.

58 Ibid., 17.

59 Emerson G. Reinsch, "New Trends in Apartment House Construction," *Journal of Property Management* 27 (December 1962): 206–8; *Journal of Property Management* 29 (May-June 1964): 228.

60 Robert Schafer, *The Suburbanization of Multifamily Housing*, 35–49.

61 *Building Management*, 9 (January, 1909): 29–30.

62 Edwin S. Jewell, "What Do Apartment Tenants Want?" *Buildings and Building Management* 35 (November 1935): 51.

63 Lovingood, *Apartment House Rental*, 16.

64 John Bodnar, Roger Simon, and Michael Weber, *Lives of Their Own: Blacks, Italians, and Poles in Pittsburgh, 1900–1960* (Urbana: University of Illinois Press, 1982), 153–4.

65 Anthony Sutcliffe, "Introduction," *Multi-Storey Living: The British Working-Class Experience* (London: Croom Helm, 1974).
66 Hamilton-Wentworth Planning and Development Department, *High Density Residential Development Study, Background Report* (November 1987).
67 Ontario, Ministry of Housing, *Municipal Building Profile Program* (January 1987); *Municipal Housing: Statement of Program* (January 1987); *Low-Rise Rehabilitation Program* (August 1987); *High-Rise Rehabilitation Program* (December 1985).

CHAPTER TEN

1 Sir Herbert Brown Ames, *The City Below the Hill, A Sociological Study of a Portion of the City of Montreal* (Toronto: University of Toronto Press, 1972, reprint of the 1897 edition).
2 David Hanna, "The Layered City: A Revolution in Housing in Mid-Nineteenth Century Montreal," *Shared Spaces: Department of Geography, McGill University*, no.6 (1986): 3–14.
3 *Hamilton Herald*, 18 April 1912.
4 HPL, Special Collections, *Beautiful Homes of Hamilton: A Souvenir Booklet Showing a Few of Hamilton's Many Tastefully Furnished and Decorated Homes* (Hamilton: Thomas C. Watkins, 1918), n.p.
5 McMaster University, Urban Documentation Centre (hereafter cited as UDC), Social Planning Council of Hamilton and District, "How to Appraise Housing Requirements," *Housing in Hamilton: A Preliminary Survey of Problem Areas: Briefs from Community Agencies and Other Background Material* (November 1965), appendix B.
6 UDC, Department of Political Economy of McMaster University, *Hamilton Housing Survey 1955: Report on Housing Conditions in the City of Hamilton* (Hamilton, 1955).
7 Canada, Dominion Bureau of Statistics, *Housing in Canada: A Study Based on the Census of 1931 and Supplementary Data* (Ottawa: King's Printer, 1941), 51.
8 Ibid., 28.
9 Ontario, Ontario Committee on Taxation, *Report, The Local Land Revenue System*, vol. 2 (Toronto: Queen's Printer, 1967), 29–31.
10 Ontario Committee on Taxation, *The Local Revenue System*, 205–7.
11 McMaster University, *Hamilton Housing Survey 1955*, 14–15.
12 Alice Kessler-Harris, *Out to Work: A History of Wage-Earning Women in the United States* (New York: Oxford University Press, 1986), 113; Allison Prentice et al., *Canadian Women: A History* (Toronto: Harcourt Brace Javonovich, 1988), 123–5.
13 Enid Gouldie, *Cruel Habitations: A History of Working Class Housing,*

1780–1918 (London: George Allen & Unwin, 1974), 92–9; John Burnett, *A Social History of Housing, 1815–1985* (London: Methuen, 2nd edition, 1986), 33, 43, 127–34.

14 Burnett, *A Social History of Housing*, 58–61, 64–9, 70–89; Lucy Caffyn, *Workers' Housing in West Yorkshire, 1750–1920*, Royal Commission on the Historical Monuments of England, supplementary series 9, and the West Yorkshire Metropolitan County Council (London: Her Majesty's Stationery Office, 1986), passim.

15 F.H.A. Aalen, "Public Housing in Ireland, 1880–1921," *Planning Perspectives* 2 (May 1987): 175–93.

16 HPL, Special Collections, M and D Papers, vol.17, Moore and Davis to Miss M. Duncan, 30 March 1910; Moore and Davis to J.S. Freeman, 7 November 1910; vol.18, Moore and Davis to "Dear Friend George" 19 May 1911; Moore and Davis to J.J. Duffy, 31 August 1911.

17 Rosemary Gagan, *Disease, Mortality and Public Health, Hamilton, Ontario, 1900–1914* (MA thesis, McMaster University, 1981), 117–30. Roberts is quoted on 137.

18 League of Nations, Economic Intelligence Service, *Urban and Rural Housing* (Geneva: League of Nations, 1939), passim.

19 United States, Department of Commerce, Bureau of Foreign and Domestic Commerce, *Real Property Inventory of 1934: Summary and Sixty-Four Cities Combined* (Washington, DC, 1934), table 2.

20 John C. Weaver, *Hamilton: An Illustrated History* (Toronto: James Lorimer and Company, 1982), 102.

21 Canada, *Report of the Royal Commission on the Relations of Capital and Labor in Canada: Evidence – Ontario* (Ottawa: Queen's Printer, 1889), 745.

22 Richard Harris, "Residential Segregation and Class Formation in Canadian Cities: A Critical Review," (unpublished paper, 1983), 23.

23 Kathleen Conzen, *Immigrant Milwaukee, 1836–1860: Accommodation and Community in a Frontier City* (Cambridge: Harvard University Press, 1976), 142–3.

24 HPL, Special Collections, Photographic Reproduction of Edwin Whitefield's sketchbook.

25 David Harvey, *Social Justice and the City* (Baltimore: The Johns Hopkins University Press, 1973), 170–1.

26 Weaver, *Hamilton*, 60, 96–9, 162–64.

27 *Hamilton Spectator*, 27 December 1877; 27 May 1879; 8 May 1882.

28 UDC, City Planning Committee of Hamilton, "Street Pattern," *Report on Existing Conditions* (Hamilton, 1945), 4.

29 Ibid., 1–8.

30 Hans Blumenfeld, *Metropolis ... and Beyond: Selected Essays*, ed. Paul D. Spreiregen (New York: John Wiley and Sons, 1979), 41–5; Canada, CMHC, *Road and Rail: Effects on Housing* (Ottawa, 1977), 18.

31 *Hamilton Times*, 21 March 1916.
32 "The Opinions and Aspirations of the Representatives of Organized Labour," *Report on Existing Conditions*, 1.
33 Ibid., 2.
34 Ibid.
35 Barry Broadfoot, *Ten Lost Years, 1929–1939; Memories of the Canadians who Survived the Depression*, (Toronto: Doubleday, 1973); Michael Bliss and L.M. Grayson, eds, *The Wretched of Canada: Letters to R.B.Bennett, 1930–1935* (Toronto: University of Toronto Press, 1971).
36 "Housing Conditions," *Report on Existing Conditions*, 16–17.
37 Leo Haak, "A Housing Survey in Hamilton," *Social Welfare* (March 1937): 6.
38 Our assessment data suggest that 172 to 180 city blocks had housing with assessed values as low as those in the 1936 study area. The percentage of truly dreadful housing units in these blocks could have varied from 25 to 70. The mean number of households per block was 27. Thus the low estimate is 172 × .25 × 27 or 1,170 households or 3.3 per cent of the city's households. The high estimate is 180 × .70 × 27 or 3,400 households or 9.6 per cent of the city's households.
39 *Hamilton Spectator*, 9 February 1949.
40 Canada, CMHC, *Public Priorities in Urban Canada; A Survey of Community Concerns* (Ottawa, 1979), table 5. Kao-Lee Liaw, *Analysis of the Local Mortality Variation* (Program for Quantitative Studies in Economics and Population, McMaster University, research report no.161, August 1986), 18, 20–1. On the variations in the levels of pollution within Hamilton, see UDC, A. C. Farhang, *The Effect of Wind on Air Quality in Hamilton, Ontario* (Downsview: Environment Canada, 1983), 13–18; Ontario Ministry of the Environment, *Air Quality Hamilton, 1970–1977* (June 1978) and subsequent reports under similar titles.
41 For comments on pollution see UDC, The Planning and Development Department of the Regional Municipality of Hamilton-Wentworth, *The Keith Neighbourhood: A Profile* (Hamilton, September 1979), n.p. The quotes are from the same series of neighbourhood profiles, *McAnulty Neighbourhood Plan and Submission* (Hamilton, September 1979), n.p. The future of the enclaves was discussed in The Planning Department of the Regional Municipality of Hamilton-Wentworth, *Review of the Residential Enclaves* (Hamilton, 1977).
42 UDC, The Planning and Development Department of the Regional Municipality of Hamilton-Wentworth, *Homewood Background Report* (Hamilton, October 1979), 10.
43 CMHC, *Public Priorities*, table 6.
44 UDC, Social Planning and Research Council of Hamilton and District, *Housing in Hamilton: A Preliminary Survey of the Problem Areas: Briefs*

from Community Agencies and Other Background Material (Hamilton, November 1965), 8.

45 UDC, Social Planning and Research Council of Hamilton and District, *Community Profile: The Gibson Planning Neighbourhood* (March 1987), 21–2.

46 Canada, Department of Health, *How to Build the Canadian House: The Little Blue Books Home Series* (Ottawa, 1923), 11.

47 There is an extensive scholarly literature on the association of amenities with property values. See, for example, J. W. Kitschen and W. S. Hendon, "Land Values Adjacent to an Urban Neighborhood Park," *Land Economics* 43 (1967): 357–60; M. R. Correll, J. H. Lillydahl, and L. D. Singell, "The Effects of Greenbelts on Residential Property Values: Some Findings on the Political Economy of Open Space," *Land Economics* 54 (1978): 207–17; V. K. Smith, "Residential Location and Environmental Amenities: A Review of Evidence," *Regional Studies* 11 (1977): 47–62; Quentin Gillard, "The Effect of Environmental Amenities on House Values: The Example of a View Lot," *Professional Geographer* 33 (May 1981): 216–20.

48 Richard Harris and Chris Hamnett, "The Myth of the Promised Land: The Social Diffusion of Home Ownership in Britain and North America," *Annals of the Association of American Geographers* 72, no. 2 (1987): 175.

49 Ibid., 176.

50 David Cannadine, *The Pleasures of the Past* (New York: W.W. Norton & Co, 1989), 237.

51 Canada, *Report of the Federal Task Force on Housing and Urban Development*, 77.

APPENDIX A

1 Examples of studies in which data have been extracted from this source include Peter G. Goheen, *Victorian Toronto, 1850 to 1900: Pattern and Process of Growth*, research paper no. 127 (Chicago: Department of Geography, University of Chicago, 1970); Robin S. Holmes, "Ownership and Migration from a Study of Rate Books," *Area* 5 (1973): 242–51, and "Identifying NineteenthCentury Properties," *Area* 6 (1974): 273–7; Michael B. Katz, *The People of Hamilton, Canada West: Family and Class in a Mid-Nineteenth-Century City* (Cambridge: Harvard University Press, 1976); Paul-André Linteau, *The Promoters' City: Building the Industrial Town of Maisonneuve 1883–1918*, translated by Robert Chodos (Toronto: James Lorimer & Co., 1985); Richard Dennis, *Landlords and Rented Housing in Toronto, 1885–1914*, research paper no. 162 (Toronto: Centre for Urban and Community Studies,

University of Toronto, 1987); Richard Harris, *The Growth of Home Ownership in Toronto, 1889–1913*, research paper no. 163 (Toronto: Centre for Urban and Community Studies, University of Toronto, 1987); Michael B. Katz, Michael J. Doucet, and Mark J. Stern, *The Social Organization of Early Industrial Capitalism* (Cambridge: Harvard University Press, 1982); Michael Doucet and John C. Weaver, "Town Fathers and Urban Continuity: The Roots of Community Power and Physical Form in Hamilton, Upper Canada in the 1830s," *Urban History Review/Revue d'histoire urbaine* 13 (1984): 75–90; A.G. Darroch, "Occupational Structure, Assessed Wealth and Homeownership during Toronto's Early Industrialization, 1861–1899," *Histoire sociale – Social History* 16 (1983): 381–410; John C. Weaver, "From Land Assembly to Social Maturity. The Suburban Life of Westdale (Hamilton), Ontario, 1911–1951," *Histoire sociale – Social History* 11 (1978): 411–40; and Michael J. Doucet, "Building the Victorian City: The Process of Land Development in Hamilton, Ontario, 1847–1881," unpublished PhD thesis, University of Toronto, 1977.

2 Goheen, Victorian Toronto, 93–4.

3 Carolyn Gray, *Historical Records of the City of Hamilton, 1847–1973* (Hamilton: Department of History, McMaster University, 1986) 38–41.

4 For a definition and discussion of Hamilton's CBD c. 1850 see Ian Davey and Michael Doucet, "The Social Geography of a Commercial City, ca. 1853," in Katz, *The People of Hamilton*, 329–34.

5 The substantial drop in cases between 1956 and 1966 appears surprising at first. Redevelopment, involving some fifteen of the centrally located blocks in the sample, explains the loss of properties. Over the decade in question urban renewal, the expansion of the CBD, the growth of institutions, industrial expansion, and the development of new parks all occurred, resulting in a net loss of assessed properties included in the sample.

6 Samples also were drawn for 1858 and 1864 but these were not used in this study.

7 This variable was added to all cross sections except for 1852, 1861, 1872, and 1881. These data sets were generated from tapes containing all assessed properties in Hamilton in each of the years and provided to us by Michael B. Katz and the York Social History Project. Obviously, the Katz files did not include this variable, and we did not feel that it would be worthwhile to redo the coding and data-entry stages for these years given the large number of cross sections that we had to produce from scratch.

8 For the reasons cited in the previous footnote, the years 1872 and 1881 were excluded.

Index